Analysis of Piezoelectric Devices

Jiashi YANG
University of Nebraska–Lincoln, USA

Analysis of Piezoelectric Devices

World Scientific

NEW JERSEY • LONDON • SINGAPORE • BEIJING • SHANGHAI • HONG KONG • TAIPEI • CHENNAI

Published by

World Scientific Publishing Co. Pte. Ltd.

5 Toh Tuck Link, Singapore 596224

USA office: 27 Warren Street, Suite 401-402, Hackensack, NJ 07601

UK office: 57 Shelton Street, Covent Garden, London WC2H 9HE

British Library Cataloguing-in-Publication Data
A catalogue record for this book is available from the British Library.

ANALYSIS OF PIEZOELECTRIC DEVICES

ISBN 981-256-861-1

Printed in Singapore by World Scientific Printers (S) Pte Ltd

Preface

This book is a natural continuation of the author's two previous books: "*An Introduction to the Theory of Piezoelectricity*" (Springer, New York, 2005) and "*The Mechanics of Piezoelectric Structures*" (World Scientific, New Jersey, 2006), which discuss the three-dimensional exact theories for piezoelectric materials and various two-, one- and zero-dimensional approximate theories for piezoelectric structures. The development of these theories was strongly influenced by the need for analyzing piezoelectric devices.

Piezoelectric materials are widely used to make various devices including transducers for converting electrical energy to mechanical energy or vice-versa, sensors, actuators, and resonators and filters for telecommunication, control and time-keeping. A few piezoelectric devices were analyzed in the above two books as examples. The present book attempts to present a systematic treatment of piezoelectric devices. However, there are many piezoelectric devices, and it is impossible to cover all of them in a single book. The present book is limited by the research of the author.

In the analysis of piezoelectric devices, very few exact solutions from the three-dimensional equations can be obtained. Numerical methods are usually needed. Another method to simplify the problems so that theoretical analyses are possible is to use lower-dimensional structural theories of plates, shells, and rods. These two approaches are both very useful in the modeling and design of piezoelectric devices; however, the author's personal experience is more on the structural side.

This book is mainly on resonant piezoelectric devices operating at a particular resonant frequency and mode of a structure. Both surface acoustic waves (SAW) and bulk acoustic waves (BAW) have been used for devices. The book is mainly on BAW devices.

Following a brief summary of the three-dimensional theories of electroelastic bodies in the first chapter, two chapters are spent on plate thickness-shear resonators. Mass sensors, fluid sensors, angular rate sensors (gyroscopes), acceleration sensors, pressure sensors, and temperature sensors are discussed in the subsequent chapters. These devices are mostly based on frequency shifts in resonators, except that for gyroscopes where angular rate induced charge, current and/or voltage are also discussed in addition to frequency shifts.

The remaining chapters are on power handling devices. These include piezoelectric generators, transformers, energy transmission through an elastic wall by acoustic waves, and acoustic wave amplifiers made from piezoelectric semiconductors.

The linear theory of piezoelectricity is sufficient for most of these devices, except that the linear theory for small fields superposed on a bias is necessary for the acceleration sensors, pressure sensors, and temperature sensors discussed in this book. The theory for small fields superposed on a bias is a consequence of the nonlinear theory, and the nonlinear theory of electroelasticity itself is employed only in a few scattered problems.

The book is strongly influenced by the author's own research. No effort is made on literature review. However, review articles known to the author on various subjects treated in this book are provided as references.

Due to the use of quite a few stress tensors and electric fields in nonlinear electroelasticity, a list of notations is provided in Appendix 1. Material constants of some common piezoelectric materials, especially those used in this book, are given in Appendix 2.

I would like to take this opportunity to thank Ms. Michelle Sitorius of the College of Engineering at UNL for her editing assistance with the book, Professor Ji Wang of Ninbo University for Figs. 3.10.2, 3.10.4, 3.11.2 and 3.11.4, and Mr. Xuechun Shen, a graduate student of the Department of Engineering Mechanics at UNL, for Figs. 2.6.2 through 2.6.7, and 3.6.2.

JSY
Lincoln, Nebraska
February, 2006

Contents

Preface v

Chapter 1: Three-Dimensional Theories 1

1.1 Nonlinear Electroelasticity for Strong Fields 1
1.2 Linear Piezoelectricity for Infinitesimal Fields 6
1.2.1 Linearization 6
1.2.2 Polarized ceramics 12
1.2.3 Quartz and langasite 13
1.3 Linear Theory for Small Fields Superposed on a Finite Bias 16
1.3.1 The reference state 16
1.3.2 The initial state 17
1.3.3 The present state 17
1.3.4 Equations for the incremental fields 18
1.3.5 Small bias 20
1.3.6 Frequency perturbation due to a small bias 21
1.4 Cubic Theory for Weak Nonlinearity 22

Chapter 2: Thickness-Shear Modes of Plate Resonators 25

2.1 Static Thickness-Shear Deformation 25
2.1.1 A plate under a mechanical load 25
2.1.2 A plate under an electrical load 28
2.2 Nonlinear Thickness-Shear Deformation 31
2.2.1 General analysis 32
2.2.2. An example 33
2.3 Effects of Initial Fields on Thickness- Shear Deformation 36
2.3.1 General analysis 37
2.3.2 An example 39
2.4 Linear Thickness-Shear Vibration 43
2.4.1 Governing equations 43
2.4.2 Free vibration 45
2.4.3 Forced vibration 47
2.4.4 An unelectroded plate 48

2.5	Effects of Electrode Inertia	49
	2.5.1 General analysis	49
	2.5.2 Identical electrodes	51
2.6	Inertial Effects of Imperfectly Bounded Electrodes	52
	2.6.1 General analysis	52
	2.6.2 Identical electrodes	54
2.7	Effects of Electrode Inertia and Shear Stiffness	60
	2.7.1 General analysis	60
	2.7.2 Special cases	62
	2.7.3 Numerical results	64
2.8	Nonlinear Thickness-Shear Vibration	66
	2.8.1 Governing equations	66
	2.8.2 Free vibration	67
	2.8.3 Forced vibration	68
2.9	Effects of Initial Fields on Thickness-Shear Vibration	70
	2.9.1 Governing equations	71
	2.9.2 Open electrodes	71
	2.9.3 Shorted electrodes	72

Chapter 3: Slowly Varying Thickness-Shear Modes **73**

3.1	Exact Waves in a Plate	73
	3.1.1 Eigenvalue problem	74
	3.1.2 An example	77
	3.1.3 A special case: Thickness-twist waves in a ceramic plate	78
	3.1.4 A special case: Thickness-twist waves in a quartz plate	85
3.2	An Approximate Equation for Thickness-Shear Waves	87
3.3	Thickness-Shear Vibration of Finite Plates	91
	3.3.1 Sinusoidal modes	92
	3.3.2 Hyperbolic modes	93
3.4	Energy Trapping in Mesa Resonators	93
3.5	Contoured Resonators	97
3.6	Energy Trapping due to Material Inhomogeneity	98
	3.6.1 General analysis	98
	3.6.2 An example	101
3.7	Energy Trapping by Electrode Mass	103
3.8	Effects of Non-uniform Electrodes	104

3.9 Effects of Electromechanical Coupling on Energy Trapping 106
3.9.1 Governing equations 107
3.9.2 Free vibration solution 108
3.9.3 Discussion 109
3.10 Coupling to Flexure 110
3.10.1 Governing equations 110
3.10.2 Waves in unbounded plates 111
3.10.3 Vibrations of finite plates 112
3.11 Coupling to Face-Shear and Flexure 114
3.11.1 Governing equations 115
3.11.2 Waves in unbounded plates 115
3.11.3 Vibrations of finite plates 117
3.12 Effects of Middle Surface Curvature 119
3.12.1 Governing equations 120
3.12.2 Thickness-shear approximation 122
3.12.3 Coupled thickness-shear and extension 124

Chapter 4: Mass Sensors **127**
4.1 Inertial Effect of a Mass Layer by Perturbation 127
4.1.1 Governing equations 127
4.1.2 Abstract notation 128
4.1.3 Perturbation 129
4.2 Thickness-Shear Modes of a Plate 131
4.3 Anti-Plane Modes of a Wedge 132
4.4 Torsional Modes of a Conical Shell 134
4.5 Effects of Inertia and Stiffness of a Mass Layer by Perturbation 135
4.5.1 Governing equations 136
4.5.2 Perturbation 138
4.6 Effects of Inertia and Stiffness of a Mass Layer by Variation 139
4.6.1 Variational formulation 140
4.6.2 Frequency shift 141
4.6.3 Discussion 142
4.7 Radial Modes of a Ring 143
4.7.1 The unperturbed mode 143
4.7.2 Perturbation solution 144
4.7.3 Variation solution 144
4.7.4 Discussion 145

4.8	Effects of Shear Deformability of a Mass Layer	146
	4.8.1 Governing equations	147
	4.8.2 Variational formulation	150
	4.8.3 Frequency shift	151
4.9	Thickness-Shear Modes of a Plate with Thick Mass Layers	153
4.10	An Ill-Posed Problem in Elasticity for Mass Sensors	157
	4.10.1 Formulation of the problem	157
	4.10.2 Known solutions	158
	4.10.3 An open problem	159
4.11	Thickness-Shear Modes of a Circular Cylinder	160
	4.11.1 Thickness-shear modes in a circular cylinder	160
	4.11.2 Mass sensitivity	161
4.12	Mass Sensitivity of Surface Waves	163
	4.12.1 Governing equations	163
	4.12.2 A half-space with an electroded surface	165
	4.12.3 A half-space with an unelectroded surface	166
4.13	Thickness-Twist Waves in a Ceramic Plate	167
	4.13.1 Governing equations	168
	4.13.2 Anti-symmetric waves	168
	4.13.3 Symmetric waves	170
4.14	Bechmann's Number for Thickness-Twist Waves	171
	4.14.1 Thickness-twist waves in an unbounded, unelectroded plate	171
	4.14.2 Thickness-twist waves in an unbounded, electroded plate	172
	4.14.3 Bechmann's number	173
4.15	Thickness-Twist Waves in a Quartz Plate	175
	4.15.1 Symmetric mass layers	175
	4.15.2 Asymmetric mass layers	176
Chapter 5: Fluid Sensors		**181**
5.1	An Ill-Posed Problem in Elasticity for Fluid Sensors	181
	5.1.1 Formulation of the problem	181
	5.1.2 Known solutions	182
	5.1.3 Boundary integral equation formulation	185
	5.1.4 Generalization to piezoelectricity	186
5.2	Perturbation Analysis	187
	5.2.1 Governing equations	188
	5.2.2 Perturbation analysis	189

5.3 Thickness-Shear Modes of a Plate 191
5.4 Torsional Modes of a Cylindrical Shell 192
5.4.1 Governing equations 193
5.4.2 Interior problem 194
5.4.3 Exterior problem 195
5.5 Thickness-Shear Modes of a Circular Cylinder 196
5.5.1 Governing equations 196
5.5.2 Interior problem 197
5.5.3 Exterior problem 198
5.5.4 Axial thickness-shear 199
5.6 Surface Wave Fluid Sensors 199
5.6.1 Governing equations 200
5.6.2 A half-space with an electroded surface 201
5.6.3 A half-space with an unelectroded surface 202
5.7 Thickness-Twist Waves in a Ceramic Plate 204
5.7.1 Governing equations 204
5.7.2 Anti-symmetric waves 205
5.7.3 Symmetric waves 206

Chapter 6: Gyroscopes — Frequency Effect 209

6.1 High Frequency Vibrations of a Small Rotating Piezoelectric Body 209
6.2 Propagation of Plane Waves 210
6.2.1 General solution 210
6.2.2 Polarized ceramics 215
6.2.3 Quartz 217
6.3 Thickness Vibrations of Plates 221
6.3.1 General solution 222
6.3.2 Ceramic plates 223
6.3.3 Quartz plates 227
6.4 Propagating Waves in a Rotating Piezoelectric Plate 228
6.4.1 General solution 228
6.4.2 Ceramic plates 232
6.5 Surface Waves over a Rotating Piezoelectric Half-Space 236
6.5.1 General solution 236
6.5.2 Ceramic half-space 238

Chapter 7: Gyroscopes — Charge Effect 245

7.1 A Rectangular Beam 245

	7.1.1	Governing equations	247
	7.1.2	Forced vibration solution	250
	7.1.3	An example	251
7.2	A Circular Tube		258
	7.2.1	Governing equations	260
	7.2.2	Forced vibration solution	261
	7.2.3	An example	263
7.3	A Beam Bimorph		264
	7.3.1	Governing equations	265
	7.3.2	Forced vibration solution	270
	7.3.3	Free vibration solution	271
	7.3.4	An example	272
7.4	An Inhomogeneous Shell		274
	7.4.1	Structure	275
	7.4.2	Governing equations	276
	7.4.3	Forced vibration solution	279
	7.4.4	Discussion	282
7.5	A Ceramic Ring		282
	7.5.1	Governing equations	284
	7.5.2	Free vibration solution	286
	7.5.3	Forced vibration solution	287
7.6	A Concentrated Mass and Ceramic Rods		288
	7.6.1	Governing equations	289
	7.6.2	Forced vibration solution	290
7.7	A Ceramic Plate by Two-Dimensional Equations		291
	7.7.1	Driving	292
	7.7.2	Sensing	294
	7.7.3	Discussion	295
7.8	A Ceramic Plate by Zero-Dimensional Equations		296
	7.8.1	Governing equations	297
	7.8.2	Free vibration solution	298
	7.8.3	Forced vibration solution	299
Chapter 8: Acceleration Sensitivity			**301**
8.1	Deformation of a Quartz Plate under Normal Acceleration		301
	8.1.1	Classical flexure	301
	8.1.2	Flexure induced in-plane extension and thickness contraction	303
	8.1.3	One-dimensional deformation	305

	8.1.4 Two-dimensional deformation	308
8.2	First-Order Acceleration Sensitivity	311
8.3	An Estimate of Second-Order Acceleration Sensitivity and its Reduction	313
	8.3.1 An estimate	314
	8.3.2 Reduction of normal acceleration sensitivity	315
8.4	Second-Order Perturbation Analysis	318
	8.4.1 Relatively large biasing deformations	318
	8.4.2 Equations for incremental vibrations	319
	8.4.3 Second-order perturbation analysis	321
8.5	Second-Order Normal Acceleration Sensitivity	324
	8.5.1 Second-order acceleration sensitivity	325
	8.5.2 One-dimensional deformation	326
	8.5.3 Two-dimensional deformation	328
8.6	Effects of Middle Surface Curvature	329
	8.6.1 Biasing deformation	329
	8.6.2 Unperturbed modes	332
	8.6.3 Frequency shift	332
8.7	Vibration Sensitivity	334
	8.7.1 Governing equations	335
	8.7.2 First-order solution	336
	8.7.3 Second-order solution	337
	8.7.4 An example	338
Chapter 9: Pressure Sensors		**341**
9.1	A Rectangular Plate in a Circular Cylindrical Shell	341
	9.1.1 Analysis	343
	9.1.2 Numerical results	345
9.2	A Circular Plate in a Circular Cylindrical Shell	347
	9.2.1 Analysis	347
	9.2.2 Numerical results	350
9.3	A Rectangular Plate in a Shallow Shell	352
	9.3.1 Analysis	352
	9.3.2 Numerical results	354
9.4	A Bimorph	355
	9.4.1 Analysis	355
	9.4.2 Numerical results	358
9.5	Surface Wave Pressure Sensors Based on Extension	361
9.6	Surface Wave Pressure Sensors Based on Flexure	364

9.6.1 Biasing deformations 364
9.6.2 Frequency shifts 368

Chapter 10: Temperature Sensors **371**

10.1 Thermoelectroelasticity 371
10.2 Linear Theory 374
10.3 Small Fields Superposed on a Thermal Bias 376
10.4 Thickness-Shear Modes of a Free Plate 379
10.5 Thickness-Shear Modes of a Constrained Plate 382
10.5.1 Analysis 382
10.5.2 An example 385

Chapter 11: Piezoelectric Generators **387**

11.1 Thickness-Stretch of a Ceramic Plate 387
11.1.1 Analysis 388
11.1.2 An example 392
11.2 A Circular Shell 395
11.3 A Beam Bimorph 399
11.3.1 Analysis 400
11.3.2 An example 406
11.4 A Spiral Bimorph 407
11.5 Nonlinear Behavior near Resonance 409
11.5.1 Analysis 410
11.5.2 Numerical results 413

Chapter 12: Piezoelectric Transformers **417**

12.1 A Thickness-Stretch Mode Plate Transformer 417
12.1.1 Governing equations 417
12.1.2 Analytical solution 419
12.1.3 Numerical results 421
12.2 Rosen Transformer 425
12.2.1 One-dimensional model 425
12.2.2 Free vibration analysis 428
12.2.3 Forced vibration analysis 429
12.2.4 Numerical results 432
12.3 A Thickness-Shear Mode Transformer — Free Vibration 434
12.3.1 Governing equations 435
12.3.2 Free vibration analysis 437
12.3.3 Ceramic transformers 439

12.4 A Thickness-Shear Mode Transformer — Forced Vibration 444
12.4.1 Governing equations 444
12.4.2 Forced vibration analysis 447

Chapter 13: Power Transmissiion through an Elastic Wall 451
13.1 Formulation of the Problem 451
13.2 Theoretical Analysis 454
13.3 Numerical Results 457

Chapter 14: Acoustic Wave Amplifiers 463
14.1 Equations for Piezoelectric Semiconductors 463
14.2 Equations for a Thin Film 464
14.3 Surface Waves 466
14.3.1 Analytical solution 466
14.3.2 Discussion 470
14.3.3 Numerical results 472
14.4 Interface Waves 474
14.4.1 Analytical solution 474
14.4.2 Discussiion 477
14.5 Waves in a Plate 478
14.5.1 Symmetric waves 479
14.5.2 Anti-symmetric waves 482
14.5.3 Discussion 483
14.6 Gap Waves 484
14.6.1 Analytical solution 484
14.6.2 Discussion 487

References 491

Appendix 1 Notation 501

Appendix 2 Electroelastic Material Constants 503

Index 517

Chapter 1
Three-Dimensional Theories

In this chapter we summarize the three-dimensional equations of the nonlinear theory of electroelasticty for large deformations and strong fields [1,2], the linear theory of piezoelectricity for infinitesimal deformation and fields [3,4], the linear theory for small fields superposed on finite biasing or initial fields [5,6,7], and the theory for weak, cubic nonlinearity [8,9]. A systematic presentation of these theories can also be found in [10]. This chapter uses the two-point Cartesian tensor notation, the summation convention for repeated tensor indices, and the convention that a comma followed by an index denotes partial differentiation with respect to the coordinate associated with the index.

1.1. Nonlinear Electroelasticity for Strong Fields

Consider a deformable continuum which, in the reference configuration at time t_0, occupies a region V with a boundary surface S (see Fig. 1.1.1). $\mathbf{N}$ is the unit exterior normal of S. In this state the body is free from deformation and fields. The position of a material point in this state is denoted by a vector $\mathbf{X} = X_K\mathbf{I}_K$ in a rectangular coordinate system X_K where X_K denotes the reference or material coordinates of the material point. They form a continuous labeling of material particles so that they are identifiable. At time t, the body occupies a region v with a boundary surface s and an exterior normal $\mathbf{n}$. The current position of the material point associated with $\mathbf{X}$ is given by $\mathbf{y} = y_k\mathbf{i}_k$, which denotes the present or spatial coordinates of the material point.

Since the coordinate systems are othogonal,

$$\mathbf{i}_k \cdot \mathbf{i}_l = \delta_{kl} \quad \text{and} \quad \mathbf{I}_K \cdot \mathbf{I}_L = \delta_{KL}, \tag{1.1.1}$$

where δ_{kl} and δ_{KL} are the Kronecker delta. In matrix notation,

$$[\delta_{kl}] = [\delta_{KL}] = \begin{bmatrix} 1 & 0 & 0 \\ 0 & 1 & 0 \\ 0 & 0 & 1 \end{bmatrix}. \tag{1.1.2}$$

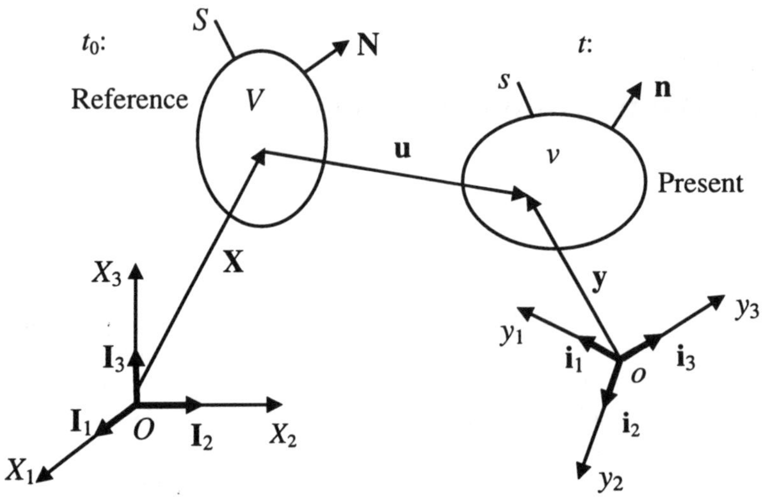

Fig. 1.1.1. Motion of a continuum and coordinate systems.

For the rest of this book the two coordinate systems are chosen to be coincident, i.e.,

$$o = O, \quad \mathbf{i}_1 = \mathbf{I}_1, \quad \mathbf{i}_2 = \mathbf{I}_2, \quad \mathbf{i}_3 = \mathbf{I}_3. \tag{1.1.3}$$

The transformation coefficients (shifters) between the two coordinate systems are denoted by

$$\mathbf{i}_k \cdot \mathbf{I}_L = \delta_{kL}. \tag{1.1.4}$$

When the two coordinate systems are coincident, δ_{kL} is simply the Kronecker delta. It is still needed for notational homogeneity. A vector can be resolved into rectangular components in different coordinate systems. For example, we can also write

$$\mathbf{y} = y_K \mathbf{I}_K, \tag{1.1.5}$$

with

$$y_M = \delta_{Mi} y_i. \tag{1.1.6}$$

The motion of the body is described by $y_i = y_i(\mathbf{X}, t)$. The equations of motion and Gauss's equation of electrostatic (the charge equation) are

$$\begin{aligned} &K_{Lj,L} + \rho_0 f_j = \rho_0 \ddot{y}_j, \\ &\mathcal{D}_{K,K} = \rho_E, \end{aligned} \tag{1.1.7}$$

where K_{Lj} is the two-point total stress tensor, ρ_0 (a scalar) is the reference mass density, f_j is the mechanical body force per unit mass, and $\mathcal{D}_K$ is the reference electric displacement vector. ρ_E, a scalar (E is not an index), is the free charge density per unit reference volume, and a superimposed dot represents the material time derivative

$$\ddot{y}_i = \frac{D^2 y_i}{Dt^2} = \left.\frac{\partial^2 y_i(\mathbf{X},t)}{\partial t^2}\right|_{\mathbf{X}\ \text{fixed}} . \tag{1.1.8}$$

In Eq. (1.1.7), K_{Lj} and $\mathcal{D}_K$ are given by:

$$\begin{aligned} &K_{Lj} = F_{Lj} + M_{Lj}, \\ &F_{Lj} = y_{j,K} T^S_{KL}, \quad M_{Lj} = JX_{L,i}\varepsilon_0 \left(E_i E_j - \frac{1}{2} E_k E_k \delta_{ij} \right), \\ &J = \det(y_{i,K}), \quad T^S_{KL} = \rho_0 \frac{\partial \psi}{\partial S_{KL}}, \quad E_i = -\phi_{,i}, \end{aligned} \tag{1.1.9}$$

and

$$\begin{aligned} &\mathcal{D}_K = \varepsilon_0 JX_{K,i} D_i = \varepsilon_0 JC^{-1}_{KL} \mathcal{E}_L + \mathcal{P}_K, \\ &C^{-1}_{KL} = X_{K,i} X_{L,i}, \\ &\mathcal{E}_K = y_{i,K} E_i = -\phi_{,K}, \quad \mathcal{P}_K = JX_{K,i} P_i = -\rho_0 \frac{\partial \psi}{\partial \mathcal{E}_K}, \end{aligned} \tag{1.1.10}$$

where ε_0 (a scalar) is the electric permittivity of free space, E_i is the electric field, P_i is the electric polarization per unit present volume, and D_i is the electric displacement vector. $\mathcal{E}_K$ is the reference electric field vector, and $\mathcal{P}_K$ is the reference electric polarization vector. ϕ is the electric potential and C^{-1}_{KL} is the inverse of the deformation tensor. $\psi = \psi(S_{KL}, \mathcal{E}_K)$ is a free energy density per unit mass, which is a function of $\mathcal{E}_K$ and the following finite strain tensor:

$$S_{KL} = (y_{i,K} y_{i,L} - \delta_{KL})/2 . \tag{1.1.11}$$

From Eqs. (1.1.9) and (1.1.10), we have

$$\begin{aligned} &K_{Lj} = y_{j,K} \rho_0 \frac{\partial \psi}{\partial S_{KL}} + JX_{L,i}\varepsilon_0 \left(E_i E_j - \frac{1}{2} E_k E_k \delta_{ij} \right), \\ &\mathcal{D}_K = \varepsilon_0 JC^{-1}_{KL} \mathcal{E}_L - \rho_0 \frac{\partial \psi}{\partial \mathcal{E}_K}. \end{aligned} \tag{1.1.12}$$

With successive substitutions from Eqs. (1.1.9) through (1.1.11), Eq. (1.1.7) can be written as four equations for the four unknowns $y_i(\mathbf{X},t)$ and $\phi(\mathbf{X},t)$.

The free energy ψ that determines the constitutive relations of nonlinear electroelastic materials may be written as

$$
\begin{aligned}
&\rho_0\psi(S_{KL},\mathcal{E}_K) \\
&= \frac{1}{2}\underset{2}{c}_{ABCD} S_{AB}S_{CD} - e_{ABC}\mathcal{E}_A S_{BC} - \frac{1}{2}\underset{2}{\chi}_{AB}\mathcal{E}_A\mathcal{E}_B \\
&+ \frac{1}{6}\underset{3}{c}_{ABCDEF} S_{AB}S_{CD}S_{EF} + \frac{1}{2}\underset{1}{k}_{ABCDE}\mathcal{E}_A S_{BC}S_{DE} \\
&- \frac{1}{2}b_{ABCD}\mathcal{E}_A\mathcal{E}_B S_{CD} - \frac{1}{6}\underset{3}{\chi}_{ABC}\mathcal{E}_A\mathcal{E}_B\mathcal{E}_C \\
&+ \frac{1}{24}\underset{4}{c}_{ABCDEFGH} S_{AB}S_{CD}S_{EF}S_{GH} + \frac{1}{6}\underset{2}{k}_{ABCDEFG}\mathcal{E}_A S_{BC}S_{DE}S_{FG} \\
&+ \frac{1}{4}\underset{1}{a}_{ABCDEF}\mathcal{E}_A\mathcal{E}_B S_{CD}S_{EF} + \frac{1}{6}\underset{3}{k}_{ABCDE}\mathcal{E}_A\mathcal{E}_B\mathcal{E}_C S_{DE} \\
&- \frac{1}{24}\underset{4}{\chi}_{ABCD}\mathcal{E}_A\mathcal{E}_B\mathcal{E}_C\mathcal{E}_D + \cdots,
\end{aligned}
\tag{1.1.13}
$$

where the material constants

$$
\begin{aligned}
&\underset{2}{c}_{ABCD},\quad e_{ABC},\quad \underset{2}{\chi}_{AB}, \\
&\underset{3}{c}_{ABCDEF},\quad \underset{1}{k}_{ABCDE},\quad b_{ABCD},\quad \underset{3}{\chi}_{ABC}, \\
&\underset{4}{c}_{ABCDEFGH},\quad \underset{2}{k}_{ABCDEFG},\quad \underset{1}{a}_{ABCDEF},\quad \underset{3}{k}_{ABCDE},\quad \underset{4}{\chi}_{ABCD}
\end{aligned}
\tag{1.1.14}
$$

are called the second-order elastic, piezoelectric, electric susceptibility, third-order elastic, first odd electroelastic, electrostrictive, third-order electric susceptibility, fourth-order elastic, second odd electroelastic, first even electroelastic, third odd electroelastic, and fourth-order electric susceptibility, respectively. The second-order constants are responsible for linear material behaviors. The third- and higher-order material constants are related to nonlinear behaviors of materials.

For mechanical boundary conditions S is partitioned into S_y and S_T, on which motion (or displacement) and traction are prescribed, respectively. Electrically S is partitioned into S_ϕ and S_D with prescribed electric potential and surface free charge, respectively, and

$$\begin{aligned} S_y \cup S_T &= S_\phi \cup S_D = S, \\ S_y \cap S_T &= S_\phi \cap S_D = 0. \end{aligned} \tag{1.1.15}$$

The usual boundary value problem for an electroelastic body consists of Eq. (1.1.7) and the following boundary conditions:

$$\begin{aligned} & y_i = \overline{y}_i \quad \text{on} \quad S_y, \\ & \phi = \overline{\phi} \quad \text{on} \quad S_\phi, \\ & K_{Lk} N_L = \overline{T}_k \quad \text{on} \quad S_T, \\ & \mathcal{D}_K N_K = -\overline{\sigma}_E \quad \text{on} \quad S_D, \end{aligned} \tag{1.1.16}$$

where $\overline{y}_i$ and $\overline{\phi}$ are the prescribed boundary motion and potential, $\overline{T}_i$ is the surface traction per unit undeformed area, and $\overline{\sigma}_E$ (a scalar) is the surface free charge per unit undeformed area.

Consider the following variational functional:

$$\begin{aligned} \Pi(\mathbf{y},\phi) = & \int_{t_0}^{t_1} dt \int_V \left[\frac{1}{2} \rho_0 \dot{y}_i \dot{y}_i - \rho_0 \psi(S_{KL}, \mathcal{E}_K) \right. \\ & \left. + \pi(S_{KL}, \mathcal{E}_K) + \rho_0 f_i y_i - \rho_E \phi \right] dV \\ & + \int_{t_0}^{t_1} dt \int_{S_T} \overline{T}_i y_i dS - \int_{t_0}^{t_1} dt \int_{S_D} \overline{\sigma}_E \phi \, dS, \end{aligned} \tag{1.1.17}$$

where

$$\pi(S_{KL}, \mathcal{E}_K) = \frac{1}{2} \varepsilon_0 J E_k E_k = \frac{1}{2} \varepsilon_0 J C_{MN}^{-1} \mathcal{E}_M \mathcal{E}_N . \tag{1.1.18}$$

The admissible y_i and ϕ for Π satisfy the following initial conditions in V and boundary conditions on S_y and S_ϕ:

$$\begin{aligned} & \delta y_i \,|_{t=t_0} = 0, \quad \delta y_i \,|_{t=t_1} = 0 \quad \text{in} \quad V, \\ & y_i = \overline{y}_i \quad \text{on} \quad S_y, \quad t_0 < t < t_1, \\ & \phi = \overline{\phi} \quad \text{on} \quad S_\phi, \quad t_0 < t < t_1. \end{aligned} \tag{1.1.19}$$

Then the first variation of Π is

$$\delta\Pi = \int_{t_0}^{t_1} dt \int_V \left[(K_{Li,L} + \rho_0 f_i - \rho_0 \ddot{y}_i)\delta y_i + (\mathcal{D}_{L,L} - \rho_E)\delta\phi\right] dV$$
$$- \int_{t_0}^{t_1} dt \int_{S_T} (K_{Li}N_L - \bar{T}_i)\delta y_i dS \qquad (1.1.20)$$
$$- \int_{t_0}^{t_1} dt \int_{S_D} (\mathcal{D}_L N_L + \bar{\sigma}_E)\delta\phi\, dS.$$

Therefore, the stationary condition of Π implies the following equations and natural boundary conditions:

$$\begin{aligned} &K_{Lk,L} + \rho_0 f_k = \rho_0 \ddot{y}_k \quad \text{in} \quad V, \\ &\mathcal{D}_{K,K} = \rho_E \quad \text{in} \quad V, \\ &K_{Lk}N_L = \bar{T}_k \quad \text{on} \quad S_T, \\ &\mathcal{D}_K N_K = -\bar{\sigma}_E \quad \text{on} \quad S_D. \end{aligned} \qquad (1.1.21)$$

1.2. Linear Piezoelectricity for Infinitesimal Fields

The equations of linear piezoelectricity are obtained by expansions and truncations of the nonlinear theory in the previous section.

1.2.1. Linearization

In linear theory, we introduce the small displacement vector $\mathbf{u} = \mathbf{y} - \mathbf{X}$ and assume infinitesimal displacement gradient and electric potential gradient. The infinitesimal strain tensor is denoted by

$$S_{kl} = \frac{1}{2}(u_{l,k} + u_{k,l}). \qquad (1.2.1)$$

The material electric field becomes

$$\mathcal{E}_K = E_i y_{i,K} \cong E_i \delta_{iK} \to E_k. \qquad (1.2.2)$$

Similarly,

$$M_{Lj} \cong 0, \quad K_{Lj} \cong F_{Lj}, \quad \mathcal{P}_K \to P_k, \quad \mathcal{D}_K \to D_k. \qquad (1.2.3)$$

Since the various stress tensors are either approximately zero (quadratic or of higher order in the infinitesimal gradients) or about the same, we use T_{ij} to denote the stress tensor that is linear in the infinitesimal gradients. This notation follows the IEEE Standard on Piezoelectricity [3]. Our notation for the rest of the linear theory will also follow the IEEE Standard. Then

$$K_{Lj} \cong F_{Lj} \to T_{lj}, \quad T^S_{KL} \to T_{kl}. \tag{1.2.4}$$

For small fields the free energy density can be approximated by

$$\begin{aligned}
&\rho_0\psi(S_{KL}, \mathcal{E}_K) - \frac{1}{2}\varepsilon_0 J E_k E_k \\
&\cong \frac{1}{2}\underset{2}{c}_{ABCD} S_{AB} S_{CD} - e_{ABC}\mathcal{E}_A S_{BC} - \frac{1}{2}\underset{2}{\chi}_{AB} \mathcal{E}_A \mathcal{E}_B - \frac{1}{2}\varepsilon_0 J E_k E_k \\
&\to \frac{1}{2}c^E_{ijkl} S_{ij} S_{kl} - e_{ijk} E_i S_{jk} - \frac{1}{2}\varepsilon^S_{ij} E_i E_j = H(S_{kl}, E_k),
\end{aligned} \tag{1.2.5}$$

where

$$\varepsilon^S_{ij} = \underset{2}{\chi}_{ij} + \varepsilon_0 \delta_{ij}. \tag{1.2.6}$$

The superscript E in c^E_{ijkl} indicates that the independent electric constitutive variable is the electric field **E**. The superscript S in ε^S_{ij} indicates that the mechanical constitutive variable is the strain tensor **S**.

In Eq. (1.2.5) we have also introduced the electric enthalpy H. The linear constitutive relations generated by H are:

$$\begin{aligned}
T_{ij} &= \frac{\partial H}{\partial S_{ij}} = c^E_{ijkl} S_{kl} - e_{kij} E_k, \\
D_i &= -\frac{\partial H}{\partial E_i} = e_{ikl} S_{kl} + \varepsilon^S_{ik} E_k.
\end{aligned} \tag{1.2.7}$$

The material constants in Eq. (1.2.7) have the following symmetries:

$$\begin{aligned}
c^E_{ijkl} &= c^E_{jikl} = c^E_{klij}, \\
e_{kij} &= e_{kji}, \\
\varepsilon^S_{ij} &= \varepsilon^S_{ji}.
\end{aligned} \tag{1.2.8}$$

We also assume that the elastic and dielectric material tensors are positive-definite in the following sense:

$$\begin{aligned}
&c^E_{ijkl} S_{ij} S_{kl} \ge 0 \quad \text{for any} \quad S_{ij} = S_{ji}, \\
&\quad \text{and} \quad c^E_{ijkl} S_{ij} S_{kl} = 0 \;\Rightarrow\; S_{ij} = 0, \\
&\varepsilon^S_{ij} E_i E_j \ge 0 \quad \text{for any} \quad E_i, \\
&\quad \text{and} \quad \varepsilon^S_{ij} E_i E_j = 0 \;\Rightarrow\; E_i = 0.
\end{aligned} \tag{1.2.9}$$

Similar to Eq. (1.2.7), linear constitutive relations can also be written as [3]

$$
\begin{aligned}
T_{ij} &= c^D_{ijkl} S_{kl} - h_{kij} D_k, \\
E_i &= -h_{ikl} S_{kl} + \beta^S_{ik} D_k,
\end{aligned}
\tag{1.2.10}
$$

$$
\begin{aligned}
S_{ij} &= s^E_{ijkl} T_{kl} + d_{kij} E_k, \\
D_i &= d_{ikl} T_{kl} + \varepsilon^T_{ik} E_k,
\end{aligned}
\tag{1.2.11}
$$

and

$$
\begin{aligned}
S_{ij} &= s^D_{ijkl} T_{kl} + g_{kij} D_k, \\
E_i &= -g_{ikl} T_{kl} + \beta^T_{ik} D_k.
\end{aligned}
\tag{1.2.12}
$$

The equations of motion and the charge equation become

$$
\begin{aligned}
T_{ji,j} + \rho f_i &= \rho \ddot{u}_i, \\
D_{i,i} &= \rho_e,
\end{aligned}
\tag{1.2.13}
$$

where ρ is the present mass density, and ρ_e (a scalar) is the free charge density per unit present volume. The difference between ρ and ρ_0, and that between ρ_E and ρ_e are neglected in Eq. (1.2.13).

In summary, the linear theory of piezoelectricity consists of the equations of motion and charge in Eq. (1.2.13), the constitutive relations

$$
T_{ij} = c_{ijkl} S_{kl} - e_{kij} E_k, \quad D_i = e_{ijk} S_{jk} + \varepsilon_{ij} E_j,
\tag{1.2.14}
$$

where the superscripts in the material constants in Eq. (1.2.7) have been dropped, and the strain-displacement and electric field-potential relations

$$
S_{ij} = (u_{i,j} + u_{j,i})/2, \quad E_i = -\phi_{,i}\,.
\tag{1.2.15}
$$

With successive substitutions from Eqs. (1.2.14) and (1.2.15), Eq. (1.2.13) can be written as four equations for $\mathbf{u}$ and ϕ:

$$
\begin{aligned}
c_{ijkl} u_{k,lj} + e_{kij} \phi_{,kj} + \rho f_i &= \rho \ddot{u}_i, \\
e_{ikl} u_{k,li} - \varepsilon_{ij} \phi_{,ij} &= \rho_e.
\end{aligned}
\tag{1.2.16}
$$

Let the region occupied by the piezoelectric body be V and its boundary surface be S as shown in Fig. 1.2.1. For linear piezoelectricity we use $\mathbf{x}$ as the independent spatial coordinates. Let the unit outward normal of S be $\mathbf{n}$.

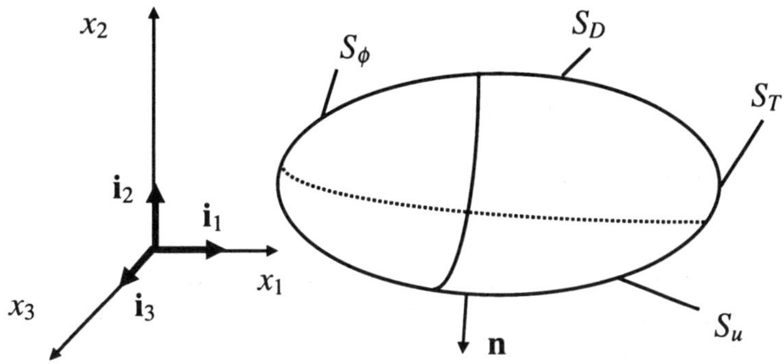

Fig. 1.2.1. A piezoelectric body and partitions of its surface.

For boundary conditions we consider the following partitions of S:

$$\begin{aligned} S_u \cup S_T = S_\phi \cup S_D = S, \\ S_u \cap S_T = S_\phi \cap S_D = 0, \end{aligned} \tag{1.2.17}$$

where S_u is the part of S on which the mechanical displacement is prescribed, and S_T is the part of S where the traction vector is prescribed. S_ϕ represents the part of S which is electroded where the electric potential is no more than a function of time, and S_D is the unelectroded part. We consider very thin electrodes whose mechanical effects can be neglected. For mechanical boundary conditions we have prescribed displacement $\bar{u}_i$

$$u_i = \bar{u}_i \quad \text{on} \quad S_u, \tag{1.2.18}$$

and prescribed traction $\bar{t}_j$

$$T_{ij} n_i = \bar{t}_j \quad \text{on} \quad S_T. \tag{1.2.19}$$

Electrically, on the electroded portion of S,

$$\phi = \bar{\phi} \quad \text{on} \quad S_\phi, \tag{1.2.20}$$

where $\bar{\phi}$ does not vary spatially. On the unelectroded part of S, the charge condition can be written as

$$D_j n_j = -\bar{\sigma}_e \quad \text{on} \quad S_D, \tag{1.2.21}$$

where $\bar{\sigma}_e$ (a scalar) is the free charge density per unit surface area.

On an electrode S_ϕ, the total free electric charge Q_e (a scalar) can be represented by

$$Q_e = \int_{S_\phi} -n_i D_i dS \,. \tag{1.2.22}$$

The electric current flowing out of the electrode is given by

$$I = -\dot{Q}_e \,. \tag{1.2.23}$$

Sometimes there are two (or more) electrodes on a body that are connected to an electric circuit. In this case, circuit equation(s) will need to be considered.

The equations and boundary conditions of linear piezoelectricity can be derived from a variational principle. Consider [4]

$$\begin{aligned} \Pi(\mathbf{u},\phi) = & \int_{t_0}^{t_1} dt \int_V \left[\frac{1}{2}\rho\dot{u}_i\dot{u}_i - H(\mathbf{S},\mathbf{E}) + \rho f_i u_i - \rho_e\phi\right] dV \\ & + \int_{t_0}^{t_1} dt \int_{S_T} \bar{t}_i u_i dS - \int_{t_0}^{t_1} dt \int_{S_D} \bar{\sigma}_e \phi \, dS. \end{aligned} \tag{1.2.24}$$

$\mathbf{u}$ and ϕ are variationally admissible if they are smooth enough and satisfy

$$\begin{aligned} & \delta u_i\,|_{t_0} = \delta u_i\,|_{t_1} = 0 \quad \text{in} \quad V, \\ & u_i = \bar{u}_i \quad \text{on} \quad S_u, \quad t_0 < t < t_1, \\ & \phi = \bar{\phi} \quad \text{on} \quad S_\phi, \quad t_0 < t < t_1. \end{aligned} \tag{1.2.25}$$

The first variation of Π is

$$\begin{aligned} \delta\Pi = & \int_{t_0}^{t_1} dt \int_V \left[(T_{ji,j} + \rho f_i - \rho\ddot{u}_i)\delta u_i + (D_{i,i} - \rho_e)\delta\phi\right] dV \\ & - \int_{t_0}^{t_1} dt \int_{S_T} (T_{ji}n_j - \bar{t}_i)\delta u_i dS - \int_{t_0}^{t_1} dt \int_{S_D} (D_i n_i + \bar{\sigma}_e)\delta\phi \, dS. \end{aligned} \tag{1.2.26}$$

Therefore, the stationary condition of Π is

$$\begin{aligned} & T_{ji,j} + \rho f_i = \rho\ddot{u}_i \quad \text{in} \quad V, \quad t_0 < t < t_1, \\ & D_{i,i} = \rho_e \quad \text{in} \quad V, \quad t_0 < t < t_1, \\ & T_{ji}n_j = \bar{t}_i \quad \text{on} \quad S_T, \quad t_0 < t < t_1, \\ & D_i n_i = -\bar{\sigma}_e \quad \text{on} \quad S_D, \quad t_0 < t < t_1. \end{aligned} \tag{1.2.27}$$

We now introduce a compact matrix notation [3,4]. This notation consists of replacing pairs of indices *ij* or *kl* by single indices *p* or *q*, where i, j, k and l take the values of 1, 2, and 3, and p and q take the values of 1, 2, 3, 4, 5, and 6 according to

$$\begin{array}{lcccccc} ij \text{ or } kl: & 11 & 22 & 33 & 23 \text{ or } 32 & 31 \text{ or } 13 & 12 \text{ or } 21 \\ p \text{ or } q: & 1 & 2 & 3 & 4 & 5 & 6 \end{array} . \qquad (1.2.28)$$

Thus

$$c_{ijkl} \to c_{pq}, \quad e_{ikl} \to e_{ip}, \quad T_{ij} \to T_p. \qquad (1.2.29)$$

For the strain tensor, we introduce S_p such that

$$\begin{aligned} &S_1 = S_{11}, \quad S_2 = S_{22}, \quad S_3 = S_{33}, \\ &S_4 = 2S_{23}, \quad S_5 = 2S_{31}, \quad S_6 = 2S_{12}. \end{aligned} \qquad (1.2.30)$$

The constitutive relations in Eq. (1.2.7) can then be written as

$$\begin{aligned} T_p &= c_{pq}^E S_q - e_{kp} E_k, \\ D_i &= e_{iq} S_q + \varepsilon_{ik}^S E_k. \end{aligned} \qquad (1.2.31)$$

In matrix form, Eq. (1.2.31) becomes

$$\begin{Bmatrix} T_1 \\ T_2 \\ T_3 \\ T_4 \\ T_5 \\ T_6 \end{Bmatrix} = \begin{pmatrix} c_{11}^E & c_{12}^E & c_{13}^E & c_{14}^E & c_{15}^E & c_{16}^E \\ c_{21}^E & c_{22}^E & c_{23}^E & c_{24}^E & c_{25}^E & c_{26}^E \\ c_{31}^E & c_{32}^E & c_{33}^E & c_{34}^E & c_{35}^E & c_{36}^E \\ c_{41}^E & c_{42}^E & c_{43}^E & c_{44}^E & c_{45}^E & c_{46}^E \\ c_{51}^E & c_{52}^E & c_{53}^E & c_{54}^E & c_{55}^E & c_{56}^E \\ c_{61}^E & c_{62}^E & c_{63}^E & c_{64}^E & c_{65}^E & c_{66}^E \end{pmatrix} \begin{Bmatrix} S_1 \\ S_2 \\ S_3 \\ S_4 \\ S_5 \\ S_6 \end{Bmatrix} - \begin{pmatrix} e_{11} & e_{21} & e_{31} \\ e_{12} & e_{22} & e_{32} \\ e_{13} & e_{23} & e_{33} \\ e_{14} & e_{24} & e_{34} \\ e_{15} & e_{25} & e_{35} \\ e_{16} & e_{26} & e_{36} \end{pmatrix} \begin{Bmatrix} E_1 \\ E_2 \\ E_3 \end{Bmatrix},$$

$$\begin{Bmatrix} D_1 \\ D_2 \\ D_3 \end{Bmatrix} = \begin{bmatrix} e_{11} & e_{12} & e_{13} & e_{14} & e_{15} & e_{16} \\ e_{21} & e_{22} & e_{23} & e_{24} & e_{25} & e_{26} \\ e_{31} & e_{32} & e_{33} & e_{34} & e_{35} & e_{36} \end{bmatrix} \begin{Bmatrix} S_1 \\ S_2 \\ S_3 \\ S_4 \\ S_5 \\ S_6 \end{Bmatrix} + \begin{pmatrix} \varepsilon_{11}^S & \varepsilon_{12}^S & \varepsilon_{13}^S \\ \varepsilon_{21}^S & \varepsilon_{22}^S & \varepsilon_{22}^S \\ \varepsilon_{31}^S & \varepsilon_{32}^S & \varepsilon_{33}^S \end{pmatrix} \begin{Bmatrix} E_1 \\ E_2 \\ E_3 \end{Bmatrix}.$$

(1.2.32)

1.2.2. Polarized ceramics

Polarized ceramics are transversely isotropic. When the poling direction is along the x_3-axis, we have

$$\begin{pmatrix} c_{11} & c_{12} & c_{13} & 0 & 0 & 0 \\ c_{21} & c_{11} & c_{13} & 0 & 0 & 0 \\ c_{31} & c_{31} & c_{33} & 0 & 0 & 0 \\ 0 & 0 & 0 & c_{44} & 0 & 0 \\ 0 & 0 & 0 & 0 & c_{44} & 0 \\ 0 & 0 & 0 & 0 & 0 & c_{66} \end{pmatrix},$$

$$\begin{pmatrix} 0 & 0 & 0 & 0 & e_{15} & 0 \\ 0 & 0 & 0 & e_{15} & 0 & 0 \\ e_{31} & e_{31} & e_{33} & 0 & 0 & 0 \end{pmatrix}, \begin{pmatrix} \varepsilon_{11} & 0 & 0 \\ 0 & \varepsilon_{11} & 0 \\ 0 & 0 & \varepsilon_{33} \end{pmatrix}, \tag{1.2.33}$$

where $c_{66} = (c_{11} - c_{12})/2$. The matrices in Eq. (1.2.33) have the same structures as those of crystals class C_{6v} (or 6mm). The constitutive relations take the following form:

$$\begin{aligned} T_{11} &= c_{11}u_{1,1} + c_{12}u_{2,2} + c_{13}u_{3,3} + e_{31}\phi_{,3}, \\ T_{22} &= c_{12}u_{1,1} + c_{22}u_{2,2} + c_{13}u_{3,3} + e_{31}\phi_{,3}, \\ T_{33} &= c_{13}u_{1,1} + c_{13}u_{2,2} + c_{33}u_{3,3} + e_{33}\phi_{,3}, \\ T_{23} &= c_{44}(u_{2,3} + u_{3,2}) + e_{15}\phi_{,2}, \\ T_{31} &= c_{44}(u_{3,1} + u_{1,3}) + e_{15}\phi_{,1}, \\ T_{12} &= c_{66}(u_{1,2} + u_{2,1}), \end{aligned} \tag{1.2.34}$$

and

$$\begin{aligned} D_1 &= e_{15}(u_{3,1} + u_{1,3}) - \varepsilon_{11}\phi_{,1}, \\ D_2 &= e_{15}(u_{2,3} + u_{3,2}) - \varepsilon_{11}\phi_{,2}, \\ D_3 &= e_{31}(u_{1,1} + u_{2,2}) + e_{33}u_{3,3} - \varepsilon_{33}\phi_{,3}. \end{aligned} \tag{1.2.35}$$

The equations of motion and charge are

$$\begin{aligned}
&c_{11}u_{1,11}+(c_{12}+c_{66})u_{2,12}+(c_{13}+c_{44})u_{3,13}+c_{66}u_{1,22}\\
&\quad +c_{44}u_{1,33}+(e_{31}+e_{15})\phi_{,13}=\rho\ddot{u}_1,\\
&c_{66}u_{2,11}+(c_{12}+c_{66})u_{1,12}+c_{11}u_{2,22}+(c_{13}+c_{44})u_{3,23}\\
&\quad +c_{44}u_{2,33}+(e_{31}+e_{15})\phi_{,23}=\rho\ddot{u}_2,\\
&c_{44}u_{3,11}+(c_{44}+c_{13})u_{1,31}+c_{44}u_{3,22}+(c_{13}+c_{44})u_{2,23}\\
&\quad +c_{33}u_{3,33}+e_{15}(\phi_{,11}+\phi_{,22})+e_{33}\phi_{,33}=\rho\ddot{u}_3,\\
&e_{15}u_{3,11}+(e_{15}+e_{31})u_{1,13}+e_{15}u_{3,22}+(e_{15}+e_{31})u_{2,32}\\
&\quad +e_{31}u_{3,33}-\varepsilon_{11}(\phi_{,11}+\phi_{,22})-\varepsilon_{33}\phi_{,33}=0.
\end{aligned} \tag{1.2.36}$$

1.2.3. Quartz and langasite

Quartz is another widely used piezoelectric crystal. It belongs to crystal class 32 (or D_3). Langasite and some of its isomorphs (langanite and langatate) are emerging piezoelectric crystals which have stronger piezoelectric coupling than quartz and also belong to crystal class 32. For such a crystal with x_3 as a trigonal axis and x_1 as a digonal axis, the material matrices are

$$\begin{pmatrix}
c_{11} & c_{12} & c_{13} & c_{14} & 0 & 0\\
c_{21} & c_{11} & c_{13} & -c_{14} & 0 & 0\\
c_{13} & c_{13} & c_{33} & 0 & 0 & 0\\
c_{14} & -c_{14} & 0 & c_{44} & 0 & 0\\
0 & 0 & 0 & 0 & c_{44} & c_{14}\\
0 & 0 & 0 & 0 & c_{14} & c_{66}
\end{pmatrix},$$

$$\begin{pmatrix}
e_{11} & -e_{11} & 0 & e_{14} & 0 & 0\\
0 & 0 & 0 & 0 & -e_{14} & -e_{11}\\
0 & 0 & 0 & 0 & 0 & 0
\end{pmatrix},
\begin{pmatrix}
\varepsilon_{11} & 0 & 0\\
0 & \varepsilon_{11} & 0\\
0 & 0 & \varepsilon_{33}
\end{pmatrix}. \tag{1.2.37}$$

The independent material constants are 6 + 2 + 2 = 10.

Quartz plates are often used to make devices. Plates taken from a bulk crystal at different orientations are referred to as plates of different cuts. A particular cut is specified by two angles, φ and θ, with respect to the crystal axes (X, Y, Z).

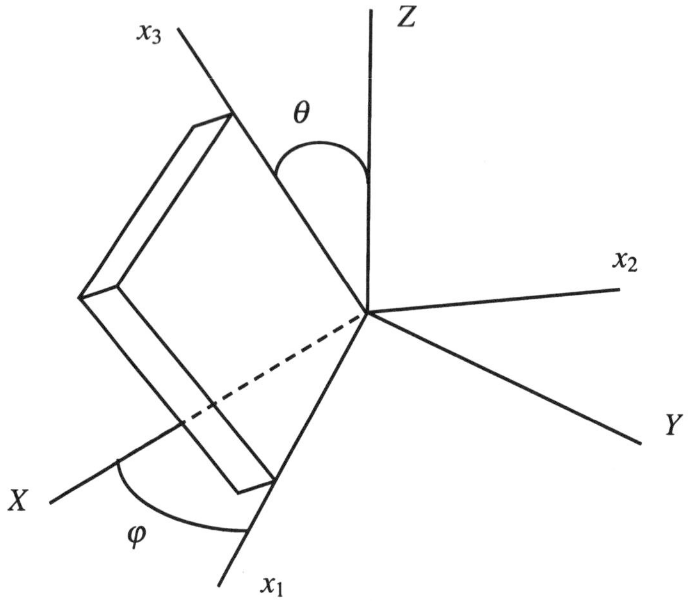

Fig. 1.2.2. A quartz plate cut from a bulk crystal.

Plates of different cuts have different material matrices with respect to coordinate axes in and normal to the plane of the plates. One class of cuts of quartz plates, called rotated Y-cuts, has $\varphi = 0$ and is particularly useful in device applications. Rotated Y-cut quartz exhibits monoclinic symmetry of class 2 (or C_2) in a coordinate system (x_1, x_2) in and normal to the plane of the plate. Therefore we list the equations for monoclinic crystals below which are useful for studying quartz devices. For monoclinic crystals, with the digonal axis along the x_1-axis, we have

$$\begin{pmatrix} c_{11} & c_{12} & c_{13} & c_{14} & 0 & 0 \\ c_{21} & c_{22} & c_{23} & c_{24} & 0 & 0 \\ c_{31} & c_{32} & c_{33} & c_{34} & 0 & 0 \\ c_{41} & c_{42} & c_{43} & c_{44} & 0 & 0 \\ 0 & 0 & 0 & 0 & c_{55} & c_{56} \\ 0 & 0 & 0 & 0 & c_{65} & c_{66} \end{pmatrix},$$

$$\begin{pmatrix} e_{11} & e_{12} & e_{13} & e_{14} & 0 & 0 \\ 0 & 0 & 0 & 0 & e_{25} & e_{26} \\ 0 & 0 & 0 & 0 & e_{35} & e_{36} \end{pmatrix}, \begin{pmatrix} \varepsilon_{11} & 0 & 0 \\ 0 & \varepsilon_{11} & \varepsilon_{23} \\ 0 & \varepsilon_{32} & \varepsilon_{33} \end{pmatrix}. \tag{1.2.38}$$

The constitutive relations are

$$\begin{aligned} T_{11} &= c_{11}u_{1,1} + c_{12}u_{2,2} + c_{13}u_{3,3} + c_{14}(u_{2,3} + u_{3,2}) + e_{11}\phi_{,1}, \\ T_{22} &= c_{12}u_{1,1} + c_{22}u_{2,2} + c_{23}u_{3,3} + c_{24}(u_{2,3} + u_{3,2}) + e_{12}\phi_{,1}, \\ T_{33} &= c_{13}u_{1,1} + c_{23}u_{2,2} + c_{33}u_{3,3} + c_{34}(u_{2,3} + u_{3,2}) + e_{13}\phi_{,1}, \\ T_{23} &= c_{14}u_{1,1} + c_{24}u_{2,2} + c_{34}u_{3,3} + c_{44}(u_{2,3} + u_{3,2}) + e_{14}\phi_{,1}, \\ T_{31} &= c_{55}(u_{3,1} + u_{1,3}) + c_{56}(u_{1,2} + u_{2,1}) + e_{25}\phi_{,2} + e_{35}\phi_{,3}, \\ T_{12} &= c_{56}(u_{3,1} + u_{1,3}) + c_{66}(u_{1,2} + u_{2,1}) + e_{26}\phi_{,2} + e_{36}\phi_{,3}, \end{aligned} \tag{1.2.39}$$

and

$$\begin{aligned} D_1 &= e_{11}u_{1,1} + e_{12}u_{2,2} + e_{13}u_{3,3} + e_{14}(u_{2,3} + u_{3,2}) - \varepsilon_{11}\phi_{,1}, \\ D_2 &= e_{25}(u_{3,1} + u_{1,3}) + e_{26}(u_{1,2} + u_{2,1}) - \varepsilon_{22}\phi_{,2} - \varepsilon_{23}\phi_{,3}, \\ D_3 &= e_{35}(u_{3,1} + u_{1,3}) + e_{36}(u_{1,2} + u_{2,1}) - \varepsilon_{23}\phi_{,2} - \varepsilon_{33}\phi_{,3}. \end{aligned} \tag{1.2.40}$$

The equations of motion and charge are

$$\begin{aligned} & c_{11}u_{1,11} + (c_{12} + c_{66})u_{2,12} + (c_{13} + c_{55})u_{3,13} + (c_{14} + c_{56})u_{2,13} \\ &\quad + (c_{14} + c_{56})u_{3,12} + 2c_{56}u_{1,23} + c_{66}u_{1,22} + c_{55}u_{1,33} \\ &\quad + e_{11}\phi_{,11} + e_{26}\phi_{,22} + (e_{36} + e_{25})\phi_{,23} + e_{35}\phi_{,33} = \rho\ddot{u}_1, \\ & c_{56}u_{3,11} + (c_{56} + c_{14})u_{1,13} + (c_{66} + c_{12})u_{1,12} + c_{66}u_{2,11} \\ &\quad + c_{22}u_{2,22} + (c_{23} + c_{44})u_{3,23} + 2c_{24}u_{2,23} + c_{24}u_{3,22} \\ &\quad + c_{34}u_{3,33} + c_{44}u_{2,33} + (e_{26} + e_{12})\phi_{,12} + (e_{36} + e_{14})\phi_{,13} = \rho\ddot{u}_2, \\ & c_{55}u_{3,11} + (c_{55} + c_{13})u_{1,13} + (c_{56} + c_{14})u_{1,12} + c_{56}u_{2,11} \\ &\quad + c_{24}u_{2,22} + 2c_{34}u_{3,23} + (c_{44} + c_{23})u_{2,23} + c_{44}u_{3,22} \\ &\quad + c_{33}u_{3,33} + c_{34}u_{2,33} + (e_{25} + e_{14})\phi_{,12} + (e_{35} + e_{13})\phi_{,13} = \rho\ddot{u}_3, \\ & e_{11}u_{1,11} + (e_{12} + e_{26})u_{2,12} + (e_{13} + e_{35})u_{3,13} + (e_{14} + e_{36})u_{2,13} \\ &\quad + (e_{14} + e_{25})u_{3,12} + (e_{25} + e_{36})u_{1,23} + e_{26}u_{1,22} \\ &\quad + e_{35}u_{1,33} - \varepsilon_{11}\phi_{,11} - \varepsilon_{22}\phi_{,22} - 2\varepsilon_{23}\phi_{,23} - \varepsilon_{33}\phi_{,33} = 0. \end{aligned} \tag{1.2.41}$$

1.3. Linear Theory for Small Fields Superposed on a Finite Bias

The theory of linear piezoelectricity assumes infinitesimal deviations from an ideal reference state of the material in which there are no pre-existing mechanical and/or electrical fields (initial or biasing fields). The presence of biasing fields makes a material apparently behave like a different material, and renders the linear theory of piezoelectricity invalid. The behavior of electroelastic bodies under biasing fields can be described by the theory for infinitesimal incremental fields superposed on finite biasing fields [5,6], which is a consequence of the nonlinear theory of electroelasticity. This section presents the theory for small fields superposed on finite biasing fields in an electroelastic body.

Consider the following three states of an electroelastic body (see Fig. 1.3.1):

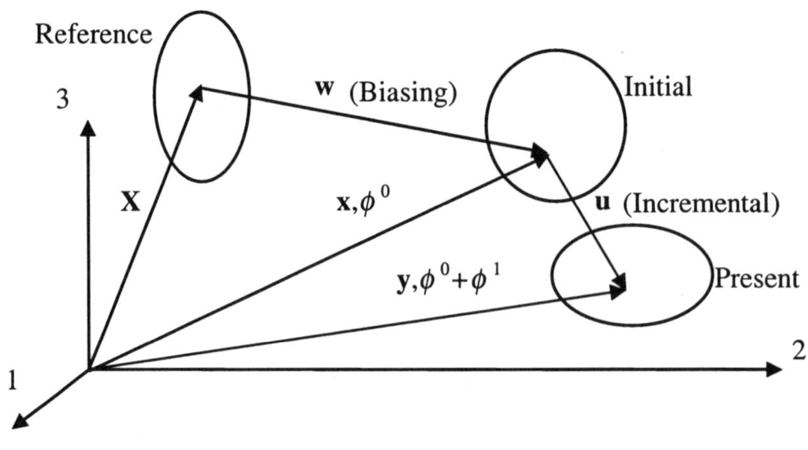

Fig. 1.3.1. Reference, initial, and present configurations of an electroelastic body.

1.3.1. The reference state

In this state the body is undeformed and is free of electric fields. A generic point at this state is denoted by **X** with Cartesian coordinates X_K. The mass density is ρ_0.

1.3.2. The initial state

In this state the body is deformed finitely and statically, and carries finite static electric fields. The body is under the action of body force f_α^0, body charge ρ_E^0, prescribed surface position $\bar{x}_\alpha$, surface traction $\bar{T}_\alpha^0$, surface potential $\bar{\phi}^0$ and surface charge $\bar{\sigma}_E^0$. The deformation and fields at this configuration are the initial or biasing fields. The position of the material point associated with **X** is given by **x** = **x**(**X**) or $x_\gamma = x_\gamma(\mathbf{X})$, with strain S_{KL}^0. Greek indices are used for the initial configuration. The electric potential in this state is denoted by $\phi^0(\mathbf{X})$, with electric field E_α^0. **x**(**X**) and $\phi^0(\mathbf{X})$ satisfy the following static equations of nonlinear electroelasticity:

$$
\begin{aligned}
&S_{KL}^0 = (x_{\alpha,K}x_{\alpha,L} - \delta_{KL})/2, \quad \mathcal{E}_K^0 = -\phi_{,K}^0, \quad E_\alpha^0 = -\phi_{,\alpha}^0, \\
&T_{KL}^0 = \rho_0 \left.\frac{\partial \psi}{\partial S_{KL}}\right|_{S_{KL}^0, \mathcal{E}_K^0}, \quad \mathcal{P}_K^0 = -\rho_0 \left.\frac{\partial \psi}{\partial \mathcal{E}_K}\right|_{S_{KL}^0, \mathcal{E}_K^0}, \\
&J^0 = \det(x_{\alpha,K}), \\
&K_{K\alpha}^0 = x_{\alpha,L}T_{KL}^0 + M_{K\alpha}^0, \quad \mathcal{D}_K^0 = \varepsilon_0 J^0 X_{K,\alpha} X_{L,\alpha} \mathcal{E}_L^0 + \mathcal{P}_K^0, \\
&M_{K\alpha}^0 = J^0 X_{K,\beta}\varepsilon_0 \left(E_\beta^0 E_\alpha^0 - \frac{1}{2} E_\gamma^0 E_\gamma^0 \delta_{\beta\alpha} \right), \\
&K_{K\alpha,K}^0 + \rho_0 f_\alpha^0 = 0, \quad \mathcal{D}_{K,K}^0 = \rho_E^0.
\end{aligned}
\tag{1.3.1}
$$

1.3.3. The present state

In this state, time-dependent, small, incremental deformations and electric fields are applied to the deformed body at the initial state. The body is under the action of f_i, ρ_E, $\bar{y}_i$, $\bar{T}_i$, $\bar{\phi}$ and $\bar{\sigma}_E$. The final position of **X** is given by **y** = **y**(**X**,t), and the final electric potential is $\phi(\mathbf{X},t)$. **y**(**X**,t) and $\phi(\mathbf{X},t)$ satisfy the dynamic equations of nonlinear electroelasticity:

$$S_{KL} = (y_{i,K}y_{i,L} - \delta_{KL})/2, \quad \mathcal{E}_K = -\phi_{,K}, \quad E_i = -\phi_{,i},$$
$$T^S_{KL} = \rho_0 \frac{\partial \psi}{\partial S_{KL}}\bigg|_{S_{KL},\mathcal{E}_K}, \quad \mathcal{P}_K = -\rho_0 \frac{\partial \psi}{\partial \mathcal{E}_K}\bigg|_{S_{KL},\mathcal{E}_K},$$
$$K_{Lj} = y_{j,K}T^S_{KL} + M_{Lj}, \quad \mathcal{D}_K = \varepsilon_0 J C^{-1}_{KL}\mathcal{E}_L + \mathcal{P}_K, \tag{1.3.2}$$
$$M_{Lj} = JX_{L,i}\varepsilon_0\left(E_iE_j - \frac{1}{2}E_kE_k\delta_{ij}\right),$$
$$K_{Lj,L} + \rho_0 f_j = \rho_0 \ddot{y}_j, \quad \mathcal{D}_{K,K} = \rho_E.$$

1.3.4. Equations for the incremental fields

Let the incremental displacement be $\mathbf{u}(\mathbf{X},t)$ and the incremental potential be $\phi^1(\mathbf{X},t)$ (see Fig. 1.3.1). $\mathbf{u}$ and ϕ^1 are assumed to be infinitesimal. We write $\mathbf{y}$ and ϕ as

$$y_i(\mathbf{X},t) = \delta_{i\alpha}[x_\alpha(\mathbf{X},t) + u_\alpha(\mathbf{X},t)],$$
$$\phi(\mathbf{X},t) = \phi^0(\mathbf{X},t) + \phi^1(\mathbf{X},t). \tag{1.3.3}$$

Then it can be shown that the equations governing the incremental fields $\mathbf{u}$ and ϕ^1 are

$$K^1_{K\alpha,K} + \rho_0 f^1_\alpha = \rho_0 \ddot{u}_\alpha,$$
$$\mathcal{D}^1_{K,K} = \rho^1_E, \tag{1.3.4}$$

where f^1_α and ρ^1_E are determined from

$$f_i = \delta_{i\alpha}(f^0_\alpha + f^1_\alpha), \quad \rho_E = \rho^0_E + \rho^1_E. \tag{1.3.5}$$

The incremental stress tensor and electric displacement vector are given by the following constitutive relations:

$$K^1_{L\gamma} = G_{L\gamma M\alpha}u_{\alpha,M} - R_{ML\gamma}\mathcal{E}^1_M,$$
$$\mathcal{D}^1_K = R_{KL\gamma}u_{\gamma,L} + L_{KL}\mathcal{E}^1_L, \tag{1.3.6}$$

where $\mathcal{E}^1_K = -\phi^1_{,K}$. Equation (1.3.6) shows that the incremental stress tensor and electric displacement vector depend linearly on the incremental displacement gradient and potential gradient. In Eq. (1.3.6),

$$
\begin{aligned}
G_{K\alpha L\gamma} &= x_{\alpha,M}\, \rho_0 \frac{\partial^2 \psi}{\partial S_{KM} \partial S_{LN}}\bigg|_{S^0_{KL},\mathcal{E}^0_K} x_{\gamma,N} + T^0_{KL}\delta_{\alpha\gamma} + g_{K\alpha L\gamma} = G_{L\gamma K\alpha}, \\
R_{KL\gamma} &= -\rho_0 \frac{\partial^2 \psi}{\partial \mathcal{E}_K \partial S_{ML}}\bigg|_{S^0_{KL},\mathcal{E}^0_K} x_{\gamma,M} + r_{KL\gamma}, \\
L_{KL} &= -\rho_0 \frac{\partial^2 \psi}{\partial \mathcal{E}_K \partial \mathcal{E}_L}\bigg|_{S^0_{KL},\mathcal{E}^0_K} + l_{KL} = L_{LK},
\end{aligned}
\tag{1.3.7}
$$

where

$$
\begin{aligned}
g_{K\alpha L\gamma} &= \varepsilon_0 J^0 [E^0_\alpha E^0_\beta (X_{K,\beta} X_{L,\gamma} - X_{K,\gamma} X_{L,\beta}) \\
&\quad - E^0_\alpha E^0_\gamma X_{K,\beta} X_{L,\beta} \\
&\quad + E^0_\beta E^0_\gamma (X_{K,\alpha} X_{L,\beta} - X_{K,\beta} X_{L,\alpha}) \\
&\quad + \frac{1}{2} E^0_\beta E^0_\beta (X_{K,\gamma} X_{L,\alpha} - X_{K,\alpha} X_{L,\gamma})], \\
r_{KL\gamma} &= \varepsilon_0 J^0 (E^0_\alpha X_{K,\alpha} X_{L,\gamma} - E^0_\alpha X_{K,\gamma} X_{L,\alpha} - E^0_\gamma X_{K,\alpha} X_{L,\alpha}), \\
l_{KL} &= \varepsilon_0 J^0 X_{K,\alpha} X_{L,\alpha}.
\end{aligned}
\tag{1.3.8}
$$

$G_{K\alpha L\gamma}$, $R_{KL\gamma}$, and L_{KL} are called the effective or apparent elastic, piezoelectric, and dielectric constants. They depend on the initial deformation $x_\alpha(\mathbf{X})$ and electric potential $\phi^0(\mathbf{X})$.

In summary, the boundary value problem for the incremental fields $\mathbf{u}$ and ϕ^1 consists of the following equations and boundary conditions:

$$
\begin{aligned}
& K^1_{K\alpha,K} + \rho_0 f^1_\alpha = \rho_0 \ddot{u}_\alpha \quad \text{in} \quad V, \\
& \mathcal{D}^1_{K,K} = \rho^1_E \quad \text{in} \quad V, \\
& K^1_{L\gamma} = G_{L\gamma M\alpha} u_{\alpha,M} + R_{ML\gamma} \phi^1_{,M} \quad \text{in} \quad V, \\
& \mathcal{D}^1_K = R_{KL\gamma} u_{\gamma,L} - L_{KL} \phi^1_{,L} \quad \text{in} \quad V, \\
& u_\alpha = \overline{u}_\alpha \quad \text{on} \quad S_y, \\
& \phi^1 = \overline{\phi}^1 \quad \text{on} \quad S_\phi, \\
& K^1_{L\alpha} N_L = \overline{T}^1_\alpha \quad \text{on} \quad S_T, \\
& \mathcal{D}^1_K N_K = -\overline{\sigma}^1_E \quad \text{on} \quad S_D.
\end{aligned}
\tag{1.3.9}
$$

Consider the following variational functional:

$$\Pi(\mathbf{u},\phi^1)=\int_{t_0}^{t_1}dt\int_V\left(\frac{1}{2}\rho_0\dot{u}_\alpha\dot{u}_\alpha-\frac{1}{2}G_{K\alpha L\gamma}u_{K,\alpha}u_{L,\gamma}\right.$$

$$\left.-R_{KL\gamma}\phi^1_{,K}u_{L,\gamma}+\frac{1}{2}L_{KL}\phi^1_{,K}\phi^1_{,L}+\rho_0 f^1_\alpha u_\alpha-\rho^1_E\phi^1\right)dV \tag{1.3.10}$$

$$+\int_{t_0}^{t_1}dt\int_{S_T}\overline{T}^1_\alpha u_\alpha dS-\int_{t_0}^{t_1}dt\int_{S_D}\overline{\sigma}^1_E\phi^1 dS.$$

The admissible $\mathbf{u}$ and ϕ^1 must satisfy

$$\begin{aligned}&\delta u_\alpha\,|_{t_0}=\delta u_\alpha\,|_{t_1}=0 \quad \text{in} \quad V,\\ &u_\alpha=\overline{u}_\alpha \quad \text{on} \quad S_u,\quad t_0<t<t_1,\\ &\phi^1=\overline{\phi}^1 \quad \text{on} \quad S_\phi,\quad t_0<t<t_1.\end{aligned} \tag{1.3.11}$$

The first variation is found to be

$$\begin{aligned}\delta\Pi(\mathbf{u},\phi^1)=&\int_{t_0}^{t_1}dt\int_V[(K^1_{L\alpha,L}+\rho_0 f^1_\alpha-\rho_0\ddot{u}_\alpha)\delta u_\alpha\\ &+(\mathcal{D}^1_{K,K}-\rho^1_E)\delta\phi^1]dV\\ &-\int_{t_0}^{t_1}dt\int_{S_T}(K^1_{L\alpha}N_L-\overline{T}^1_\alpha)\delta u_\alpha dS\\ &-\int_{t_0}^{t_1}dt\int_{S_D}(\mathcal{D}^1_K N_K+\overline{\sigma}^1_E)\delta\phi^1 dS.\end{aligned} \tag{1.3.12}$$

Therefore, the stationary condition of the functional gives the following governing equations and boundary conditions:

$$\begin{aligned}&K^1_{K\alpha,K}+\rho_0 f^1_\alpha=\rho_0\ddot{u}_\alpha \quad \text{in} \quad V,\\ &\mathcal{D}^1_{K,K}=\rho^1_E \quad \text{in} \quad V,\\ &K^1_{L\alpha}N_L=\overline{T}^1_\alpha \quad \text{on} \quad S_T,\\ &\mathcal{D}^1_K N_K=-\overline{\sigma}^1_E \quad \text{on} \quad S_D.\end{aligned} \tag{1.3.13}$$

1.3.5. Small bias

In many applications, the biasing deformations and fields are also infinitesimal. In this case, usually only their first-order effects on the incremental fields need to be considered. Then the following energy density of a cubic polynomial is sufficient:

$$\rho_0\psi(S_{KL},\mathcal{E}_K)=\frac{1}{2}c_{ABCD}S_{AB}S_{CD}-e_{ABC}\mathcal{E}_A S_{BC}-\frac{1}{2}\chi_{AB}\mathcal{E}_A\mathcal{E}_B$$
$$+\frac{1}{6}c_{ABCDEF}S_{AB}S_{CD}S_{EF}+\frac{1}{2}k_{ABCDE}\mathcal{E}_A S_{BC}S_{DE} \quad (1.3.14)$$
$$-\frac{1}{2}b_{ABCD}\mathcal{E}_A\mathcal{E}_B S_{CD}-\frac{1}{6}\chi_{ABC}\mathcal{E}_A\mathcal{E}_B\mathcal{E}_C,$$

where the subscripts indicating the orders of the material constants have been dropped. For small biasing fields it is convenient to introduce the small displacement vector **w** of the initial deformation (see Fig. 1.3.1), given as

$$x_\alpha=\delta_{\alpha K}X_K+w_\alpha\,. \quad (1.3.15)$$

Then, neglecting the quadratic terms of the gradients of **w** and ϕ^0, the effective material constants take the following form [5,6]:

$$\begin{aligned}G_{K\alpha L\gamma}&=c_{K\alpha L\gamma}+\hat{c}_{K\alpha L\gamma},\\ R_{KL\gamma}&=e_{KL\gamma}+\hat{e}_{KL\gamma},\\ L_{KL}&=\varepsilon_{KL}+\hat{\varepsilon}_{KL},\end{aligned} \quad (1.3.16)$$

where

$$\begin{aligned}\hat{c}_{K\alpha L\gamma}&=T^0_{KL}\delta_{\alpha\gamma}+c_{K\alpha LN}w_{\gamma,N}+c_{KNL\gamma}w_{\alpha,N}\\ &\quad+c_{K\alpha L\gamma AB}S^0_{AB}+k_{AK\alpha L\gamma}\mathcal{E}^0_A,\\ \hat{e}_{KL\gamma}&=e_{KLM}w_{\gamma,M}-k_{KL\gamma AB}S^0_{AB}+b_{AKL\gamma}\mathcal{E}^0_A\\ &\quad+\varepsilon_0(\mathcal{E}^0_K\delta_{L\gamma}-\mathcal{E}^0_L\delta_{K\gamma}-\mathcal{E}^0_M\delta_{M\gamma}\delta_{KL}),\\ \hat{\varepsilon}_{KL}&=b_{KLAB}S^0_{AB}+\chi_{KLA}\mathcal{E}^0_A+\varepsilon_0(S^0_{MM}\delta_{KL}-2S^0_{KL}),\\ T^0_{KL}&=c_{KLMN}S^0_{MN}-e_{AKL}\mathcal{E}^0_A\\ S^0_{AB}&\cong(w_{A,B}+w_{B,A})/2,\\ \mathcal{E}^0_K&=-\phi^0_{,K}\,.\end{aligned} \quad (1.3.17)$$

1.3.6. Frequency perturbation due to a small bias

For many applications to be discussed later, we are interested in the effect of biasing fields on resonant frequencies of an electroelastic body.

Consider the following eigenvalue problem for free vibrations of an electroelastic body under small biasing fields:

$$-[(c_{L\gamma M\alpha}+\hat{c}_{L\gamma M\alpha})u_{\alpha,M}+(e_{ML\gamma}+\hat{e}_{ML\gamma})\phi^1_{,M}]_{,L}=\rho_0\omega^2 u_\gamma,$$
$$[(e_{KL\gamma}+\hat{e}_{KL\gamma})u_{\gamma,L}-(\varepsilon_{KL}+\hat{\varepsilon}_{KL})\phi^1_{,L}]_{,K}=0, \tag{1.3.18}$$

with appropriate boundary conditions. In Eq. (1.3.18), ω and u_α are the resonant frequency and the corresponding mode, respectively, when the biasing fields are present and may be called a perturbed frequency and mode. Then it can be shown [7] that, to the first-order of the biasing fields,

$$\Delta\omega=$$
$$\frac{1}{2\omega_0}\frac{\int_V [\hat{c}_{L\gamma M\alpha}U_{\gamma,L}U_{\alpha,M}+2\hat{e}_{ML\gamma}\Phi_{,M}U_{\gamma,L}-\hat{\varepsilon}_{ML}\Phi_{,M}\Phi_{,L}]dV}{\int_V \rho_0 U_\alpha U_\alpha dV}, \tag{1.3.19}$$

where $\Delta\omega=\omega-\omega_0$. ω_0, U_α and Φ are the frequency and mode when the biasing fields are not present, or the unperturbed frequency and mode. They are governed by

$$-(c_{L\gamma M\alpha}U_{\alpha,M}+e_{ML\gamma}\Phi_{,M})_{,L}=\rho_0\omega_0^2 U_\gamma,$$
$$(e_{KL\gamma}U_{\gamma,L}-\varepsilon_{KL}\Phi_{,L})_{,K}=0, \tag{1.3.20}$$

with appropriate boundary conditions. Equation (1.3.19) is the result of a first-order perturbation analysis. It breaks a complicated eigenvalue problem of Eq. (1.3.18) for vibrations under a bias into two simpler problems of Eq. (1.3.19) for vibrations without a bias, and the biasing field problem.

1.4. Cubic Theory for Weak Nonlinearity

By cubic theory we mean that effects of all terms up to the third power of the displacement and potential gradients or their products are included [8]. Cubic theory is an approximate theory for relatively weak nonlinearities, and can be obtained by expansions and truncations from the nonlinear theory in the first section of this chapter. The resulting equations are:

$$
\begin{aligned}
F_{Lj} \cong \delta_{jM} \Big[& \underset{2}{c}_{LMAB} u_{A,B} + e_{ALM} \phi_{,A} + \frac{1}{2} \underset{2}{c}_{LMAB} u_{K,A} u_{K,B} \\
& + \underset{2}{c}_{LKAB} u_{M,K} u_{A,B} + \frac{1}{2} \underset{3}{c}_{LMABCD} u_{A,B} u_{C,D} \\
& + e_{ALK} u_{M,K} \phi_{,A} - \underset{1}{k}_{ABCLM} u_{B,C} \phi_{,A} - \frac{1}{2} b_{ABLM} \phi_{,A} \phi_{,B} \\
& + \frac{1}{2} \underset{2}{c}_{LRAB} u_{M,R} u_{K,A} u_{K,B} + \frac{1}{2} \underset{3}{c}_{LKABCD} u_{M,K} u_{A,B} u_{C,D} \\
& + \frac{1}{2} \underset{3}{c}_{LMABcD} u_{A,B} u_{K,C} u_{K,D} + \frac{1}{6} \underset{4}{c}_{LMABCDEF} u_{A,B} u_{C,D} u_{E,F} \\
& - \underset{1}{k}_{ABCLK} u_{B,C} u_{M,K} \phi_{,A} - \frac{1}{2} \underset{1}{k}_{ABCLM} u_{K,B} u_{K,C} \phi_{,A} \\
& - \frac{1}{2} \underset{2}{k}_{ABCDELM} u_{B,C} u_{D,E} \phi_{,A} - \frac{1}{2} b_{ABLK} u_{M,K} \phi_{,A} \phi_{,B} \\
& + \frac{1}{2} \underset{1}{a}_{ABCDLM} u_{c,D} \phi_{,A} \phi_{,B} + \frac{1}{6} \underset{3}{k}_{ABCLM} \phi_{,A} \phi_{,B} \phi_{,C} \Big],
\end{aligned}
\tag{1.4.1}
$$

$$
\begin{aligned}
\mathcal{P}_L \cong\ & e_{LBC} u_{B,C} - \underset{2}{\chi}_{AL} \phi_{,A} + \frac{1}{2} e_{LBC} u_{K,B} u_{K,C} - \frac{1}{2} \underset{1}{k}_{LBCDE} u_{B,C} u_{D,E} - b_{ALCD} u_{C,D} \phi_{,A} \\
& + \frac{1}{2} \underset{3}{\chi}_{ABL} \phi_{,A} \phi_{,B} - \frac{1}{2} \underset{1}{k}_{LBCDE} u_{B,C} u_{K,D} u_{K,E} \\
& - \frac{1}{6} \underset{2}{k}_{LBCDEFG} u_{B,C} u_{D,E} u_{F,G} - \frac{1}{2} b_{ALCD} u_{K,C} u_{K,D} \phi_{,A} \\
& + \frac{1}{2} \underset{1}{a}_{ALCDEF} u_{C,D} u_{E,F} \phi_{,A} + \frac{1}{2} \underset{3}{k}_{ABLDE} u_{D,E} \phi_{,A} \phi_{,B} - \frac{1}{6} \underset{4}{\chi}_{ABCL} \phi_{,A} \phi_{,B} \phi_{,C},
\end{aligned}
\tag{1.4.2}
$$

$$
\begin{aligned}
M_{Lj} \cong \varepsilon_0 \delta_{jM} \Big[& \phi_{,L} \phi_{,M} - \frac{1}{2} \phi_{,K} \phi_{,K} \delta_{LM} - \phi_{,K} \phi_{,M} u_{K,L} \\
& - \phi_{,K} \phi_{,M} u_{L,K} + \phi_{,L} \phi_{,M} u_{K,K} - \phi_{,L} \phi_{,K} u_{K,M} \\
& + \phi_{,K} \phi_{,R} u_{R,K} \delta_{LM} + \frac{1}{2} \phi_{,K} \phi_{,K} u_{L,M} - \frac{1}{2} \phi_{,R} \phi_{,R} u_{K,K} \delta_{LM} \Big],
\end{aligned}
\tag{1.4.3}
$$

$$\varepsilon_0 J C_{KL}^{-1} \mathcal{E}_K \cong \varepsilon_0 \Big[-\phi_{,L} + \phi_{,K} u_{L,K} - \phi_{,L} u_{K,K} + \phi_{,K} u_{K,L} - \phi_{,M} u_{L,K} u_{K,M} + \phi_{,K} u_{M,M} u_{L,K} - \frac{1}{2}\phi_{,L} u_{K,K} u_{M,M} + \frac{1}{2}\phi_{,L} u_{K,M} u_{M,K} - \phi_{,M} u_{L,K} u_{M,K} + \phi_{,M} u_{M,L} u_{K,K} - \phi_{,M} u_{M,K} u_{K,L} \Big]. \tag{1.44}$$

A special case of the cubic theory is the case of relatively large mechanical deformations and weak electric fields [9]. In this case all electrical nonlinearities can be neglected. The following energy density is sufficient:

$$\rho_0 \Psi = \frac{1}{2} c_{ABCD} S_{AB} S_{CD} - e_{ABC} \mathcal{E}_A S_{BC} - \frac{1}{2} \chi_{AB} \mathcal{E}_A \mathcal{E}_B + \frac{1}{6} c_{ABCDEF} S_{AB} S_{CD} S_{EF} + \frac{1}{24} c_{ABCDEFGH} S_{AB} S_{CD} S_{EF} S_{GH}. \tag{1.4.5}$$

Keeping the linear terms of the electric potential gradient and up to cubic terms of the displacement gradient, we obtain

$$\begin{aligned} K_{LM} &= c_{LMRS} u_{R,S} + e_{KLM} \phi_{,K} \\ &\quad + c^e_{LMRSKN} u_{R,S} u_{K,N} + c^e_{LMRSKNIJ} u_{R,S} u_{K,N} u_{I,J}, \\ \mathcal{D}_K &= e_{KRS} u_{R,S} - \varepsilon_{KL} \phi_{,L}, \end{aligned} \tag{1.4.6}$$

where

$$\begin{aligned} c^e_{LMRSKN} &= \frac{1}{2}(c_{LMRSKN} + c_{LMNS}\delta_{KR} + c_{LNRS}\delta_{KM}), \\ c^e_{LMRSKNIJ} &= \frac{1}{6} c_{LMRSKNIJ} \\ &\quad + \frac{1}{2}(c_{LMKNSJ}\delta_{RI} + c_{LNSJ}\delta_{MK}\delta_{RI} + c_{LNRSIJ}\delta_{MK}). \end{aligned} \tag{1.4.7}$$

Chapter 2

Thickness-Shear Modes of Plate Resonators

Thickness modes of a plate are modes of unbounded plates. They do not depend on the in-plane coordinates of the plate and only vary along the plate thickness. Material particles in a plane parallel to the surfaces of the plate have the same motion. They are classified into thickness-shear and thickness-stretch modes, depending on whether the material particles are moving parallel to the plate surfaces or perpendicular to them. Thickness modes are the so-called high frequency modes whose frequencies are determined by the plate thickness, the smallest dimension. This is in contrast to the low frequency modes of extension and flexure in traditional structural engineering, whose frequencies depend strongly on the length and/or width of the plate. We mainly study thickness-shear modes which are the most common operating modes for piezoelectric resonators.

2.1. Static Thickness-Shear Deformation

First we examine static thickness-shear deformations of a plate. We consider two cases of a plate under a mechanical load or an electrical load separately.

2.1.1. A plate under a mechanical load

Consider an unbounded plate of crystals of monoclinic symmetry (see Fig. 2.1.1). The major surfaces of the plate are under a tangential traction p and are electroded. Two cases of shorted and open electrodes will be considered.

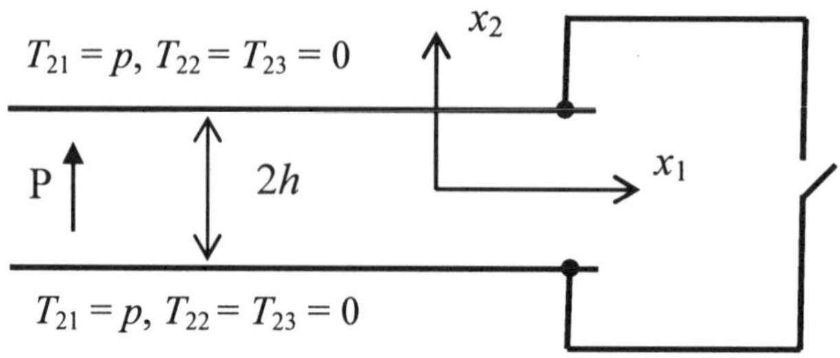

Fig. 2.1.1. An electroded crystal plate under mechanical loads.

2.1.1.1. Governing equations

The boundary-value problem is:

$$
\begin{aligned}
&T_{ji,j} = 0, \quad D_{i,i} = 0, \quad |x_2| < h,\\
&T_{ij} = c_{ijkl}S_{kl} - e_{kij}E_k, \quad D_i = e_{ikl}S_{kl} + \varepsilon_{ik}E_k, \quad |x_2| < h,\\
&S_{ij} = (u_{i,j} + u_{j,i})/2, \quad E_i = -\phi_{,i}, \quad |x_2| < h,\\
&T_{ji}n_j = p\delta_{2i}, \quad x_2 = \pm h,\\
&\phi(x_2 = h) = \phi(x_2 = -h), \quad \text{if the electrodes are shorted,}\\
&\text{or} \quad D_2(x_2 = \pm h) = 0, \quad \text{if the electrodes are open.}
\end{aligned}
\tag{2.1.1}
$$

Consider the possibility of the following displacement and potential fields:

$$u_1 = u_1(x_2), \quad u_2 = u_3 = 0, \quad \phi = \phi(x_2). \tag{2.1.2}$$

From Eqs. (1.2.39) and (1.2.40), we have the following non-vanishing components of strain, electric field, stress, and electric displacement:

$$
\begin{aligned}
&S_6 = u_{1,2}, \quad E_2 = -\phi_{,2},\\
&T_{31} = c_{56}u_{1,2} + e_{25}\phi_{,2}, \quad T_{21} = c_{66}u_{1,2} + e_{26}\phi_{,2},\\
&D_2 = e_{26}u_{1,2} - \varepsilon_{22}\phi_{,2}, \quad D_3 = e_{36}u_{1,2} - \varepsilon_{23}\phi_{,2}.
\end{aligned}
\tag{2.1.3}
$$

The equations left to be satisfied by u_1 and ϕ are

$$
\begin{aligned}
&T_{21,2} = c_{66}u_{1,22} + e_{26}\phi_{,22} = 0,\\
&D_{2,2} = e_{26}u_{1,22} - \varepsilon_{22}\phi_{,22} = 0.
\end{aligned}
\tag{2.1.4}
$$

Hence,

$$u_{1,22} = 0, \quad \phi_{,22} = 0, \tag{2.1.5}$$

unless

$$k_{26}^2 = \frac{e_{26}^2}{\varepsilon_{22} c_{66}} = 1, \tag{2.1.6}$$

which we do not consider because usually $k_{26}^2 < 1$. Equation (2.1.5) implies that all the strain, stress, electric field, and electric displacement components are constants. The mechanical displacement and electric potential are no more than linear functions of x_2.

2.1.1.2. *Shorted electrodes*

When the electrodes are shorted, the potential at the two electrodes are equal. Since E_2 is a constant, we must have

$$E_2 = 0. \tag{2.1.7}$$

The mechanical boundary conditions require that $T_{21} = p$. Then

$$S_6 = u_{1,2} = \frac{p}{c_{66}}. \tag{2.1.8}$$

The work done by p to the plate per unit volume is

$$W_1 = \frac{1}{2} T_6 S_6 = \frac{p^2}{2c_{66}}. \tag{2.1.9}$$

2.1.1.3. *Open electrodes*

In the case of open electrodes, the boundary conditions require that

$$\begin{aligned} T_{21} &= c_{66} u_{1,2} + e_{26} \phi_{,2} = p, \\ D_2 &= e_{26} u_{1,2} - \varepsilon_{22} \phi_{,2} = 0, \end{aligned} \tag{2.1.10}$$

which implies that

$$S_6 = u_{1,2} = \frac{p}{c_{66}(1 + k_{26}^2)}. \tag{2.1.11}$$

The work done by p to the plate per unit volume is

$$W_2 = \frac{1}{2} T_6 S_6 = \frac{p^2}{2c_{66}(1 + k_{26}^2)}. \tag{2.1.12}$$

Clearly, from Eqs. (2.1.9) and (2.1.12) (also see Fig. 2.1.2),

$$W_2 < W_1. \tag{2.1.13}$$

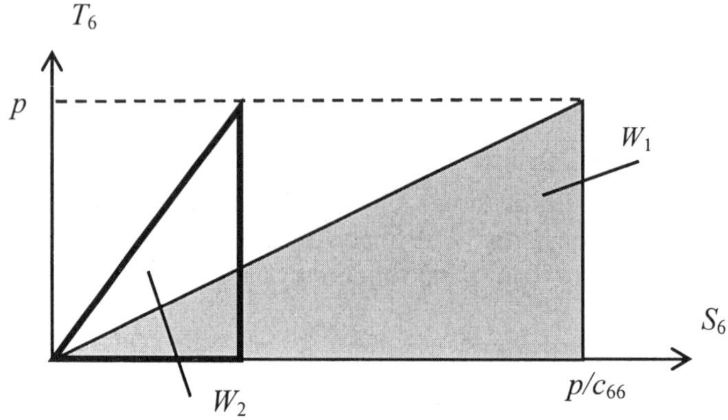

Fig. 2.1.2. Work done to the plate in different processes.

2.1.1.4. Electromechanical coupling factor

Electromechanical coupling factors are important indicators for the electrical-mechanical energy converting capability of piezoelectric materials. The electromechanical coupling factor for the thickness-shear deformation of a monoclinic plate is

$$\bar{k}_{26}^2 = \frac{W_1 - W_2}{W_1} = 1 - \frac{1}{1 + k_{26}^2} = \frac{k_{26}^2}{1 + k_{26}^2}. \tag{2.1.14}$$

2.1.2. A plate under an electrical load

Consider the same monoclinic crystal plate under an applied voltage V across the plate thickness (see Fig. 2.1.3). Two cases of mechanical boundary conditions will be considered.

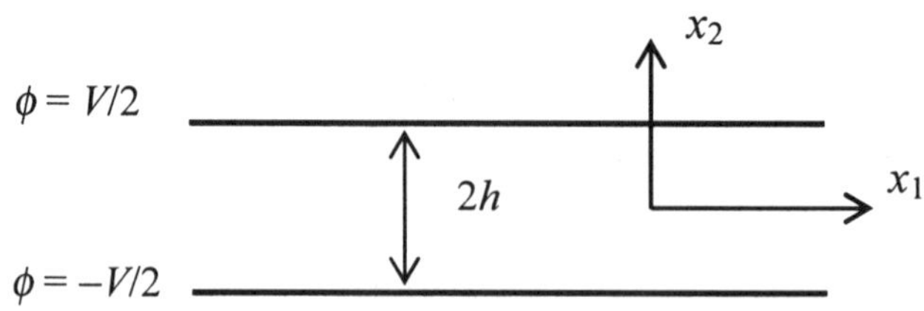

Fig. 2.1.3. A crystal plate under a voltage V.

2.1.2.1. Governing equations

The boundary-value problem is:

$$\begin{aligned}
&T_{ji,j} = 0, \quad D_{i,i} = 0, \quad |x_2| < h,\\
&T_{ij} = c_{ijkl}S_{kl} - e_{kij}E_k, \quad D_i = e_{ikl}S_{kl} + \varepsilon_{ik}E_k, \quad |x_2| < h,\\
&S_{ij} = (u_{i,j} + u_{j,i})/2, \quad E_i = -\phi_{,i} \quad |x_2| < h,\\
&\phi = \pm V/2, \quad x_2 = \pm h,\\
&T_{ji}n_j = 0, \quad x_2 = \pm h, \quad \text{if the surfaces are traction-free,}\\
&\text{or} \quad u_j = 0, \quad x_2 = \pm h, \quad \text{if the surfaces are clamped (fixed).}
\end{aligned} \tag{2.1.15}$$

Consider the possibility of a solution in the form of Eq. (2.1.2). Then Eqs. (2.1.3) through (2.1.5) still apply. All the strain, stress, electric field, and electric displacement components are constants, and the mechanical displacement and electric potential are no more than linear functions of x_2. In particular,

$$E_2 = -\frac{V}{2h}. \tag{2.1.16}$$

2.1.2.2. Free surfaces

When the surfaces are traction free, we have

$$T_{12} = c_{66}u_{1,2} + e_{26}\phi_{,2} = 0\,. \tag{2.1.17}$$

Then,

$$\begin{aligned}
&S_{12} = u_{1,2} = -\frac{e_{26}}{c_{66}}\frac{V}{2h},\\
&T_{31} = -c_{56}\frac{e_{26}}{c_{66}}\frac{V}{2h} + e_{25}\frac{V}{2h} = \left(e_{25} - e_{26}\frac{c_{56}}{c_{66}}\right)\frac{V}{2h},\\
&D_2 = -e_{26}\frac{e_{26}}{c_{66}}\frac{V}{2h} - \varepsilon_{22}\frac{V}{2h} = -(1 + k_{26}^2)\varepsilon_{22}\frac{V}{2h},\\
&D_3 = -e_{36}\frac{e_{26}}{c_{66}}\frac{V}{2h} - \varepsilon_{23}\frac{V}{2h} = -\left(\varepsilon_{23} + \frac{e_{36}e_{26}}{c_{66}}\right)\frac{V}{2h}.
\end{aligned} \tag{2.1.18}$$

The free charge per unit area on the electrode at $x_2 = h$ is

$$\sigma_e = -D_2 = (1+k_{26}^2)\varepsilon_{22}\frac{V}{2h}. \qquad (2.1.19)$$

Hence the capacitance of the plate per unit area is

$$\frac{\sigma_e}{V} = (1+k_{26}^2)\frac{\varepsilon_{22}}{2h}. \qquad (2.1.20)$$

Equation (2.1.20) shows that the effect of piezoelectric coupling enhances the capacitance by a portion of k_{26}^2. The electrical energy stored in the plate capacitor per unit area is

$$W_1 = \frac{1}{2}\sigma_e V = \frac{1}{2}(1+k_{26}^2)\varepsilon_{22}\frac{V^2}{2h}. \qquad (2.1.21)$$

2.1.2.3. Clamped surfaces

When the plate surfaces are clamped, the strain S_{12} is still a constant and u_1 is still no more than a linear function of x_2. The displacement boundary conditions require this linear function to vanish at two points. Hence

$$u_1 = 0, \qquad (2.1.22)$$

which implies that

$$S_{12} = u_{1,2} = 0, \quad T_{31} = e_{25}\frac{V}{2h}, \quad T_{12} = e_{26}\frac{V}{2h},$$
$$D_2 = -\varepsilon_{22}\frac{V}{2h}, \quad D_3 = -\varepsilon_{23}\frac{V}{2h}. \qquad (2.1.23)$$

The free charge per unit area on the electrode at $x_2 = h$ is

$$\sigma_e = -D_2 = \varepsilon_{22}\frac{V}{2h}. \qquad (2.1.24)$$

Hence the capacitance of the plate per unit area is

$$\frac{\sigma_e}{V} = \frac{\varepsilon_{22}}{2h}. \qquad (2.1.25)$$

The electrical energy stored in the capacitor per unit area is

$$W_2 = \frac{1}{2}\sigma_e V = \frac{1}{2}\varepsilon_{22}\frac{V^2}{2h}. \qquad (2.1.26)$$

2.1.2.4. Electromechanical coupling factor

We calculate the following:

$$\frac{W_1 - W_2}{W_1} = \frac{k_{26}^2}{1 + k_{26}^2} = \bar{k}_{26}^2, \tag{2.1.27}$$

which is the same as the electromechanical coupling factor in Eq. (2.1.14).

A rotated Y-cut quartz plate of $\theta = 32.5°$ is called an AT-cut and is widely used in devices. From the material constants in Appendix 2,

$$k_{26}^2 = \frac{0.095^2}{(39.8 \times 10^{-12})(29.0 \times 10^9)} = 0.0078,$$
$$k_{26} = 0.088, \tag{2.1.28}$$

which is a very small number. Quartz is often used for signal processing in telecommunication or sensing rather than for power handling. Therefore a small electromechanical coupling coefficient is usually sufficient.

2.2. Nonlinear Thickness-Shear Deformation

In this section we consider nonlinear, static thickness deformations of a plate (see Fig. 2.2.1) [11]. The bottom of the plate is fixed as a reference to eliminate rigid-body displacement. A uniform surface traction t_i is applied at the top. For pure thickness deformations the top surface is

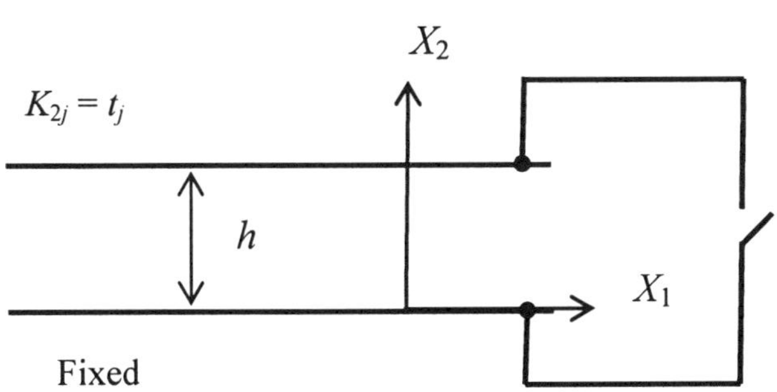

Fig. 2.2.1. Reference configuration of an electroded electroelastic plate under a uniform surface load.

displaced but is undeformed. Therefore t_i can be considered as either the traction per unit present area or unit reference area. The plate is electroded at its top and bottom surfaces. The electrodes can be shorted or opened.

2.2.1. General analysis

Let the displacement field be defined by $\mathbf{u} = \mathbf{y} - \mathbf{X}$. In pure thickness deformations the equilibrium equation and the equation of charge can be satisfied by the following linear displacement and electric potential which produce constant stress and electric displacement:

$$u_1 = a_1 X_2, \quad u_2 = a_2 X_2, \quad u_3 = a_3 X_2, \quad \phi = b X_2, \tag{2.2.1}$$

where a_i and b are undetermined constants. The mechanical boundary conditions at the fixed bottom are already satisfied by Eq. (2.2.1). a_i and b are determined by the mechanical boundary conditions at the top and electrical boundary conditions. We consider two separate cases below.

2.2.1.1. Shorted electrodes

First consider the case of shorted electrodes. In this case $b = 0$. There is no electric field and the related piezoelectric stiffening effect. a_i are determined from the traction boundary conditions at $X_2 = h$:

$$K_{Lj} N_L = K_{2j} = t_j, \tag{2.2.2}$$

which are three nonlinear equations for a_i and can be solved for a_i in terms of t_i. Equation (2.2.2) can also be viewed as functions of t_i versus a_i. Let t_i be slowly increased from zero to a final value T_i. The mechanical work done by t_i per unit area is given by

$$W_1 = \int_0^{T_i} \left[K_{2j} du_j \right]_{X_2 = h}, \tag{2.2.3}$$

where the integration limits are symbolic, meaning that the integration is for the loading process beginning with zero surface traction to the final value T_i.

2.2.1.2. Open electrodes

Next consider the case when the electrodes are open. In this case a_i and b are determined by the following boundary conditions as functions of t_i:

$$K_{2j} = t_j, \quad \mathcal{D}_2 = 0. \tag{2.2.4}$$

Let t_i increase from zero to the same T_i. The work done by t_i in this process is denoted by

$$W_2 = \int_0^{T_i} \left[K_{2j} du_j\right]_{X_2=h}. \tag{2.2.5}$$

In this case, since electric field and piezoelectric stiffening are present, the material usually appears to be stiffer. We expect a smaller displacement field and $W_2 < W_1$.

2.2.1.3. Electromechanical coupling factor

We define the electromechanical coupling factor as

$$k^2 = \frac{W_1 - W_2}{W_1}. \tag{2.2.6}$$

2.2.2. An example

The above procedure can be seen specifically through an example. Based on the availability of nonlinear material constants, we consider thickness-shear deformation in the X_1 direction of an AT-cut quartz plate under the following boundary load at the upper surface:

$$K_{21} = t, \quad u_2 = u_3 = 0, \quad X_2 = h. \tag{2.2.7}$$

The applied traction t causes a thickness-shear in the X_1 direction with K_{21} as the corresponding shear stress. In addition, K_{23} and K_{22} may be present due to anisotropy and/or nonlinearity. They are needed to maintain the displacement boundary conditions. In particular, the presence of the normal stress K_{22} due to a nonlinear shear is called the Poynting effect. We look for a solution in the following form:

$$u_1 = aX_2, \quad u_2 = u_3 = 0, \quad \phi = bX_2. \tag{2.2.8}$$

Quartz has very weak piezoelectric coupling. Under a mechanical load, the nonlinearity is mainly mechanical. For cubic nonlinear effects with weak electric fields, the relevant stress component and electric displacement component are [9]

$$\begin{aligned} K_{21} &= c_{66}u_{1,2} + e_{26}\phi_{,2} + \gamma(u_{1,2})^3 = ca + eb + \gamma a^3, \\ \mathcal{D}_2 &= e_{26}u_{1,2} - \varepsilon_{22}\phi_{,2} = ea - \varepsilon b, \end{aligned} \tag{2.2.9}$$

where we have denoted $c = c_{66}$, $e = e_{26}$, $\varepsilon = \varepsilon_{22}$, and, under the compact matrix notation, we have

$$\begin{aligned} &c = 29.01\times10^{9}\,\mathrm{N/m^2}, \quad e = -0.095\,\mathrm{C/m^2}, \\ &\varepsilon = 39.82\times10^{-12}\,\mathrm{C/Vm}, \\ &\gamma = \frac{1}{2}c_{22} + c_{266} + \frac{1}{6}c_{6666} = 13.7\times10^{11}\,\mathrm{N/m^2}. \end{aligned} \tag{2.2.10}$$

One immediate observation from Eqs. (2.2.9) and (2.2.10) is that γ is positive, which represents a nonlinear stiffening effect. The implication on the coupling factor is that W_1 with a larger strain than W_2 is affected more by this nonlinear stiffening. Therefore we expect the coupling factor to be lowered by this nonlinear effect.

2.2.2.1. *Shorted electrodes*

First consider the case of a plate with shorted electrodes. In this case $b = 0$ and Eq. $(2.2.7)_1$ requires that

$$ca + \gamma a^3 = t\,. \tag{2.2.11}$$

Although the exact solution to Eq. (2.2.11) can be obtained, for our purpose the following simple, approximate solution is sufficient. Since the nonlinear term is a small modification of the linear term, we write the solution to Eq. (2.2.11) as the sum of the linear solution and a small perturbation Δ_1

$$a \cong \frac{t}{c}(1 - \Delta_1)\,. \tag{2.2.12}$$

Substituting Eq. (2.2.12) into Eq. (2.2.11), keeping up to linear terms of Δ_1, we obtain a linear equation for Δ_1. The solution is

$$\begin{aligned} &a = f_1(t) \cong \frac{t}{c}(1 - \Delta_1) = \frac{t}{c}\left(1 - \frac{\delta_1}{1 + 3\delta_1}\right), \\ &\delta_1 = \frac{\gamma}{c}\frac{t^2}{c^2}. \end{aligned} \tag{2.2.13}$$

When t increases from zero to its final value T, we denote

$$A_1 = f_1(T)\,. \tag{2.2.14}$$

The work done by t from zero to T can be written as

$$\begin{aligned} W_1 &= \int_0^T t du_1 |_{X_2=h} \\ &= \int_0^{A_1} (ca+\gamma a^3) d(ah) = h\left(\frac{1}{2}cA_1^2 + \frac{1}{4}\gamma A_1^4\right). \end{aligned} \tag{2.2.15}$$

2.2.2.2. Open electrodes

Next consider the case of open electrodes. In this case the boundary conditions require that

$$\begin{aligned} ca + eb + \gamma a^3 &= t, \\ ea - \varepsilon b &= 0, \end{aligned} \tag{2.2.16}$$

which implies, by eliminating b, that

$$\bar{c}a + \gamma a^3 = t, \quad \bar{c} = c + \frac{e^2}{\varepsilon}. \tag{2.2.17}$$

Equation $(2.2.17)_1$ determines

$$\begin{aligned} a &= f_2(t) \cong \frac{t}{\bar{c}}(1-\Delta_2) = \frac{t}{\bar{c}}\left(1 - \frac{\delta_2}{1+3\delta_2}\right), \\ \delta_2 &= \frac{\gamma}{\bar{c}}\frac{t^2}{\bar{c}^2}. \end{aligned} \tag{2.2.18}$$

When t reaches its final value T, we denote

$$A_2 = f_2(T). \tag{2.2.19}$$

The work done by t can be written as

$$\begin{aligned} W_2 &= \int_0^T t du_1 |_{X_2=h} \\ &= \int_0^{A_2} (\bar{c}a+\gamma a^3) d(ah) = h\left(\frac{1}{2}\bar{c}A_2^2 + \frac{1}{4}\gamma A_2^4\right). \end{aligned} \tag{2.2.20}$$

2.2.2.3. Electromechanical coupling factor

The coupling factor is given by

$$k^2 = \frac{W_1 - W_2}{W_1} = \frac{cA_1^2 + \frac{1}{2}\gamma A_1^4 - \left(\bar{c}A_2^2 + \frac{1}{2}\gamma A_2^4\right)}{cA_1^2 + \frac{1}{2}\gamma A_1^4} . \quad (2.2.21)$$

We make the following observations from Eq. (2.2.21):

(i) When nonlinearity is neglected, Eq. (2.2.21) reduces to

$$k^2 = \frac{cA_1^2 - \bar{c}A_2^2}{cA_1^2} = \frac{c\left(\frac{T}{c}\right)^2 - \bar{c}\left(\frac{T}{\bar{c}}\right)^2}{c\left(\frac{T}{c}\right)^2} = \frac{k_{26}^2}{1+k_{26}^2} = \bar{k}_{26}^2 , \quad (2.2.22)$$

which is the linear electromechanical coupling factor in Eq. (2.1.14).

(ii) The nonlinear coupling factor depends on the load level T, or equivalently the deformation amplitude A.

We plot k^2 versus the load level T in Fig. 2.2.2. The figure shows that for small loads the coupling factor is a constant which is the linear coupling factor. For relatively large loads the factor becomes load dependent and is smaller than the linear coupling factor as expected.

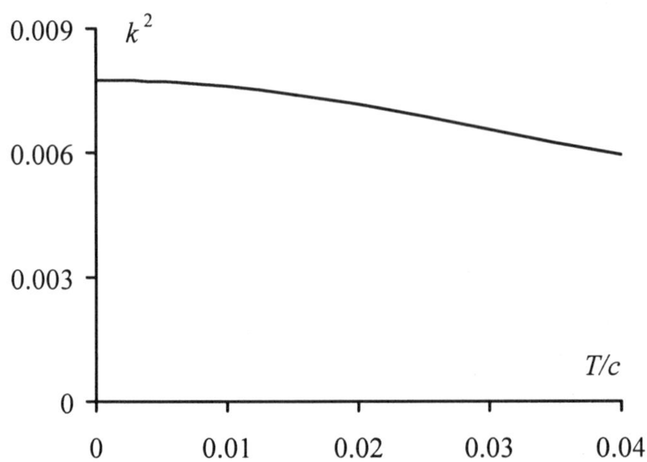

Fig. 2.2.2. Electromechanical coupling factor versus load level.

2.3. Effects of Initial Fields on Thickness-Shear Deformation

We now consider static thickness deformations of a plate under biasing fields [12]. The plate is electroded at its top and bottom surfaces. The electrodes can be shorted or opened. Figure 2.3.1 is the reference configuration of the plate.

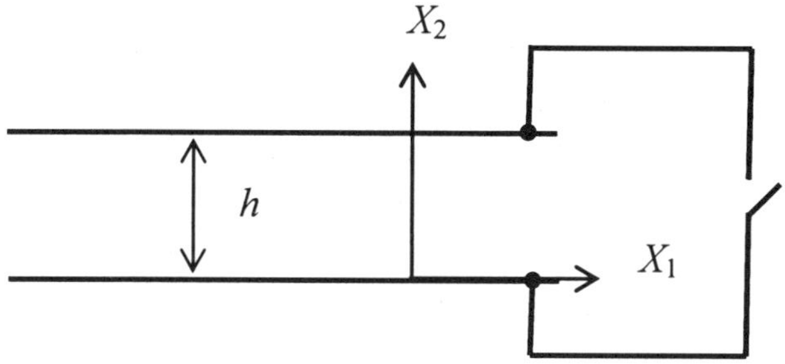

Fig. 2.3.1. Reference configuration of an electroded electroelastic plate.

2.3.1. General analysis

The plate is pre-loaded and uniform biasing fields are produced. Then the bottom of the plate is fixed as a reference to eliminate rigid-body incremental displacement. A uniform incremental surface traction t_i per unit undeformed area is then applied at the top. In pure incremental thickness deformations the displacement and electric potential can be represented by linear functions of X_2:

$$\begin{aligned} &u_1 = a_1 X_2, \quad u_2 = a_2 X_2, \quad u_3 = a_3 X_2, \\ &\phi^1 = b X_2, \end{aligned} \tag{2.3.1}$$

where a_i and b are undetermined constants. The mechanical boundary conditions at the fixed bottom are already satisfied by Eq. (2.3.1). Equation (2.3.1) implies that K^1_{Lj} and $\mathcal{D}^1_L$ are constants and hence the equation of motion and the charge equation are satisfied. a_i and b are determined by the mechanical boundary conditions at the top and electrical boundary conditions. We consider two separate cases below.

2.3.1.1 . Shorted electrodes

First consider the case of shorted electrodes. In this case $b = 0$. There is no electric field. a_i are determined from the traction boundary conditions at $X_2 = h$:

$$\begin{aligned} K_{21}^1 &= G_{2121}u_{1,2} + G_{2122}u_{2,2} + G_{2123}u_{3,2} = t_1, \\ K_{22}^1 &= G_{2221}u_{1,2} + G_{2222}u_{2,2} + G_{2223}u_{3,2} = t_2, \\ K_{23}^1 &= G_{2321}u_{1,2} + G_{2322}u_{2,2} + G_{2323}u_{3,2} = t_3, \end{aligned} \tag{2.3.2}$$

which are three equations for a_i and can be solved for a_i in terms of t_i. Equation (2.3.2) can also be viewed as functions of t_i versus a_i. Let t_i be slowly increased from zero to a final value T_i. Then the mechanical work done by t_i per unit area is given by

$$W_1 = \int_0^{T_i} [K_{2j}^1 du_j]_{X_2=h}. \tag{2.3.3}$$

2.3.1.2. Open electrodes

Next consider the case when the electrodes are open. In this case a_i and b are determined by the following boundary conditions as functions of t_i:

$$\begin{aligned} K_{21}^1 &= G_{2121}u_{1,2} + G_{2122}u_{2,2} + G_{2123}u_{3,2} + R_{221}\phi_{,2}^1 = t_1, \\ K_{22}^1 &= G_{2221}u_{1,2} + G_{2222}u_{2,2} + G_{2223}u_{3,2} + R_{222}\phi_{,2}^1 = t_2, \\ K_{23}^1 &= G_{2321}u_{1,2} + G_{2322}u_{2,2} + G_{2323}u_{3,2} + R_{223}\phi_{,2}^1 = t_3, \\ \mathcal{D}_2^1 &= R_{221}u_{1,2} + R_{222}u_{2,2} + R_{223}u_{3,2} - L_{22}\phi_{,2}^1 = 0, \end{aligned} \tag{2.3.4}$$

which are four equations for a_i and b. Let t_i increase from zero to the same final value T_i. The work done by t_i in this process is denoted by

$$W_2 = \int_0^{T_i} [K_{2j}^1 du_j]_{X_2=h}. \tag{2.3.5}$$

In this case, since an electric field is present, it usually causes piezoelectric stiffening and the material appears to be stiffer. We expect a smaller displacement field and, correspondingly, $W_2 < W_1$.

2.3.1.3. Electromechanical coupling factor

We define the electromechanical coupling factor as

$$k^2 = \frac{W_1 - W_2}{W_1}. \tag{2.3.6}$$

2.3.2. An example

The above procedure can be seen specifically through a simple example. Based on the availability of nonlinear material constants necessary in calculating the effects of biasing fields, we consider a Y-cut quartz plate and a Y-cut langasite plate below. Motions with $w_3 = 0$, $u_3 = 0$ and $\partial/\partial X_3 = 0$ are sufficient for our purpose. For biasing fields consider a uniform extension in the X_1 direction (see Fig. 2.3.2).

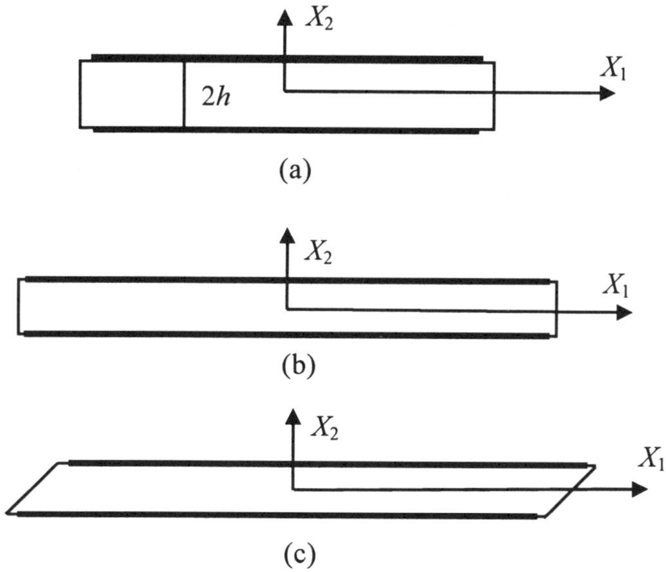

Fig. 2.3.2. (a) Reference, (b) initial, and (c) present configurations of a crystal plate.

We first determine the biasing state. The biasing strain components are:

$$S_1^0 = S^0, \quad S_2^0 = -\frac{c_{21}}{c_{22}} S^0, \tag{2.3.7}$$

where S_2^0 is due to Poisson's effect. Electrically, for the biasing state, the electrodes are shorted and there is no biasing electric field.

For the incremental motion we consider thickness-shear in the X_1 direction:

$$u_1 = u_1(X_2), \; u_2 = u_3 = 0, \quad \phi^1 = \phi^1(X_2). \tag{2.3.8}$$

Under a mechanical load, mechanical nonlinearities are the major nonlinearities. Therefore, as an approximation, we only consider the effects of the biasing fields through the elastic constants, linear and nonlinear. The relevant incremental constitutive relations are

$$\begin{aligned} K_{21}^1 &= c_{66}(1+cS_1^0)u_{1,2} + e_{26}\phi_{,2}^1, \\ K_{22}^1 &= (c_{261}S_1^0 + c_{262}S_2^0)u_{1,2}, \\ K_{23}^1 &= (c_{461}S_1^0 + c_{462}S_2^0)u_{1,2}, \\ \mathcal{D}_2^1 &= e_{26}u_{1,2} - \varepsilon_{22}\phi_{,2}^1, \end{aligned} \tag{2.3.9}$$

where

$$c = 2 + \frac{c_{661}}{c_{66}} - \frac{c_{662}c_{21}}{c_{66}c_{22}}, \tag{2.3.10}$$

and, under the compact matrix notation, c_{PQR} are the third-order elastic constants. For Y-cut quartz and langasite:

$$c_{261} = c_{262} = c_{461} = c_{462} = 0, \tag{2.3.11}$$

and hence $K_{22}^1 = K_{23}^1 = 0$. The equations of equilibrium and charge are

$$K_{21,2}^1 = 0, \quad K_{22,2}^1 = 0, \quad K_{23,2}^1 = 0, \quad \mathcal{D}_{2,2}^1 = 0. \tag{2.3.12}$$

Equations $(2.3.12)_{2,3}$ are trivially satisfied. Equations $(2.3.12)_{1,4}$ take the following form:

$$\begin{aligned} c_{66}(1+cS^0)u_{1,22} + e_{26}\phi_{,22}^1 &= 0, \\ e_{26}u_{1,22} - \varepsilon_{22}\phi_{,22}^1 &= 0. \end{aligned} \tag{2.3.13}$$

Equation (2.3.13) shows that both $u_1(X_2)$ and $\phi^1(X_2)$ depend on X_2 linearly. We therefore take

$$u_1(X_2) = aX_2, \qquad u_2 = u_3 = 0, \qquad \phi^1(X_2) = bX_2. \tag{2.3.14}$$

The relevant non-zero stress component and electric displacement component are

$$\begin{aligned} K_{21}^1 &= c_{66}(1+cS^0)a + e_{26}b, \\ \mathcal{D}_2^1 &= e_{26}a - \varepsilon_{22}b. \end{aligned} \tag{2.3.15}$$

2.3.2.1. Shorted electrodes

First we consider the case of a plate with shorted electrodes. In this case $b = 0$ in Eq. (2.3.15). When t_1 increases from zero to its final value T, a goes from zero to

$$A_1 = \frac{T}{c_{66}(1+cS^0)}. \tag{2.3.16}$$

The work done by t_1 during this process can be written as

$$\begin{aligned} W_1 &= \int_0^T [K_{21}^1 du_1]_{X_2=h} = \int_0^{A_1} c_{66}(1+cS^0)ad(ah) \\ &= \frac{T^2 h}{2c_{66}(1+cS^0)}. \end{aligned} \tag{2.3.17}$$

2.3.2.2. Open electrodes

When the two electrodes are open, the electrical boundary condition $\mathcal{D}_2^1 = 0$ requires that $b = e_{26}\varepsilon_{22}^{-1}a$ from Eq. $(2.3.15)_2$. Thus,

$$K_{21}^1 = c_{66}\left[(1+cS^0)+k_{26}^2\right]a, \quad k_{26}^2 = \frac{e_{26}^2}{\varepsilon_{22}c_{66}}. \tag{2.3.18}$$

When t_1 increases from zero to its final value T, a goes from zero to

$$A_2 = \frac{T}{c_{66}\left[(1+cS^0)+k_{26}^2\right]}. \tag{2.3.19}$$

The work done by t_1 during this process can be written as

$$\begin{aligned} W_2 &= \int_0^T [K_{21}^1 du_1]_{X_2=h} = \int_0^{A_2} c_{66}(1+cS^0+k_{26}^2)ad(ah) \\ &= \frac{T^2 h}{2c_{66}(1+cS^0+k_{26}^2)}. \end{aligned} \tag{2.3.20}$$

2.3.2.3. Electromechanical coupling factor

Substitution of Eqs. (2.3.17) and (2.3.20) into Eq. (2.3.6), we obtain the electromechanical coupling factor as

$$k^2 = \frac{k_{26}^2}{1 + cS^0 + k_{26}^2}. \tag{2.3.21}$$

Obviously, when the biasing field is absent, i.e. $S^0 = 0$, Eq. (2.3.21) reduces to

$$k^2 = \frac{k_{26}^2}{1 + k_{26}^2} = \overline{k}_{26}^2, \tag{2.3.22}$$

where $\overline{k}_{26}^2$ is the linear electromechanical coupling factor without biasing fields.

To show the quantitative effect of the biasing fields on the electromechanical coupling factor, two numerical examples are calculated below. We plot k^2 versus S^0 for quartz in Fig. 2.3.3 and for langasite in Fig. 2.3.4, respectively. It follows from Fig. 2.3.3 that the electromechanical coupling factor of a Y-cut quartz plate decreases monotonically. Therefore a biasing compression enlarges k^2 and a biasing extension reduces k^2. Figure 2.3.4 shows that a Y-cut langasite plate has the opposite behavior. An extension raises k^2 and a compression lowers it.

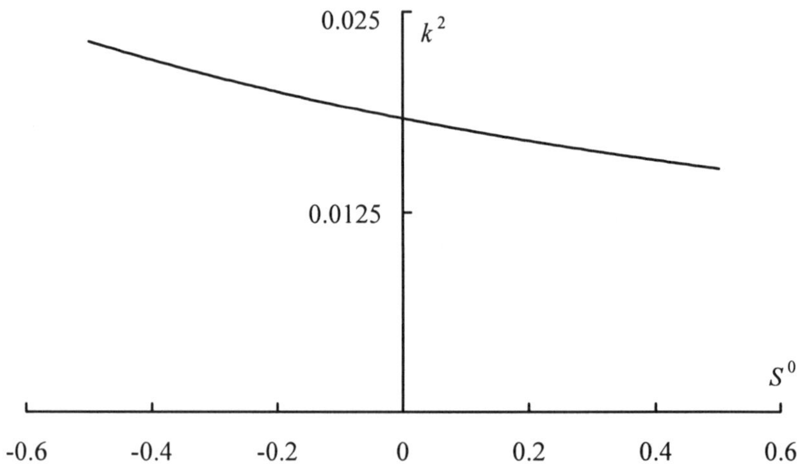

Fig. 2.3.3. Electromechanical coupling factor of a Y-cut quartz plate.

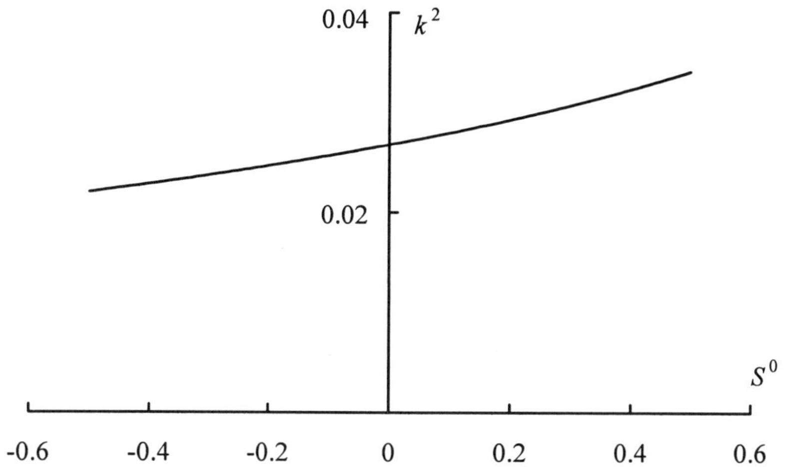

Fig. 2.3.4. Electromechanical coupling factor of a Y-cut langasite plate.

2.4. Linear Thickness-Shear Vibration

Thickness vibrations of piezoelectric plates were analyzed in [13] in general. We study the special case of thickness-shear vibration of a quartz plate. Consider an unbounded, rotated Y-cut quartz plate (see Fig. 2.4.1). The two major surfaces are traction-free and are electroded, with a driving voltage $V\exp(i\omega t)$ across the thickness.

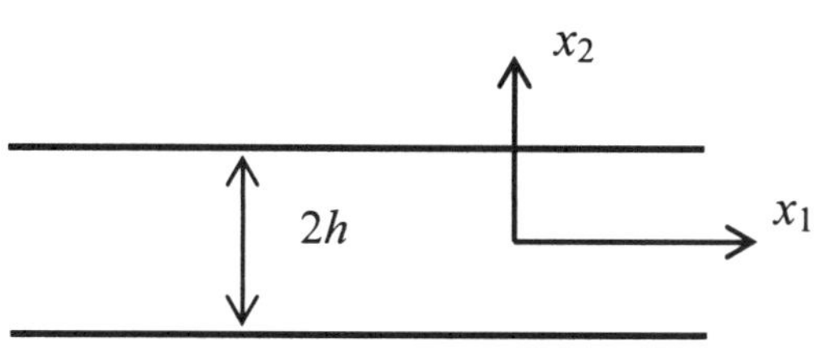

Fig. 2.4.1. An electroded quartz plate.

2.4.1. *Governing equations*

The governing equations and boundary conditions are:

$$T_{ji,j} = \rho \ddot{u}_i, \quad D_{i,i} = 0, \quad |x_2| < h,$$
$$T_{ij} = c_{ijkl} S_{kl} - e_{kij} E_k, \quad D_i = e_{ikl} S_{kl} + \varepsilon_{ik} E_k, \quad |x_2| < h, \tag{2.4.1}$$
$$S_{ij} = (u_{i,j} + u_{j,i})/2, \quad E_i = -\phi_{,i}, \quad |x_2| < h,$$

and

$$T_{2j} = 0, \quad x_2 = \pm h,$$
$$\phi(x_2 = h) - \phi(x_2 = -h) = V \exp(i\omega t). \tag{2.4.2}$$

Consider the possibility of the following displacement and potential fields:

$$u_1 = u_1(x_2)\exp(i\omega t), \quad u_2 = u_3 = 0, \quad \phi = \phi(x_2)\exp(i\omega t). \tag{2.4.3}$$

The non-trivial components of strain, electric field, stress, and electric displacement are

$$2S_{12} = u_{1,2}, \quad E_2 = -\phi_{,2}, \tag{2.4.4}$$

and

$$T_{31} = c_{56} u_{1,2} + e_{25}\phi_{,2}, \quad T_{12} = c_{66} u_{1,2} + e_{26}\phi_{,2},$$
$$D_2 = e_{26} u_{1,2} - \varepsilon_{22}\phi_{,2}, \quad D_3 = e_{36} u_{1,2} - \varepsilon_{23}\phi_{,2}, \tag{2.4.5}$$

where the time-harmonic factor has been dropped. The equation of motion and the charge equation require that

$$T_{21,2} = c_{66} u_{1,22} + e_{26}\phi_{,22} = -\rho\omega^2 u_1,$$
$$D_{2,2} = e_{26} u_{1,22} - \varepsilon_{22}\phi_{,22} = 0. \tag{2.4.6}$$

Equation $(2.4.6)_2$ can be integrated to yield

$$\phi = \frac{e_{26}}{\varepsilon_{22}} u_1 + B_1 x_2 + B_2, \tag{2.4.7}$$

where B_1 and B_2 are integration constants and B_2 is immaterial. Substituting Eq. (2.4.7) into the expressions for T_{21}, D_2, and Eq. $(2.4.6)_1$, we obtain

$$T_{21} = \bar{c}_{66} u_{1,2} + e_{26} B_1, \quad D_2 = -\varepsilon_{22} B_1, \tag{2.4.8}$$

$$\bar{c}_{66} u_{1,22} = -\rho\,\omega^2 u_1, \tag{2.4.9}$$

where

$$\bar{c}_{66} = c_{66}(1 + k_{26}^2), \quad k_{26}^2 = \frac{e_{26}^2}{\varepsilon_{22} c_{66}}. \tag{2.4.10}$$

The general solution to Eq. (2.4.9) and the corresponding expression for the electric potential are

$$u_1 = A_1 \sin \xi x_2 + A_2 \cos \xi x_2,$$
$$\phi = \frac{e_{26}}{\varepsilon_{22}}(A_1 \sin \xi x_2 + A_2 \cos \xi x_2) + B_1 x_2 + B_2, \tag{2.4.11}$$

where A_1 and A_2 are integration constants, and

$$\xi^2 = \frac{\rho}{\bar{c}_{66}}\omega^2. \tag{2.4.12}$$

Then the expression for the stress component relevant to boundary conditions is

$$T_{21} = \bar{c}_{66}(A_1 \xi \cos \xi x_2 - A_2 \xi \sin \xi x_2) + e_{26} B_1. \tag{2.4.13}$$

The boundary conditions require that

$$\bar{c}_{66} A_1 \xi \cos \xi h - \bar{c}_{66} A_2 \xi \sin \xi h + e_{26} B_1 = 0,$$
$$\bar{c}_{66} A_1 \xi \cos \xi h + \bar{c}_{66} A_2 \xi \sin \xi h + e_{26} B_1 = 0, \tag{2.4.14}$$
$$2\frac{e_{26}}{\varepsilon_{22}} A_1 \sin \xi h + 2B_1 h = V.$$

We can add Eqs. $(2.14)_{1,2}$, and also subtract them from each other. The Eqs. (2.4.14) become:

$$\bar{c}_{66} A_1 \xi \cos \xi h + e_{26} B_1 = 0,$$
$$\bar{c}_{66} A_2 \xi \sin \xi h = 0, \tag{2.4.15}$$
$$2\frac{e_{26}}{\varepsilon_{22}} A_1 \sin \xi h + 2B_1 h = V.$$

2.4.2. Free vibration

First, consider free vibrations with $V = 0$. Equation (2.4.15) decouples into two sets of equations. For symmetric modes,

$$\bar{c}_{66} A_2 \xi \sin \xi h = 0. \tag{2.4.16}$$

Non-trivial solutions may exist if

$$\sin \xi h = 0, \tag{2.4.17}$$

or

$$\xi^{(n)} h = \frac{n\pi}{2}, \quad n = 0, 2, 4, 6, \ldots, \tag{2.4.18}$$

which determines the following resonant frequencies:

$$\omega^{(n)} = \frac{n\pi}{2h}\sqrt{\frac{\bar{c}_{66}}{\rho}}, \quad n = 0, 2, 4, 6, \ldots. \tag{2.4.19}$$

Equation (2.4.17) implies that $B_1 = 0$ and $A_1 = 0$. The corresponding modes are

$$u_1 = \cos \xi^{(n)} x_2, \quad \phi = \frac{e_{26}}{\varepsilon_{22}} \cos \xi^{(n)} x_2, \tag{2.4.20}$$

where $n = 0$ represents a rigid-body mode.

For anti-symmetric modes,

$$\begin{aligned} &\bar{c}_{66} A_1 \xi \cos \xi h + e_{26} B_1 = 0, \\ &2 \frac{e_{26}}{\varepsilon_{22}} A_1 \sin \xi h + 2 B_1 h = 0. \end{aligned} \tag{2.4.21}$$

The resonance frequencies are determined by

$$\begin{vmatrix} \bar{c}_{66} \xi \cos \xi h & e_{26} \\ \dfrac{e_{26}}{\varepsilon_{22}} \sin \xi h & h \end{vmatrix} = \bar{c}_{66} \xi h \cos \xi h - \frac{e_{26}^2}{\varepsilon_{22}} \sin \xi h = 0, \tag{2.4.22}$$

or

$$\cot \xi h = \frac{\bar{k}_{26}^2}{\xi h}, \tag{2.4.23}$$

where

$$\bar{k}_{26}^2 = \frac{e_{26}^2}{\varepsilon_{22} \bar{c}_{66}} = \frac{e_{26}^2}{\varepsilon_{22} c_{66} (1 + k_{26}^2)} = \frac{k_{26}^2}{1 + k_{26}^2}. \tag{2.4.24}$$

Equations (2.4.23) and (2.4.21) determine the resonant frequencies and modes.

If the small piezoelectric coupling for quartz is neglected in Eq. (2.4.23), a set of frequencies similar to Eq. (2.4.19) with n equals odd numbers can be determined for a set of modes with sine dependence on the thickness coordinate. Static thickness-shear deformation and the first few thickness-shear modes in a plate are shown in Fig. 2.4.2.

For quartz $\bar{k}_{26}^2$ is very small. Equation (2.4.23) can be solved approximately. For example, for the most widely used fundamental mode of $n = 1$, we can write [4]

$$\xi h = \frac{\pi}{2} - \Delta, \tag{2.4.25}$$

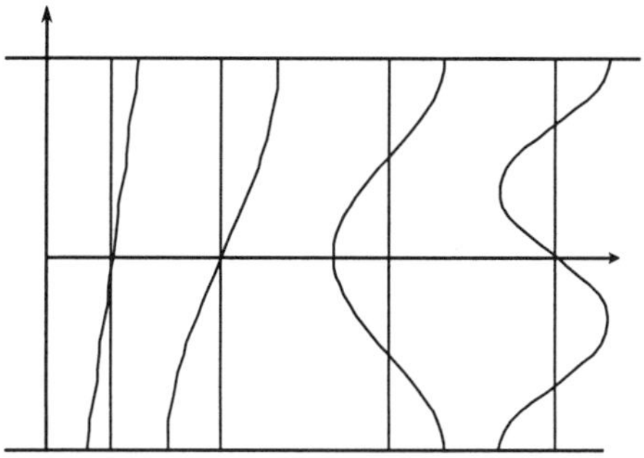

Fig. 2.4.2. Thickness-shear deformation and modes in a plate.

where Δ is a small number. Substituting Eq. (2.4.25) into Eq. (2.4.23), for small $\bar{k}_{26}^2$ and small Δ, we obtain

$$\Delta \cong \frac{2}{\pi}\bar{k}_{26}^2, \tag{2.4.26}$$

or

$$\begin{aligned}\omega^{(1)} &\cong \frac{\pi}{2h}\sqrt{\frac{\bar{c}_{66}}{\rho}}\left(1-\frac{4}{\pi^2}\bar{k}_{26}^2\right) = \frac{\pi}{2h}\sqrt{\frac{c_{66}(1+k_{26}^2)}{\rho}}\left(1-\frac{4}{\pi^2}\bar{k}_{26}^2\right)\\ &\cong \frac{\pi}{2h}\sqrt{\frac{c_{66}}{\rho}}\left(1+\frac{1}{2}k_{26}^2\right)\left(1-\frac{4}{\pi^2}\bar{k}_{26}^2\right)\\ &\cong \frac{\pi}{2h}\sqrt{\frac{c_{66}}{\rho}\left(1+\frac{1}{2}k_{26}^2-\frac{4}{\pi^2}k_{26}^2\right)} > \frac{\pi}{2h}\sqrt{\frac{c_{66}}{\rho}}.\end{aligned} \tag{2.4.27}$$

Equation (2.4.27) shows that piezoelectric coupling raises the frequency. This effect is called piezoelectric stiffening.

2.4.3. Forced vibration

For forced vibration we have $A_2 = 0$ and

$$A_1 = \frac{\begin{vmatrix} 0 & e_{26} \\ V & 2h \end{vmatrix}}{\begin{vmatrix} \bar{c}_{66}\xi\cos\xi h & e_{26} \\ 2\dfrac{e_{26}}{\varepsilon_{22}}\sin\xi h & 2h \end{vmatrix}} = \frac{-e_{26}V}{2\bar{c}_{66}\xi h\cos\xi h - 2\dfrac{e_{26}^2}{\varepsilon_{22}}\sin\xi h}, \tag{2.4.28}$$

$$B_1 = \frac{\begin{vmatrix} \bar{c}_{66}\xi\cos\xi h & 0 \\ 2\dfrac{e_{26}}{\varepsilon_{22}}\sin\xi h & V \end{vmatrix}}{\begin{vmatrix} \bar{c}_{66}\xi\cos\xi h & e_{26} \\ 2\dfrac{e_{26}}{\varepsilon_{22}}\sin\xi h & 2h \end{vmatrix}} = \frac{V\bar{c}_{66}\xi\cos\xi h}{2\bar{c}_{66}\xi h\cos\xi h - 2\dfrac{e_{26}^2}{\varepsilon_{22}}\sin\xi h}. \tag{2.4.29}$$

Hence

$$D_2 = -\varepsilon_{22}B_1 = -\varepsilon_{22}\frac{V}{2h}\frac{\xi h}{\xi h - \bar{k}_{26}^2\tan\xi h} = -\sigma_e, \tag{2.4.30}$$

where σ_e is the surface free charge per unit area on the electrode at $x_2 = h$. The capacitance per unit area is then

$$C = \frac{\sigma_e}{V} = \frac{\varepsilon_{22}}{2h}\frac{\xi h}{\xi h - \bar{k}_{26}^2\tan\xi h}. \tag{2.4.31}$$

Note the following limits:

$$\lim_{e_{26}\to 0} C = \frac{\varepsilon_{22}}{2h},$$
$$\lim_{\omega\to 0} C = \frac{\varepsilon_{22}}{2h}\frac{1}{1 - \dfrac{k_{26}^2}{1+k_{26}^2}} = \frac{\varepsilon_{22}}{2h}(1 + k_{26}^2). \tag{2.4.32}$$

2.4.4. An unelectroded plate

Finally, for free vibrations of an unelectroded plate (or equivalently an electroded plate with open electrodes), $D_2 = 0$ at the plate surfaces implies from Eq. $(2.4.8)_2$ that $B_1 = 0$. For symmetric modes $A_1 = 0$ and

$$\omega^{(n)} = \frac{n\pi}{2h}\sqrt{\frac{\bar{c}_{66}}{\rho}}, \quad n = 0, 2, 4, 6, \ldots. \tag{2.4.33}$$

For anti-symmetric modes $A_2 = 0$ and

$$\omega^{(n)} = \frac{n\pi}{2h}\sqrt{\frac{\bar{c}_{66}}{\rho}}, \quad n = 1, 3, 5, \ \ldots. \tag{2.4.34}$$

We note that the $\omega^{(1)}$ from Eq. (2.4.34) is higher than that in Eq. (2.4.27) because for an unelectroded plate or an electroded plate with open electrodes there is more piezoelectric stiffening and hence higher frequency.

2.5. Effects of Electrode Inertia

The electrodes on the crystal plate in the previous section were assumed to be very thin. Their mechanical effects were neglected. When the electrodes are not very thin, their mechanical effects, e.g. inertia and stiffness, need to be considered. In this section we study the inertial effect of the electrode mass only. In resonator manufacturing, one electrode is deposited first with a pre-determined thickness. Then the electrode on the other side of the crystal plate has a thickness that is determined by that the electroded plate having a desired frequency. This usually results in a crystal plate with two electrodes of slightly different thickness. Consider a plate of monoclinic crystals with electrodes of unequal thickness as shown in Fig. 2.5.1. The electrodes are shorted. The case of electrodes of equal thickness was treated in [4,14], and the case of electrodes of unequal thickness was studied in [15].

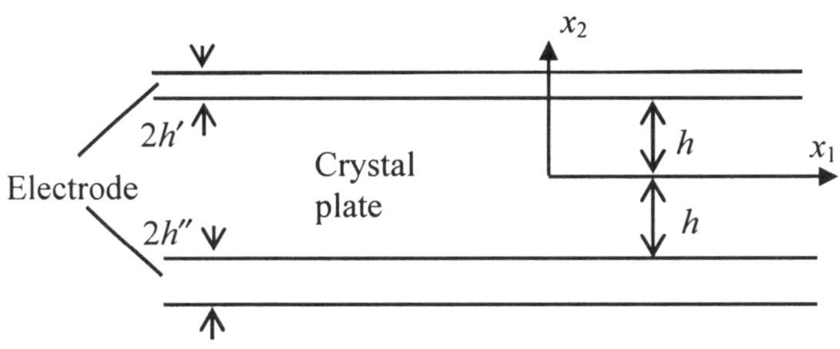

Fig. 2.5.1. A crystal plate with electrodes of different thickness.

2.5.1 General analysis

We have the following equations and boundary conditions:

$$\begin{aligned}
&T_{2i,2} = \rho \ddot{u}_i, \quad D_{2,2} = 0, \quad |x_2| < h, \\
&T_{ij} = c_{ijkl} S_{kl} - e_{kij} E_k, \quad D_i = e_{ijk} S_{jk} + \varepsilon_{ij} E_j, \quad |x_2| < h, \\
&S_{ij} = (u_{i,j} + u_{j,i})/2, \quad E_i = -\phi_{,i}, \quad |x_2| < h, \\
&-T_{2j} = 2\rho' h' \ddot{u}_j, \quad x_2 = h, \\
&T_{2j} = 2\rho' h'' \ddot{u}_j, \quad x_2 = -h, \\
&\phi(x_2 = h) = \phi(x_2 = -h),
\end{aligned} \tag{2.5.1}$$

where ρ' is the electrode mass density. For monoclinic crystals, thickness-shear modes described by the following displacement and potential fields are allowed:

$$u_1 = u_1(x_2)\exp(i\omega t), \quad u_2 = u_3 = 0, \quad \phi = \phi(x_2)\exp(i\omega t). \tag{2.5.2}$$

From Eqs. (2.4.8) through (2.4.13) we still have

$$\begin{aligned}
&\xi^2 = \rho\omega^2 / \bar{c}_{66}, \\
&u_1 = A_1 \sin\xi x_2 + A_2 \cos\xi x_2, \\
&\phi = \frac{e_{26}}{\varepsilon_{22}}(A_1 \sin\xi x_2 + A_2 \cos\xi x_2) + B_1 x_2 + B_2, \\
&T_{21} = \bar{c}_{66}(A_1 \xi \cos\xi x_2 - A_2 \xi \sin\xi x_2) + e_{26} B_1, \\
&D_2 = -\varepsilon_{22} B_1.
\end{aligned} \tag{2.5.3}$$

Substituting Eq. (2.5.3) into the boundary conditions in Eq. (2.5.1), we obtain

$$\begin{aligned}
&-\bar{c}_{66}\xi(A_1 \cos\xi h - A_2 \sin\xi h) - e_{26} B_1 \\
&\quad = -\omega^2 2\rho' h'(A_1 \sin\xi h + A_2 \cos\xi h), \\
&\bar{c}_{66}\xi(A_1 \cos\xi h + A_2 \sin\xi h) + e_{26} B_1 \\
&\quad = -\omega^2 2\rho' h''(-A_1 \sin\xi h + A_2 \cos\xi h), \\
&\frac{e_{26}}{\varepsilon_{22}} A_1 \sin\xi h + B_1 h = 0.
\end{aligned} \tag{2.5.4}$$

For non-trivial solutions of the undetermined constants, the determinant of the coefficient matrix of Eq. (2.5.4) has to vanish. This results in the following frequency equation:

$$\begin{vmatrix} -\bar{c}_{66}\xi\cos\xi h+\omega^2 2\rho' h'\sin\xi h & \bar{c}_{66}\xi\sin\xi h+\omega^2 2\rho' h'\cos\xi h & -e_{26} \\ \bar{c}_{66}\xi\cos\xi h-\omega^2 2\rho' h''\sin\xi h & \bar{c}_{66}\xi\sin\xi h+\omega^2 2\rho' h''\cos\xi h & e_{26} \\ \frac{e_{26}}{\varepsilon_{22}}\sin\xi h & 0 & h \end{vmatrix}=0. \quad (2.5.5)$$

2.5.2. *Identical electrodes*

We examine the special case of symmetric electrodes when $h''=h'$. The modes can be classified as symmetric and anti-symmetric and Eq. (2.5.5) splits into two factors. For symmetric modes $A_1=0$ and $B_1=0$. The frequency equation is

$$\tan\xi h=-R\xi h, \quad (2.5.6)$$

where R is the electrode-crystal mass ratio given by

$$R=\frac{2\rho' h'}{\rho h}. \quad (2.5.7)$$

For anti-symmetric modes $A_2=0$. The frequency equation is

$$\cot\xi h-\frac{\bar{k}_{26}^2}{\xi h}=R\xi h. \quad (2.5.8)$$

For the fundamental anti-symmetric mode, if we write [4]

$$\xi h=\frac{\pi}{2}-\Delta, \quad (2.5.9)$$

we obtain

$$\Delta\cong\frac{2}{\pi}\bar{k}_{26}^2+\frac{\pi}{2}R, \quad (2.5.10)$$

or

$$\omega^{(1)}\cong\frac{\pi}{2h}\sqrt{\frac{\bar{c}_{66}}{\rho}}\left(1-\frac{4}{\pi^2}\bar{k}_{26}^2-R\right). \quad (2.5.11)$$

Obviously, R lowers the frequency. When $R=0$, Eqs. (2.5.6), (2.5.8) and (2.5.11) reduce to Eqs. (2.4.17), (2.4.23) and (2.4.27), respectively.

2.6. Inertial Effects of Imperfectly Bounded Electrodes

The electrodes in the previous section follow the crystal surfaces perfectly with the same displacements as the crystal surfaces. In this section, we analyze thickness-shear vibration of a crystal plate with imperfectly bonded electrodes [16]. The interfaces between the crystal and the electrodes are described by the so-called shear-lag model. Consider a plate of monoclinic crystals with imperfectly bonded electrodes of unequal thickness as shown in Fig. 2.6.1.

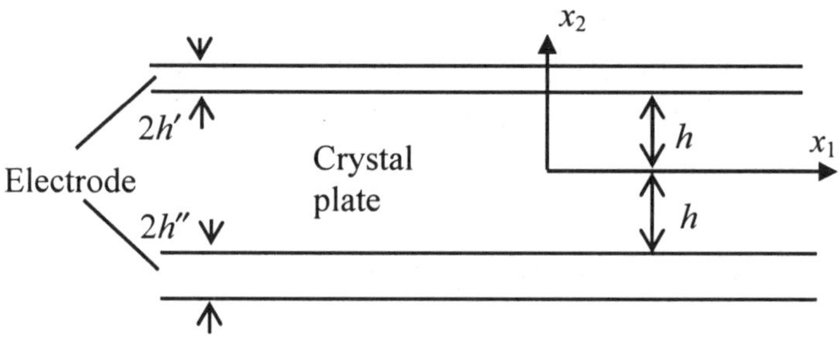

Fig. 2.6.1. A crystal plate with imperfectly bonded electrodes.

2.6.1. General analysis

For free vibration frequencies we have the following equations and boundary conditions:

$$\begin{aligned}
&T_{2i,2} = \rho \ddot{u}_i, \quad D_{2,2} = 0, \quad |x_2| < h, \\
&T_{ij} = c_{ijkl}S_{kl} - e_{kij}E_k, \quad D_i = e_{ijk}S_{jk} + \varepsilon_{ij}E_j, \quad |x_2| < h, \\
&S_{ij} = (u_{i,j} + u_{j,i})/2, \quad E_i = -\phi_{,i}, \quad |x_2| < h, \\
&-T_{2j} = 2\rho' h' \ddot{u}'_j, \quad x_2 = h, \\
&T_{2j} = 2\rho'' h'' \ddot{u}''_j, \quad x_2 = -h, \\
&\phi(x_2 = h) = \phi(x_2 = -h),
\end{aligned} \tag{2.6.1}$$

where ρ' and ρ'' are the mass densities of the top and bottom electrodes, which are allowed to be different. u'_j and u''_j are the displacements of the top and bottom electrodes. According to the shear-lag model, u'_j and u''_j are allowed to be different from the crystal surface displacements and the following constitutive relations describe the behavior of the interfaces between the crystal and the electrodes:

$$\begin{aligned} T_{2j} &= k'(u'_j - u_j), \quad x_2 = h, \\ T_{2j} &= k''(u_j - u''_j), \quad x_2 = -h, \end{aligned} \tag{2.6.2}$$

where k' and k'' are elastic constants of the interfaces. With k' and k'', the interfaces are allowed to possess strain energies. Substituting Eq. (2.6.2) into the boundary conditions in Eq. (2.6.1), we have

$$\begin{aligned} &-k'(u'_j - u_j) = 2\rho' h' \ddot{u}'_j, \quad x_2 = h, \\ &k''(u_j - u''_j) = 2\rho'' h'' \ddot{u}''_j, \quad x_2 = -h, \\ &\phi(x_2 = h) = \phi(x_2 = -h). \end{aligned} \tag{2.6.3}$$

For monoclinic crystals, thickness-shear modes described by the following displacement and potential fields are allowed:

$$\begin{aligned} &u_1 = u_1(x_2)\exp(i\omega t), \quad u_2 = u_3 = 0, \quad \phi = \phi(x_2)\exp(i\omega t), \\ &u'_1 = A' \exp(i\omega t), \quad u''_1 = A'' \exp(i\omega t). \end{aligned} \tag{2.6.4}$$

From Eqs. (2.4.8) through (2.4.13) we still have

$$\begin{aligned} &\xi^2 = \rho\omega^2 / \bar{c}_{66}, \\ &u_1 = A_1 \sin\xi x_2 + A_2 \cos\xi x_2, \\ &\phi = \frac{e_{26}}{\varepsilon_{22}}(A_1 \sin\xi x_2 + A_2 \cos\xi x_2) + B_1 x_2 + B_2, \\ &T_{21} = \bar{c}_{66}(A_1\xi \cos\xi x_2 - A_2\xi \sin\xi x_2) + e_{26} B_1 \\ &D_2 = -\varepsilon_{22} B_1. \end{aligned} \tag{2.6.5}$$

Substituting Eqs. (2.6.4) and (2.6.5) into Eqs. (2.6.2) and (2.6.3), we obtain

$$\begin{aligned} &\bar{c}_{66}\xi(A_1 \cos\xi h - A_2 \sin\xi h) + e_{26} B_1 \\ &\quad = k'[A' - (A_1 \sin\xi h + A_2 \cos\xi h)], \\ &\bar{c}_{66}\xi(A_1 \cos\xi h + A_2 \sin\xi h) + e_{26} B_1 \\ &\quad = k''[-A_1 \sin\xi h + A_2 \cos\xi h - A''], \\ &-k'[A' - (A_1 \sin\xi h + A_2 \cos\xi h)] = -\omega^2 2\rho' h' A', \\ &k''[-A_1 \sin\xi h + A_2 \cos\xi h - A''] = -\omega^2 2\rho'' h'' A'', \\ &\frac{e_{26}}{\varepsilon_{22}} A_1 \sin\xi h + B_1 h = 0. \end{aligned} \tag{2.6.6}$$

For non-trivial solutions of the undetermined constants, the determinant of the coefficient matrix of Eq. (2.6.6) has to vanish. This results in the following frequency equation:

$$\begin{vmatrix} \bar{c}_{66}\xi\cos\xi h + k'\sin\xi h & -\bar{c}_{66}\xi\sin\xi h + k'\cos\xi h & e_{26} & -k' & 0 \\ \bar{c}_{66}\xi\cos\xi h + k''\sin\xi h & \bar{c}_{66}\xi\sin\xi h - k''\cos\xi h & e_{26} & 0 & k'' \\ k'\sin\xi h & k'\cos\xi h & 0 & \omega^2 2\rho' h' - k' & 0 \\ -k''\sin\xi h & k''\cos\xi h & 0 & 0 & \omega^2 2\rho'' h'' - k'' \\ \frac{e_{26}}{\varepsilon_{22}}\sin\xi h & 0 & h & 0 & 0 \end{vmatrix} = 0 \tag{2.6.7}$$

2.6.2. *Identical electrodes*

We examine the special case of symmetric electrodes. In this case $h'' = h'$, $\rho'' = \rho'$, and $k'' = k'$.

2.6.2.1. *Identical electrodes in general*

The modes can be classified as symmetric and anti-symmetric and Eq. (2.6.7) has two factors. For symmetric modes $A_1 = B_1 = 0$ and $A'' = A'$. The frequency equation is

$$\bar{c}_{66}\xi h\tan\xi h = k'h + \frac{k'^2 h}{2\rho' h'\omega^2 - k'}. \tag{2.6.8}$$

For anti-symmetric modes $A_2 = 0$ and $A'' = -A'$. The frequency equation is

$$\bar{c}_{66}\xi h\left(\cot\xi h - \frac{\bar{k}_{26}^2}{\xi h}\right) = -k'h - \frac{k'^2 h}{2\rho' h'\omega^2 - k'}. \tag{2.6.9}$$

With

$$X = \xi h, \quad \alpha = \frac{\bar{c}_{66}}{k'h}, \tag{2.6.10}$$

Equations (2.6.8) and (2.6.9) can be written as

$$\tan X = \frac{1}{\alpha X}\left(1 + \frac{1}{\alpha R X^2 - 1}\right), \tag{2.6.11}$$

and

$$\cot X = \frac{\bar{k}_{26}^2}{X} - \frac{1}{\alpha X}\left(1 + \frac{1}{\alpha R X^2 - 1}\right), \tag{2.6.12}$$

where

$$R = \frac{2\rho' h'}{\rho h}. \tag{2.6.13}$$

The roots of Eq. (2.6.11) are determined by the intersections of the following two families of curves:

$$Y_1 = \tan X, \quad Y_2 = \frac{1}{\alpha X}\left(1 + \frac{1}{\alpha R X^2 - 1}\right), \tag{2.6.14}$$

which are shown in Fig. 2.6.2.

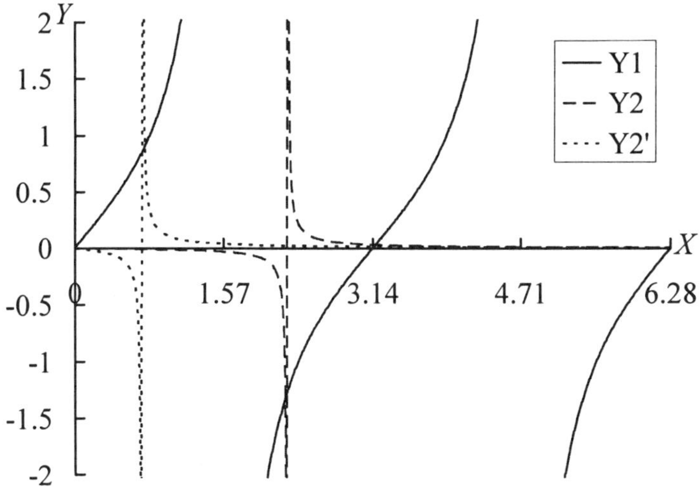

Fig. 2.6.2. Roots of Eq. (2.6.11) for symmetric modes, $\alpha = 20$.
Y2: $R = 0.01$. Y2′: $R = 0.1$.

Equation (2.6.12) can be treated in the same way with

$$Y_1 = \cot X, \quad Y_2 = \frac{\bar{k}_{26}^2}{X} - \frac{1}{\alpha X}\left(1 + \frac{1}{\alpha R X^2 - 1}\right), \tag{2.6.15}$$

and is plotted in Fig. 2.6.3 for Y-cut quartz.

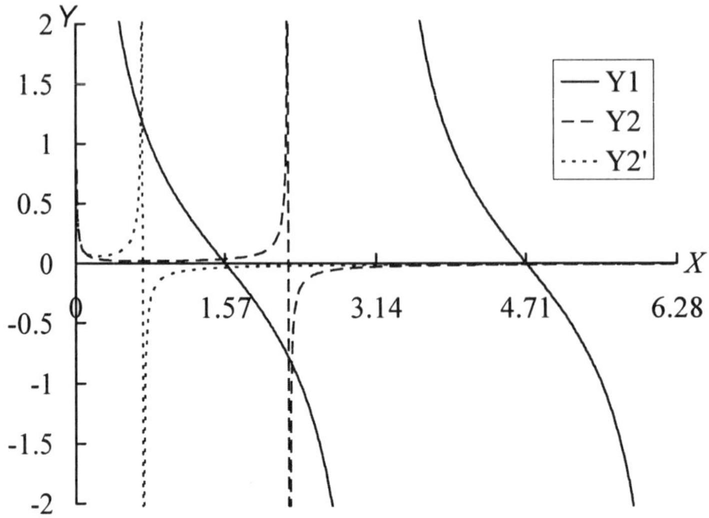

Fig. 2.6.3. Roots of Eq. (6.12) for anti-symmetric modes (quartz), $\alpha = 20$. Y2: $R = 0.01$. Y2′: $R = 0.1$.

2.6.2.2. Large k′ limit

We examine a limit case of Eqs. (2.6.11) and (2.6.12). For large k', $\alpha \to 0$. To the lowest order, Eqs. (2.6.11) and (2.6.12) are asymptotic to

$$\tan X = -RX \tag{2.6.16}$$

and

$$\cot X - \frac{\bar{k}_{26}^2}{X} = RX, \tag{2.6.17}$$

which are the results for perfectly bonded electrodes in Eqs. (2.5.6) and (2.5.8).

2.6.2.3. Small k′ limit

Next we examine another limit case of Eqs. (2.6.11) and (2.6.12). For loosely bonded electrodes, in the limit of $k' \to 0$ or $\alpha \to \infty$, to the lowest order, Eqs. (2.6.11) and (2.6.12) are asymptotic to

$$\tan X = \frac{1}{\alpha X}, \tag{2.6.18}$$

and

$$\cot X = \frac{\bar{k}_{26}^2 - \dfrac{1}{\alpha}}{X}, \tag{2.6.19}$$

respectively. If α is set to infinity, Eqs. (2.6.18) and (2.6.19) reduce to Eqs. (2.4.17) and (2.4.23) for a plate with very thin electrodes of negligible mass, as expected.

The roots of Eq. (2.6.18) are determined by the intersections of the curves in Fig. 2.6.4 (for two values of α). Compared to the frequencies of a traction-free plate or when the mass layers are not moving ($k' = 0$ or $\alpha = \infty$), which are determined by $\sin X = 0$, from the figure it can be easily concluded that in this case the mass layers raise the resonant frequencies. This is not surprising because for small k' the mass layers do not follow the crystal surface much and the interface shear then contributes to the restoring force on the vibrating crystal. From an energy point of view, the resonant frequencies are determined by the Rayleigh quotient of strain energy over kinetic energy. For small k' the mass layers do not move much and contribute little to the kinetic energy, and at the same time the interfaces are in large shears with considerable strain energy. Therefore the change of strain energy dominates and the resonant frequencies of the crystal become higher. The figure also shows that a relatively smaller k' corresponds to a larger α and a lower frequency, as expected.

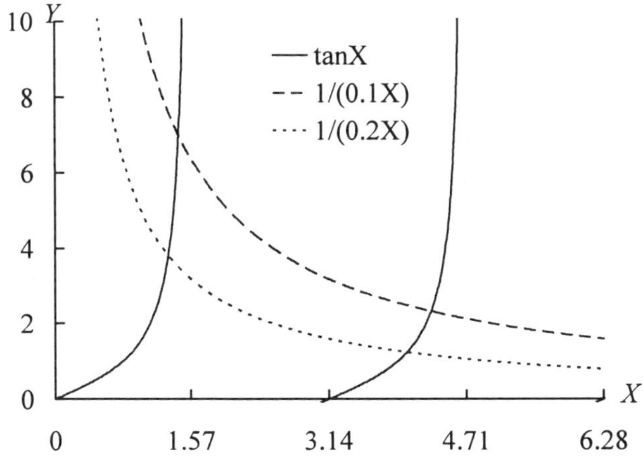

Fig. 2.6.4. Roots of Eq. (2.6.18) for symmetric modes.

The roots of Eq. (2.6.19) are determined by the intersections of the two families of curves in Fig. 2.6.5. This case is slightly more complicated due to piezoelectric coupling. There are two types of curves and intersections depending on the sign of $\bar{k}_{26}^2 - 1/\alpha$. Numerical results show that it is easier to plot the curves for Y-cut langasite with a larger electromechanical coupling coefficient $\bar{k}_{26}^2 \cong 0.0268$ than Y-cut quartz with $\bar{k}_{26}^2 \cong 0.0078$. The case of $\bar{k}_{26}^2 - 1/\alpha < 0$ is shown in Fig. 2.6.5. For the case of $\bar{k}_{26}^2 - 1/\alpha > 0$, two curves corresponding to two different values of α are shown in Figs. 2.6.6 and 2.6.7 respectively. The curve for $\alpha = 5000$ is slightly higher than the curve for $\alpha = 500$. In this case it can also be concluded that a larger α leads to a lower frequency.

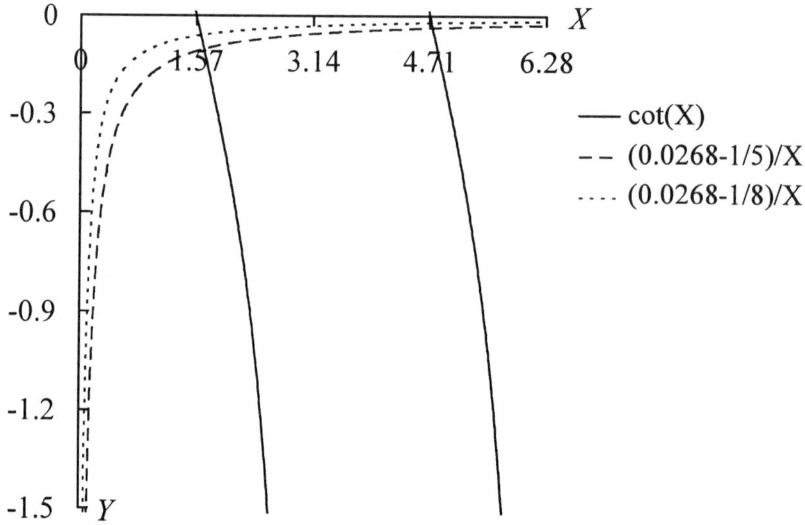

Fig. 2.6.5. Roots of Eq. (2.6.19) for anti-symmetric modes (langasite, $\bar{k}_{26}^2 - 1/\alpha < 0$).

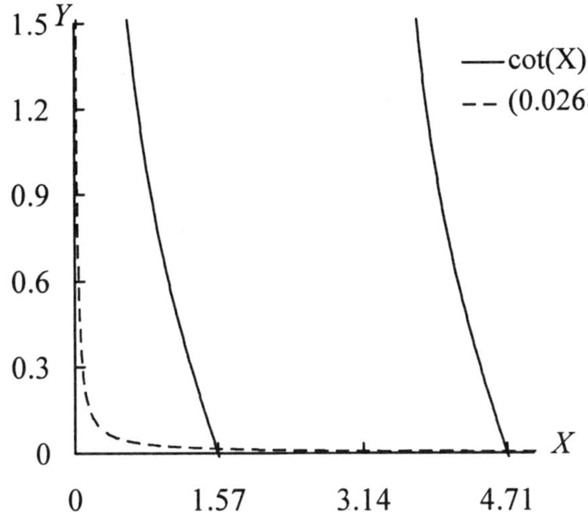

Fig. 2.6.6. Roots of Eq. (2.6.19) for anti-symmetric modes (langasite, $\bar{k}_{26}^2 - 1/\alpha > 0$).

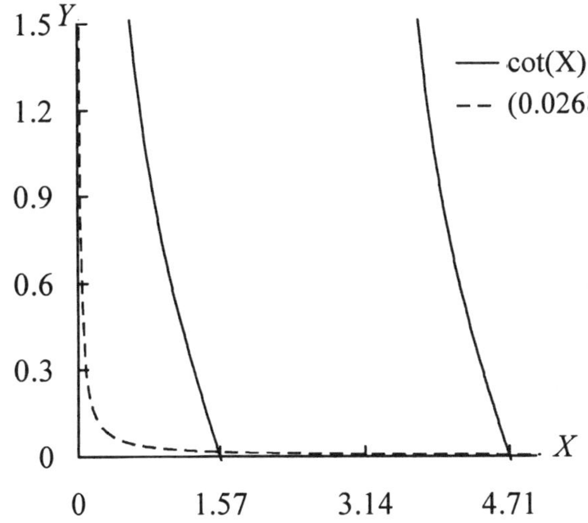

Fig. 2.6.7. Roots of Eq. (2.6.19) for anti-symmetric modes (langasite, $\bar{k}_{26}^2 - 1/\alpha > 0$).

2.7 Effects of Electrode Inertia and Shear Stiffness

When the electrodes are not very thin compared to the crystal plate, the shear stiffness of the electrodes cannot always be neglected. For symmetric electrodes the shear stiffness of the electrodes was considered in [17], and for asymmetric electrodes in [18]. In this section, we analyze thickness-shear vibration of a crystal plate with electrodes of unequal thickness. Both the mass effect and the effect of the shear stiffness of the electrodes are considered. The electrodes are modeled by the three-dimensional equations of elasticity. Consider a plate of monoclinic crystals with electrodes of unequal thickness as shown in Fig. 2.7.1. The electrodes are shorted.

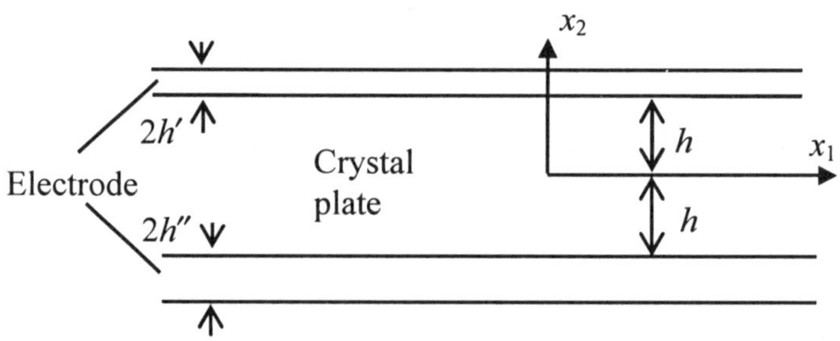

Fig. 2.7.1. A crystal plate with perfectly bonded, thick electrodes.

2.7.1. General analysis

We have the following boundary and continuity conditions:

$$\begin{aligned} &T_{2j} = 0, \quad x_2 = h + 2h', \quad x_2 = -h - 2h'', \\ &u_j(x_2 = h^-) = u_j(x_2 = h^+), \\ &T_{2j}(x_2 = h^-) = T_{2j}(x_2 = h^+), \\ &u_j(x_2 = -h^+) = u_j(x_2 = -h^-), \\ &T_{2j}(x_2 = -h^+) = T_{2j}(x_2 = -h^-), \\ &\phi(x_2 = h) = \phi(x_2 = -h). \end{aligned} \tag{2.7.1}$$

Consider the following thickness-shear modes:

$$u_1 = u_1(x_2)\exp(i\omega t), \quad u_2 = u_3 = 0, \quad \phi = \phi(x_2)\exp(i\omega t). \tag{2.7.2}$$

For fields in the crystal plate, from Eqs. (2.4.8) through (2.4.13) we still have

$$\begin{aligned}
&\xi^2 = \rho\omega^2 / \bar{c}_{66}, \\
&u_1 = A_1 \sin \xi x_2 + A_2 \cos \xi x_2, \\
&\phi = \frac{e_{26}}{\varepsilon_{22}}(A_1 \sin \xi x_2 + A_2 \cos \xi x_2) + B_1 x_2 + B_2, \\
&T_{21} = \bar{c}_{66}(A_1 \xi \cos \xi x_2 - A_2 \xi \sin \xi x_2) + e_{26} B_1, \\
&D_2 = -\varepsilon_{22} B_1.
\end{aligned} \tag{2.7.3}$$

The electrodes are isotropic, elastic, perfect conductors. Consider the upper electrode first. We have

$$T_{21,2} = c'_{66} u_{1,22} = -\rho'\omega^2 u_1 , \tag{2.7.4}$$

$$u_1 = A'_1 \sin \xi'(x_2 - h) + A'_2 \cos \xi'(x_2 - h), \tag{2.7.5}$$

$$T_{21} = c'_{66}[A'_1 \xi' \cos \xi'(x_2 - h) - A'_2 \xi' \sin \xi'(x_2 - h)], \tag{2.7.6}$$

where A'_1 and A'_2 are integration constants, ρ' and c'_{66} are the density and shear constants of the electrodes, and

$$(\xi')^2 = \frac{\rho'}{c'_{66}}\omega^2 . \tag{2.7.7}$$

Similarly, for the lower electrode we have

$$u_1 = A''_1 \sin \xi'(x_2 + h) + A''_2 \cos \xi'(x_2 + h), \tag{2.7.8}$$

$$T_{21} = c'_{66}[A''_1 \xi' \cos \xi'(x_2 + h) - A''_2 \xi' \sin \xi'(x_2 + h)], \tag{2.7.9}$$

where A''_1 and A''_2 are integration constants.

Substituting the relevant fields into the boundary and continuity conditions in Eq. (2.7.1), we have

$$\begin{aligned}
&A_1 \sin \xi h + A_2 \cos \xi h = A'_2, \\
&-A_1 \sin \xi h + A_2 \cos \xi h = A''_2, \\
&\bar{c}_{66}\xi(A_1 \cos \xi h - A_2 \sin \xi h) + e_{26} B_1 = c'_{66}\xi' A'_1, \\
&\bar{c}_{66}\xi(A_1 \cos \xi h + A_2 \sin \xi) h + e_{26} B_1 = c'_{66}\xi' A''_1, \\
&A'_1 \cos \xi' 2h' - A'_2 \sin \xi 2h' = 0, \\
&A''_1 \cos \xi 2h'' + A''_2 \sin \xi 2h'' = 0, \\
&\frac{e_{26}}{\varepsilon_{22}} A_1 \sin \xi h + B_1 h = 0.
\end{aligned} \tag{2.7.10}$$

For non-trivial solutions of the undetermined constants, the determinant of the coefficient matrix of Eq. (2.7.10) has to vanish. This results in the following frequency equation:

$$\begin{aligned}\left(1-\bar{k}_{26}^2\frac{\tan\xi h}{\xi h}\right)&\left[2\tan\xi h+\sqrt{\frac{\rho'c'_{66}}{\rho\bar{c}_{66}}}\left(\tan\xi'2h'+\tan\xi'2h''\right)\right]\\ &=\sqrt{\frac{\rho'c'_{66}}{\rho\bar{c}_{66}}}\tan\xi h\left[\tan\xi h\left(\tan\xi'2h'+\tan\xi'2h''\right)\right.\\ &\qquad\left.+2\sqrt{\frac{\rho'c'_{66}}{\rho\bar{c}_{66}}}\tan\xi'2h'\tan\xi'2h''\right].\end{aligned} \tag{2.7.11}$$

2.7.2. Special cases

We examine the following special cases of Eq. (2.7.11).

2.7.2.1. Very thin electrodes

When $h'=0$ and $h''=0$, i.e., the mechanical effects of the electrodes are neglected, Eq. (2.7.11) reduces to

$$\left(1-\bar{k}_{26}^2\frac{\tan\xi h}{\xi h}\right)\tan\xi h=0, \tag{2.7.12}$$

which is the frequency equation of both the symmetric and anti-symmetric modes in Eqs. (2.4.17) and (2.4.23).

2.7.2.2. Identical electrodes

When $h'=h''$, i.e. the electrodes are of the same thickness, Eq. (2.7.11) reduces to

$$\begin{aligned}&\left(1-\bar{k}_{26}^2\frac{\tan\xi h}{\xi h}-\sqrt{\frac{\rho'c'_{66}}{\rho\bar{c}_{66}}}\tan\xi h\tan\xi'2h'\right)\times\\ &\times\left(\tan\xi h+\sqrt{\frac{\rho'c'_{66}}{\rho\bar{c}_{66}}}\tan\xi'2h'\right)=0.\end{aligned} \tag{2.7.13}$$

The first factor of Eq. (2.7.13) is the frequency equation for the anti-symmetric modes treated in [17]. The second factor is for symmetric modes.

For very small h', i.e. very thin and equal electrodes, we approximately have

$$\tan\xi'2h' \cong \xi'2h', \quad \sqrt{\frac{\rho' c'_{66}}{\rho\bar{c}_{66}}}\tan\xi'2h' \cong R\xi h, \quad R = \frac{\rho' 2h'}{\rho h}. \tag{2.7.14}$$

In this case Eq. (2.7.13) reduces to

$$\tan\xi h = \frac{\xi h}{\bar{k}_{26}^2 + R(\xi h)^2}, \quad \tan\xi h = -R\xi h, \tag{2.7.15}$$

which are Eqs. (2.5.8) and (2.5.6).

2.7.2.3. *Thin electrodes*

When h' and h'' are both small, i.e. thin and unequal electrodes, Eq. (2.7.11) reduces to

$$\begin{aligned}&\left(1 - \bar{k}_{26}^2\frac{\tan\xi h}{\xi h}\right)\left[2\tan\xi h + (R' + R'')\xi h\right] \\ &\quad = \xi h \tan\xi h\left[(R' + R'')\tan\xi h + 2R'R''\xi h\right],\end{aligned} \tag{2.7.16}$$

where we have denoted

$$R' = \frac{\rho' 2h'}{\rho h}, \quad R'' = \frac{\rho' 2h''}{\rho h}. \tag{2.7.17}$$

To the lowest (first) order of the mass effect, the $R'R''$ term on the right-hand side of Eq. (2.7.16) can be dropped.

2.7.2.4. *Nonpiezoelectric plates*

For a three-layered elastic plate, $k_{26} = 0$. Eq. (2.7.11) reduces to

$$\begin{aligned}&2\tan\xi h + \sqrt{\frac{\rho' c'_{66}}{\rho\bar{c}_{66}}}\left(\tan\xi'2h' + \tan\xi'2h''\right) = \sqrt{\frac{\rho' c'_{66}}{\rho\bar{c}_{66}}}\tan\xi h \\ &\times\left[\tan\xi h\left(\tan\xi'2h' + \tan\xi'2h''\right) + 2\sqrt{\frac{\rho' c'_{66}}{\rho\bar{c}_{66}}}\tan\xi'2h'\tan\xi'2h''\right].\end{aligned} \tag{2.7.18}$$

For a two-layered elastic plate, we set $h'' = 0$ in Eq. (2.7.18) and obtain

$$2\tan\xi h + \sqrt{\frac{\rho' c'_{66}}{\rho \bar{c}_{66}}}\tan\xi' 2h' = \sqrt{\frac{\rho' c'_{66}}{\rho \bar{c}_{66}}}\tan\xi h \tan\xi h \tan\xi' 2h'. \quad (2.7.19)$$

2.7.3. Numerical results

Introduce a normalized frequency

$$\Omega = \frac{\omega}{\omega_0}, \quad \omega_0^2 = \frac{\pi^2 \bar{c}_{66}}{\rho h^2}. \quad (2.7.20)$$

Then Eq. (2.7.11) can be written as an equation for Ω:

$$\left(1 - \bar{k}_{26}^2 \frac{\tan\frac{\pi}{2}\Omega}{\frac{\pi}{2}\Omega}\right)\left[2\tan\frac{\pi}{2}\Omega + \alpha\left(\tan\beta'\Omega + \tan\beta''\Omega\right)\right]$$

$$= \alpha\tan\frac{\pi}{2}\Omega\left[\tan\frac{\pi}{2}\Omega\left(\tan\beta'\Omega + \tan\beta''\Omega\right) + 2\alpha\tan\beta'\Omega\tan\beta''\Omega\right], \quad (2.7.21)$$

where

$$\alpha = \sqrt{\frac{\rho' c'_{66}}{\rho \bar{c}_{66}}}, \quad \beta' = \sqrt{\frac{\rho' c'_{66}}{\rho \bar{c}_{66}}}\frac{2h'}{h}\frac{\pi}{2}, \quad \beta'' = \sqrt{\frac{\rho' c'_{66}}{\rho \bar{c}_{66}}}\frac{2h''}{h}\frac{\pi}{2}. \quad (2.7.22)$$

Consider an AT-cut quartz plate. For the electrodes we consider the following materials:

	ρ'(kg/m^3)	c'_{66} (10^{10} N/m^2)
Gold	19 300	2.85
Silver	10 490	2.7

The values of α and k_{26} are determined by material constants. We plot the normalized fundamental thickness-shear frequency, the one most important in applications, from Eq. (2.7.21) in Figs. 2.7.2 and 2.7.3 as a function of β' for different values of β'' which are relatively small or the electrodes are relatively thin. For thin electrodes the inertial effect dominates which is essentially linear. It tends to lower the resonance frequencies as expected. Gold is heavier than silver and has a larger effect on frequencies.

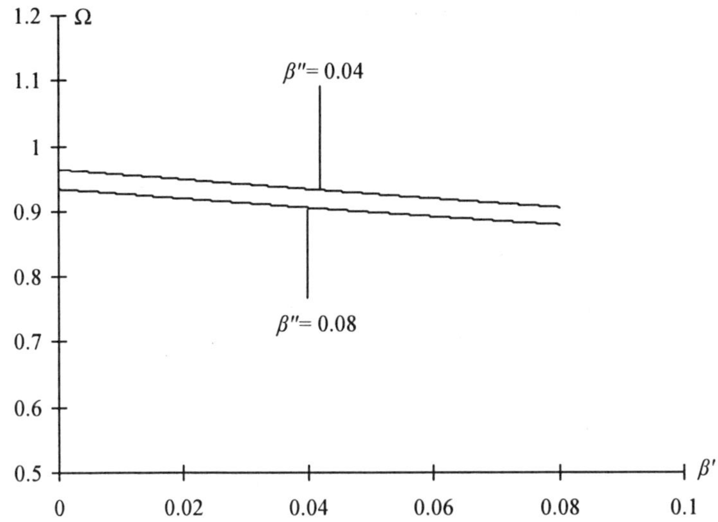

Fig. 2.7.2. Effects of gold electrodes on resonant frequency.

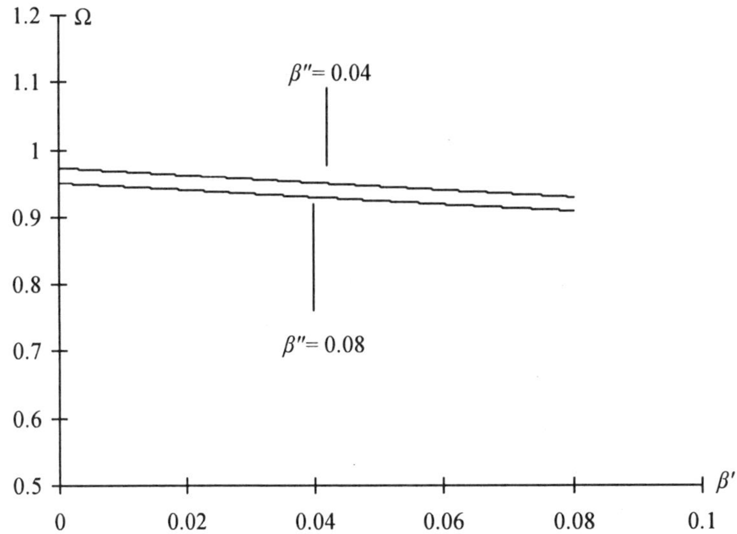

Fig. 2.7.3. Effects of silver electrodes on resonant frequency.

2.8. Nonlinear Thickness-Shear Vibration

In this section we analyze nonlinear thickness-shear vibrations of a crystal plate with relatively large amplitude. The plate is of monoclinic crystals, electroded at its two major faces (see Fig. 2.8.1). A voltage is applied across the thickness of the plate with $\phi(\pm h) = \pm 0.5 V \cos\omega t$.

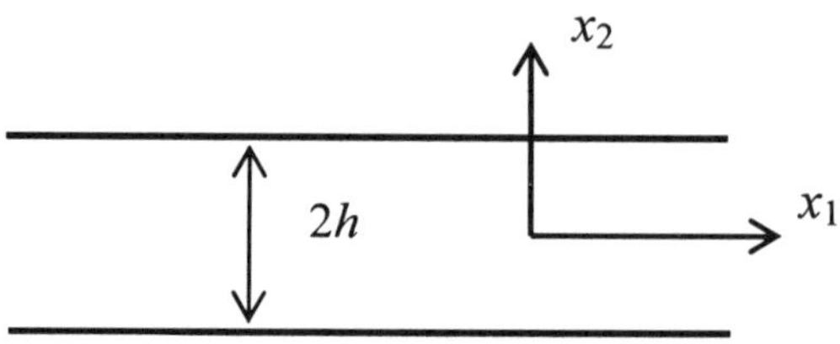

Fig. 2.8.1. An electroded crystal plate.

2.8.1. Governing equations

We use the two-dimensional equations of an electroelastic plate [19,20]. The displacement and potential are approximated by

$$
\begin{aligned}
&u_1 \cong x_2 u_1^{(1)}(t), \quad u_2 \cong 0, \quad u_3 \cong 0, \\
&\phi \cong \phi^{(0)}(t) + x_2 \phi^{(1)}(t), \quad \phi^{(0)} = 0, \quad \phi^{(1)} = \frac{V}{2h} \cos\omega t.
\end{aligned}
\tag{2.8.1}
$$

Then the relevant plate equations are

$$
\begin{aligned}
-K_{21}^{(0)} &= \frac{2\rho_0 h^3}{3} \ddot{u}_1^{(1)}, \\
K_{21}^{(0)} &= 2h(\bar{c}'_{2112} U_{12}^{(0)} - \bar{e}'_{221} \mathcal{E}_2^{(0)} \\
&\quad + \bar{c}_{211212} u_1^{(1)} u_1^{(1)} + \bar{c}_{21121212} u_1^{(1)} u_1^{(1)} u_1^{(1)}), \\
\mathcal{D}_2^{(0)} &= 2h(\bar{e}'_{212} U_{12}^{(0)} + \bar{\varepsilon}_{22} \mathcal{E}_2^{(0)}), \\
U_{12}^{(0)} &= u_1^{(1)}, \\
\mathcal{E}_2^{(0)} &= -\phi^{(1)},
\end{aligned}
\tag{2.8.2}
$$

where $K_{21}^{(0)}$ is the shear resultant, $\mathcal{D}_2^{(0)}$ is the resultant of the electric displacement, and $\bar{c}'_{2121}$, $\bar{c}_{211212}$, $\bar{c}_{21121212}$, $\bar{e}'_{221}$ and $\bar{\varepsilon}_{22}$ are plate material constants. From Eqs. (2.8.2)$_{2,4,5}$ the equation of motion in Eq. (2.8.2)$_1$ can be written as

$$\ddot{u}+\omega_0^2 u+\beta u^2+\gamma u^3=\alpha\phi^{(1)}. \tag{2.8.3}$$

In Eq. (2.8.3) we have denoted

$$\begin{aligned}
&u=u_1^{(1)},\quad \omega_0^2=3\bar{c}'_{2112}/(\rho_0 h^2),\quad \alpha=-3\bar{e}'_{221}/(\rho_0 h^2),\\
&\bar{c}'_{2112}=\kappa_1^2\bar{c}_{2112},\quad \bar{e}'_{221}=\kappa_1\bar{e}_{221},\\
&\kappa_1^2=\frac{\pi^2}{12}(1-\frac{8\bar{k}_{26}^2}{\pi^2}),\quad \bar{k}_{26}^2=\frac{e_{26}^2}{\varepsilon_{22}(c_{66}+e_{26}^2/\varepsilon_{22})},\\
&\beta=3\bar{c}_{211212}/(\rho_0 h^2),\quad \gamma=3\bar{c}_{21121212}/(\rho_0 h^2),
\end{aligned} \tag{2.8.4}$$

where κ_1 is a thickness-shear correction factor [4]. In Eq. (2.8.4), ω_0 is the fundamental thickness-shear frequency from a linear solution. We study free and forced vibrations near ω_0.

2.8.2. Free vibration

For free vibrations we look for a periodic solution with undetermined frequency ω and amplitude A to the homogeneous form of Eq. (2.8.3). Consider AT-cut quartz for which β = 0. Therefore the quadratic nonlinear term in Eq. (2.8.3) disappears. Substituting $u = A\cos\omega t$ into Eq. (2.8.3) (with $V = 0$), and neglecting the $\cos 3\omega t$ term, we obtain

$$-\omega^2 A+\omega_0^2 A+\frac{3}{4}\gamma A^3=0. \tag{2.8.5}$$

Equation (2.8.5) yields the following expression for the nonlinear resonant frequency:

$$\omega=\sqrt{\omega_0^2+\frac{3}{4}\gamma A^2}\cong\omega_0\left(1+\frac{3\gamma A^2}{8\omega_0^2}\right), \tag{2.8.6}$$

and the corresponding free vibration mode:

$$u=A\cos\left[\omega_0\left(1+\frac{3\gamma A^2}{8\omega_0^2}\right)t\right]. \tag{2.8.7}$$

Equation (2.8.6) shows that for large amplitude vibrations the frequency becomes amplitude dependent.

2.8.3. Forced vibration

Next consider electrically forced vibrations under $\phi(\pm h) = \pm 0.5V\cos\omega t$. Equation (2.8.3) can be written as

$$\ddot{u}+\omega_0^2 u+c\dot{u}+\gamma u^3=\frac{\alpha V}{2h}\cos\omega t\,, \tag{2.8.8}$$

where we have also introduced a damping term with a damping coefficient

$$c = 2\omega_0\zeta = \omega_0/Q. \tag{2.8.9}$$

ζ is the relative damping coefficient and Q is the quality factor. We look for a solution to Eq. (2.8.8) in the following form:

$$u = A\cos(\omega t+\psi)\,, \tag{2.8.10}$$

where A and ψ are undetermined constants. Substituting Eq. (2.8.10) into Eq. (2.8.8), neglecting the $\cos3(\omega t+\psi)$ term, and collecting the coefficients of $\sin\omega t$ and $\cos\omega t$, we obtain

$$\begin{cases}\left[(\omega_0^2-\omega^2)A+\dfrac{3}{4}\gamma A^3\right]\cos\psi - cA\omega\sin\psi=\dfrac{\alpha V}{2h}, \\ \left[(\omega_0^2-\omega^2)A+\dfrac{3}{4}\gamma A^3\right]\sin\psi + cA\omega\cos\psi=0. \end{cases} \tag{2.8.11}$$

Multiplying Eqs. $(2.8.11)_{1,2}$ by $\cos\psi$ and $\sin\psi$ respectively and then adding them, and multiplying Eq. $(2.8.11)_{1,2}$ by $\sin\psi$ and $\cos\psi$ respectively and then subtracting one from the other, we have

$$\begin{cases}(\omega_0^2-\omega^2)A+\dfrac{3}{4}\gamma A^3=\dfrac{\alpha V}{2h}\cos\psi, \\ cA\omega=-\dfrac{\alpha V}{2h}\sin\psi. \end{cases} \tag{2.8.12}$$

Squaring both sides of Eq. $(2.8.12)_{1,2}$ and then adding them yields

$$\left[(\omega_0^2-\omega^2)A+\frac{3}{4}\gamma A^3\right]^2+(cA\omega)^2=\left(\frac{\alpha V}{2h}\right)^2. \tag{2.8.13}$$

We are interested in resonant behaviors near ω_0. Therefore we denote $\omega=\omega_0+\Delta\omega$. Then $\omega_0^2-\omega^2 \cong -2\omega_0\Delta\omega$, and Eq. (2.8.13) can be written as

$$\left(\Delta\omega - \frac{3\gamma A^2}{8\omega_0}\right)^2 + \frac{c^2}{4} = \left(\frac{\alpha V}{4h\omega_0 A}\right)^2, \tag{2.8.14}$$

from which we can solve for $\Delta\omega$:

$$\Delta\omega = \frac{3\gamma A^2}{8\omega_0} \pm \frac{1}{2}\left[\left(\frac{\alpha V}{2h\omega_0 A}\right)^2 - c^2\right]^{\frac{1}{2}}. \tag{2.8.15}$$

We calculate the electric current flowing out of the driving electrodes, which is useful in resonator design. From Eqs. $(2.8.2)_3$, we have, for the free electric charge per unit undeformed area of the electrode at $X_2 = h$,

$$\begin{aligned} Q_e &= -\frac{\mathcal{D}_2^{(0)}}{2h} = -(\kappa_1 \bar{e}_{26} u - \bar{\varepsilon}_{22}\phi^{(1)}) \\ &\cong -\kappa_1 \bar{e}_{26} u = -\kappa_1 \bar{e}_{26} A\cos(\omega t + \psi), \end{aligned} \tag{2.8.16}$$

where, for near resonance behavior, we have neglected the electrostatic term in the expression of Q_e which is much smaller than the piezoelectric term near resonance. Then the current flows out of the electrode is given by

$$\begin{aligned} -\dot{Q}_e &= -\kappa_1 \bar{e}_{26} A\omega \sin(\omega t + \psi) = I\sin(\omega t + \psi), \\ I &= -\kappa_1 \bar{e}_{26} A\omega \cong -\kappa_1 \bar{e}_{26} A\omega_0. \end{aligned} \tag{2.8.17}$$

From Eqs. $(2.8.17)_2$ and (2.8.15), we obtain the frequency-current amplitude relation

$$\Delta\omega = \frac{3\gamma}{8\omega_0}\left(\frac{I}{\kappa_1 \bar{e}_{26}\omega_0}\right)^2 \pm \frac{1}{2}\left[\left(\frac{\alpha V \kappa_1 \bar{e}_{26}}{2hI}\right)^2 - c^2\right]^{1/2}. \tag{2.8.18}$$

Quartz is a material with very little damping. We choose the quality factor to be $Q = 10^5$. Consider a fundamental mode resonator with $h =$ 0.8273 mm. The frequency-current amplitude relation predicted by Eq. (2.8.18) is plotted in Fig. 2.8.2, which is typical for nonlinear resonance.

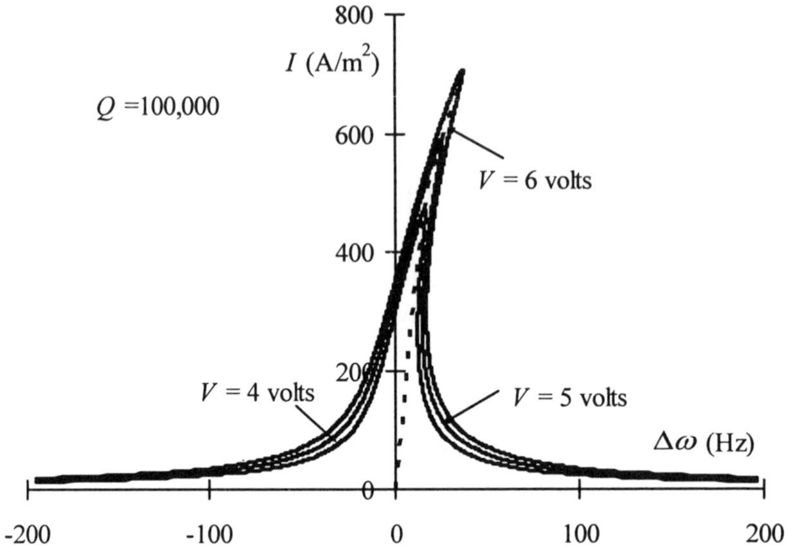

Fig. 2.8.2. Nonlinear amplitude-frequency behavior near resonance.

2.9. Effects of Initial Fields on Thickness-Shear Vibration

The effects of biasing fields on resonant frequencies represent a problem of fundamental importance in resonator frequency stability and acoustic wave sensors. The effects of various biasing fields on frequencies of thickness modes were studied in, e.g. [21,22]. In this section we analyze the effects of a uniform extensional strain on thickness-shear frequencies of an electroded monoclinic crystal plate (see Fig. 2.9.1).

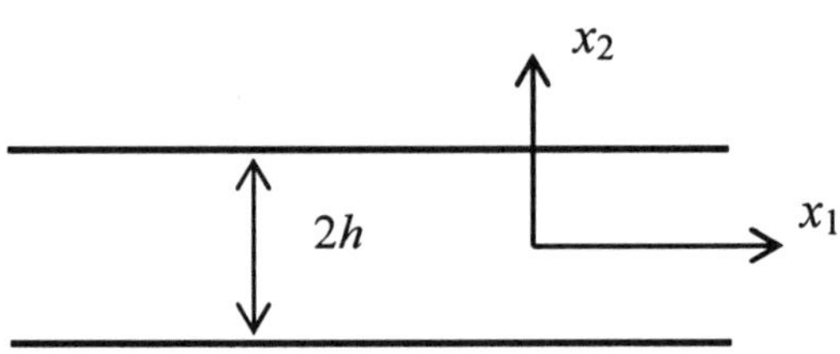

Fig. 2.9.1. An electroded crystal plate.

2.9.1. Governing equations

The biasing fields consist of an extensional strain $S_1^0 = S^0$ and the accompanying Poisson's contraction in the plate thickness

$$S_2^0 = -\frac{c_{21}}{c_{22}} S^0 . \tag{2.9.1}$$

For the biasing fields the electrodes are shorted. Therefore the biasing electric field is zero.

For the incremental motion we consider

$$u_1 = u_1(X_2,t), \quad u_2 = u_3 = 0, \quad \phi^1 = \phi^1(X_2,t) . \tag{2.9.2}$$

From Eqs. (2.3.9) through (2.3.11) we have

$$\begin{aligned} K_{21}^1 &= c_{66}(1+cS^0)u_{1,2} + e_{26}\phi_{,2}^1 , \\ K_{22}^1 &= 0, \quad K_{23}^1 = 0, \\ \mathcal{D}_2^1 &= e_{26}u_{1,2} - \varepsilon_{22}\phi_{,2}^1 , \\ c &= 2 + \frac{c_{661}}{c_{66}} - \frac{c_{662}c_{21}}{c_{66}c_{22}} . \end{aligned} \tag{2.9.3}$$

The equations of motion are

$$\begin{aligned} K_{21,2}^1 &= \rho_0 \ddot{u}_1 , \quad K_{22,2} = 0, \quad K_{23,2} = 0, \\ \mathcal{D}_{2,2} &= 0. \end{aligned} \tag{2.9.4}$$

Equations $(2.9.4)_{2,3}$ are trivially satisfied. Equations $(2.9.4)_{1,4}$ take the following form:

$$\begin{aligned} c_{66}(1+cS^0)u_{1,22} + e_{26}\phi_{,22}^1 &= \rho_0 \ddot{u}_1 , \\ e_{26}u_{1,22} - \varepsilon_{22}\phi_{,22}^1 &= 0. \end{aligned} \tag{2.9.5}$$

2.9.2. Open electrodes

For the incremental motion consider open electrodes so that

$$\begin{aligned} \mathcal{D}_2^1 &= e_{26}u_{1,2} - \varepsilon_{22}\phi_{,2}^1 = 0, \\ K_{21}^1 &= c_{66}(1+k_{26}^2+cS^0)u_{1,2} . \end{aligned} \tag{2.9.6}$$

Then Eq. (2.9.5) can be written as

$$c_{66}(1+k_{26}^2+cS^0)u_{1,22}=\rho_0\ddot{u}_1\,. \tag{2.9.7}$$

Consider odd thickness-shear modes which are excitable by a thickness electric field. Let

$$u_1(X_2,t)=u(t)\sin\frac{n\pi}{2h}X_2,\quad n=1,3,5\ldots, \tag{2.9.8}$$

which satisfies the boundary conditions

$$K_{21}^1(X_2=\pm h)=0\,. \tag{2.9.9}$$

Substitution of Eq. (2.9.8) into Eq. (2.9.7) results in the following ordinary differential equation for $u(t)$,

$$\ddot{u}+\omega_0^2(1+k_{26}^2+cS^0)u=0\,, \tag{2.9.10}$$

where

$$\omega_0^2=\frac{n^2\pi^2c_{66}}{4h^2\rho_0}\,. \tag{2.9.11}$$

Letting

$$u(t)=\exp(i\omega t) \tag{2.9.12}$$

in Eq. (2.9.10), we obtain the frequency

$$\omega=\omega_0\sqrt{1+k_{26}^2+cS^0}\cong\omega_0\left(1+\frac{1}{2}k_{26}^2+\frac{c}{2}S^0\right). \tag{2.9.13}$$

2.9.3. Shorted electrodes

If, for the incremental motion, the electrodes are shorted, we have $\phi^1=0$. Then, from Eq. (2.9.5),

$$c_{66}(1+cS^0)u_{1,22}=\rho_0\ddot{u}_1\,, \tag{2.9.14}$$

which leads to

$$\omega\cong\omega_0\left(1+\frac{c}{2}S^0\right). \tag{2.9.15}$$

Chapter 3
Slowly Varying Thickness-Shear Modes

From the viewpoint of the three-dimensional theory of elasticity or piezoelectricity, the pure thickness modes of a plate treated in the previous chapter are waves with wave vectors along the thickness direction of the plate. The waves are bounced back and forth between the two major faces of the plate. These pure thickness waves can exist only in infinite plates. They are the idealized operating modes of many resonant piezoelectric devices. In reality, due to the finite size of a plate-like device, pure thickness modes are not possible. The operating modes of these devices are in fact related to three-dimensional waves whose wave vectors are slightly off the thickness direction of the plate, with a small component within the plane of the plate. From the view point of plate theories, these waves are long plate waves with in-plane variations that are slow as compared to the plate thickness. They will be called slowly varying thickness modes in this book. In this chapter we analyze slowly varying thickness modes in plates. Due to the in-plane variation of the modes, the problem becomes more complicated. Approximations are often needed in theoretical analyses. The in-plane variation of the modes results in many interesting and useful phenomena.

3.1. Exact Waves in a Plate

First we study exact waves in a plate. Knowledge of these exact waves will serve as guidance and criteria for approximations to be made later. Exact waves in monoclinic crystal plates were analyzed in [23–26], in ceramic plates in [27], and in general in [28,29]. Consider an unbounded piezoelectric plate of thickness $2h$ as schematically illustrated in Fig. 3.1.1. The major surfaces of the plate are traction-free and are electroded. The electrodes are shorted.

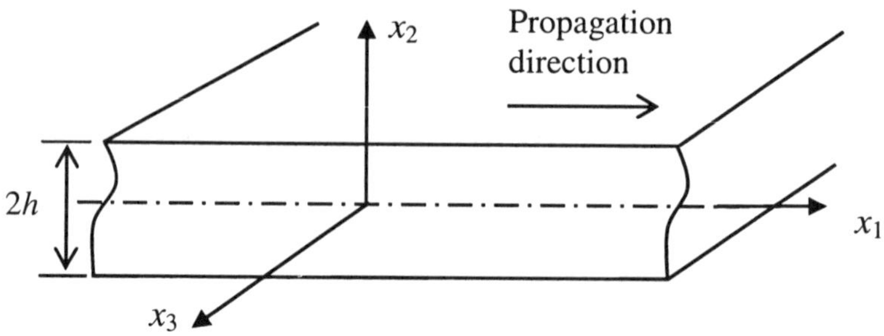

Fig. 3.1.1. Propagating waves in a piezoelectric plate.

3.1.1. *Eigenvalue problem*

We study the so-called straight-crested waves without x_3 dependence. Then the homogeneous form of Eq. (1.2.16) takes the following form:

$$\begin{aligned}
&c_{11}u_{1,11} + c_{12}u_{2,21} + c_{14}u_{3,21} + c_{15}u_{3,11} + c_{16}(u_{1,21} + u_{2,11}) \\
&\quad + c_{16}u_{1,12} + c_{26}u_{2,22} + c_{46}u_{3,22} + c_{56}u_{3,12} + c_{66}(u_{1,22} + u_{2,12}) \\
&\quad + e_{11}\phi_{,11} + e_{21}\phi_{,21} + e_{16}\phi_{,12} + e_{26}\phi_{,22} = \rho\ddot{u}_1, \\
&c_{16}u_{1,11} + c_{26}u_{2,21} + c_{46}u_{3,21} + c_{56}u_{3,11} + c_{66}(u_{1,21} + u_{2,11}) \\
&\quad + c_{12}u_{1,12} + c_{22}u_{2,22} + c_{24}u_{3,22} + c_{25}u_{3,12} + c_{26}(u_{1,22} + u_{2,12}) \\
&\quad + e_{16}\phi_{,11} + e_{26}\phi_{,21} + e_{12}\phi_{,12} + e_{22}\phi_{,22} = \rho\ddot{u}_2, \\
&c_{15}u_{1,11} + c_{25}u_{2,21} + c_{45}u_{3,21} + c_{55}u_{3,11} + c_{56}(u_{1,21} + u_{2,11}) \\
&\quad + c_{14}u_{1,12} + c_{24}u_{2,22} + c_{44}u_{3,22} + c_{45}u_{3,12} + c_{46}(u_{1,22} + u_{2,12}) \\
&\quad + e_{15}\phi_{,11} + e_{25}\phi_{,21} + e_{14}\phi_{,12} + e_{24}\phi_{,22} = \rho\ddot{u}_3, \\
&e_{11}u_{1,11} + e_{12}u_{2,21} + e_{14}u_{3,21} + e_{15}u_{3,11} + e_{16}(u_{1,21} + u_{2,11}) \\
&\quad + e_{21}u_{1,12} + e_{22}u_{2,22} + e_{24}u_{3,22} + e_{25}u_{3,12} + e_{26}(u_{1,22} + u_{2,12}) \\
&\quad - \varepsilon_{11}\phi_{,11} - \varepsilon_{12}\phi_{,21} - \varepsilon_{12}\phi_{,12} - \varepsilon_{22}\phi_{,22} = 0.
\end{aligned} \tag{3.1.1}$$

We seek solutions representing waves propagating in the x_1 direction:

$$\begin{aligned}
u_j(\mathbf{x},t) &= A_j \exp(k\eta x_2)\exp[i(kx_1 - \omega t)], \\
\phi(\mathbf{x},t) &= A_4 \exp(k\eta x_2)\exp[i(kx_1 - \omega t)],
\end{aligned} \tag{3.1.2}$$

where k and ω are the wave number in the x_1 direction and the frequency, respectively. ηk is related to the wave number in the x_2 direction. A_j (j = 1, 2, 3) and A_4 are complex constants, representing the wave amplitude. Substitution of Eq. (3.1.2) into Eq. (3.1.1) leads to the following four linear algebraic equations for A_j and A_4:

$$\begin{aligned}
&[\rho\omega^2 + k^2(\eta^2 c_{66} + i2\eta c_{16} - c_{11})]A_1 \\
&\quad + k^2(\eta^2 c_{26} + i\eta c_{66} + i\eta c_{12} - c_{16})A_2 \\
&\quad + k^2(\eta^2 c_{46} + i\eta c_{56} + i\eta c_{14} - c_{15})A_3 \\
&\quad + k^2(\eta^2 e_{26} + i\eta e_{21} + i\eta e_{16} - e_{11})A_4 = 0, \\
&k^2(\eta^2 c_{26} + i\eta c_{21} + i\eta c_{66} - c_{16})A_1 \\
&\quad + [\rho\omega^2 + k^2(\eta^2 c_{22} + i2\eta c_{26} - c_{66})]A_2 \\
&\quad + k^2(\eta^2 c_{24} + i\eta c_{25} + i\eta c_{46} - c_{56})A_3 \\
&\quad + k^2(\eta^2 e_{22} + i\eta e_{26} + i\eta e_{12} - e_{16})A_4 = 0, \\
&k^2(\eta^2 c_{46} + i\eta c_{14} + i\eta c_{56} - c_{15})A_1 \\
&\quad + k^2(\eta^2 c_{24} + i\eta c_{46} + i\eta c_{25} - c_{56})A_2 \\
&\quad + [\rho\omega^2 + k^2(\eta^2 c_{44} + i2\eta c_{45} - c_{55})]A_3 \\
&\quad + k^2(\eta^2 e_{24} + i\eta e_{25} + i\eta e_{14} - e_{15})A_4 = 0, \\
&(\eta^2 e_{26} + i\eta e_{21} + i\eta e_{16} - e_{11})A_1 \\
&\quad + (\eta^2 e_{22} + i\eta e_{26} + i\eta e_{12} - e_{16})A_2 \\
&\quad + (\eta^2 e_{24} + i\eta e_{25} + i\eta e_{14} - e_{15})A_3 \\
&\quad - (\eta^2 \varepsilon_{22} + i2\eta \varepsilon_{12} - \varepsilon_{11})A_4 = 0.
\end{aligned} \tag{3.1.3}$$

For non-trivial solutions of A_j and/or A_4, the determinant of the coefficient matrix of the above equations must vanish. This leads to a polynomial equation of degree eight for η. We denote the eight roots of the equation by $\eta_{(m)}$, and the corresponding eigenvectors by ($\overline{A}_j^{(m)}$, $\overline{A}_4^{(m)}$), m = 1, 2, …, 8. Thus, the general wave solution to Eq. (3.1.1) in the form of Eq. (3.1.2) can be written as

$$u_i = \sum_{m=1}^{8} C_{(m)} \overline{A}_i^{(m)} \exp(k\eta_{(m)} x_2) \exp[i(kx_1 - \omega t)],$$
$$\phi = \sum_{m=1}^{8} C_{(m)} \overline{A}_4^{(m)} \exp(k\eta_{(m)} x_2) \exp[i(kx_1 - \omega t)], \tag{3.1.4}$$

where the constants $C_{(m)}$ (m = 1, 2, ..., 8) are to be determined. Substituting Eq. (3.1.4) into the boundary conditions

$$T_{2j}(x_2 = \pm h) = [c_{2jkl} u_{k,l} + e_{k2j} \phi_{,k}]_{x_2 = \pm h} = 0 \tag{3.1.5}$$

and

$$\phi(x_2 = \pm h) = 0 \tag{3.1.6}$$

yields the following eight linear algebraic equations for $C_{(m)}$:

$$\begin{aligned}
&\sum_{m=1}^{8} (c_{12} i\overline{A}_1^{(m)} + c_{22}\eta_{(m)}\overline{A}_2^{(m)} + c_{24}\eta_{(m)}\overline{A}_3^{(m)} \\
&\quad + c_{25} i\overline{A}_3^{(m)} + c_{26}\eta_{(m)}\overline{A}_1^{(m)} + c_{26} i\overline{A}_2^{(m)} \\
&\quad + e_{12} i\overline{A}_4^{(m)} + e_{22}\eta_{(m)}\overline{A}_4^{(m)}) \exp[\pm k\eta_{(m)} h] C_{(m)} = 0, \\
&\sum_{m=1}^{8} (c_{16} i\overline{A}_1^{(m)} + c_{26}\eta_{(m)}\overline{A}_2^{(m)} + c_{46}\eta_{(m)}\overline{A}_3^{(m)} \\
&\quad + c_{56} i\overline{A}_3^{(m)} + c_{66}\eta_{(m)}\overline{A}_1^{(m)} + c_{66} i\overline{A}_2^{(m)} \\
&\quad + e_{16} i\overline{A}_4^{(m)} + e_{26}\eta_{(m)}\overline{A}_4^{(m)}) \exp[\pm k\eta_{(m)} h] C_{(m)} = 0, \\
&\sum_{m=1}^{8} (c_{14} i\overline{A}_1^{(m)} + c_{24}\eta_{(m)}\overline{A}_2^{(m)} + c_{44}\eta_{(m)}\overline{A}_3^{(m)} \\
&\quad + c_{45} i\overline{A}_3^{(m)} + c_{46}\eta_{(m)}\overline{A}_1^{(m)} + c_{46} i\overline{A}_2^{(m)} \\
&\quad + e_{14} i\overline{A}_4^{(m)} + e_{24}\eta_{(m)}\overline{A}_4^{(m)}) \exp[\pm k\eta_{(m)} h] C_{(m)} = 0, \\
&\sum_{j=1}^{8} \overline{A}_4^{(m)} \exp[\pm k\eta_{(m)} h] C_{(m)} = 0.
\end{aligned} \tag{3.1.7}$$

For non-trivial solutions of $C_{(m)}$, the determinant of the coefficient matrix of Eq. (3.1.7) has to vanish, which yields the frequency equation that contains ω and k. The above derivation is for materials with general anisotropy.

3.1.2. An example

As a numerical example, consider a plate of polarized ceramics poled in the x_3 direction. In this case, u_1 is coupled with u_2, and u_3 is coupled with ϕ, but the two groups do not couple to each other. The dispersion relations (ω versus k) for PZT-5H are plotted in Fig. 3.1.2.

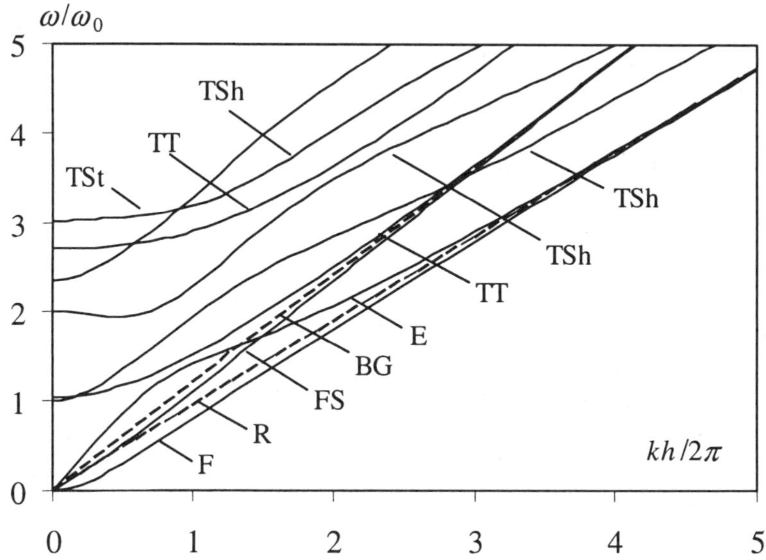

Fig. 3.1.2. Dispersion relations of waves in an electroded ceramic plate poled along the x_3 direction.

There exist an infinite number of branches of dispersion relations. Only the first nine branches are shown. The wave frequencies are normalized by

$$\omega_0 = \frac{\pi}{2h}\sqrt{\frac{c_{44}}{\rho}} . \tag{3.1.8}$$

The dispersion relations are labeled as follows, along with the dominant displacement component at small wave numbers:

E = extension (u_1),

F = flexure (u_2),

FS = face-shear (u_3),

TSh = thickness-shear (u_1),

TSt = thickness-stretch (u_2),

TT = thickness-twist (u_3),

R = Rayleigh surface wave (u_1 and u_2),

BG = Bleustein-Gulyaev surface wave (u_3).

In the figure, the three branches passing through the origin are the so-called low frequency branches. They represent the extensional, flexural, and face-shear waves. The other six branches are high-frequency branches, which have finite intercepts with the ω-axis. These intercepts are called cutoff frequencies below which the corresponding waves cannot propagate. Cutoff frequencies are, in fact, the frequencies of pure thickness modes. The six high frequency branches shown represent three thickness-shear waves, one thickness-stretch wave, and two thickness-twist waves. One of the two dotted lines in the figure is the well-known Rayleigh surface wave, which can propagate over an elastic half-space and is not dispersive. The other dotted line is the well known Bleustein-Gulyaev surface wave which has only one displacement component u_3 and can propagate over a piezoelectric half-space but does not have an elastic counterpart. These two surface waves are included as references. The figure shows that for short waves with larger k, the frequencies of the extensional and flexural waves approach that of the Rayleigh surface wave. Similarly, for short waves, the frequencies of the face-shear wave approach that of the Bleustein-Gulyaev wave.

3.1.3. A special case: Thickness-twist waves in a ceramic plate

As a special case, consider a ceramic plate poled in the x_3 direction (see Fig. 3.1.3). The major surfaces of the plate are traction-free and electroded, and the electrodes are grounded. (The case of an unelectroded plate will be discussed at the end of this subsection).

We are interested in faces-shear and thickness-twist waves described by $u_3(x_1,x_2,t)$ and $\phi(x_1,x_2,t)$. They are also called anti-plane or shear horizontal (SH) waves. With the notation in [27,10], the governing equations are

$$\bar{c}\nabla^2 u = \rho\ddot{u}, \quad \nabla^2\psi = 0, \quad \phi = \psi + \frac{e}{\varepsilon}u, \quad -h < x_2 < h, \tag{3.1.9}$$

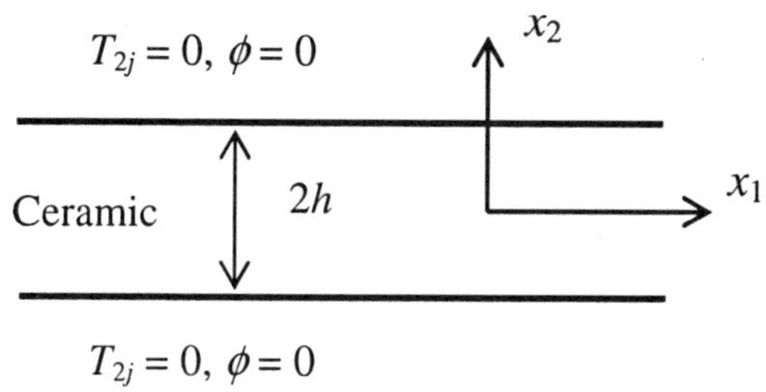

Fig. 3.1.3. An electroded ceramic plate.

where $u = u_3$, $c = c_{44}$, $e = e_{15}$, $\varepsilon = \varepsilon_{11}$, and $\bar{c} = c(1 + e^2 / c\varepsilon)$. The boundary conditions are

$$\begin{aligned} T_{23} &= 0, \quad x_2 = \pm h, \\ \phi &= 0, \quad x_2 = \pm h, \end{aligned} \tag{3.1.10}$$

or, in terms of u and ψ,

$$\begin{aligned} &\bar{c}u_{,2} + e\psi_{,2} = 0, \quad x_2 = \pm h, \\ &\psi + \frac{e}{\varepsilon}u = 0, \quad x_2 = \pm h. \end{aligned} \tag{3.1.11}$$

The waves that we are interested in may be classified as symmetric and anti-symmetric. We discuss them separately below.

3.1.3.1. Anti-symmetric waves (electroded plate)

For anti-symmetric waves we consider the possibility of

$$\begin{aligned} u &= A \sin \xi_2 x_2 \cos(\xi_1 x_1 - \omega t), \\ \psi &= B \sinh \xi_1 x_2 \cos(\xi_1 x_1 - \omega t), \end{aligned} \tag{3.1.12}$$

where A and B are constants. For Eq. (3.1.12) to satisfy Eq. (3.1.9), we have

$$\begin{aligned} &\xi_2^2 = \frac{\rho\omega^2}{\bar{c}} - \xi_1^2 = \xi_1^2\left(\frac{v^2}{v_T^2} - 1\right), \\ &v^2 = \frac{\omega^2}{\xi_1^2}, \quad v_T^2 = \frac{\bar{c}}{\rho}. \end{aligned} \tag{3.1.13}$$

Substitution of Eq. (3.1.12) into Eq. (3.1.11) leads to

$$\begin{aligned} &\bar{c}A\xi_2\cos\xi_2 h + eB\xi_1\cosh\xi_1 h = 0, \\ &\frac{e}{\varepsilon}A\sin\xi_2 h + B\sinh\xi_1 h = 0, \end{aligned} \qquad (3.1.14)$$

which is a system of linear, homogeneous equations for A and B. For nontrivial solutions we must have

$$\frac{\tan\xi_2 h}{\tanh\xi_1 h} = \frac{\xi_2 h}{\xi_1 h}\frac{\varepsilon\bar{c}}{e^2}\,. \qquad (3.1.15)$$

Equation (3.1.15) can be written as [27]

$$\frac{\tan\frac{\pi}{2}(\Omega^2 - Z^2)^{1/2}}{\tanh\frac{\pi}{2}Z} = \frac{(\Omega^2 - Z^2)^{1/2}}{\bar{k}^2 Z}, \qquad (3.1.16)$$

where the dimensionless frequency and the dimensionless wave number in the x_1 direction are defined by

$$\Omega^2 = \omega^2\Big/\left(\frac{\pi^2\bar{c}}{4\rho h^2}\right), \quad Z = \xi_1\Big/\left(\frac{\pi}{2h}\right). \qquad (3.1.17)$$

In the limit when $Z \to 0$, Eq. (3.1.16) reduces to

$$\tan\frac{\pi}{2}\Omega = \frac{\pi}{2}\frac{\Omega}{\bar{k}^2}, \qquad (3.1.18)$$

which is the frequency equation for anti-symmetric thickness-shear modes in a ceramic plate.

3.1.3.2. Symmetric waves (electroded plate)

For Symmetric waves we consider

$$\begin{aligned} u &= A\cos\xi_2 x_2\cos(\xi_1 x_1 - \omega t), \\ \psi &= B\cosh\xi_1 x_2\cos(\xi_1 x_1 - \omega t), \end{aligned} \qquad (3.1.19)$$

where A and B are constants. For Eq. (3.1.19) to satisfy Eq. (3.1.9), we still have Eq. (3.1.13). Substitution of Eq. (3.1.19) into Eq. (3.1.11) leads to

$$
\begin{aligned}
&-\bar{c}A\xi_2 \sin\xi_2 h + eB\xi_1 \sinh\xi_1 h = 0, \\
&\frac{e}{\varepsilon}A\cos\xi_2 h + B\cosh\xi_1 h = 0.
\end{aligned}
\tag{3.1.20}
$$

For non-trivial solutions of A and/or B, we must have

$$
\frac{\tan\xi_2 h}{\tanh\xi_1 h} = -\frac{\xi_1 h}{\xi_2 h}\frac{e^2}{\varepsilon\bar{c}},
\tag{3.1.21}
$$

or

$$
\frac{\tan\frac{\pi}{2}(\Omega^2 - Z^2)^{1/2}}{\tanh\frac{\pi}{2}Z} = -\frac{\bar{k}^2 Z}{(\Omega^2 - Z^2)^{1/2}}.
\tag{3.1.22}
$$

The first few branches of Eqs. (3.1.16) and (3.1.22) are plotted in Fig. 3.1.4. The dotted straight line represents Bleustein-Gulyaev surface waves in a ceramic half-space with an electroded surface, which is plotted for comparison.

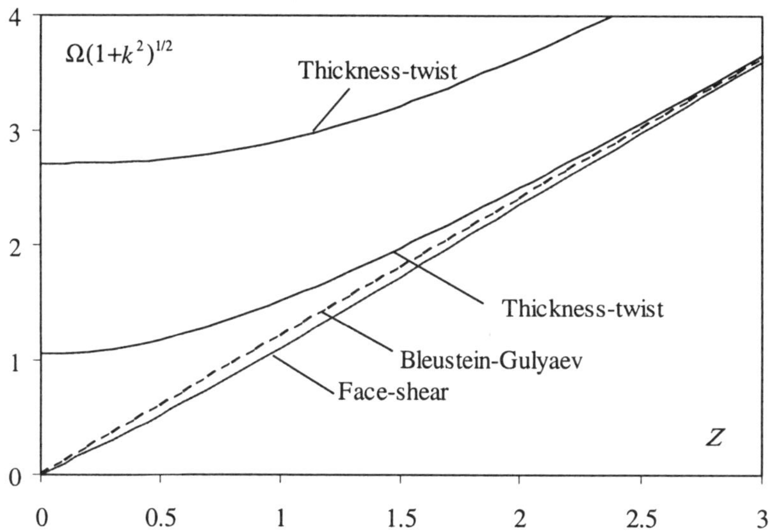

Fig. 3.1.4. Dispersion relations for anti-plane waves in a plate.

3.1.3.3. An unelectroded plate

For an unelectroded plate, electric fields may exist in the surrounding free space. Continuity of the electric potential and the normal component of the electric displacement vector need to be imposed on the place surfaces, in addition to the traction-free boundary conditions. The resulting frequency equations that determine the dispersion relations of the anti-symmetric and symmetric waves are, respectively [27],

$$\begin{aligned}&\bar{k}^2 Z \tan\frac{\pi}{2}(\Omega^2 - Z^2)^{1/2}\\&\quad = (\Omega^2 - Z^2)^{1/2}\left(\tanh\frac{\pi}{2}Z + \frac{\varepsilon_{11}}{\varepsilon_0}\right),\\&(\Omega^2 - Z^2)^{1/2}\left(1 + \frac{\varepsilon_{11}}{\varepsilon_0}\tanh\frac{\pi}{2}Z\right)\tan\frac{\pi}{2}(\Omega^2 - Z^2)^{1/2}\\&\quad = -\bar{k}^2 Z \tanh\frac{\pi}{2}Z.\end{aligned} \tag{3.1.23}$$

If the electric field in the free space is neglected and the normal component of the electric displacement at the plate surfaces are set to zero as boundary conditions, Eq. (3.1.23) reduces to $\cos\pi(\Omega^2 - Z^2)^{1/2}/2 = 0$ and $\sin\pi(\Omega^2 - Z^2)^{1/2}/2 = 0$. These can be formally obtained by setting $\varepsilon_0/\varepsilon_{11} = 0$ in Eq. (3.1.23).

3.1.3.4. Vibrations of a finite plate

Consider a plate finite in the x_1 direction whose cross-section is shown in Fig. 3.1.5.

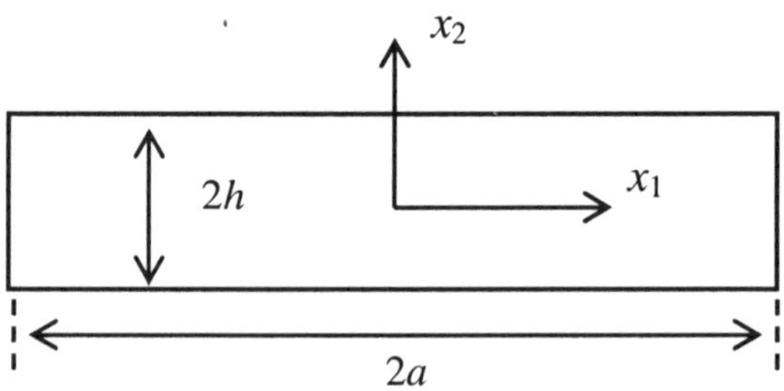

Fig. 3.1.5 A rectangular piezoelectric plate.

First consider an unelectroded plate. The boundary conditions are

$$T_{23}=0,\quad D_2=0,\quad x_2=\pm h,$$
$$T_{13}=0,\quad D_1=0,\quad x_1=\pm a, \tag{3.1.24}$$

or, equivalently,

$$u_{,2}=0,\quad \psi_{,2}=0,\quad x_2=\pm h,$$
$$u_{,1}=0,\quad \psi_{,1}=0,\quad x_1=\pm a. \tag{3.1.25}$$

Consider the possibility of

$$u=A\cos\xi_2 x_2\cos\xi_1 x_1\exp(i\omega t),$$
$$\psi=B\cosh\xi_1 x_2\cos\xi_1 x_1\exp(i\omega t), \tag{3.1.26}$$

where A and B are constants. Note that Eq. $(3.1.26)_2$ satisfies Eq. $(3.1.9)_2$. For Eq. (3.1.26) to satisfy Eq. (3.1.25), we must have

$$\xi_1 a=m\pi/2,\quad m=0,2,4,\cdots,$$
$$\xi_2 h=n\pi/2,\quad n=0,2,4,\cdots, \tag{3.1.27}$$
$$B=0.$$

For Eq. $(3.1.26)_1$ to satisfy Eq. $(3.1.9)_1$ we must have

$$\rho\omega^2=\bar{c}(\xi_1^2+\xi_2^2)=\bar{c}\left[\left(\frac{m\pi}{2a}\right)^2+\left(\frac{n\pi}{2h}\right)^2\right]. \tag{3.1.28}$$

By similar considerations, it can be concluded that for an unelectroded plate the following are all allowable modes:

$$u=\begin{cases} A_1\cos\dfrac{n\pi}{2h}x_2\cos\dfrac{m\pi}{2a}x_1\exp(i\omega t), \\ A_2\sin\dfrac{n\pi}{2h}x_2\cos\dfrac{m\pi}{2a}x_1\exp(i\omega t), \\ A_3\cos\dfrac{n\pi}{2h}x_2\sin\dfrac{m\pi}{2a}x_1\exp(i\omega t), \\ A_4\sin\dfrac{n\pi}{2h}x_2\sin\dfrac{m\pi}{2a}x_1\exp(i\omega t), \end{cases} \tag{3.1.29}$$
$$\psi=0,\quad \phi=eu/\varepsilon,$$

where, corresponding to A_1, A_2, A_3 and A_4, we have (m, n) = (even, even), (even, odd), (odd, even) and (odd, odd), and the frequencies are given by Eq. (3.1.28).

Next consider another case of practical interest in which the boundaries at $x_2 = \pm h$ are electroded and the electrodes are shorted. The boundary conditions are

$$\begin{aligned} &T_{23} = 0, \quad \phi = \psi + eu/\varepsilon = 0, \quad x_2 = \pm h, \\ &T_{13} = 0, \quad D_1 = 0, \quad x_1 = \pm a. \end{aligned} \tag{3.1.30}$$

In this case it is convenient to classify the modes as symmetric and anti-symmetric according to whether their x_2 dependence is even or odd, and discuss symmetric and anti-symmetric modes separately.

For symmetric modes consider the possibility of

$$\begin{aligned} &u = A\cos\xi_2 x_2 \cos\xi_1 x_1 \exp(i\omega t), \\ &\psi = B\cosh\xi_1 x_2 \cos\xi_1 x_1 \exp(i\omega t), \\ &\xi_1 = \frac{m\pi}{2a}, \quad m = 0, 2, 4, \ldots, \end{aligned} \tag{3.1.31}$$

which satisfy Eq. (3.1.9)$_2$ and the boundary conditions at $x_1 = \pm a$. For Eq. (3.1.31)$_1$ to satisfy Eq. (3.1.9)$_1$ we must have

$$\rho\omega^2 = \bar{c}\left[\left(\frac{m\pi}{2a}\right)^2 + \xi_2^2\right]. \tag{3.1.32}$$

For Eq. (3.1.31) to satisfy the boundary conditions at $x_2 = \pm h$, we must have

$$\begin{aligned} &-\bar{c}A\xi_2 \sin\xi_2 h + eB\xi_1 \sinh\xi_1 h = 0, \\ &\frac{e}{\varepsilon}A\cos\xi_2 h + B\cosh\xi_1 h = 0. \end{aligned} \tag{3.1.33}$$

For non-trivial solutions, the determinant of the coefficient matrix of Eq. (3.1.33) must vanish, which yields

$$\tan\xi_2 h + \bar{k}^2 \frac{\xi_1}{\xi_2}\tanh\xi_1 h = 0, \tag{3.1.34}$$

where $\bar{k}^2 = e^2/(\bar{c}\varepsilon)$. With

$$\xi_1 = \frac{m\pi}{2a}, \quad \xi_2^2 = \frac{\rho\omega^2}{\bar{c}} - \left(\frac{m\pi}{2a}\right)^2, \tag{3.1.35}$$

Eq. (3.1.34) can be written as a frequency equation for ω. Similarly,

$$
\begin{aligned}
&u = A\cos\xi_2 x_2 \sin\xi_1 x_1 \exp(i\omega t),\\
&\psi = B\cosh\xi_1 x_2 \sin\xi_1 x_1 \exp(i\omega t),\\
&\xi_1 = \frac{m\pi}{2a}, \quad m = 1,3,5,\ldots,
\end{aligned}
\tag{3.1.36}
$$

are also modes that are even in x_2. The frequency equation is still Eq. (3.1.34).

For modes anti-symmetric in x_2, consider

$$
\begin{aligned}
&u = A\sin\xi_2 x_2 \cos\xi_1 x_1 \exp(i\omega t),\\
&\psi = B\sinh\xi_1 x_2 \cos\xi_1 x_1 \exp(i\omega t),\\
&\xi_1 = \frac{m\pi}{2a}, \quad m = 0,2,4,\ldots,
\end{aligned}
\tag{3.1.37}
$$

which satisfy Eq. $(3.1.9)_2$ and the boundary conditions at $x_1 = \pm a$. For Eq. $(3.1.37)_1$ to satisfy Eq. $(3.1.9)_1$ we still have Eq. (3.1.32). For Eq. (3.1.37) to satisfy the boundary conditions at $x_2 = \pm h$, we must have

$$
\begin{aligned}
&\bar{c}A\xi_2 \cos\xi_2 h + eB\xi_1 \cosh\xi_1 h = 0,\\
&\frac{e}{\varepsilon} A\sin\xi_2 h + B\sinh\xi_1 h = 0.
\end{aligned}
\tag{3.1.38}
$$

For non-trivial solutions,

$$
\tanh\xi_1 h - \bar{k}^2 \frac{\xi_1}{\xi_2} \tan\xi_2 h = 0\,, \tag{3.1.39}
$$

which, with Eq. (3.1.35), represents the frequency equation. Similarly,

$$
\begin{aligned}
&u = A\sin\xi_2 x_2 \sin\xi_1 x_1 \exp(i\omega t),\\
&\psi = B\sinh\xi_1 x_2 \sin\xi_1 x_1 \exp(i\omega t),\\
&\xi_1 = \frac{m\pi}{2a}, \quad m = 1,3,5,\ldots,
\end{aligned}
\tag{3.1.40}
$$

are also anti-symmetric modes whose frequency equation is also Eq. (3.1.39).

3.1.4. A special case: Thickness-twist waves in a quartz plate

A simple solution can be obtained for thickness-twist waves in a quartz plate [26]. Consider a plate of rotated Y-cut quartz as shown in Fig. 3.1.6.

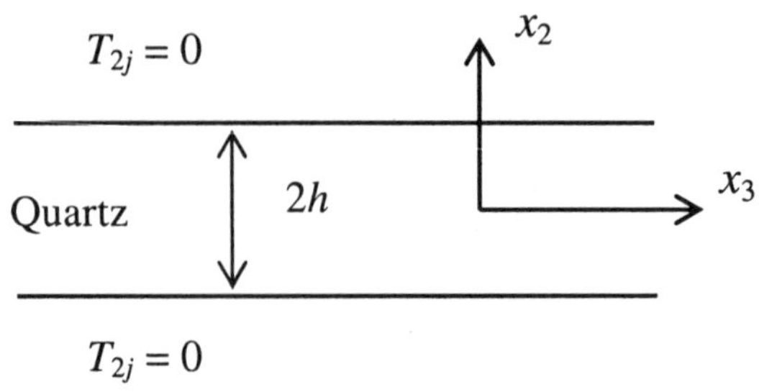

Fig. 3.1.6. An infinite plate of rotated Y-cut quartz.

We neglect the weak piezoelectric coupling in quartz. Thickness-twist modes are described by

$$u_1 = u_1(y, z, t), \tag{3.1.41}$$

where $y = x_2$ and $z = x_3$. From Eqs. (1.2.41) and (1.2.39) we have the governing equation and the stress component relevant to boundary conditions as

$$\begin{aligned} &c_{66}u_{1,22} + c_{55}u_{1,33} + 2c_{56}u_{1,23} = \rho\ddot{u}_1, \\ &T_{21} = c_{56}u_{1,3} + c_{66}u_{1,2}. \end{aligned} \tag{3.1.42}$$

Consider

$$u_1 = U(y, z)\exp(i\omega t). \tag{3.1.43}$$

Then Eq. (3.1.42) becomes

$$\begin{aligned} &c_{66}U_{,yy} + c_{55}U_{,zz} + 2c_{56}U_{,yz} = -\rho\omega^2 U, \\ &T_{21} = c_{56}U_{,z} + c_{66}U_{,y}. \end{aligned} \tag{3.1.44}$$

Let

$$U = C_1 \sin(\eta_1 y + \varsigma z) + C_2 \sin(\eta_2 y - \varsigma z), \tag{3.1.45}$$

where C_1 and C_2 are undetermined constants. For Eq. (3.1.45) to satisfy Eq. $(3.1.44)_1$, we must have

$$\rho\omega^2 = c_{66}\eta_1^2 + c_{55}\varsigma^2 + 2c_{56}\eta_1\varsigma = c_{66}\eta_2^2 + c_{55}\varsigma^2 - 2c_{56}\eta_2\varsigma, \tag{3.1.46}$$

which implies that

$$c_{66}\eta_2 - c_{56}\varsigma = c_{66}\eta_1 + c_{56}\varsigma \,. \tag{3.1.47}$$

Substituting Eq. (3.1.45) into the boundary conditions

$$T_{21} = c_{56}U_{,z} + c_{66}U_{,y} = 0, \quad x_2 = \pm h \,, \tag{3.1.48}$$

with the use of Eq. (3.1.47), we obtain

$$\frac{C_2}{C_1} = -\frac{\cos\eta_1 h}{\cos\eta_2 h} = \frac{\sin\eta_1 h}{\sin\eta_2 h}, \tag{3.1.49}$$

hence

$$\sin(\eta_1 + \eta_2)h = 0 \,, \tag{3.1.50}$$

or

$$(\eta_1 + \eta_2)h = m\pi, \quad m = 0, 1, 2, \ldots . \tag{3.1.51}$$

From Eqs. (3.1.47) and (3.1.51),

$$\eta_1 = \frac{m\pi}{2h} - \frac{c_{56}}{c_{66}}\varsigma, \quad \eta_2 = \frac{m\pi}{2h} + \frac{c_{56}}{c_{66}}\varsigma \,. \tag{3.1.52}$$

Then Eq. (3.1.46) becomes

$$\omega^2 = \frac{c_{66}}{\rho}\left(\frac{\pi}{2h}\right)^2\left[m^2 + \frac{\bar{c}_{55}}{c_{66}}\left(\frac{2\varsigma h}{\pi}\right)^2\right], \tag{3.1.53}$$

where

$$\bar{c}_{55} = c_{55} - c_{56}^2 / c_{66} \,. \tag{3.1.54}$$

3.2. An Approximate Equation for Thickness-Shear Waves

As seen from the previous section, waves in plates are relatively complicated. In this section we develop a simple, approximate description for slowly varying thickness-shear waves. For device applications, we are interested in long waves slowly varying in the plane of a plate in the sense that the wavelength is much larger than the plate thickness. Approximations can be made for these slowly varying waves.

Most of the wave behaviors we need to study are mechanical behaviors not relying on piezoelectric coupling. Therefore in this chapter we analyze elastic waves most of the time. We are mainly interested in anti-symmetric thickness-shear waves. Consider a plate of thickness $2h$ with traction-free surfaces as shown in Fig. 3.2.1. The plate is made of monoclinic crystals.

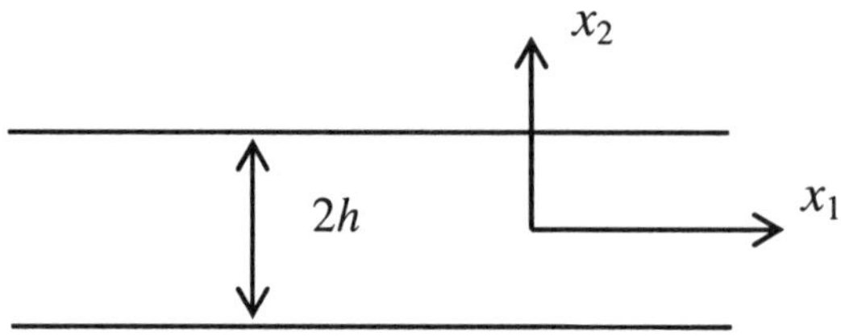

Fig. 3.2.1. An elastic plate of monoclinic crystals.

We study thickness-shear waves described by [30]

$$\begin{aligned} u_1 &\cong f(x_1,t)\sin\frac{n\pi x_2}{2h}, \quad n=1,3,5,\ldots, \\ u_2 &\cong 0, \quad u_3 \cong 0, \end{aligned} \tag{3.2.1}$$

where f is an amplitude function which is slowly varying with x_1. We want to derive an equation for f using the variational principle in Eq. (1.2.26). The strain components determined from Eq. (3.2.1) are

$$\begin{aligned} S_1 &= u_{1,1} = f'\sin\frac{n\pi x_2}{2h}, \\ S_2 &= S_3 = S_4 = S_5 = 0, \\ S_6 &= u_{1,2} = \frac{n\pi}{2h} f\cos\frac{n\pi x_2}{2h}, \end{aligned} \tag{3.2.2}$$

where a prime is a derivative with respect to x_1. We also assume that h can be a slowly varying function of x_1, and its derivative with respect to x_1 can be neglected, as in the expression of S_1. Basically we assume that $h' << f'$. For monoclinic crystals,

$$
\begin{aligned}
T_1 &= c_{11}S_1 = c_{11}f'\sin\frac{n\pi x_2}{2h},\\
T_2 &= c_{21}S_1 = c_{21}f'\sin\frac{n\pi x_2}{2h},\\
T_3 &= c_{31}S_1 = c_{31}f'\sin\frac{n\pi x_2}{2h},\\
T_4 &= c_{41}S_1 = c_{31}f'\sin\frac{n\pi x_2}{2h},\\
T_5 &= c_{56}S_6 = c_{56}\frac{n\pi}{2h}f\cos\frac{n\pi x_2}{2h},\\
T_6 &= c_{66}S_6 = c_{66}\frac{n\pi}{2h}f\cos\frac{n\pi x_2}{2h}.
\end{aligned}
\tag{3.2.3}
$$

For the fundamental mode with $n = 1$, over a cross section of the plate, T_6 does not change sign along the plate thickness and is the major stress over the cross section. At the same time T_1 changes sign once over the cross section. Therefore T_1 produces a bending moment but does not produce a net extensional force over the cross section. T_1 is relatively small compared to T_6 because it depends on f' which is the derivative of a slowly varying function. We note that the T_6 in Eq. (3.2.3) satisfies traction-free boundary conditions at $x_2 = \pm h$ although it is not required by the variational principle. The energy densities relevant to the variational principle are

$$
\begin{aligned}
\frac{1}{2}\rho\dot{u}_i\dot{u}_i &= \frac{1}{2}\rho(\dot{f})^2\sin^2\frac{n\pi x_2}{2h},\\
H &= \frac{1}{2}c_{11}(f')^2\sin^2\frac{n\pi x_2}{2h}+\frac{1}{2}c_{66}(f)^2\left(\frac{n\pi}{2h}\right)^2\cos^2\frac{n\pi x_2}{2h}.
\end{aligned}
\tag{3.2.4}
$$

Substituting the above into the variational formulation in Eq. (1.2.26), for an arbitrary portion of the plate between $a < x_1 < b$ with unit thickness in the x_3 direction, we obtain

$$
\Pi(f) = \int_a^b \left[\frac{1}{2}\rho(\dot{f})^2 h - \frac{1}{2}c_{11}(f')^2 h - \frac{1}{2}c_{66}(f)^2\left(\frac{n\pi}{2h}\right)^2 h\right]dx_1, \tag{3.2.5}
$$

where boundary terms are not included. The first variation is

$$\delta\Pi = \int_a^b \left[-\rho \ddot{f} h + (c_{11} f' h)' - c_{66} f \left(\frac{n\pi}{2h}\right)^2 h \right] \delta f \, dx_1 \qquad (3.2.6)$$
$$- [c_{11} f' h \delta f]_a^b + \cdots .$$

Hence the equation for f is

$$(c_{11} f' h)' - \left(\frac{n\pi}{2h}\right)^2 c_{66} f h = \rho h \ddot{f} \,. \qquad (3.2.7)$$

For boundary conditions we may prescribe

$$f \quad \text{or} \quad c_{11} f' h \,. \qquad (3.2.8)$$

As a special case we consider a homogeneous plate with constant c_{11}, c_{66}, h and ρ. Then Eq. (3.2.7) can be written as

$$\omega_0^2 \beta^2 h^2 f'' - \omega_0^2 f = \ddot{f} \,, \qquad (3.2.9)$$

where

$$\omega_0^2 = \frac{c_{66} n^2 \pi^2}{\rho 4 h^2}, \quad \omega_0^2 \beta^2 h^2 = \frac{c_{11}}{\rho},$$
$$\beta^2 = \frac{c_{11}}{\rho} \frac{1}{\omega_0^2 h^2} = \frac{c_{11}}{\rho} \frac{\rho 4 h^2}{c_{66} n^2 \pi^2} \frac{1}{h^2} = \frac{4 c_{11}}{n^2 \pi^2 c_{66}}. \qquad (3.2.10)$$

Consider the propagation of the following thickness-shear wave in Eq. (3.2.9):

$$f = \exp(i\xi x_1 - \omega t) \,. \qquad (3.2.11)$$

Substitution of Eq. (3.2.11) into Eq. (3.2.9) yields the following dispersion relation:

$$\left(\frac{\omega}{\omega_0}\right)^2 = 1 + \beta^2 (\xi h)^2 \,. \qquad (3.2.12)$$

The dispersion curve described by Eq. (3.2.12) is qualitatively shown in Fig. 3.2.2 for $n = 1$. It is the long wave approximation of the first branch of the dispersion curves of thickness-shear waves in Fig. 3.1.2.

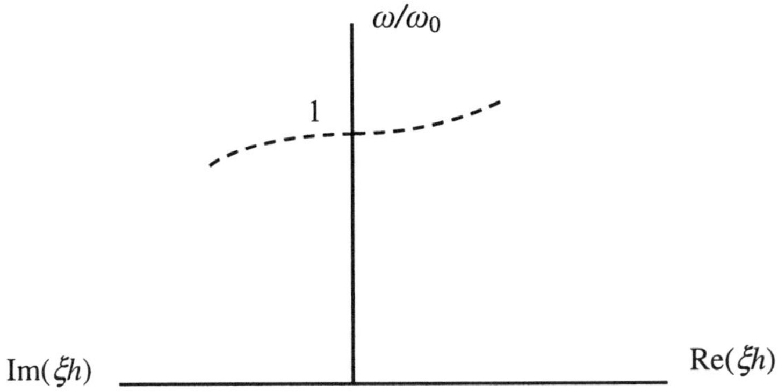

Fig. 3.2.2. Dispersion curve of Eq. (3.2.9).

3.3. Thickness-Shear Vibration of Finite Plates

So far the modes or waves we have studied are for unbounded plates. With the approximate equation developed in the previous section, in this section we study thickness-shear vibration of finite plates. Consider a finite plate of monoclinic crystals as shown in Fig. 3.3.1.

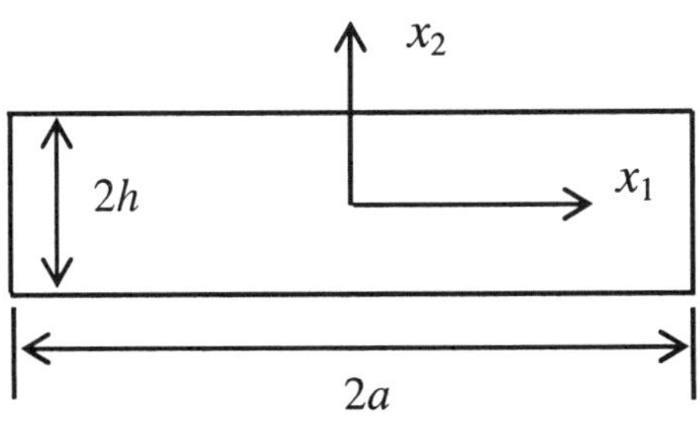

Fig. 3.3.1. A finite plate of monoclinic crystals.

The two edges of the plate are assumed to be traction-free. For free vibration with a frequency ω, we have the following eigenvalue problem:

$$-\omega_0^2\beta^2h^2f'' + \omega_0^2 f = \omega^2 f, \quad |x_1| < a$$
$$f' = 0, \quad x_1 = \pm a, \tag{3.3.1}$$
$$\omega_0^2 = \frac{c_{66}n^2\pi^2}{\rho 4h^2}, \quad \beta^2 = \frac{4c_{11}}{n^2\pi^2 c_{66}}.$$

Eq. $(3.3.1)_2$ dictates that the shear stress T_6 vanishes at both ends of a free plate, which is equivalent to vanishing shear strains at both ends.

3.3.1. Sinusoidal modes

We try $f = \cos\xi x_1$ in Eq. $(3.3.1)_1$, which leads to

$$\xi = \frac{1}{\beta h}\sqrt{\left(\frac{\omega}{\omega_0}\right)^2 - 1}, \tag{3.3.2}$$

and we assume

$$\left(\frac{\omega}{\omega_0}\right)^2 - 1 > 0 \tag{3.3.3}$$

for the moment. Then the boundary conditions in Eq. $(3.3.1)_2$ require that

$$\cos\xi a = 0. \tag{3.3.4}$$

Equation (3.3.4) determines the resonant frequencies as

$$\xi a = \frac{m\pi}{2}, \quad m = 1, 3, 5, \ldots, \tag{3.3.5}$$

or

$$\left(\frac{\omega}{\omega_0}\right)^2 = 1 + \left(\frac{\beta m\pi h}{2a}\right)^2, \quad m = 1, 3, 5, \ldots. \tag{3.3.6}$$

These modes may be called symmetric modes with respect to x_1.

Similarly, there exist anti-symmetric modes with

$$f = \sin\xi x_1,$$
$$\xi = \frac{1}{\beta h}\sqrt{\left(\frac{\omega}{\omega_0}\right)^2 - 1}, \tag{3.3.7}$$
$$\left(\frac{\omega}{\omega_0}\right)^2 = 1 + \left(\frac{\beta m\pi h}{2a}\right)^2, \quad m = 0, 2, 4, \ldots.$$

3.3.2. Hyperbolic modes

For the case of

$$\left(\frac{\omega}{\omega_0}\right)^2 - 1 \le 0, \tag{3.3.8}$$

we can try a symmetric solution in the form

$$f = \cosh \xi x_1 . \tag{3.3.9}$$

Equation $(3.3.1)_1$ implies that

$$\xi = \frac{1}{\beta h}\sqrt{1 - \left(\frac{\omega}{\omega_0}\right)^2} . \tag{3.3.10}$$

Then the boundary conditions in Eq. $(3.3.1)_2$ require that

$$\cosh \xi a = 0 , \tag{3.3.11}$$

which has no solution.

Similarly, if we look for

$$f = \sinh \xi x_1 , \tag{3.3.12}$$

then from Eq. (3.3.1) we must have

$$\xi = \frac{1}{\beta h}\sqrt{1 - \left(\frac{\omega}{\omega_0}\right)^2} , \quad \sinh \xi a = 0. \tag{3.3.13}$$

Equation (3.3.13) does allow a root unless $\xi = 0$, which represents the infinite plate thickness-shear mode

$$f = 1, \quad \omega = \omega_0 . \tag{3.3.14}$$

3.4. Energy Trapping in Mesa Resonators

Mesa resonators are plates with inhomogeneous thickness (see Fig. 3.4.1).

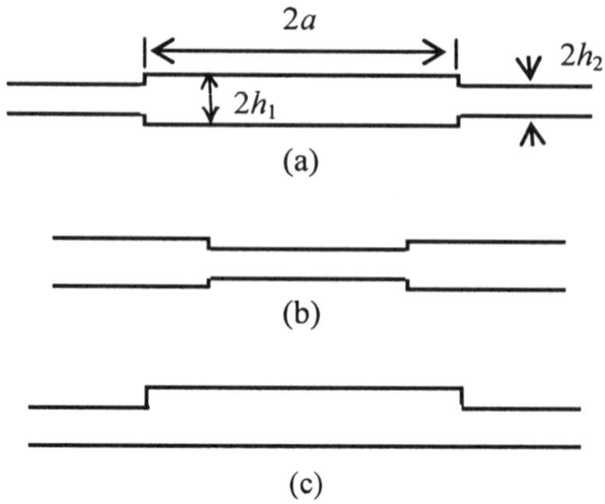

Fig. 3.4.1. Mesa resonators. (a) Bi-mesa. (b) Inverted mesa. (c) Plano-mesa.

Due to the inhomogeneous thickness, certain thickness-shear modes in mesa resonators exhibit the so-called energy trapping phenomenon in which the vibration is confined to a portion of the plate. In the rest of the plate there is essentially no vibration so that mounting and wiring can be achieved there without affecting the vibration. This can be qualitatively seen as follows. Consider the bi-mesa structure shown in Fig. 3.4.1(a). For long, thickness-shear waves with an $\exp i(\xi x_1 - \omega t)$ dependence on x_1 and t, the dispersion relations for two unbounded plates with thickness $2h_1$ and $2h_2$ respectively are as shown in Fig. 3.4.2, assuming $h_1 > h_2$.

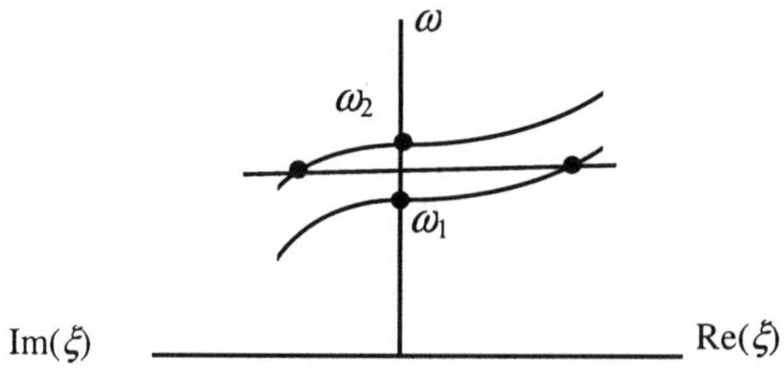

Fig. 3.4.2. Dispersion curves of thickness-shear waves in plates with thickness $2h_1$ and $2h_2$. $h_1 > h_2$.

We have

$$\omega_1^2 = \frac{c_{66} n^2 \pi^2}{\rho 4 h_1^2} < \frac{c_{66} n^2 \pi^2}{\rho 4 h_2^2} = \omega_2^2 . \tag{3.4.1}$$

Then for a mesa resonator important resonances may lie between ω_1 and ω_2. When $\omega_1 < \omega < \omega_2$, the wave number in the thin outer region is imaginary but in the thick central region it is real (see the intersections of the horizontal line with the dispersion curves). Therefore the modes have sinusoidal behavior in the central region, and decay exponentially in the outer region. Thus the vibration energy is essentially trapped in the central region.

Quantitatively, for free vibration modes in a mesa resonator, the eigenvalue problem is

$$\begin{aligned}
&-\omega_1^2 \beta^2 h_1^2 f'' + \omega_1^2 f = \omega^2 f, \quad |x_1| < a, \\
&-\omega_2^2 \beta^2 h_2^2 f'' + \omega_2^2 f = \omega^2 f, \quad |x_1| > a, \\
&f(\pm a^+) = f(\pm a^-), \\
&c_{11} f'(a^+) h_2 = c_{11} f'(a^-) h_1, \\
&c_{11} f'(-a^-) h_2 = c_{11} f'(-a^+) h_1, \\
&f = 0, \quad x_1 = \pm\infty.
\end{aligned} \tag{3.4.2}$$

Equation $(3.4.2)_{1,2}$ can be written as

$$\begin{aligned}
&f'' + \delta_1^2 f = 0, \quad |x_1| < a, \\
&f'' - \delta_2^2 f = 0, \quad |x_1| > a,
\end{aligned} \tag{3.4.3}$$

where

$$\begin{aligned}
\delta_1^2 &= \frac{\omega^2 - \omega_1^2}{\omega_1^2 \beta^2 h_1^2} = (\omega^2 - \omega_1^2)\frac{\rho}{c_{11}} > 0, \\
\delta_2^2 &= \frac{\omega_2^2 - \omega^2}{\omega_2^2 \beta^2 h_2^2} = (\omega_2^2 - \omega^2)\frac{\rho}{c_{11}} > 0.
\end{aligned} \tag{3.4.4}$$

We are interested in symmetric modes that are even functions of x_1. Therefore we consider half of the structure with $x_1 > 0$:

$$f = \begin{cases} \cos(\delta_1 x_1), & 0 < x_1 < a, \\ \cos(\delta_1 a)\exp[-\delta_2 (x_1 - a)], & a < x_1. \end{cases} \tag{3.4.5}$$

The only condition in Eq. (3.4.2) left for Eq. (3.4.5) to satisfy is Eq. $(3.4.2)_4$, which yields the following frequency equation:

$$\tan(\delta_1 a) = \frac{\delta_2 h_2}{\delta_1 h_1}. \tag{3.4.6}$$

The first symmetric mode is qualitatively shown in Fig. 3.4.3 below.

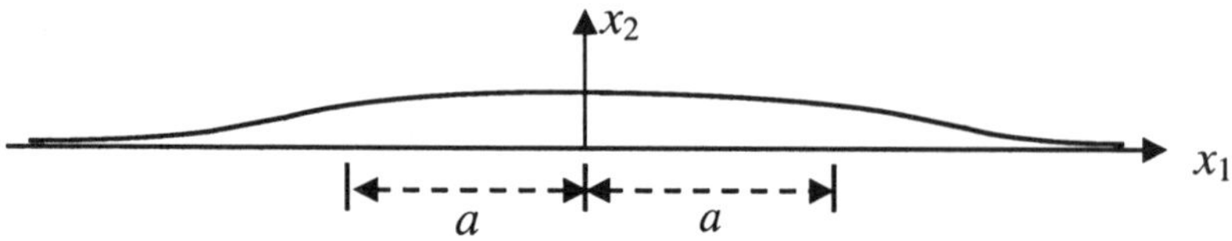

Fig. 3.4.3. A trapped thickness-shear mode.

For a mesa plate, there may exist several trapped modes which are sinusoidal in the thick central region and are exponentially decaying in the thin outer region. Figure 3.4.4 shows another possible trapped mode. These higher-order modes may have nodal points in the x_1 direction. The frequencies of these trapped modes are closely spaced in a small interval (ω_1, ω_2).

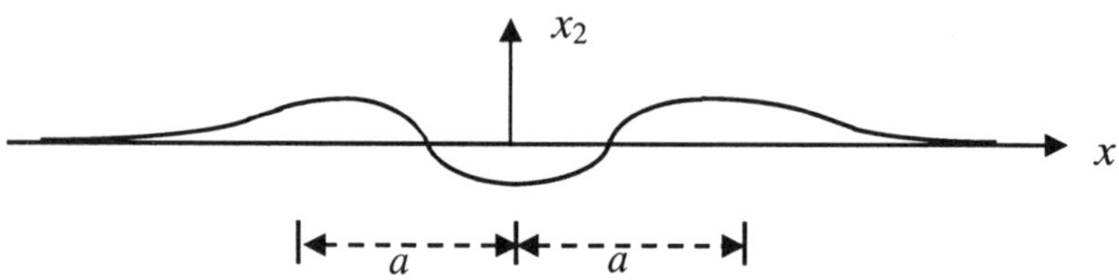

Fig. 3.4.4. Another trapped thickness-shear mode.

In applications usually only the first mode without a nodal point in the x_1 direction is important. Bechmann observed that reduction of the length of the thick central region eliminates the undesirable modes one by one until only the desired mode remains. Bechmann's number [31] is defined as the ratio of the length of the thick central region over the plate thickness below which only the mode without a nodal point in the x_1 direction exists. Bechmann's number was originally defined for a homogeneous plate with symmetric electrodes at the central portion, which effectively is a plate thicker in the central region.

For the inverted mesa in Fig. 3.4.1(b), thickness-shear modes are trapped in the outer thicker regions.

3.5. Contoured Resonators

Energy trapping in bi-mesa resonators is due to the thick central region. It can be expected that a plate with a varying thickness (wedge or contoured resonators, see Fig. 3.5.1) can produce stronger energy trapping.

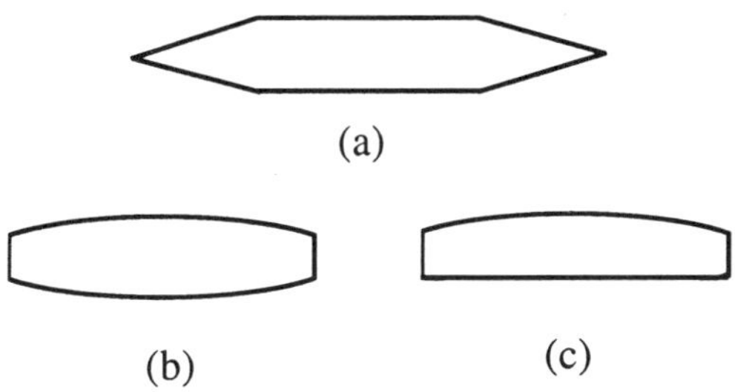

Fig. 3.5.1. Wedge and contoured resonators for strong energy trapping. (a) Wedge. (b) Bi-convex. (c) Plano-convex.

If we allow the plate thickness h in Eq. (3.2.7) to be a slowly varying function of x_1, and assume that h varies much more slowly than f (or $h' << f'$), then Eq. (3.2.9) is still approximately valid. We have, for contoured resonators,

$$f'' + \left[\omega^2 - \frac{c_{66} n^2 \pi^2}{\rho 4 h^2(x_1)} \right] \frac{\rho}{c_{11}} f = 0, \tag{3.5.1}$$

which is an equation with a variable coefficient. Solutions to these equations are usually associated with various special functions. Quite a few contoured and wedge resonators were analyzed in [32–41] using the structural equations of plates and beams as well as the three-dimensional equations.

3.6. Energy Trapping due to Material Inhomogeneity

In Eq. (3.2.7), the elastic constants and mass density can also be considered as slowly varying functions of x_1. Plates of inhomogeneous materials are called plates of functionally graded materials (FGM). Time-harmonic thickness motions with frequency ω in an FGM plate are governed by Eq. (3.2.7) in the following form:

$$-\left[c_{11}(x_1)f'\right]' + c_{66}(x_1)\left(\frac{n\pi}{2h}\right)^2 f = \lambda\rho(x_1)f\,, \qquad (3.6.1)$$

where $\lambda = \omega^2$.

3.6.1. General analysis

In an FGM plate energy trapping can be produced by material inhomogeneity [30]. Consider the inhomogeneous plate in Fig. 3.6.1.

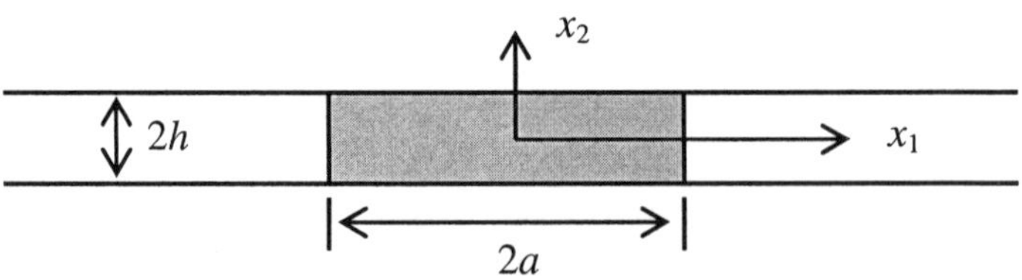

Fig. 3.6.1. An inhomogeneous plate.

The plate has the following piecewise constant material constants:

$$c_{11}(x_1), c_{66}(x_1), \rho(x_1) = \begin{cases} c_{11}, c_{66}, \rho, & |x_1| < a, \\ \hat{c}_{11}, \hat{c}_{66}, \hat{\rho}, & |x_1| > a. \end{cases} \qquad (3.6.2)$$

The governing equations are

$$\begin{aligned} f'' + \alpha_n^2 f &= 0, \quad |x_1| < a, \\ f'' - \beta_n^2 f &= 0, \quad |x_1| > a, \end{aligned} \qquad (3.6.3)$$

where

$$\alpha_n^2 = \frac{\rho}{c_{11}}\left(\omega^2 - \omega_n^2\right), \quad \beta_n^2 = \frac{\hat{\rho}}{\hat{c}_{11}}\left(\hat{\omega}_n^2 - \omega^2\right). \qquad (3.6.4)$$

$n = 1, 3, 5, \ldots$ for odd behavior in the plate thickness direction. We are only interested in trapped modes with frequencies satisfying

$$\frac{n^2\pi^2}{4h^2}\frac{c_{66}}{\rho} = \omega_n^2 < \omega^2 < \hat{\omega}_n^2 = \frac{n^2\pi^2}{4h^2}\frac{\hat{c}_{66}}{\hat{\rho}}, \tag{3.6.5}$$

for which the material constants have to satisfy

$$\frac{c_{66}}{\rho} < \frac{\hat{c}_{66}}{\hat{\rho}}. \tag{3.6.6}$$

Otherwise trapping cannot happen. Equations (3.6.3) allow sinusoidal behavior for $|x_1| < a$ and exponential behavior for $|x_1| > a$. For the applications we are interested in we look for solutions that are even functions of x_1, or

$$f = \begin{cases} \cos(\alpha_n x_1), & |x_1| < a, \\ \cos(\alpha_n a)\exp[-\beta_n(|x_1| - a)], & |x_1| > a. \end{cases} \tag{3.6.7}$$

The f in Eq. (3.6.7) is continuous at $|x_1| = a$. For a plate with piecewise constant material constants, we need to impose the continuity of $c_{11}f'$ at $|x_1| = a$, which yields the following frequency equation for ω:

$$\tan(\alpha_n a) = \frac{\hat{c}_{11}\beta_n}{c_{11}\alpha_n}. \tag{3.6.8}$$

As an equation for $\lambda = \omega^2$, in the interval of $(\omega_n^2, \hat{\omega}_n^2)$, Eq. (3.6.8) may have more than one root, which can be labeled as λ_{nm}. Since we are considering modes that are even in x_1, it is proper to allow m to assume 0, 2, 4, … . Equation (3.6.8) can be further written as

$$\tan X = \frac{\sqrt{\delta_n - \nu X^2}}{X}, \tag{3.6.9}$$

where

$$X = \alpha_n a, \quad \delta_n = \frac{\hat{\rho}\,\hat{c}_{11}a^2}{c_{11}^2}\left(\hat{\omega}_n^2 - \omega_n^2\right), \quad \nu = \frac{\hat{\rho}\,\hat{c}_{11}}{\rho c_{11}}. \tag{3.6.10}$$

The roots of Eq. (3.6.9) can be determined graphically by the intersections of the following two families of curves:

$$Y_1 = \tan X, \quad Y_2 = \frac{\sqrt{\delta_n - \nu X^2}}{X}. \tag{3.6.11}$$

The case of $\delta_n = (2.5\pi)^2$ and $\nu = 1$ is shown in Fig. 3.6.2.

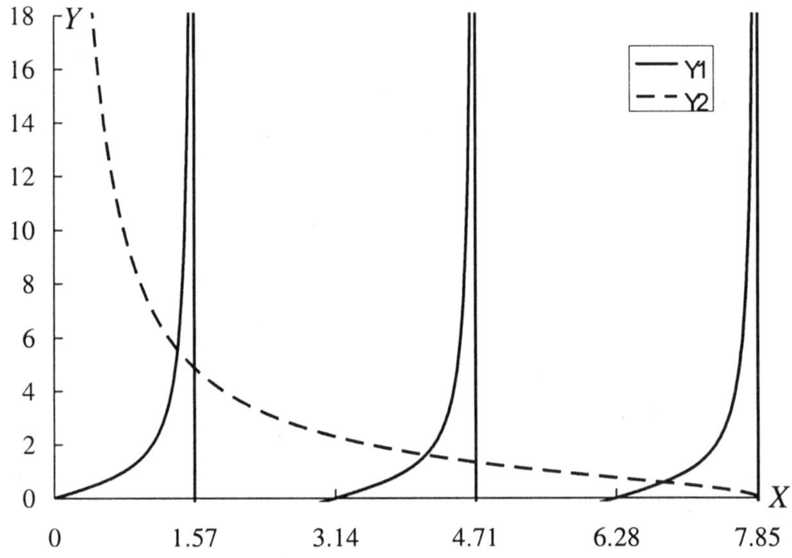

Fig. 3.6.2. Intersections of $Y_1(X)$ and $Y_2(X)$.

Bechmann's number is the ratio of *a/h* below which only one trapped mode exists. From Eq. (3.6.11), the domain of $Y_2(X)$ is $\delta_n - \nu X^2 > 0$. For Eq. (3.6.9) to have one root only, we must have

$$X = \sqrt{\frac{\delta_n}{\nu}} \le \pi, \quad \Rightarrow \quad \delta_n = \frac{\hat{\rho}\hat{c}_{11} a^2}{c_{11}^2}\left(\hat{\omega}_n^2 - \omega_n^2\right) \le \nu\pi^2, \tag{3.6.12}$$

which gives an upper bound of *a/h* and thus determines Bechmann's number.

3.6.2. An example

As an example, we choose the material properties in the outer region as

$$\begin{aligned}\hat{c}_{11} &= c_{11}(1+\varepsilon),\\ \hat{c}_{66} &= c_{66}(1+\varepsilon),\\ \hat{\rho} &= \rho(1-\varepsilon),\end{aligned} \tag{3.6.13}$$

where ε is a small number and may be called the material property grading index. With the above material properties, we calculate the resonant frequencies and plot them versus the grading index ε in Fig. 3.6.3. The figure shows that larger ε leads to more trapped modes with higher frequencies, as expected.

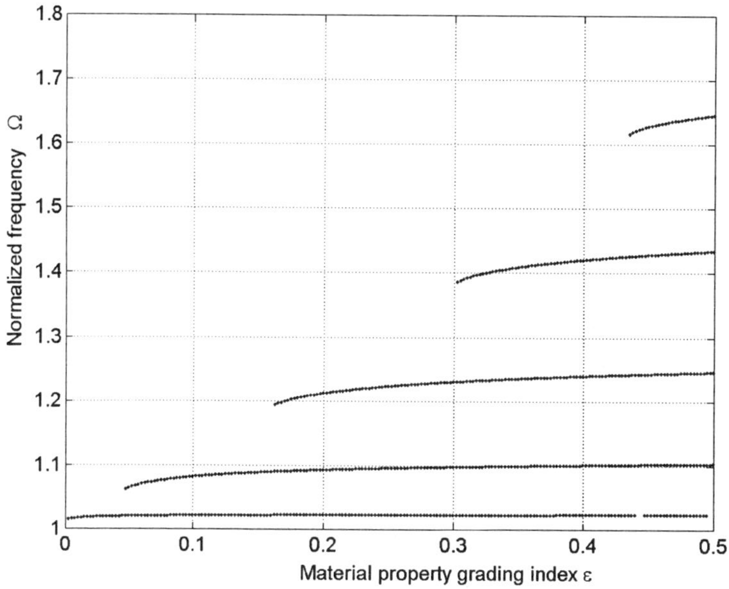

Fig. 3.6.3. Frequency versus the grading index ε.

When $\varepsilon = 0.20$, the first three trapped modes symmetric in x_1 are shown in Fig. 3.6.4. Due to symmetry only the region with $x_1 > 0$ is shown. The modes are labeled as TSh-1, TSh-2, and TSh-3 corresponding to $m = 0$, 2 and 4, respectively. m represents the number of nodal points along the x_1-axis.

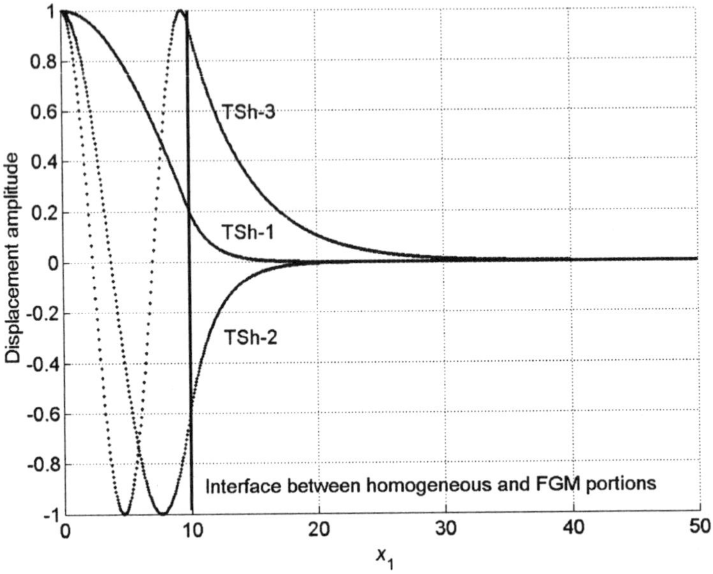

Fig. 3.6.4. The first three thickness-shear modes when $\varepsilon = 0.2$, $a = 10$.

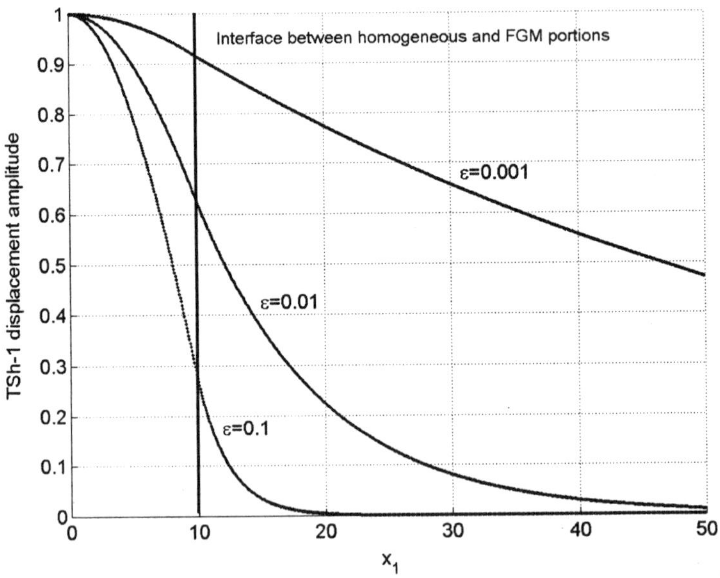

Fig. 3.6.5. The TSh-1 mode for different ε. $a = 10$.

The first mode (TSh-1) is plotted in Fig. 3.6.5 for different values of ε. Clearly, larger ε yields stronger energy trapping, as expected.

3.7. Energy Trapping by Electrode Mass

Historically, energy trapping was first observed and analyzed in partially electroded plates (see Fig. 3.7.1) as a result of the inertia of the electrode mass [31, 42].

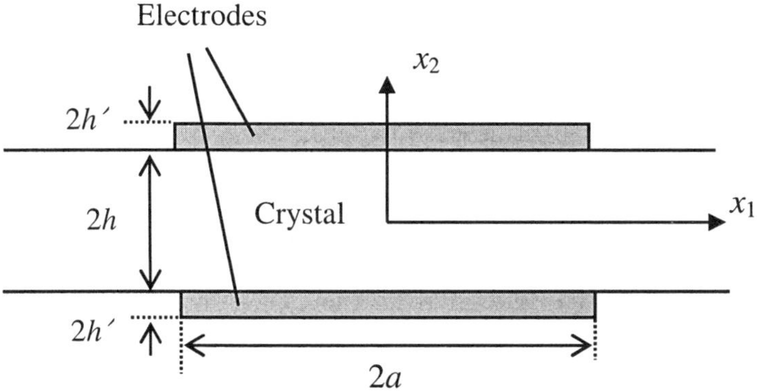

Fig. 3.7.1. A partially electroded crystal plate.

We use approximate plate equations to study this effect. The thickness-shear displacement is approximated by

$$u_1(x_1, x_2, t) \cong x_2 u_1^{(1)}(x_1, t). \quad (3.7.1)$$

The equation governing $u_1^{(1)}$ is [43]

$$\gamma_{11} u_{1,11}^{(1)} - \rho\omega_0^2 u_1^{(1)} + \frac{3}{2h^3}[hT_{21}(h) + hT_{21}(-h)] = \rho\ddot{u}_1^{(1)}, \quad (3.7.2)$$

where, for rotated Y-cut crystals,

$$\gamma_{11} = c_{11} - \frac{c_{12}^2 c_{44} + c_{14}^2 c_{22} - 2c_{12}c_{24}c_{14}}{c_{22}c_{44} - c_{24}^2} = \frac{s_{33}}{s_{11}s_{33} - s_{13}^2},$$
$$\omega_0^2 = \frac{\pi^2}{4h^2}\frac{c_{66}}{\rho}. \quad (3.7.3)$$

The electrodes are very thin with negligible stiffness. Their motions are governed by

$$-T_{21}(h)=\rho'2h'h\ddot{u}_1^{(1)},\quad T_{21}(-h)=\rho'2h'(-h\ddot{u}_1^{(1)})\,. \tag{3.7.4}$$

Substituting Eq. (3.7.4) into Eq. (3.7.2), we obtain the following equation for the electroded plate:

$$\gamma_{11}u_{1,11}^{(1)}-\rho\omega_0^2u_1^{(1)}=\rho(1+3R)\ddot{u}_1^{(1)}\,, \tag{3.7.5}$$

where the mass ratio R is given by

$$R=\frac{2\rho'h'}{\rho h}\,. \tag{3.7.6}$$

For time-harmonic motions with frequency ω, Eq. (3.7.5) takes the following form:

$$\gamma_{11}u_{1,11}^{(1)}+\rho[(1+3R)\omega^2-\omega_0^2]u_1^{(1)}=0\,. \tag{3.7.7}$$

For the partially electroded plate in Fig. 3.7.1, the free vibration eigenvalue problem is

$$\begin{aligned}&\gamma_{11}u_{1,11}^{(1)}+\rho[(1+3R)\omega^2-\omega_0^2]u_1^{(1)}=0,\quad |x_1|<a,\\&\gamma_{11}u_{1,11}^{(1)}-\rho(\omega_0^2-\omega^2)u_1^{(1)}=0,\quad |x_1|>a,\end{aligned} \tag{3.7.8}$$

plus boundary and continuity conditions. In the interval of

$$\frac{\omega_0^2}{1+3R}<\omega^2<\omega_0^2\,, \tag{3.7.9}$$

the shear deformation in the central region is sinusoidal and that in the outer region is exponential. Therefore we have trapped modes.

3.8. Effects of Non-uniform Electrodes

Energy trapping by electrode mass is usually a relatively weak effect when the electrodes are thin. Contoured resonators with varying thickness can produce stronger energy trapping. However, the manufacturing process of contoured resonators is technically complicated. On the other hand, it is not hard to deposit electrodes with varying thickness on a flat resonator. This offers another way of producing resonators with strong energy trapping. In this section, we analyze energy trapping of thickness-shear modes in a monoclinic crystal resonator with varying electrodes (see Fig. 3.8.1) [44].

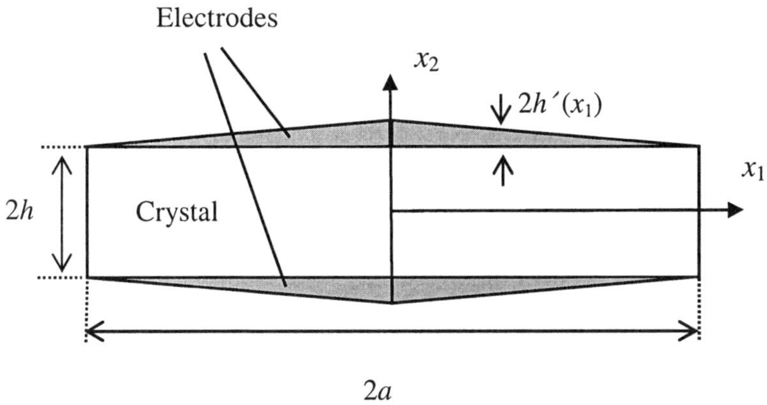

Fig. 3.8.1. A crystal plate with electrodes of varying thickness.

For varying electrodes we need to allow R in Eq. (3.7.5) to be a slowly varying function of x_1:

$$\gamma_{11}u_{1,11}^{(1)} - \rho\omega_0^2 u_1^{(1)} = \rho[1+3R(x_1)]\ddot{u}_1^{(1)}, \quad (3.8.1)$$

where

$$R(x_1) = \frac{2\rho' h'(x_1)}{\rho h}, \quad \omega_0^2 = \frac{\pi^2}{4h^2}\frac{c_{66}}{\rho}. \quad (3.8.2)$$

For time-harmonic motions with frequency ω, Eq. (3.8.2) takes the following form:

$$\gamma_{11}u_{1,11}^{(1)} + \rho\{[1+3R(x_1)]\omega^2 - \omega_0^2\}u_1^{(1)} = 0. \quad (3.8.3)$$

Based on our experience from contoured resonators, we study electrodes with a thickness given by

$$h'(x_1) = h_0[1 - f(x_1)],$$
$$R(x_1) = \frac{2\rho' h_0(1-f)}{\rho h} = R_0(1-f), \quad R_0 = \frac{2\rho' h_0}{\rho h}, \quad (3.8.4)$$

where h_0 and R_0 are constants. f is a positive, dimensionless, and increasing function. Then Eq. (3.8.3) can be written as

$$u_{1,11}^{(1)} + [\alpha - \beta f(x_1)]u_1^{(1)} = 0, \quad (3.8.5)$$

where

$$\alpha = \rho \frac{(1+3R_0)\omega^2 - \omega_0^2}{\gamma_{11}}, \quad \beta = \rho \frac{3R_0\omega^2}{\gamma_{11}}. \tag{3.8.6}$$

Eq. (3.8.5) has variable coefficients. The mathematical challenge is the same as that of contoured resonators. Since electrodes are usually made from heavy metals (gold and silver, etc.), it can be expected that the effect of varying electrodes on energy trapping is strong.

3.9. Effects of Electromechanical Coupling on Energy Trapping

In our analysis of energy trapping so far, we have been considering pure elastic plates and the mechanical effects of electrodes only. From Eq. (2.5.11), it can be seen that electromechanical coupling also affects thickness-shear resonant (cutoff) frequencies, in addition to electrode mass. Therefore electromechanical coupling also affects energy trapping which depends on the difference of cutoff frequencies (see Fig. 3.4.2). Recently, new crystals of the langasite family with electromechanical coupling stronger than quartz have been developed and used to make thickness-shear resonators. In addition, other crystals with relatively strong piezoelectric coupling, e.g. lithium niobate and lithium tantalate, have also been used to make thickness-shear devices. For thickness-shear devices made from materials with strong piezoelectric coupling, the

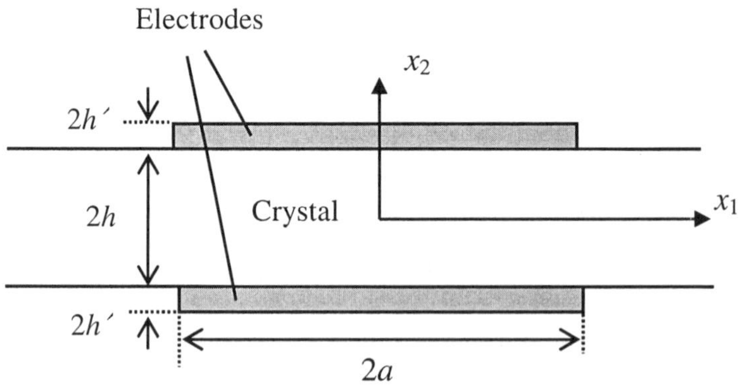

Fig. 3.9.1. A partially electroded crystal plate.

understanding of energy trapping based on an elastic analysis may be inadequate. An analysis with piezoelectric coupling may be necessary to quantify the energy trapping phenomenon. In this section we analyze energy trapping of thickness-shear modes in a partially electroded crystal plate of monoclinic symmetry. Piezoelectric coupling is included in the analysis [45]. Consider an unbounded, partially electroded piezoelectric plate of crystals of monoclinic symmetry (see Fig. 3.9.1).

3.9.1. Governing equations

When piezoelectric coupling is considered, the equation governing the thickness-shear displacement $u_1^{(1)}(x_1)$ in the electroded region is given by [46]

$$\bar{\gamma} u_{1,11}^{(1)} + \bar{\rho}(\omega^2 - \bar{\omega}_0^2) u_1^{(1)} = 0, \tag{3.9.1}$$

where

$$\begin{aligned} &\bar{\gamma} = \gamma_{11} + \frac{c_{66}\bar{\kappa}_6^2}{r}, \quad \gamma_{11} = c_{11} - \frac{c_{12}^2}{c_{22}} - \frac{(c_{14} - c_{12}c_{24}/c_{22})^2}{c_{44} - c_{24}^2/c_{22}}, \\ &\bar{\kappa}_6^2 = \frac{\pi^2}{12}\left(1 + R - \frac{8k_{26}^2}{\pi^2}\right)\frac{\bar{c}_{66}}{c_{66}}, \quad \bar{\rho} = \rho(1+3R), \quad R = \frac{2\rho' h'}{\rho h}, \\ &k_{26}^2 = \frac{e_{26}^2}{\varepsilon_{22}\bar{c}_{66}}, \quad \bar{c}_{66} = c_{66} + \frac{e_{26}^2}{\varepsilon_{22}}, \quad r = \frac{1+R}{1+3R}, \\ &\bar{\omega}_0^2 = \frac{\pi^2}{4h^2}\left(1 - R - \frac{4k_{26}^2}{\pi^2}\right)^2 \frac{\bar{c}_{66}}{\rho}. \end{aligned} \tag{3.9.2}$$

The most basic effects of mass loading and piezoelectric coupling are represented by the presence of R and k_{26}^2.

Similarly, in an unelectroded region of the plate we have [46]

$$\gamma u_{1,11}^{(1)} + \rho(\omega^2 - \omega_0^2) u_1^{(1)} = 0, \tag{3.9.3}$$

where

$$\gamma = \gamma_{11} + c_{66}\kappa_6^2, \quad \kappa_6^2 = \frac{\pi^2}{12}\left(1 + k_{26}^2\right), \quad \omega_0^2 = \frac{\pi^2}{4h^2}\frac{\bar{c}_{66}}{\rho}. \tag{3.9.4}$$

At the junctions between the electroded and unelectroded regions, the continuity of $u_1^{(1)}$ and $u_{1,1}^{(1)}$ needs to be imposed.

3.9.2. Free vibration solution

We look for trapped thickness-shear modes in the following form:

$$u_1^{(1)} = \begin{cases} A\exp\xi(a+x_1), & x_1 \le -a, \\ \bar{A}\cos\bar{\xi}x_1, & |x_1| \le a, \\ A\exp\xi(a-x_1), & x_1 \ge a, \end{cases} \tag{3.9.5}$$

where A, $\bar{A}$, ξ and $\bar{\xi}$ are undetermined constants. Modes represented by Eq. (3.9.5) are symmetric about x_1. The anti-symmetric modes will not be considered. Substituting Eq. (3.9.5) into Eqs. .9.1) and (3.9.3), we obtain

$$\begin{aligned} \bar{\xi}^2 &= \frac{\bar{\rho}}{\bar{\gamma}}(\omega^2 - \bar{\omega}_0^2) > 0, \\ \xi^2 &= \frac{\rho}{\gamma}(\omega_0^2 - \omega^2) > 0. \end{aligned} \tag{3.9.6}$$

The inequalities hold because $\bar{\omega}_0^2 < \omega_0^2$. The frequencies we are looking for are in the interval $\bar{\omega}_0^2 < \omega^2 < \omega_0^2$. The continuity of $u_1^{(1)}$ and $u_{1,1}^{(1)}$ at $x_1 = \pm a$ implies

$$\begin{aligned} \bar{A}\cos\bar{\xi}a &= A, \\ \bar{A}\bar{\xi}\sin\bar{\xi}a &= A\xi. \end{aligned} \tag{3.9.7}$$

For non-rivial solutions of A and $\bar{A}$, the determinant of coefficient matrix of Eq. (3.9.7) has to vanish, which yields

$$\tan\bar{\xi}a = \frac{\xi}{\bar{\xi}}. \tag{3.9.8}$$

With Eq. (3.9.6), Eq. (3.9.8) can be written as an equation for ω. From Eq. $(3.9.7)_1$, the modes can be written as

$$u_1^{(1)} = \begin{cases} \bar{A}\cos\bar{\xi}a\exp\xi(a+x_1), & x_1 \le -a, \\ \bar{A}\cos\bar{\xi}x_1, & |x_1| \le a, \\ \bar{A}\cos\bar{\xi}a\exp\xi(a-x_1), & x_1 \ge a. \end{cases} \tag{3.9.9}$$

3.9.3. Discussion

Consider the first resonant frequency above $\overline{\omega}_0$, which is the one most important in applications. When $a >> h$, the first mode has an essentially uniform thickness-shear deformation (small $\overline{\xi}$) in the long, electroded region, with an exponential decay right outside the electroded region. A limit solution to Eq. (3.9.8) for this case of small $\overline{\xi}$ and large a is

$$\overline{\xi} \cong 0, \quad \overline{\xi} a \cong \frac{\pi}{2}. \tag{3.9.10}$$

From Eq. $(3.9.6)_1$ and Eq. $(3.9.10)_1$ we have the resonant frequency for this mode approximately as

$$\omega \cong \overline{\omega}_0 \,. \tag{3.9.11}$$

Then, from Eq. $(3.9.6)_2$,

$$\xi^2 \cong \frac{\rho}{\gamma}(\omega_0^2 - \overline{\omega}_0^2)\,. \tag{3.9.12}$$

The decay coefficient of the mode, ξa, characterizes how the mode is trapped. From Eqs. (3.9.12), (3.9.2) and (3.9.4), the decay coefficient for this mode is approximately

$$\begin{aligned}
\xi^2 a^2 &\cong \frac{\rho}{\gamma}(\omega_0^2 - \overline{\omega}_0^2)a^2 \\
&= \frac{\rho}{\gamma}\left[\frac{\pi^2}{4h^2}\frac{\overline{c}_{66}}{\rho} - \frac{\pi^2}{4h^2}\left(1 - R - \frac{4k_{26}^2}{\pi^2}\right)^2 \frac{\overline{c}_{66}}{\rho}\right]a^2 \\
&= \frac{\rho}{\gamma}\frac{\pi^2}{4h^2}\frac{\overline{c}_{66}}{\rho}a^2\left[1 - \left(1 - R - \frac{4k_{26}^2}{\pi^2}\right)^2\right] \\
&\cong \frac{\pi^2 a^2}{2h^2}\frac{\overline{c}_{66}}{\gamma}\left(R + \frac{4k_{26}^2}{\pi^2}\right).
\end{aligned} \tag{3.9.13}$$

We make the following observations from Eq. (3.9.13):

(i) Both mass ratio and piezoelectric coupling contribute to energy trapping;

(ii) The contribution from mass ratio is linear. The contribution from piezoelectric coupling is quadratic in the electromechanical coupling coefficient k_{26} or the piezoelectric constant e_{26};

(iii) Energy trapping is stronger with increasing *a/h* ratio;

(iv) To get some quantitative estimates, we calculate the electromechanical coupling factors of a few common crystals below:

Plate	Y - cut quartz	AT - cut quartz	Y - cut langasite
k_{26}	13.7%	8.8%	17.2%
$\frac{4k_{26}^2}{\pi^2}$	0.76%	0.32%	1.20%

We observe that even for a relatively large mass loading on the order of $R = 1\%$, the piezoelectric coupling contribution to energy trapping is comparable to the mass effect and needs to be considered.

3.10. Coupling to Flexure

For modes in finite plates or slowly varying waves in unbounded plates, thickness-shear and flexure are more or less coupled [47–49]. An accurate prediction of this coupling is important in design for reducing or avoiding the undesirable flexure. Consider the monoclinic crystal plate in Fig. 3.10.1.

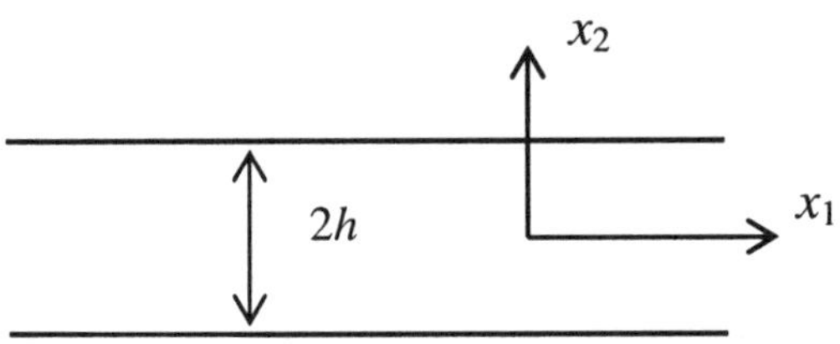

Fig. 3.10.1. A monoclinic crystal plate.

3.10.1. Governing equations

The displacement fields are approximated by

$$u_1(x_1,x_2,t) \cong x_2 u_1^{(1)}(x_1,t),$$
$$u_2(x_1,x_2,t) \cong u_2^{(0)}(x_1,t), \qquad (3.10.1)$$
$$u_3 \cong 0.$$

The governing equations for the flexure $u_2^{(0)}$ and the thickness-shear $u_1^{(1)}$ are

$$\begin{aligned} &\kappa_1^2 c_{66} u_{2,11}^{(0)} + \kappa_1^2 c_{66} u_{1,1}^{(1)} = \rho \ddot{u}_2^{(0)}, \\ &\gamma_{11} u_{1,11}^{(1)} - 3h^{-2}\kappa_1^2 c_{66}(u_{2,1}^{(0)} + u_1^{(1)}) = \rho \ddot{u}_1^{(1)}, \end{aligned} \qquad (3.10.2)$$

where γ_{11} is as in Eq. (3.7.3), $\kappa_1^2 = \pi^2/12$. The shear force and bending moments are given by

$$\begin{aligned} &T_{12}^{(0)} = 2h\kappa_1^2 c_{66}(u_{2,1}^{(0)} + u_1^{(1)}), \\ &T_1^{(1)} = \frac{2h^3}{3}\gamma_{11} u_{1,1}^{(1)}. \end{aligned} \qquad (3.10.3)$$

3.10.2. Waves in unbounded plates

Consider the following coupled thickness-shear and flexural waves:

$$u_1^{(1)} = A_1 \exp i(\xi x_1 - \omega t), \quad u_2^{(0)} = A_2 \exp i(\xi x_1 - \omega t). \qquad (3.10.4)$$

Substitution of Eq. (3.10.4) into Eq. (3.10.2) results in two linear equations for A_1 and A_2

$$\begin{aligned} &\kappa_1^2 c_{66} i\xi A_1 + (\rho\omega^2 - \kappa_1^2 c_{66}\xi^2)A_2 = 0, \\ &(\rho\omega^2 - \gamma_{11}\xi^2 - 3h^{-2}\kappa_1^2 c_{66})A_1 - 3h^{-2}\kappa_1^2 c_{66} i\xi A_2 = 0. \end{aligned} \qquad (3.10.5)$$

For non-trivial solutions the determinant of the coefficient matrix of the linear equations has to vanish. This yields the following equation that determines the dispersion relations for coupled flexural and thickness-shear waves:

$$\begin{vmatrix} \kappa_1^2 c_{66} i\xi & \rho\omega^2 - \kappa_1^2 c_{66}\xi^2 \\ \rho\omega^2 - \gamma_{11}\xi^2 - 3h^{-2}\kappa_1^2 c_{66} & -3h^{-2}\kappa_1^2 c_{66} i\xi \end{vmatrix} = 0. \qquad (3.10.6)$$

Equation (3.10.6) predicts two branches of dispersion curves. One is mainly flexure (F), the other is mainly thickness-shear (TSh). Given a value of ω, we denote the two corresponding wave numbers determined from Eq. (3.10.6) by $\xi^{(1)}(\omega)$ and $\xi^{(2)}(\omega)$, and the corresponding eigenvectors from Eq. (3.10.5) by $\{\beta_1^{(1)},\beta_2^{(1)}\}$ and $\{\beta_1^{(2)},\beta_2^{(2)}\}$. Then the general right-traveling wave solution to Eq. (3.10.2) can be written as

$$\begin{aligned} u_1^{(1)} &= C_1\beta_1^{(1)}\exp(i\xi^{(1)}x_1-\omega t)+C_2\beta_1^{(2)}\exp(i\xi^{(2)}x_1-\omega t), \\ u_2^{(0)} &= C_1\beta_2^{(1)}\exp(i\xi^{(1)}x_1-\omega t)+C_2\beta_2^{(2)}\exp(i\xi^{(2)}x_1-\omega t), \end{aligned} \tag{3.10.7}$$

where C_1 and C_2 are undetermined constants. The dispersion curves determined by Eq. (3.10.6) are plotted in Fig. 3.10.2, where

$$\Omega=\frac{\omega}{\omega_0},\quad \omega_0^2=\frac{\pi^2 c_{66}}{4h^2\rho},\quad Z=\xi\Big/\frac{\pi}{2b}\,. \tag{3.10.8}$$

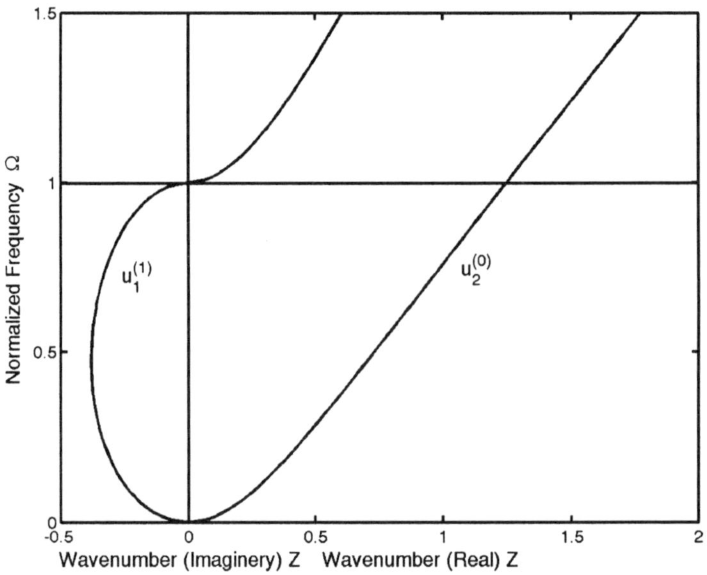

Fig. 3.10.2. Dispersion curves for coupled thickness-shear and flexure along with the dominant displacements.

3.10.3. Vibrations of finite plates

Next consider a bounded plate in $|x_1| < a$ (see Fig. 3.10.3).

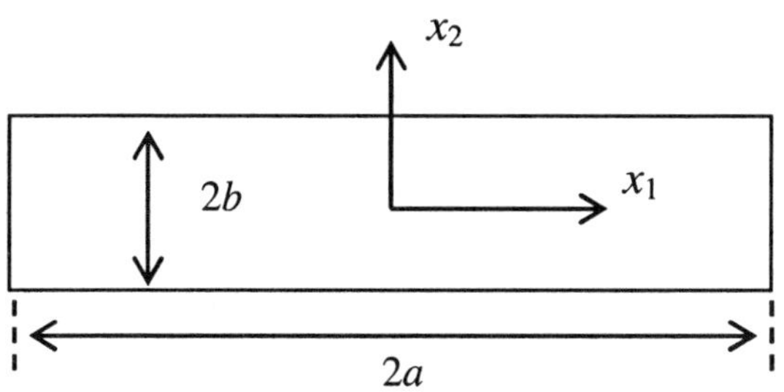

Fig. 3.10.3. A finite plate.

Consider stationary waves:

$$u_1^{(1)} = A_1 \cos \xi x_1 \exp(i\omega t), \quad u_2^{(0)} = A_2 \sin \xi x_1 \exp(i\omega t). \tag{3.10.9}$$

They may be considered as symmetric modes in view of that the shear displacement is an even function of x_1. Then

$$\begin{aligned} &-\kappa_1^2 c_{66} \xi A_1 + (\rho\omega^2 - \kappa_1^2 c_{66} \xi^2) A_2 = 0, \\ &(\rho\omega^2 - \gamma_{11}\xi^2 - 3h^{-2}\kappa_1^2 c_{66}) A_1 - 3h^{-2}\kappa_1^2 c_{66} \xi A_2 = 0, \end{aligned} \tag{3.10.10}$$

and

$$\begin{vmatrix} -\kappa_1^2 c_{66} \xi & \rho\omega^2 - \kappa_1^2 c_{66} \xi^2 \\ \rho\omega^2 - \gamma_{11}\xi^2 - 3h^{-2}\kappa_1^2 c_{66} & -3h^{-2}\kappa_1^2 c_{66} \xi \end{vmatrix} = 0. \tag{3.10.11}$$

We still denote the two dispersion relations by $\xi^{(1)}(\omega)$ and $\xi^{(2)}(\omega)$, and the corresponding eigenvectors by $\{\beta_1^{(1)}, \beta_2^{(1)}\}$ and $\{\beta_1^{(2)}, \beta_2^{(2)}\}$. Then the general symmetric stationary wave solution to Eq. (3.10.2) can be written as

$$\begin{aligned} u_1^{(1)} &= C_1 \beta_1^{(1)} \cos \xi^{(1)} x_1 \exp(i\omega t) + C_2 \beta_1^{(2)} \cos \xi^{(2)} x_1 \exp(i\omega t), \\ u_2^{(0)} &= C_1 \beta_2^{(1)} \sin \xi^{(1)} x_1 \exp(i\omega t) + C_2 \beta_2^{(2)} \sin \xi^{(2)} x_1 \exp(i\omega t). \end{aligned} \tag{3.10.12}$$

Consider the case of free edges. The boundary conditions are

$$\begin{aligned} T_{12}^{(0)} &= 2h\kappa_1^2 c_{66} (u_{2,1}^{(0)} + u_1^{(1)}) = 0, \quad x_1 = \pm a, \\ T_1^{(1)} &= \frac{2h^3}{3} \gamma_{11} u_{1,1}^{(1)} = 0, \quad x_1 = \pm a. \end{aligned} \tag{3.10.13}$$

Substitution of Eqs. (3.10.12) into (3.10.13) results in two homogeneous equations for C_1 and C_2. For non-trivial solutions the determinant of the coefficient matrix has to vanish, which yields the frequency equation for symmetric modes. The solution to the frequency equation is plotted in Fig. 3.10.4 as a function of the aspect ratio a/b. This is called a spectral plot and is very useful in resonator design.

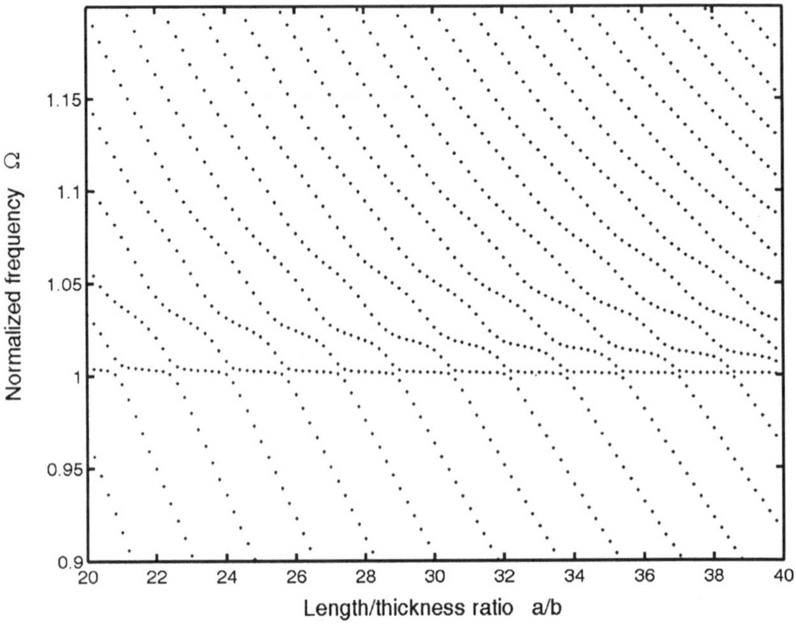

Fig. 3.10.4. Frequency versus aspect ratio.

For each aspect ratio there are infinitely many frequencies. At the relatively flat portions of the curves the modes are essentially thickness-shear. Similarly, anti-symmetric modes with the shear displacement as an odd function of x_1 can be obtained.

3.11. Coupling to Face-Shear and Flexure

In addition to coupling to flexure, the thickness-shear operating mode of a resonator may also couple to face-shear due to anisotropy. Consider the monoclinic crystal plate in Fig. 3.11.1.

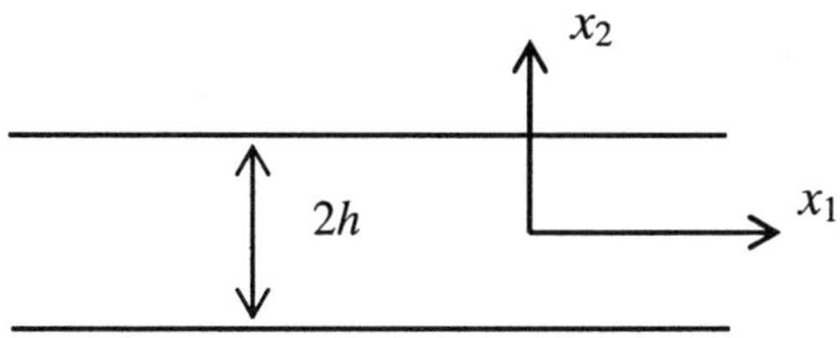

Fig. 3.11.1. A monoclinic crystal plate.

3.11.1. Governing equations

The displacement fields are approximated by

$$\begin{aligned} &u_1(x_1,x_2,t) \cong x_2 u_1^{(1)}(x_1,t), \\ &u_2(x_1,x_2,t) \cong u_2^{(0)}(x_1,t), \\ &u_3 \cong u_3^{(0)}(x_1,t). \end{aligned} \tag{3.11.1}$$

The governing equations for the thickness-shear $u_1^{(1)}$, the flexure $u_2^{(0)}$ and the face-shear $u_3^{(0)}$ are

$$\begin{aligned} &\gamma_{11} u_{1,11}^{(1)} - 3h^{-2}[\kappa_1 c_{56} u_{3,1}^{(0)} + \kappa_1^2 c_{66}(u_{2,1}^{(0)} + u_1^{(1)})] = \rho \ddot{u}_1^{(1)}, \\ &\kappa_1^2 c_{66} u_{1,1}^{(1)} + \kappa_1^2 c_{66} u_{2,11}^{(0)} + \kappa_1 c_{56} u_{3,11}^{(0)} = \rho \ddot{u}_2^{(0)}, \\ &\kappa_1 c_{56} u_{1,1}^{(1)} + \kappa_1 c_{56} u_{2,11}^{(0)} + c_{55} u_{3,11}^{(0)} = \rho \ddot{u}_3^{(0)}. \end{aligned} \tag{3.11.2}$$

The shear forces and bending moments are given by

$$\begin{aligned} &T_1^{(1)} = \frac{2h^3}{3} \gamma_{11} u_{1,1}^{(1)}, \\ &T_5^{(0)} = 2h[c_{55} u_{3,1}^{(0)} + \kappa_1 c_{56}(u_{2,1}^{(0)} + u_1^{(1)})], \\ &T_6^{(0)} = 2h[\kappa_1 c_{56} u_{3,1}^{(0)} + \kappa_1^2 c_{66}(u_{2,1}^{(0)} + u_1^{(1)})]. \end{aligned} \tag{3.11.3}$$

3.11.2. Waves in unbounded plates

Consider stationary waves:

$$\begin{aligned} u_1^{(1)} &= A_1 \cos\xi x_1 \exp(i\omega t), \\ u_2^{(0)} &= A_2 \sin\xi x_1 \exp(i\omega t), \\ u_3^{(0)} &= A_3 \sin\xi x_1 \exp(i\omega t). \end{aligned} \tag{3.11.4}$$

They may be considered as symmetric modes in view of the fact that the shear displacement is an even function of x_1. Substitution of Eq. (3.11.4) into Eq. (3.11.2) results in three linear equations for A_1, A_2 and A_3

$$\begin{aligned} &(\rho\omega^2 - \gamma_{11}\xi^2 - 3h^{-2}\kappa_1^2 c_{66})A_1 \\ &\quad - 3h^{-2}\kappa_1^2 c_{66}\xi A_2 - 3h^{-2}\kappa_1 c_{56}\xi A_3 = 0 \\ &-\kappa_1^2 c_{66}\xi A_1 + (\rho\omega^2 - \kappa_1^2 c_{66}\xi^2)A_2 - \kappa_1 c_{56}\xi^2 A_3 = 0, \\ &-\kappa_1 c_{56}\xi A_5 - \kappa_1 c_{56}\xi^2 A_2 + (\rho\omega^2 - c_{55}\xi^2)A_3 = 0. \end{aligned} \tag{3.11.5}$$

For non-trivial solutions the determinant of the coefficient matrix of the linear equations has to vanish. This yields the following equation that determines the dispersion relations for coupled flexural, thickness-shear and face-shear waves:

$$\begin{vmatrix} \rho\omega^2 - \gamma_{11}\xi^2 - 3h^{-2}\kappa_1^2 c_{66} & -3h^{-2}\kappa_1^2 c_{66}\xi & -3h^{-2}\kappa_1 c_{56}\xi \\ -\kappa_1^2 c_{66}\xi & \rho\omega^2 - \kappa_1^2 c_{66}\xi^2 & -\kappa_1 c_{56}\xi^2 \\ -\kappa_1 c_{56}\xi & -\kappa_1 c_{56}\xi^2 & \rho\omega^2 - c_{55}\xi^2 \end{vmatrix} = 0\,. \tag{3.11.6}$$

Equation (3.11.6) is a cubic equation for ξ^2. It predicts three branches of dispersion curves. Given a value of ω, we denote the three corresponding wave numbers by $\xi^{(i)}(\omega)$, $i = 1, 2, 3$, and the corresponding eigenvectors by $\{\beta_1^{(i)}, \beta_2^{(i)}, \beta_3^{(i)}\}$. Then the general symmetric wave solution to Eq. (3.11.2) can be written as

$$\begin{aligned} u_1^{(1)} &= \sum_{i=1}^{3} C_i\beta_1^{(i)} \cos\xi^{(i)} x_1 \exp(i\omega t), \\ u_2^{(0)} &= \sum_{i=1}^{3} C_i\beta_2^{(i)} \sin\xi^{(i)} x_1 \exp(i\omega t), \\ u_3^{(0)} &= \sum_{i=1}^{3} C_i\beta_3^{(i)} \sin\xi^{(i)} x_1 \exp(i\omega t), \end{aligned} \tag{3.11.7}$$

where C_i are undetermined constants. The dispersion curves determined by Eq. (3.11.6) are plotted in Fig. 3.11.2, where

$$\Omega = \frac{\omega}{\omega_0}, \quad \omega_0^2 = \frac{\pi^2 c_{66}}{4h^2 \rho}, \quad Z = \xi \Big/ \frac{\pi}{2h} . \tag{3.11.8}$$

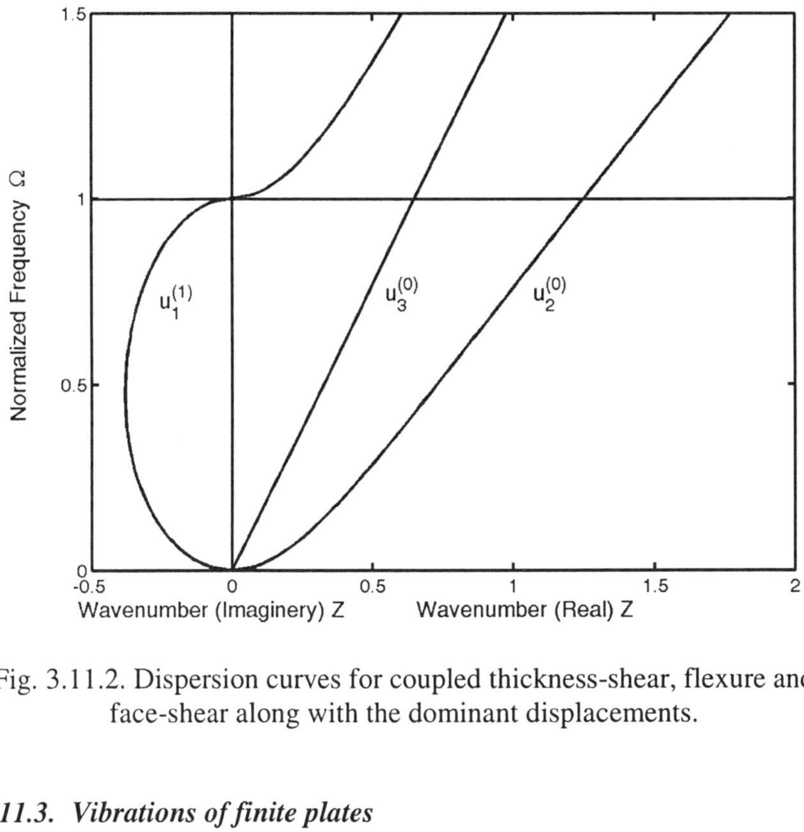

Fig. 3.11.2. Dispersion curves for coupled thickness-shear, flexure and face-shear along with the dominant displacements.

3.11.3. Vibrations of finite plates

Next consider a bounded plate in $|x_1| < a$ (see Fig. 3.11.3).

Consider the case of free edges. The boundary conditions are

$$T_{11}^{(1)} = 0, \quad T_{12}^{(0)} = 0, \quad T_{13}^{(0)} = 0, \quad x_1 = \pm a . \tag{3.11.9}$$

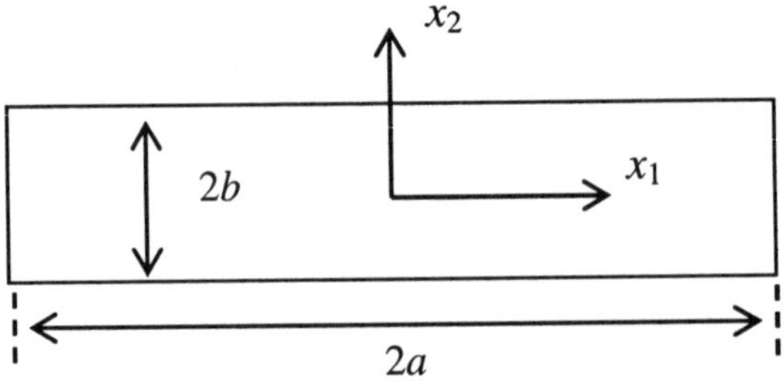

Fig. 3.11.3. A finite plate.

Substitution of Eq. (3.11.7) into Eq. (3.11.9) results in three homogeneous equations for C_i. For non-trivial solutions the determinant of the coefficient matrix has to vanish, which yields the frequency equation for symmetric modes. The solution to the frequency equation is plotted in Fig. 3.11.4 as a function of the aspect ratio a/b.

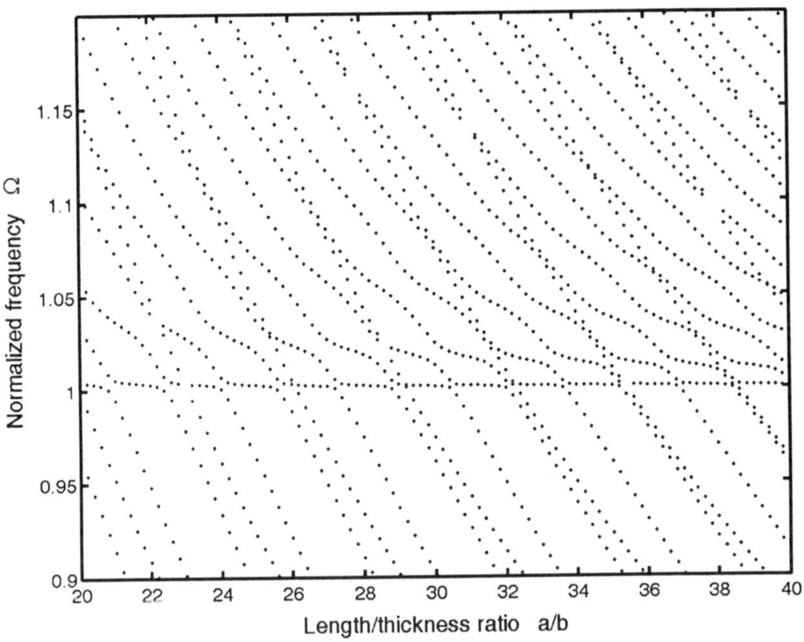

Fig. 3.11.4. Frequency versus aspect ratio.

Similarly, anti-symmetric modes with the shear displacement as an odd function of x_1 can be obtained.

3.12. Effects of Middle Surface Curvature

Plano-convex plates (see Fig. 3.12.1) are widely used for crystal resonators. From a structural point of view, it is obvious that the middle surface (the dotted line in the figure, consisting of points at equal distance to the top and bottom surfaces) of such a resonator is curved. By definition this makes the resonator structure a shell rather than a plate with a flat middle surface. For a plano-convex resonator the curvature of the middle surface is usually small and therefore the resonator is what is called a shallow shell in theories of structures. A plano-convex resonator is in fact a shallow shell with a particular thickness variation such that its bottom surface is flat.

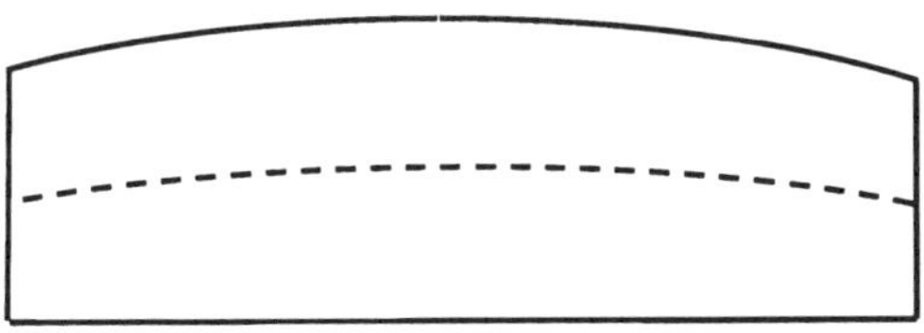

Fig. 3.12.1. A plano-convex resonator.

A fundamental difference between a shell and a plate is that in a shell flexural and thickness-shear vibrations are inherently coupled to extensional vibration due to the curvature of the middle surface. This coupling exists irrespective of whether the shell has a uniform or varying thickness. In this section we study the effects of the middle surface curvature on the vibration characteristics of a thickness-shear resonator [50]. For the most basic effects it is sufficient to study a shallow shell of uniform thickness (see Fig. 3.12.2) .

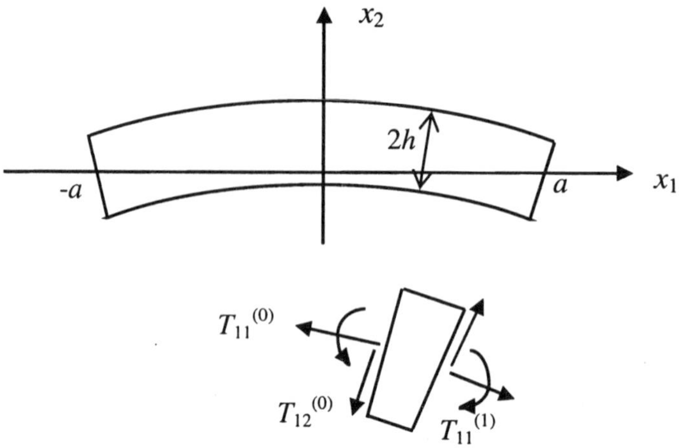

Fig. 3.12.2. A shallow shell (concave-convex resonator) and resultants.

3.12.1. Governing equations

We study motions with $u_3 = 0$ and $\partial/\partial x_3 = 0$ by shell equations [20]. Let the equation for the middle surface be $x_2 = f(x_1)$, and Y be the local coordinate along the shell thickness direction measured from the middle surface. Let $u_1^{(0)}(x_1,t)$ and $u_2^{(0)}(x_1,t)$ be the extensional and flexural displacements of the middle surface, and $u_1^{(1)}(x_1,t)$ the shell thickness-shear displacement. Then the three-dimensional displacement fields are approximately

$$u_1 \cong u_1^{(0)} + Yu_1^{(1)}, \quad u_2 \cong u_2^{(0)}. \tag{3.12.1}$$

From the shell point of view the middle surface ($Y = 0$) is where the shear displacement $Yu_1^{(1)}$ vanishes, but in general it is not where the u_1 displacement vanishes because the middle surface may be in extension for a shell. The equations for coupled extensional, flexural and thickness-shear vibration under the shallow shell assumptions are

$$T_{11,1}^{(0)} = 2h\rho_0 \ddot{u}_1^{(0)},$$
$$-k_1 T_{11}^{(0)} + T_{12,1}^{(0)} + F_2^{(0)} = 2h\rho_0 \ddot{u}_2^{(0)}, \quad (3.12.2)$$
$$T_{11,1}^{(1)} - T_{21}^{(0)} = \frac{2h^3}{3}\rho_0 \ddot{u}_1^{(1)},$$

where

$$k_1 = -\frac{\partial^2 f}{\partial x_1^2} \quad (3.12.3)$$

is the curvature of the middle surface, and $F_2^{(0)}$ is the load in the x_2 direction per unit area of the middle surface. We consider a thin and shallow shell with $k_1 h << k_1 a << 1$. The extensional, shear and bending resultants are given by

$$T_{11}^{(0)} = \int_{-h}^{h} T_{11} dY = 2h\bar{c}_{11}(u_{1,1}^{(0)} + k_1 u_2^{(0)}),$$
$$T_{12}^{(0)} = T_{21}^{(0)} = \int_{-h}^{h} T_{12} dY = 2h\kappa^2 c_{66}(u_1^{(1)} + u_{2,1}^{(0)} - k_1 u_1^{(0)}), \quad (3.12.4)$$
$$T_{11}^{(1)} = \int_{-h}^{h} T_{11} Y dY = \frac{2h^3}{3}\bar{c}_{11} u_{1,1}^{(1)},$$

where

$$\bar{c}_{11} = c_{11} - \frac{c_{12}^2}{c_{22}} \quad (3.12.5)$$

is the extensional elastic stiffness in the x_1 direction after the stress relaxation in the x_2 direction. κ is a thickness-shear correction factor. We will use $\kappa^2 = \pi^2/12$ so that in the limit of $k_1 \to 0$ the exact frequency for the fundamental pure thickness-shear motion independent of x_1 will be obtained. Due to the curvature of the middle surface, the extensional resultant $T_{11}{}^{(0)}$ is present in Eq. $(3.12.2)_2$ for flexure, the flexural displacement $u_2{}^{(0)}$ has a contribution to the extensional strain in Eq. $(3.12.4)_1$, and the extensional displacement $u_1{}^{(0)}$ affects the shear strain in Eq. $(3.12.4)_2$. Consider the case that the middle surface has a constant curvature. In this case the equations have constant coefficients. With substitutions from Eq. (3.12.4), Eq. (3.12.2) can be written as three equations for $u_1^{(0)}$, $u_2^{(0)}$ and $u_1^{(1)}$:

$$\begin{aligned}
&2h\bar{c}_{11}(u_{1,11}^{(0)} + k_1 u_{2,1}^{(0)}) = 2h\rho_0 \ddot{u}_1^{(0)}, \\
&-k_1 2h\bar{c}_{11}(u_{1,1}^{(0)} + k_1 u_2^{(0)}) \\
&\quad + 2h\kappa^2 c_{66}(u_{1,1}^{(1)} + u_{2,11}^{(0)} - k_1 u_{1,1}^{(0)}) + F_2^{(0)} = 2h\rho_0 \ddot{u}_2^{(0)}, \\
&\frac{2h^3}{3}\bar{c}_{11} u_{1,11}^{(1)} - 2h\kappa^2 c_{66}(u_1^{(1)} + u_{2,1}^{(0)} - k_1 u_1^{(0)}) = \frac{2h^3}{3}\rho_0 \ddot{u}_1^{(1)}.
\end{aligned} \tag{3.12.6}$$

3.12.2. Thickness-shear approximation

For the simplest model that has an effect of the middle surface curvature, we extend the thickness-shear approximation for coupled thickness-shear and flexural vibrations of plates in [46] to our case of shells with extension. This approximation is based on two observations. One is that we are interested in thickness-shear waves or modes long and slowly varying in the x_1 direction, hence certain terms with higher order derivatives with respect to x_1 can be neglected. The other is that we are considering frequencies very close to the fundamental plate thickness-shear frequency

$$\omega_0^2 = \frac{\pi^2 c_{66}}{4\rho_0 h^2} = \frac{3\kappa^2 c_{66}}{\rho_0 h^2}. \tag{3.12.7}$$

Therefore, Eq. (3.12.6)$_{1,2}$, for $F_2^{(0)} = 0$, can be approximated by

$$\begin{aligned}
&2h\bar{c}_{11} k_1 u_{2,1}^{(0)} = 2h\rho_0(-\omega_0^2) u_1^{(0)}, \\
&-k_1 2h\bar{c}_{11} k_1 u_2^{(0)} + 2h\kappa^2 c_{66} u_{1,1}^{(1)} = 2h\rho_0(-\omega_0^2) u_2^{(0)}.
\end{aligned} \tag{3.12.8}$$

Solving Eq. (3.12.8) approximately under the small curvature assumption, up to k_1^2, we have

$$\begin{aligned}
u_2^{(0)} &\cong -\left(1 + \frac{4k_1^2 h^2 \bar{c}_{11}}{\pi^2 c_{66}}\right)\frac{h^2}{3} u_{1,1}^{(1)}, \\
u_1^{(0)} &\cong \frac{4k_1 h^4 \bar{c}_{11}}{3\pi^2 c_{66}} u_{1,11}^{(1)}.
\end{aligned} \tag{3.12.9}$$

Substituting Eq. (3.12.9) into Eq. (3.12.6)$_3$, we obtain

$$\frac{2h^3}{3}\left(\bar{c}_{11} + \kappa^2 c_{66} + \frac{2}{3}k_1^2 h^2 \bar{c}_{11}\right) u_{1,11}^{(1)} - 2h\kappa^2 c_{66} u_1^{(1)} = \frac{2h^3}{3}\rho_0 \ddot{u}_1^{(1)}. \tag{3.12.10}$$

Under Eq. (3.12.9), the resultants are approximated by

$$\begin{aligned} T_{11}^{(0)} &= -2h\bar{c}_{11}k_1\left(1+\frac{4k_1^2h^2\bar{c}_{11}}{\pi^2 c_{66}}\right)\frac{h^2}{3}u_{1,1}^{(1)}, \\ T_{12}^{(0)} &= 2h\kappa^2 c_{66}u_1^{(1)}, \\ T_{11}^{(1)} &= \frac{2h^3}{3}\bar{c}_{11}u_{1,1}^{(1)}. \end{aligned} \tag{3.12.11}$$

Consider the thickness-shear modes given by

$$\begin{aligned} &u_1^{(1)} = \cos\frac{n\pi}{2a}x_1\exp(i\omega t), \quad n=1,3,\ldots 5,\ldots, \\ \text{or}\quad &u_1^{(1)} = \sin\frac{n\pi}{2a}x_1\exp(i\omega t), \quad n=2,4,6,\ldots, \end{aligned} \tag{3.12.12}$$

which satisfy the boundary conditions that $u_1^{(1)}(x_1 = \pm a) = 0$. Substitution of Eq. (3.12.12) into Eq. (3.12.10) yields

$$\frac{\omega^2}{\omega_0^2} = 1 + n^2\frac{h^2}{a^2}\left(\frac{\bar{c}_{11}}{c_{66}} + \kappa^2 + \frac{2k_1^2h^2}{3}\frac{\bar{c}_{11}}{c_{66}}\right), \quad n=1,2,3,\ldots, \tag{3.12.13}$$

where the last term in the parentheses is due to the curvature. Equation (3.12.13) shows that the curvature raises the thickness-shear frequency.

For some numerical estimates we consider Y-cut and AT-cut quartz and Y-cut langasite. For the size and curvature of the resonator, we choose k_1 = 1/(20 cm), a = 1 cm, and h = 0.5 mm. For an AT-cut resonator these data yield ω = 10.45 MHz. For Y-cut langasite ω = 8.54 MHz. From Eq. (3.12.13), for n = 1, we obtain the contribution of curvature to frequency as

$$\frac{1}{2}\frac{h^2}{a^2}\frac{2k_1^2h^2}{3}\frac{\bar{c}_{11}}{c_{66}}\Big/\omega_0 \cong \begin{cases} 5.96\times10^{-15} & (\text{AT}-\text{cut quartz}), \\ 3.69\times10^{-15} & (\text{Y}-\text{cut quartz}), \\ 7.63\times10^{-15} & (\text{Y}-\text{cut langasite}), \end{cases} \tag{3.12.14}$$

which gives an idea of the order of magnitude of the effect of the middle surface curvature on thickness-shear frequencies. This effect seems to be small enough to be neglected in most applications.

3.12.3. Coupled thickness-shear and extension

The middle surface curvature does more than the small frequency shift shown in Eq. (3.12.14). The related coupling to extension can become strong when the frequency of a higher order extensional mode is the same as or close to the operating frequency of the fundamental thickness-shear mode (i.e., the modes are degenerate). Since thickness-shear can be trapped but extension cannot, this mode coupling is highly undesirable because it affects energy trapping and mounting, and should be avoided in design. Therefore, it is of practical importance to predict when this coupling occurs. In order to do so we use Eq. $(3.12.9)_1$ to eliminate the flexure in Eq. $(3.12.6)_{1,3}$ and obtain the following two equations for coupled thickness-shear and extension:

$$\begin{aligned} &\bar{c}_{11}u_{1,11}^{(0)} - \frac{1}{3}k_1 h^2 \bar{c}_{11} u_{1,11}^{(1)} = \rho_0 \ddot{u}_1^{(0)}, \\ &\left(\bar{c}_{11} + \kappa^2 c_{66} + \frac{1}{3}k_1^2 h^2 \bar{c}_{11}\right) u_{1,11}^{(1)} - \rho\omega_0^2 u_1^{(1)} + k_1 \rho_0 \omega_0^2 u_1^{(0)} = \rho_0 \ddot{u}_1^{(1)}. \end{aligned} \tag{3.12.15}$$

The resultants are approximated by

$$\begin{aligned} T_{11}^{(0)} &= 2h\bar{c}_{11}u_{1,1}^{(0)} - 2h\bar{c}_{11}k_1\left(1 + \frac{4k_1^2 h^2 \bar{c}_{11}}{\pi^2 c_{66}}\right)\frac{h^2}{3}u_{1,1}^{(1)}, \\ T_{12}^{(0)} &= 2h\kappa^2 c_{66}(u_1^{(1)} - k_1 u_1^{(0)}), \\ T_{11}^{(1)} &= \frac{2h^3}{3}\bar{c}_{11}u_{1,1}^{(1)}. \end{aligned} \tag{3.12.16}$$

For boundary conditions we consider the case that

$$u_1^{(0)}(x_1 = \pm a) = 0, \quad u_1^{(1)}(x_1 = \pm a) = 0. \tag{3.12.17}$$

Consider the following modes

$$u_1^{(0)} = A\cos\xi x_1 \exp(i\omega t), \quad u_1^{(1)} = B\cos\xi x_1 \exp(i\omega t), \tag{3.12.18}$$

where A and B are constants, and ξ is the wave number in the x_1 direction. For Eq. (3.12.18) to satisfy Eq. (3.12.17) we must have

$$\cos\xi a = 0, \quad \xi = \frac{n\pi}{2a}, \quad n = 1,3,5,\cdots. \tag{3.12.19}$$

Substituting Eq. (3.12.18) into Eq. (3.12.15), for non-trivial solutions of A and B, we obtain the following frequency equation:

$$\begin{vmatrix} \rho_0\omega^2 - \bar{c}_{11}\xi^2 & \frac{1}{3}k_1 h^2 \bar{c}_{11}\xi^2 \\ k_1\rho\omega_0^2 & \rho_0\omega^2 - \left(\bar{c}_{11} + \kappa^2 c_{66} + \frac{1}{3}k_1^2 h^2 \bar{c}_{11}\right)\xi^2 - \rho_0\omega_0^2 \end{vmatrix} = 0\,. \quad (3.12.20)$$

Similarly, there exist another set of modes

$$u_1^{(0)} = C\sin\xi x_1 \exp(i\omega t), \quad u_1^{(1)} = D\sin\xi x_1 \exp(i\omega t), \quad (3.12.21)$$

where C and D are constants, and

$$\sin\xi a = 0, \quad \xi = \frac{n\pi}{2a}, \quad n = 2, 4, 6, \ldots. \quad (3.12.22)$$

The frequency equation is still given by Eq. (3.12.20).

For the special case of $k_1 = 0$, Eq. (3.12.20) reduces to two sets of frequencies for uncoupled extension and thickness-shear. When $k_1 \neq 0$ but is small, Eq. (3.12.20) yields two sets of frequencies. One represents essentially extensional modes. The other is essentially thickness-shear modes. We plot the two sets of frequencies when $k_1 = 0$ versus the aspect ration a/h in Fig. 3.12.3 for AT-cut quartz. The thickness-shear modes

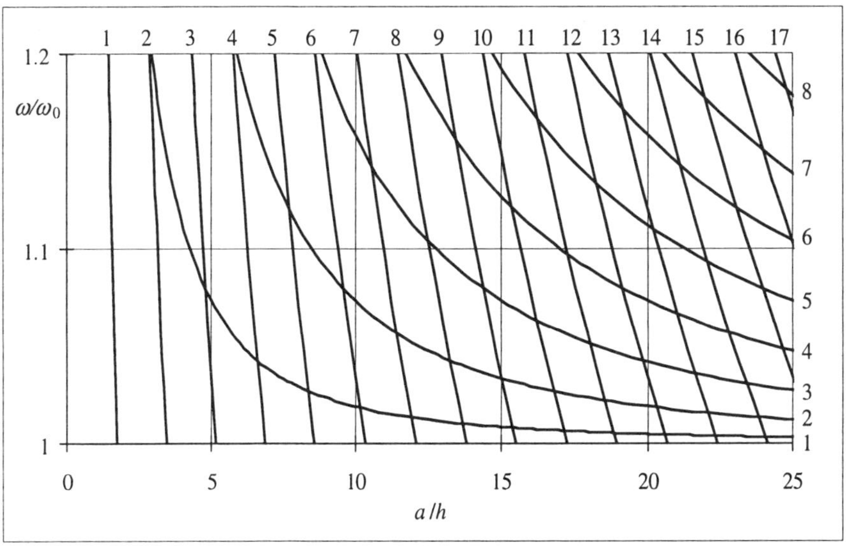

Fig. 3.12.3. Frequency spectra for coupled thickness-shear and extension in AT-cut quartz.

are numbered on the right. The extensional modes are numbered on the top. The Fig. shows that for certain values of a/h the two sets of curves intersect. These are the aspect ratios at which a thickness-shear mode and an extensional mode have the same resonant frequency and should be avoided in resonator design.

The extensional and thickness-shear frequency spectra for Y-cut langasite are shown in Fig. 3.12.4.

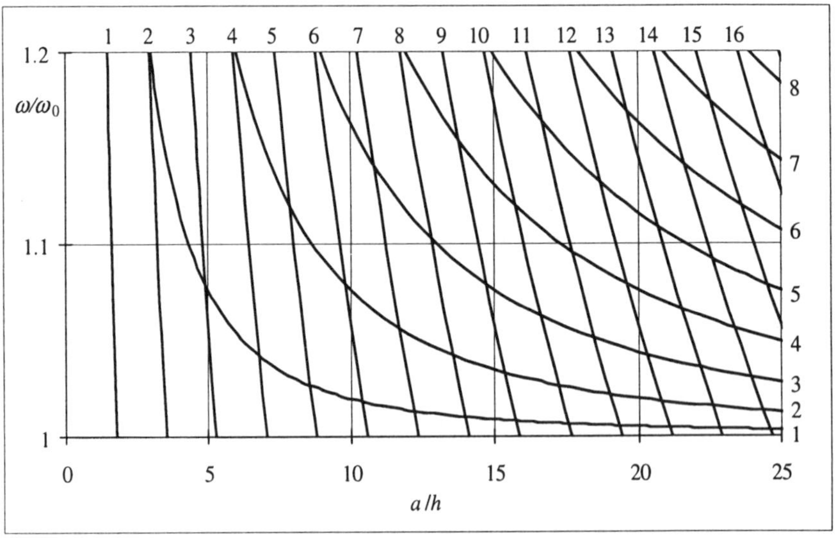

Fig. 3.12.4. Frequency spectra for coupled thickness-shear and extension in Y-cut langasite.

Chapter 4

Mass Sensors

Frequency shifts in a crystal resonator due to a thin layer of mass added to part of the crystal surface is the foundation of many chemical and biological acoustic wave sensors. These sensors detect a substance through the mass-frequency effect of the substance accumulated on the crystal surface by some chemically or biologically active films. In this chapter, we analyze frequency shifts in a three-dimensional body due to a thin layer of surface mass.

4.1. Inertial Effect of a Mass Layer by Perturbation

Consider a piezoelectric body with a thin mass layer of thickness h' and density ρ' on part of its surface (see Fig. 4.1.1). We study the mass-frequency effect by a perturbation method [51].

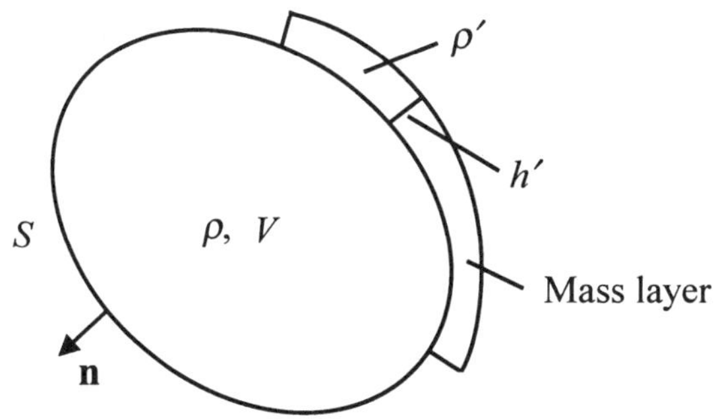

Fig. 4.1.1. A piezoelectric body with a surface mass layer.

4.1.1. Governing equations

Let the region occupied by the piezoelectric body be V and its boundary surface be S. The unit outward normal of S is $\mathbf{n}$. The mass layer is assumed to be very thin. For the lowest order effect of the mass layer, only the inertial effect of the layer is to be considered; its stiffness is neglected. For free vibration with a frequency ω, by Newton's second law, the traction boundary condition on the surface area with the mass layer is

$$-T_{ji}n_j = \rho' h' \ddot{u}_i = -\rho' h' \omega^2 u_i \,. \tag{4.1.1}$$

From Eqs. (1.2.16) through (1.2.21) and Eq. (4.1.1), the eigenvalue problem for the resonant frequencies and modes of the crystal with the mass layer is

$$\begin{aligned}
&-c_{jikl}u_{k,lj} - e_{kji}\phi_{,kj} = \rho\lambda u_i \quad \text{in} \quad V, \\
&-e_{ikl}u_{k,li} + \varepsilon_{ik}\phi_{,ki} = 0 \quad \text{in} \quad V, \\
&u_i = 0 \quad \text{on} \quad S_u, \\
&T_{ji}n_j = (c_{jikl}u_{k,l} + e_{kji}\phi_{,k})n_j = \varepsilon\lambda\rho' h' u_i \quad \text{on} \quad S_T, \\
&\phi = 0 \quad \text{on} \quad S_\phi, \\
&D_i n_i = (e_{ikl}u_{k,l} - \varepsilon_{ik}\phi_{,k})n_i = 0 \quad \text{on} \quad S_D,
\end{aligned} \tag{4.1.2}$$

where we have denoted

$$\lambda = \omega^2 \,. \tag{4.1.3}$$

In Eq. $(4.1.2)_4$ we have artificially introduced a dimensionless number ε to be used in our perturbation analysis. The real physical problem is represented by $\varepsilon = 1$.

4.1.2. Abstract notation

Eq. (4.1.2) can be written in a more compact form as

$$\begin{aligned}
&\mathbf{AU} = \lambda\mathbf{BU} \quad \text{in} \quad V, \\
&u_i = 0 \quad \text{on} \quad S_u, \\
&T_{ji}(\mathbf{U})n_j = \varepsilon\lambda\rho' h' u_i \quad \text{on} \quad S_T, \\
&\phi = 0 \quad \text{on} \quad S_\phi, \\
&D_i(\mathbf{U})n_i = 0 \quad \text{on} \quad S_D,
\end{aligned} \tag{4.1.4}$$

where $\mathbf{U} = \{u_k, \phi\}$ is a 4-vector. The differential operators $\mathbf{A}$ and $\mathbf{B}$ are defined by

$$\begin{aligned}
&\mathbf{AU} = \{-c_{jikl}u_{k,lj} - e_{kji}\phi_{,kj}, -e_{ikl}u_{k,li} + \varepsilon_{ik}\phi_{,ki}\}, \\
&\mathbf{BU} = \{\rho u_i, 0\}.
\end{aligned} \tag{4.1.5}$$

$T_{ji}(\mathbf{U})$ and $D_i(\mathbf{U})$ are the stress tensor and electric displacement vector in terms of the 4-vector.

4.1.3. Perturbation

We make the following perturbation expansions:

$$\lambda \cong \lambda^{(0)} + \varepsilon \lambda^{(1)},$$
$$\mathbf{U} = \begin{Bmatrix} u_i \\ \phi \end{Bmatrix} \cong \begin{Bmatrix} u_i^{(0)} \\ \phi^{(0)} \end{Bmatrix} + \varepsilon \begin{Bmatrix} u_i^{(1)} \\ \phi^{(1)} \end{Bmatrix} = \mathbf{U}^{(0)} + \varepsilon \mathbf{U}^{(1)}. \tag{4.1.6}$$

Substituting Eq. (4.1.6) into Eq. (4.1.4), collecting terms of equal powers of ε, we obtain a series of perturbation problems of successive orders. We are interested in the lowest order effect of the mass layer. Therefore we collect coefficients of terms with powers of ε^0 and ε^1 only. The zero-order problem is

$$\begin{aligned}
&-c_{jikl}u_{k,lj}^{(0)} - e_{kji}\phi_{,kj}^{(0)} = \rho\lambda^{(0)}u_i^{(0)} \quad \text{in} \quad V, \\
&-e_{ikl}u_{k,li}^{(0)} + \varepsilon_{ik}\phi_{,ki}^{(0)} = 0 \quad \text{in} \quad V, \\
&u_i^{(0)} = 0 \quad \text{on} \quad S_u, \\
&(c_{jikl}u_{k,l}^{(0)} + e_{kji}\phi_{,k}^{(0)})n_j = 0 \quad \text{on} \quad S_T, \\
&\phi^{(0)} = 0 \quad \text{on} \quad S_\phi, \\
&(e_{ikl}u_{k,l}^{(0)} - \varepsilon_{ik}\phi_{,k}^{(0)})n_i = 0 \quad \text{on} \quad S_D.
\end{aligned} \tag{4.1.7}$$

This represents free vibrations of the body without the surface mass layer. The solution to the zero-order problem, $\lambda^{(0)}$ and $\mathbf{U}^{(0)}$, is assumed known as usual in a perturbation analysis. $\lambda^{(0)}$ is real and $\mathbf{U}^{(0)}$ can also be written as real functions [10]. The following first-order problem is to be solved:

$$\begin{aligned}
&-c_{jikl}u_{k,lj}^{(1)} - e_{kji}\phi_{,kj}^{(1)} = \rho\lambda^{(1)}u_i^{(0)} + \rho\lambda^{(0)}u_i^{(1)} \quad \text{in} \quad V, \\
&-e_{ikl}u_{k,li}^{(1)} + \varepsilon_{ik}\phi_{,ki}^{(1)} = 0 \quad \text{in} \quad V, \\
&u_i^{(1)} = 0 \quad \text{on} \quad S_u, \\
&(c_{jikl}u_{k,l}^{(1)} + e_{kji}\phi_{,k}^{(1)})n_j = \rho'h'\lambda^{(0)}u_i^{(0)} \quad \text{on} \quad S_T, \\
&\phi^{(1)} = 0 \quad \text{on} \quad S_\phi, \\
&(e_{ikl}u_{k,l}^{(1)} - \varepsilon_{ik}\phi_{,k}^{(1)})n_i = 0 \quad \text{on} \quad S_D.
\end{aligned} \tag{4.1.8}$$

The equations for the first-order problem, Eq. $(4.1.8)_{1,2}$, can be written as

$$\mathbf{AU}^{(1)} = \lambda^{(0)}\mathbf{BU}^{(1)} + \lambda^{(1)}\mathbf{BU}^{(0)}. \tag{4.1.9}$$

Multiplying both sides of Eq. (4.1.9) by $\mathbf{U}^{(0)}$ and integrating the resulting equation over V, we have

$$\begin{aligned} &< \mathbf{AU}^{(1)};\mathbf{U}^{(0)} > \\ &= \int_V [(-c_{jikl}u_{k,lj}^{(1)} - e_{kji}\phi_{,kj}^{(1)})u_i^{(0)} + (-e_{ikl}u_{k,li}^{(1)} + \varepsilon_{ik}\phi_{,ki}^{(1)})\phi^{(0)}]dV \\ &= \lambda^{(0)} < \mathbf{BU}^{(1)};\mathbf{U}^{(0)} > + \lambda^{(1)} < \mathbf{BU}^{(0)};\mathbf{U}^{(0)} >, \end{aligned} \tag{4.1.10}$$

where, for simplicity, we have used $<;>$ to represent the product of two 4-vectors and the integration over V. With integration by parts, it can be shown that

$$\begin{aligned} &< \mathbf{AU}^{(0)};\mathbf{U}^{(1)} > \\ &= -\int_S [T_{ji}(\mathbf{U}^{(0)})n_j u_i^{(1)} + D_i(\mathbf{U}^{(0)})n_i\phi^{(1)}]dS \\ &+ \int_S [T_{kl}(\mathbf{U}^{(1)})n_l u_k^{(0)} + D_k(\mathbf{U}^{(1)})n_k\phi^{(0)}]dS + < \mathbf{U}^{(0)};\mathbf{AU}^{(1)} >. \end{aligned} \tag{4.1.11}$$

With the boundary conditions in Eqs. $(4.1.7)_{3\text{-}6}$ and $(4.1.8)_{3\text{-}6}$, Eq. (4.1.11) becomes

$$\begin{aligned} &< \mathbf{AU}^{(0)};\mathbf{U}^{(1)} > \\ &= \int_{S_T} \rho' h' \lambda^{(0)} u_k^{(0)} u_k^{(0)} dS + < \mathbf{U}^{(0)};\mathbf{AU}^{(1)} >. \end{aligned} \tag{4.1.12}$$

Substitution of Eq. (4.1.12) into Eq. (4.1.10) gives

$$\begin{aligned} &< \mathbf{AU}^{(0)};\mathbf{U}^{(1)} > - \int_{S_T} \rho' h' \lambda^{(0)} u_k^{(0)} u_k^{(0)} dS \\ &= \lambda^{(0)} < \mathbf{BU}^{(1)};\mathbf{U}^{(0)} > + \lambda^{(1)} < \mathbf{BU}^{(0)};\mathbf{U}^{(0)} >, \end{aligned} \tag{4.1.13}$$

which can be further written as

$$\begin{aligned} &< \mathbf{AU}^{(0)} - \lambda^{(0)}\mathbf{BU}^{(0)};\mathbf{U}^{(1)} > - \int_{S_T} \rho' h' \lambda^{(0)} u_k^{(0)} u_k^{(0)} dS \\ &= \lambda^{(1)} < \mathbf{BU}^{(0)};\mathbf{U}^{(0)} >. \end{aligned} \tag{4.1.14}$$

With Eq. $(4.1.7)_{1,2}$, we obtain, from Eq. (4.1.14),

$$\lambda^{(1)} = -\frac{\int_{S_T} \rho' h' \lambda^{(0)} u_k^{(0)} u_k^{(0)} dS}{< \mathbf{BU}^{(0)};\mathbf{U}^{(0)} >} = -\lambda^{(0)} \frac{\int_{S_T} \rho' h' u_k^{(0)} u_k^{(0)} dS}{\int_V \rho u_i^{(0)} u_i^{(0)} dV}. \tag{4.1.15}$$

The above expression is for the perturbation of the eigenvalue $\lambda = \omega^2$. For ω we make the following expansion:

$$\omega \cong \omega^{(0)} + \varepsilon\omega^{(1)}. \tag{4.1.16}$$

Then,

$$\lambda = \omega^2 \cong (\omega^{(0)} + \varepsilon\omega^{(1)})^2 \cong (\omega^{(0)})^2 + 2\varepsilon\omega^{(0)}\omega^{(1)} \cong \lambda^{(0)} + \varepsilon\lambda^{(1)}. \tag{4.1.17}$$

Hence,

$$\frac{\varepsilon\omega^{(1)}}{\omega^{(0)}} \cong \frac{1}{2(\omega^{(0)})^2}\varepsilon\lambda^{(1)} = -\frac{1}{2(\omega^{(0)})^2}\varepsilon\lambda^{(0)}\frac{\int_{S_T} \rho' h' u_k^{(0)} u_k^{(0)} dS}{\int_V \rho u_i^{(0)} u_i^{(0)} dV}. \tag{4.1.18}$$

Finally, setting $\varepsilon = 1$ in Eq. (4.1.18), we obtain

$$\frac{\omega - \omega^{(0)}}{\omega^{(0)}} \cong -\frac{1}{2}\frac{\int_{S_T} \rho' h' u_k^{(0)} u_k^{(0)} dS}{\int_V \rho u_i^{(0)} u_i^{(0)} dV}. \tag{4.1.19}$$

Eq. (4.1.19) is linear in the density and thickness of the mass layer. It shows that the inertia of the mass layer lowers the frequency, as expected. It is also related to the kinetic energies of the mass layer and the crystal. This will become clear later in an variational analysis.

4.2. Thickness-Shear Modes of a Plate

Consider an unbounded plate of rotated Y-cut quartz (see Fig. 4.2.1). The plate carries a thin mass layer on both of its surfaces.

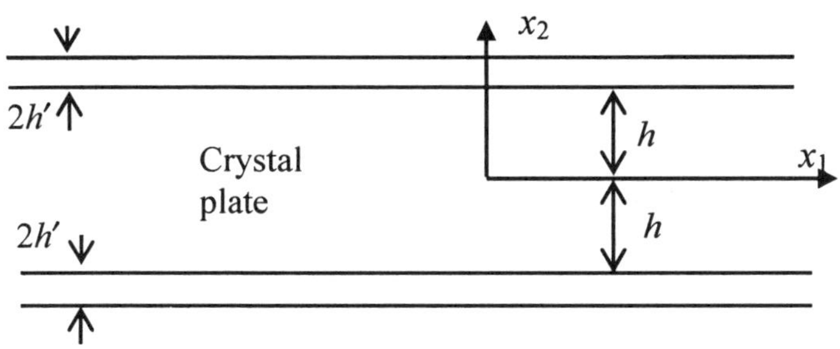

Fig. 4.2.1. An elastic plate with mass layers.

Quartz is a material with very weak piezoelectric coupling. For a frequency analysis the weak piezoelectric coupling can usually be neglected and an elastic analysis is sufficient. We are interested in the mass sensitivity of the fundamental thickness-shear mode. When the surface mass is not present, the frequency of interest is given by Eq. (2.4.34) for $n = 1$, which, along with the mode, in the notation of the present chapter, takes the following form:

$$\omega^{(0)} = \frac{\pi}{2h}\sqrt{\frac{c_{66}}{\rho}}, \quad u_1^{(0)} = \sin\frac{\pi}{2h}x_2, \quad u_2^{(0)} = u_3^{(0)} = 0 . \tag{4.2.1}$$

When the mass layers are present, from Eq. (4.1.19) we obtain

$$\frac{\Delta\omega}{\omega^{(0)}} = -\frac{2\rho' h'}{\rho h} = -R . \tag{4.2.2}$$

The exact fundamental thickness-shear frequency for the plate and the mass layers is given by Eq. (2.5.8):

$$\cot\left(\omega h\sqrt{\frac{\rho}{c_{66}}}\right) = \frac{2\rho' h'}{\rho h}\omega h\sqrt{\frac{\rho}{c_{66}}} . \tag{4.2.3}$$

For small h', to the first order effect of h', the ω determined by Eq. (4.2.3) is approximately given by Eq. (2.5.11):

$$\omega = \frac{\pi}{2h}\sqrt{\frac{c_{66}}{\rho}}(1-R) = \omega^{(0)}(1-R) . \tag{4.2.4}$$

Thus the perturbation integral predicts the lowest order inertial effect of the mass layer.

4.3. Anti-Plane Modes of a Wedge

In this section we study the mass sensitivity of anti-plane or shear-horizontal (SH) modes in an isotropic elastic wedge [52]. Anti-plane wedge modes are face-shear and thickness-twist modes with material particles of the wedge surfaces moving tangentially within the surfaces, and are ideal for mass sensors. In addition, as modes of a perfect semi-infinite structure, wedge modes do not have the edge effect associated with using modes of an infinite plate to make a finite size sensor. Consider a semi-infinite wedge with mass layers as shown in Fig. 4.3.1.

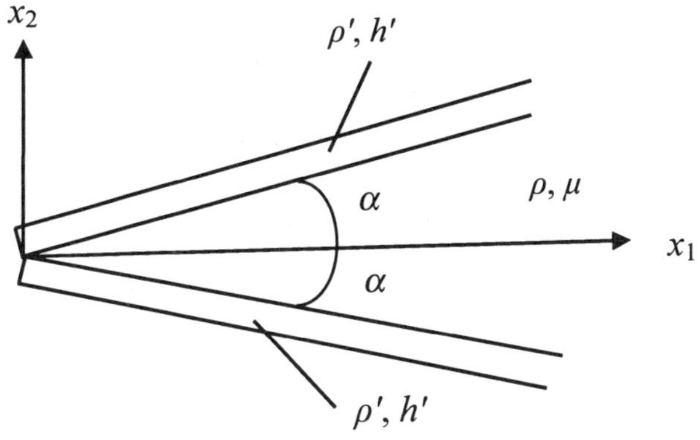

Fig. 4.3.1. An elastic wedge with mass layers.

When the mass layers are not present, the wedge supports the following anti-plane modes in a polar coordinate system defined by $x_1 = r\cos\theta$ and $x_2 = r\sin\theta$ [41]:

$$u_3 = J_\nu(\eta r)\sin\nu\theta\exp(-i\omega t), \quad n = 1,3,5,\ldots, \\ u_3 = J_\nu(\eta r)\cos\nu\theta\exp(-i\omega t), \quad n = 0,2,4,\ldots, \tag{4.3.1}$$

where

$$\eta = \frac{\omega}{c_2}, \quad c_2^2 = \frac{\mu}{\rho}, \quad \nu = \frac{n\pi}{2\alpha}. \tag{4.3.2}$$

μ is the shear elastic constant and ρ is the mass density of the wedge. J_ν is the first kind Bessel function of order ν. $n = 0$ is the face-shear mode and the rest are thickness-twist modes. As a semi-infinite structure, a wedge has a continuous spectrum and ω can be arbitrary.

Substitution of Eq. (4.3.1) into Eq. (4.1.19) yields

$$\frac{\Delta\omega}{\omega} \cong -\frac{\omega\rho' h'}{\alpha\sqrt{\rho\mu}}\gamma(n)B(\nu), \tag{4.3.3}$$

where

$$\gamma(n) = \begin{cases} 1/2, & n = 0, \\ 1, & n = 1,2,3,4,\ldots, \end{cases} \quad B(\nu) = \frac{\int_{-\infty}^{\infty} J_\nu^2(\xi)d\xi}{\int_{-\infty}^{\infty} \xi J_\nu^2(\xi)d\xi}. \tag{4.3.4}$$

Equation (4.3.3) shows that high frequency modes, narrow wedges, and light and compliant wedge materials imply higher mass sensitivity, as expected. Since the mass layer has been assumed to be much thinner than the wedge, α in Eq. (4.3.3) cannot go to zero.

4.4. Torsional Modes of a Conical Shell

In this section we study the mass sensitivity of torsional modes in a circular conical shell [53]. Consider a conical shell with a mass layer as shown in Fig. 4.4.1. The shell has density ρ, thickness h and shear elastic constant μ. The mass layer is assumed to be much thinner than the shell, with density ρ' and thickness h'.

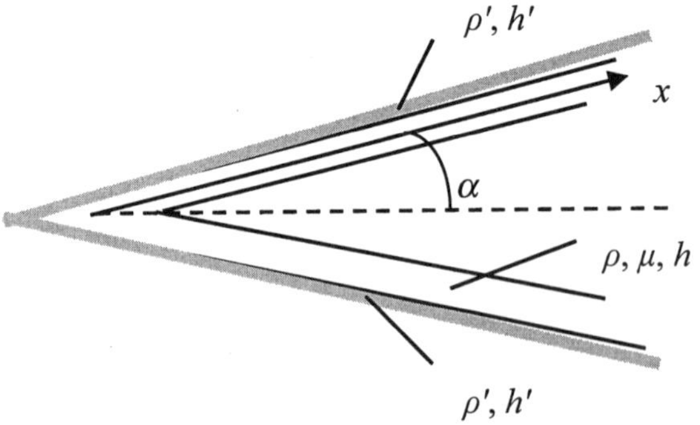

Fig. 4.4.1. An elastic circular conical shell with a mass layer.

When the mass layer is not present, the torsional modes of the shell is governed by [54]

$$\begin{aligned} &\frac{\partial(xN_{x\theta})}{\partial x} + N_{x\theta} = \rho h \ddot{u}_\theta, \\ &N_{x\theta} = \mu h \gamma_{x\theta}, \\ &\gamma_{x\theta} = \frac{\partial u_\theta}{\partial x} - \frac{u_\theta}{x}, \end{aligned} \tag{4.4.1}$$

where u_θ is the torsional displacement, $N_{x\theta}$ is the shear force for torsion, and $\gamma_{x\theta}$ is the shear strain for torsion. With successive substitutions, Eq. (4.4.1) can be written as

$$\mu h\left(\frac{\partial^2 u_\theta}{\partial x^2}+\frac{1}{x}\frac{\partial u_\theta}{\partial x}-\frac{1}{x^2}u_\theta\right)=\rho h\ddot{u}_\theta\,, \tag{4.4.2}$$

where, as expected, h and α in fact do not affect the torsional modes within a shell theory. For time-harmonic motion with a frequency ω, Eq. (4.4.2) can be written as

$$\frac{d^2u_\theta}{dX^2}+\frac{1}{X}\frac{du_\theta}{dX}+\left(1-\frac{1}{X^2}\right)u_\theta=0\,, \tag{4.4.3}$$

where

$$X=\xi x,\quad \xi^2=\frac{\rho\omega^2}{\mu}\,. \tag{4.4.4}$$

Eq. (4.4.3) is Bessel equation of order one. Its solution bounded at $X = 0$ is given by

$$u_\theta=J_1(X)=J_1(\xi x)\,, \tag{4.4.5}$$

where J_1 is the first kind Bessel function of order one. As a semi-infinite structure, a conical shell has a continuous spectrum and ω can be arbitrary. $J_1(X)$ decays like $1/\sqrt{X}$ for large X.

Substitution of Eq. (4.4.5) into Eq. (4.1.19) yields the following simple expression for the frequency shift:

$$\frac{\Delta\omega}{\omega}\cong-\frac{1}{2}\frac{\rho'h'}{\rho h}\,. \tag{4.4.6}$$

4.5. Effects of Inertia and Stiffness of a Mass Layer by Perturbation

In the first section of this chapter, only the inertial effect of the mass layer was considered, its stiffness neglected. When the mass layer is not very thin, its stiffness may need to be considered. In this section, we analyze frequency shifts due to a mass layer with both the inertial effect and the stiffness effect of the layer included in the analysis [55]. Consider a piezoelectric body with a thin mass layer of thickness h' and density ρ' on part of its surface (see Fig. 4.5.1).

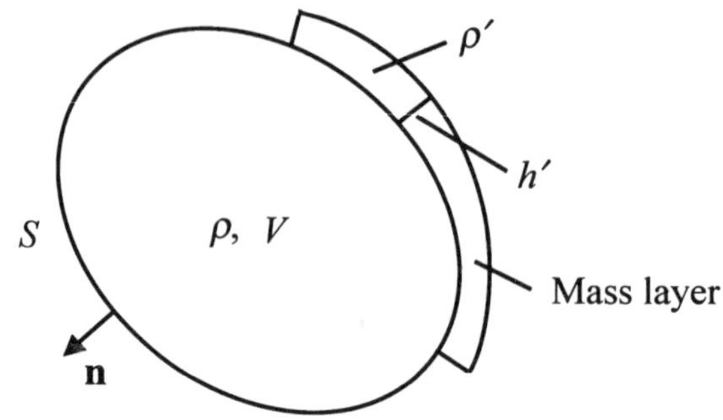

Fig. 4.5.1. A piezoelectric body with a surface mass layer.

4.5.1. Governing equations

For free vibrations, from Eqs. (1.2.16) through (1.2.21), the governing equations and boundary conditions of the body are

$$\begin{aligned}
&c_{jikl}u_{k,lj} + e_{kji}\phi_{,kj} = \rho\ddot{u}_i \quad \text{in} \quad V, \\
&-e_{ikl}u_{k,li} + \varepsilon_{ik}\phi_{,ki} = 0 \quad \text{in} \quad V, \\
&u_i = 0 \quad \text{on} \quad S_u, \\
&T_{ji}n_j = (c_{jikl}u_{k,l} + e_{kji}\phi_{,k})n_j = t_i \quad \text{on} \quad S_T, \\
&\phi = 0 \quad \text{on} \quad S_\phi, \\
&D_i n_i = (e_{ikl}u_{k,l} - \varepsilon_{ik}\phi_{,k})n_i = 0 \quad \text{on} \quad S_D,
\end{aligned} \tag{4.5.1}$$

where t_i is the traction vector between the body and the film. For the film, we use the two-dimensional equations of an elastic shell [54]. For the lowest order effect of the stiffness of the mass layer, we use the membrane theory for a thin shell in extension, which describes a film that does not resist bending. The shell displacement vector is given by $u_\beta = u_\beta(\alpha_1, \alpha_2, t)$, where α_1 and α_2 are the shell middle surface principal coordinates, and α_3 is the shell thickness coordinate. β = 1, 2, 3 is associated with the shell principle coordinates (α_1,α_2,α_3). Then the shell membrane strains are

$$S_1 = \frac{1}{A_1}\frac{\partial u_1}{\partial \alpha_1} + \frac{u_2}{A_1 A_2}\frac{\partial A_1}{\partial \alpha_2} + \kappa_1 u_3,$$
$$S_2 = \frac{1}{A_2}\frac{\partial u_2}{\partial \alpha_2} + \frac{u_1}{A_1 A_2}\frac{\partial A_2}{\partial \alpha_1} + \kappa_2 u_3, \tag{4.5.2}$$
$$S_6 = \frac{A_2}{A_1}\frac{\partial}{\partial \alpha_1}\left(\frac{u_2}{A_2}\right) + \frac{A_1}{A_2}\frac{\partial}{\partial \alpha_2}\left(\frac{u_1}{A_1}\right),$$

where A_1 and A_2 are the Lamè coefficients corresponding to α_1 and α_2, and κ_1 and κ_2 are the principal curvatures of the middle surface of the shell. The mass layer is assumed isotropic. The membrane stress resultants are given by the following membrane constitutive relations:

$$N_1 = h'\gamma_{11}S_1 + h'\gamma_{12}S_2,$$
$$N_2 = h'\gamma_{12}S_1 + h'\gamma_{11}S_2, \tag{4.5.3}$$
$$N_6 = h'\gamma_{66}S_6,$$

where γ_{11}, γ_{12} and γ_{66} are the shell membrane elastic constants [20]. The membrane equations of motion are

$$\frac{\partial}{\partial \alpha_1}(A_2 N_1) + \frac{\partial}{\partial \alpha_2}(A_1 N_6) + N_6 \frac{\partial A_1}{\partial \alpha_2} - N_2 \frac{\partial A_2}{\partial \alpha_1} + A_1 A_2 f_1 = A_1 A_2 \rho' h' \ddot{u}_1,$$
$$\frac{\partial}{\partial \alpha_1}(A_2 N_6) + \frac{\partial}{\partial \alpha_2}(A_1 N_2) + N_6 \frac{\partial A_2}{\partial \alpha_1} - N_1 \frac{\partial A_1}{\partial \alpha_2} + A_1 A_2 f_2 = A_1 A_2 \rho' h' \ddot{u}_2, \tag{4.5.4}$$
$$-A_1 A_2 \kappa_1 N_1 - A_1 A_2 \kappa_2 N_2 + A_1 A_2 f_3 = A_1 A_2 \rho' h' \ddot{u}_3,$$

where f_i is the load per unit middle surface area. With successive substitutions from Eqs. (4.5.2) and (4.5.3), Eq. (4.5.4) can be written in the following compact form

$$-L_\beta(\mathbf{u}) + f_\beta = \rho' h' \ddot{u}_\beta, \tag{4.5.5}$$

where L_β are linear differential operators. At every point of the middle surface of the shell, there exists a set of transformation coefficients $\delta_{i\beta}$ between the local shell principal coordinates and the global Cartesian coordinates. Multiplying Eq. (4.5.5) by the transformation coefficients we have

$$-L_i(\mathbf{u})+f_i=\rho'h'\ddot{u}_i\,. \tag{4.5.6}$$

The t_i in Eq. $(4.5.1)_4$ and the f_i in Eq. (4.5.6) are actions and reactions, equal in magnitude and opposite in direction (f_i = $-t_i$). Substituting Eq. (4.5.6) into Eq. $(4.5.1)_4$, considering time-harmonic motions with $\exp(i\omega t)$ time dependence, the eigenvalue problem for the free vibration of the body with the mass layer can be written as

$$\begin{aligned}
&-c_{jikl}u_{k,lj}-e_{kji}\phi_{,kj}=\rho\lambda u_i \quad \text{in} \quad V,\\
&-e_{ikl}u_{k,li}+\varepsilon_{ik}\phi_{,ki}=0 \quad \text{in} \quad V,\\
&u_i=0 \quad \text{on} \quad S_u,\\
&T_{ji}n_j=(c_{jikl}u_{k,l}+e_{kji}\phi_{,k})n_j\\
&\quad=\varepsilon[\rho'h'\lambda u_i-L_i(\mathbf{u})] \quad \text{on} \quad S_T,\\
&\phi=0 \quad \text{on} \quad S_\phi,\\
&D_in_i=(e_{ikl}u_{k,l}-\varepsilon_{ik}\phi_{,k})n_i=0 \quad \text{on} \quad S_D,
\end{aligned} \tag{4.5.7}$$

where we have denoted $\lambda=\omega^2$. ε is a parameter introduced.

4.5.2. Perturbation

For a perturbation solution we make the following expansions:

$$\begin{aligned}
&\lambda\cong\lambda^{(0)}+\varepsilon\,\lambda^{(1)},\\
&\mathbf{U}=\begin{Bmatrix}u_i\\ \phi\end{Bmatrix}\cong\begin{Bmatrix}u_i^{(0)}\\ \phi^{(0)}\end{Bmatrix}+\varepsilon\begin{Bmatrix}u_i^{(1)}\\ \phi^{(1)}\end{Bmatrix}=\mathbf{U}^{(0)}+\varepsilon\mathbf{U}^{(1)},
\end{aligned} \tag{4.5.8}$$

where $\omega^{(0)}$, $\mathbf{u}^{(0)}$ and $\phi^{(0)}$ are the frequency and modes when the mass layer is not present. They are supposed to be known and can be written as real numbers and real functions. Comparison of Eq. (4.5.7) to Eq. (4.1.2) shows that the only difference is the term with L_i. Therefore, the perturbation procedure is very similar to that in the first section of this chapter; the only modification needed is the L_i term. The perturbation procedure will not be repeated. It is easy to see that the result is simply

$$\lambda^{(1)} = \frac{\int_{S_T} [L_k(\mathbf{u}^{(0)})u_k^{(0)} - \rho' h' \lambda^{(0)} u_k^{(0)} u_k^{(0)}] dS}{\int_V \rho u_i^{(0)} u_i^{(0)} dV}. \tag{4.5.9}$$

The above expression is for the eigenvalue $\lambda = \omega^2$. For ω, from the first section of this chapter, we have

$$\omega^{(1)} = \lambda^{(1)} / (2\omega^{(0)}). \tag{4.5.10}$$

Hence,

$$\frac{\varepsilon\omega^{(1)}}{\omega^{(0)}} \cong \frac{1}{2(\omega^{(0)})^2} \varepsilon\lambda^{(1)}. \tag{4.5.11}$$

Finally, substituting Eq. (4.5.9) into Eq. (4.5.11), setting $\varepsilon = 1$, we obtain

$$\frac{\omega - \omega^{(0)}}{\omega^{(0)}} \cong \frac{1}{2(\omega^{(0)})^2} \times \frac{\int_{S_T} [L_k(\mathbf{u}^{(0)})u_k^{(0)} - \rho' h' (\omega^{(0)})^2 u_k^{(0)} u_k^{(0)}] dS}{\int_V \rho u_i^{(0)} u_i^{(0)} dV}. \tag{4.5.12}$$

The stiffness effect of the film is in L_k. When $L_k = 0$, Eq. (4.5.12) reduces to Eq. (4.1.19).

4.6. Effects of Inertia and Stiffness of a Mass Layer by Variation

In this section we study the inertial and stiffness effects of a mass layer on a crystal by variational method [56]. Since the inertial and stiffness effects are both mechanical, for simplicity we consider an elastic rather than a piezoelectric body with a thin elastic mass layer of thickness h' and density ρ' on part of its surface (see Fig. 4.6.1).

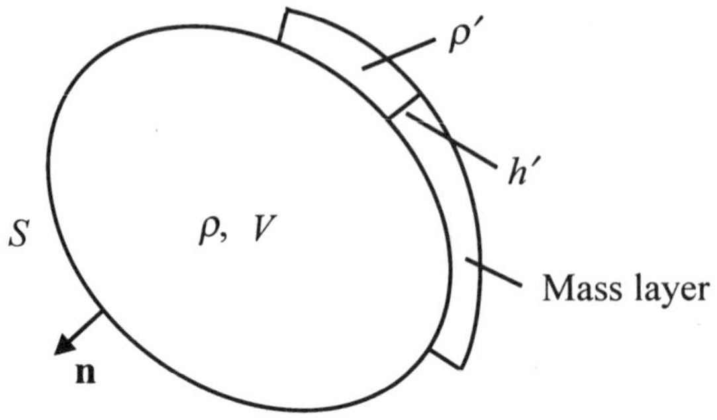

Fig. 4.6.1. An elastic body with a surface mass layer.

4.6.1. Variational formulation

For free vibrations, the governing equations and boundary conditions of the body are

$$
\begin{aligned}
&c_{jikl}u_{k,lj} = \rho\ddot{u}_i \quad \text{in} \quad V,\\
&u_i = 0 \quad \text{on} \quad S_u,\\
&T_{ji}n_j = c_{jikl}S_{kl}n_j = c_{jikl}u_{k,l}n_j = t_i \quad \text{on} \quad S_T,
\end{aligned}
\tag{4.6.1}
$$

where t_i is the traction vector between the body and the mass layer. For the mass layer, Eqs. (4.5.2) through (4.5.6) remain the same for this section. The eigenvalue problem for the free vibration of the body with the mass layer can be written as

$$
\begin{aligned}
&-c_{jikl}u_{k,lj} = \rho\lambda u_i \quad \text{in} \quad V,\\
&u_i = 0 \quad \text{on} \quad S_u,\\
&T_{ji}n_j = c_{jikl}u_{k,l}n_j = \rho' h'\lambda u_i - L_i(\mathbf{u}) \quad \text{on} \quad S_T,
\end{aligned}
\tag{4.6.2}
$$

where we have denoted $\lambda = \omega^2$.

The variational formulation of the eigenvalue problem of an elastic body is given by the following functional (Rayleigh quotient) [10]:

$$
\lambda = R(\mathbf{u}) = \frac{U(\mathbf{u})}{T(\mathbf{u})}, \tag{4.6.3}
$$

where U is the potential or strain energy and T is the kinetic energy. Using three-dimensional equations for the body and shell equations for the mass layer, we can write the Rayleigh quotient of the whole body including the mass layer as

$$\lambda = \frac{\int_V \frac{1}{2} c_{ijkl} S_{ij} S_{kl} dV + \int_{S_T} \frac{1}{2} \gamma_{ab} S_a S_b h' dS}{\int_V \frac{1}{2} \rho u_i u_i dV + \int_{S_T} \frac{1}{2} \rho' u_i u_i h' dS}, \tag{4.6.4}$$

where a, b = 1, 2, 6 are indices in the local shell coordinate system.

4.6.2. *Frequency shift*

From Eq. (4.6.3), a perturbation of the eigenvalue can be written as

$$\delta\lambda = \frac{T\delta U - U\delta T}{T^2}. \tag{4.6.5}$$

It is informative to compare Eq. (4.6.5) with Eq. (4.1.15). Denote the frequency, displacement and strains in the crystal body when the mass layer is not present by $\lambda^{(0)}$, $\mathbf{u}^{(0)}$ and $\mathbf{S}^{(0)}$. From a perturbation point of view, $\lambda^{(0)}$, $\mathbf{u}^{(0)}$ and $\mathbf{S}^{(0)}$ are assumed known and are perturbed slightly when a thin mass layer is added. Therefore, as an approximation, we use $\mathbf{u}^{(0)}$ and $\mathbf{S}^{(0)}$ for $\mathbf{u}$ and $\mathbf{S}$ in Eq. (4.6.5) for both the crystal and the mass layer. Denoting

$$\begin{aligned} &T^{(0)} = \int_V \frac{1}{2} \rho u_i^{(0)} u_i^{(0)} dV, \quad U^{(0)} = \int_V \frac{1}{2} c_{ijkl} S_{ij}^{(0)} S_{kl}^{(0)} dV, \\ &T' = \int_{S_T} \frac{1}{2} \rho' u_i^{(0)} u_i^{(0)} h' dS, \quad U' = \int_{S_T} \frac{1}{2} \gamma_{ab} S_a^{(0)} S_b^{(0)} h' dS, \\ &\lambda^{(0)} = \frac{U^{(0)}}{T^{(0)}}, \end{aligned} \tag{4.6.6}$$

we write Eq. (4.6.5) as

$$\begin{aligned} \lambda - \lambda^{(0)} &\cong \frac{U' - \lambda^{(0)} T'}{T^{(0)}} \\ &= \frac{\int_{S_T} (\gamma_{ab} S_a^{(0)} S_b^{(0)} - \lambda^{(0)} \rho' u_i^{(0)} u_i^{(0)}) h' dS}{\int_V \rho u_i^{(0)} u_i^{(0)} dV}, \end{aligned} \tag{4.6.7}$$

where it has been assumed that T' and U' are small for a thin mass layer. Equation (4.6.7) gives the perturbation of the eigenvalue in terms of the energies of the unperturbed frequencies and modes. The stiffness effect of the film is represented by the term with γ_{ab}. The inertial effect is represented by the ρ' term. Equation (4.6.7) can be further written as

$$\frac{\omega-\omega^{(0)}}{\omega^{(0)}} \cong \frac{1}{2(\omega^{(0)})^2} \times \frac{\int_{S_T} (\gamma_{ab} S_a^{(0)} S_b^{(0)} - \lambda^{(0)} \rho' u_i^{(0)} u_i^{(0)}) h' dS}{\int_V \rho u_i^{(0)} u_i^{(0)} dV}. \tag{4.6.8}$$

4.6.3. Discussion

Since the shell strain energy density $\gamma_{ab} S_a^{(0)} S_b^{(0)}$ is positive definite, one immediate observation from Eq. (4.6.8) is that the film stiffness tends to increase the frequency, and the film inertia lowers the frequency as expected. Since Eq. (4.6.8) is in terms of energies, it is more convenient to use than Eq. (4.5.12) which is in terms of the second-order differential operator **L**. Eq. (4.6.8) is formally valid for a piezoelectric body with an elastic film although it was derived for an elastic body with an elastic film. This is because the numerator of Eq. (4.6.8) depends on the film only, and the denominator depends on the kinetic energy of the body only.

To see the relation between Eq. (4.6.7) with Eq. (4.5.9), we proceed as follows. With Eq. (4.5.2) through (4.5.6), it can be verified that

$$\begin{aligned} U' &= \int_{S_T} \frac{1}{2} \gamma_{ab} S_a^{(0)} S_b^{(0)} h' dS \\ &= \int_{S_T} \frac{1}{2} N_a^{(0)} S_a^{(0)} dS = \int_{S_T} \frac{1}{2} L_k(\mathbf{u}^{(0)}) u_k^{(0)} dS, \end{aligned} \tag{4.6.9}$$

where integration by parts and the fact the edge of the mass layer is traction-free as a shell boundary have been used. With Eq. (4.6.9) we can write Eq. (4.6.7) as

$$\lambda - \lambda^{(0)} = \frac{\int_{S_T} [L_k(\mathbf{u}^{(0)}) u_k^{(0)} - \lambda^{(0)} \rho' h' u_k^{(0)} u_k^{(0)}] dS}{\int_V \rho u_i^{(0)} u_i^{(0)} dV}, \tag{4.6.10}$$

which is exactly Eq. (4.5.9).

4.7. Radial Modes of a Ring

As an example of the applications of Eqs. (4.5.12) and (4.6.8), consider the radial vibration of a thin elastic ring (see Fig. 4.7.1) of a mean radius R, thickness h, Young's modulus E, Poisson's ratio ν, and mass density ρ. The mass layer has a thickness h', Young's modulus E' and mass density ρ'.

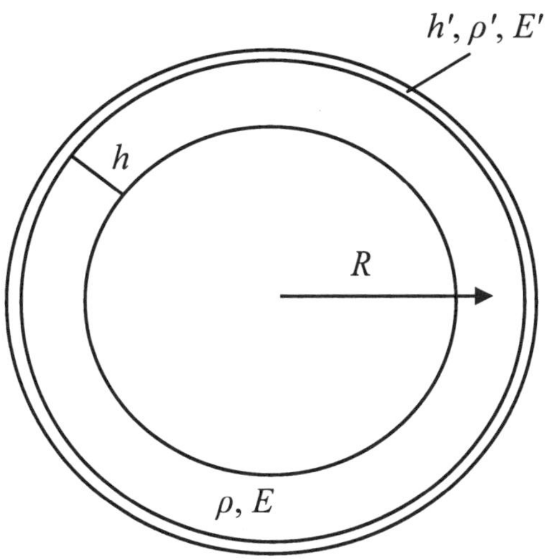

Fig. 4.7.1. A thin elastic ring with a mass layer.

4.7.1. The unperturbed mode

When the mass layer is not present, in polar coordinates, the lowest radial mode is given by [10]

$$(\omega^{(0)})^2 = \frac{E}{R^2 \rho},$$
$$u_r^{(0)} = C, \quad u_\theta^{(0)} = 0, \quad u_z^{(0)} = -\nu \frac{C}{R} z \cong 0, \tag{4.7.1}$$

where C is an arbitrary constant. $u_z^{(0)}$ is due to Poisson's effect, and is relatively small and is neglected as an approximation. For the mass layer, the equation of motion in the radial direction is [20]

$$-\frac{h'E'}{R^2}u_r + f_r = \rho' h' \ddot{u}_r \,, \tag{4.7.2}$$

from which, by comparison with Eq. (4.5.5), we identify

$$L_r(\mathbf{u}) = \frac{h'E'}{R^2}u_r, \quad L_\theta(\mathbf{u}) = 0, \quad L_z(\mathbf{u}) = 0 \,. \tag{4.7.3}$$

4.7.2. Perturbation solution

Substituting Eqs. (4.7.1) and (4.7.3) into the perturbation integral in Eq. (4.5.12), we obtain the frequency shift due to the mass layer as

$$\begin{aligned}\frac{\omega - \omega^{(0)}}{\omega^{(0)}} &\cong \frac{1}{2(\omega^{(0)})^2}\frac{1}{\rho h}\left[\frac{h'E'}{R^2} - \rho' h'(\omega^{(0)})^2\right] \\ &= \frac{1}{2(\omega^{(0)})^2}\frac{1}{\rho h}\left(\frac{h'E'}{R^2} - \rho' h' \frac{E}{R^2 \rho}\right) \\ &= \frac{1}{2}\frac{h'}{h}\left(\frac{E'}{E} - \frac{\rho'}{\rho}\right).\end{aligned} \tag{4.7.4}$$

Eq. (4.7.4) shows that the inertial effect of the mass layer lowers the frequency, and the stiffness of the mass layer raises the frequency as expected. It also shows that compared to the inertial effect, the stiffness effect is not necessarily small, depending on the materials involved.

4.7.3. Variation solution

From Eq. (4.7.1), the major strain component is

$$S_\theta^{(0)} = \frac{C}{R} \,. \tag{4.7.5}$$

Substituting Eqs. (4.7.1) and (4.7.5) into Eq. (4.6.8), we obtain the frequency shift due to the mass layer as

$$\begin{aligned}\frac{\omega-\omega^{(0)}}{\omega^{(0)}} &\cong \frac{1}{2(\omega^{(0)})^2}\frac{1}{\rho h C^2}\left[E'\left(\frac{C}{R}\right)^2 h' - (\omega^{(0)})^2 \rho' C^2 h'\right] \\ &= \frac{1}{2(\omega^{(0)})^2}\frac{1}{\rho h}\left(\frac{h'E'}{R^2} - \rho' h' \frac{E}{R^2\rho}\right) \\ &= \frac{1}{2}\frac{h'}{h}\left(\frac{E'}{E} - \frac{\rho'}{\rho}\right),\end{aligned} \tag{4.7.6}$$

which is the same as Eq. (4.7.4).

4.7.4. Discussion

From the approximate one-dimensional equations for a composite ring, we can also obtain the frequency of the ring with a thin mass layer as

$$\begin{aligned}\omega^2 &= \frac{Eh+E'h'}{R^2(\rho h+\rho' h')} \\ &= \frac{Eh(1+E'h'/Eh)}{R^2\rho h(1+\rho' h'/\rho h)} \cong (\omega^{(0)})^2\left(1+\frac{E'h'}{Eh} - \frac{\rho' h'}{\rho h}\right).\end{aligned} \tag{4.7.7}$$

Hence,

$$\begin{aligned}\frac{\omega^2-(\omega^{(0)})^2}{(\omega^{(0)})^2} &\cong \frac{2\omega^{(0)}(\omega-\omega^{(0)})}{(\omega^{(0)})^2} = \frac{2(\omega-\omega^{(0)})}{\omega^{(0)}} \\ &= \frac{E'h'}{Eh} - \frac{\rho' h'}{\rho h},\end{aligned} \tag{4.7.8}$$

or

$$\frac{\omega-\omega^{(0)}}{\omega^{(0)}} \cong \frac{1}{2}\frac{h'}{h}\left(\frac{E'}{E} - \frac{\rho'}{\rho}\right), \tag{4.7.9}$$

which is the same as Eqs. (4.7.4) and (4.7.6).

The result of Eq. (4.7.9) and the result from Eq. (4.1.19) are plotted in Fig. 4.7.2 using $E = s_{11}$ and ρ from quartz. Equation (4.1.19) is a special case of Eq. (4.7.9) by neglecting the layer stiffness ($E' = 0$). For E' and ρ' we consider two cases: gold and aluminum. The figure shows that for a heavy mass layer of gold the inertial effect of the mass layer dominates. However, for a light mass layer of aluminum, the stiffness effect is also important.

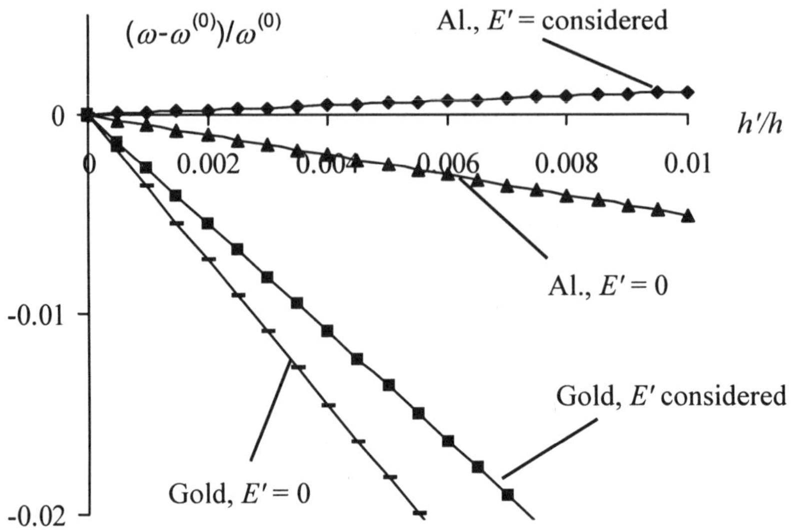

Fig. 4.7.2. Frequency shifts of a ring due to a mass layer.

4.8. Effects of Shear Deformability of a Mass Layer

In the analysis of the mass-frequency effects in the previous sections, it was assumed that the mass layer follows the motion of the crystal surfaces perfectly, with the same displacement as the crystal surface. Most of the mass sensors operate with the vibration modes of a crystal for which the material particles of the crystal surface move tangentially to the crystal surface, e.g. thickness-shear modes of a plate. For these modes, the phenomenon that the mass layer does not perfectly follow the motion of the crystal surface partially exhibits itself as thickness-shear deformation of the mass layer, which can be described by the shear deformation plate or shell theory [20]. This is a different mechanism or approach from the shear-lag interface model in the sixth section of the second chapter. In this section we study the effect of mass layer shear deformability [57]. Consider the crystal body with a surface bonded mass layer shown in Fig. 4.8.1.

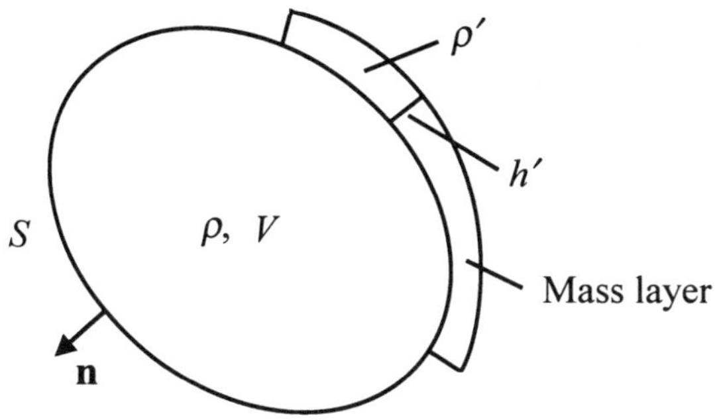

Fig. 4.8.1. An elastic body with a surface mass layer.

4.8.1. Governing equations

For free vibrations, the governing equations and boundary conditions of the elastic body are

$$\begin{aligned} &c_{jikl}u_{k,lj} = \rho\ddot{u}_i \quad \text{in} \quad V, \\ &u_i = 0 \quad \text{on} \quad S_u, \\ &T_{ji}n_j = c_{jikl}S_{kl}n_j = c_{jikl}u_{k,l}n_j = t_i \quad \text{on} \quad S_T. \end{aligned} \tag{4.8.1}$$

For the mass layer, we use the two-dimensional equations of an elastic shell for coupled extensional, flexural and thickness-shear deformations [20] which include the effects of extensional, flexural and thickness-shear stiffness, and the extensional, flexural and rotatory inertia. Let α_1 and α_2 be the shell middle surface principal coordinates, and α_3 be the shell thickness coordinate. Let the Greek indices β, γ = 1, 2, 3 be associated with the shell principle coordinates $(\alpha_1,\alpha_2,\alpha_3)$. The shell displacement vector is given by [20]

$$u_\beta = v_\beta(\alpha_1,\alpha_2,t) + \delta_{\beta 1}\alpha_3 w_1(\alpha_1,\alpha_2,t) + \delta_{\beta 2}\alpha_3 w_2(\alpha_1,\alpha_2,t), \tag{4.8.2}$$

where v_1 and v_2 are the middle surface extensional displacements, v_3 is the flexural displacement, and w_1 and w_2 are the thickness-shear displacements of the shell. $\delta_{\beta\gamma}$ is the Kronecker delta. The equations of motion are

$$\begin{aligned}
&\frac{\partial(N_{11}A_2)}{\partial\alpha_1}+\frac{\partial(N_{21}A_1)}{\partial\alpha_2}+N_{12}\frac{\partial A_1}{\partial\alpha_2}-N_{22}\frac{\partial A_2}{\partial\alpha_1}\\
&\quad+N_{13}A_1A_2\frac{1}{R_1}+F_1^{(0)}=\rho' h' A_1A_2\ddot{v}_1,\\
&\frac{\partial(N_{12}A_2)}{\partial\alpha_1}+\frac{\partial(N_{22}A_1)}{\partial\alpha_2}+N_{21}\frac{\partial A_2}{\partial\alpha_1}-N_{11}\frac{\partial A_1}{\partial\alpha_2}\\
&\quad+N_{23}A_1A_2\frac{1}{R_2}+F_2^{(0)}=\rho' h' A_1A_2\ddot{v}_2,\\
&\frac{\partial}{\partial\alpha_1}(N_{13}A_2)+\frac{\partial}{\partial\alpha_2}(N_{23}A_1)-N_{11}\frac{A_1A_2}{R_1}-N_{22}\frac{A_1A_2}{R_2}\\
&\quad+F_3^{(0)}=\rho' h' A_1A_2\ddot{v}_3
\end{aligned} \tag{4.8.3}$$

for extensions and flexure, and

$$\begin{aligned}
&\frac{\partial(M_{11}A_2)}{\partial\alpha_1}+\frac{\partial(M_{21}A_1)}{\partial\alpha_2}+M_{12}\frac{\partial A_1}{\partial\alpha_2}-M_{22}\frac{\partial A_2}{\partial\alpha_1}\\
&\quad-N_{31}A_1A_2+F_1^{(1)}=\frac{\rho' h'^3}{12}A_1A_2\ddot{w}_1,\\
&\frac{\partial(M_{12}A_2)}{\partial\alpha_1}+\frac{\partial(M_{22}A_1)}{\partial\alpha_2}+M_{21}\frac{\partial A_2}{\partial\alpha_1}-M_{11}\frac{\partial A_1}{\partial\alpha_2}\\
&\quad-N_{32}A_1A_2+F_2^{(1)}=\frac{\rho' h'^3}{12}A_1A_2\ddot{w}_2
\end{aligned} \tag{4.8.4}$$

for thickness-shears, where A_1 and A_2 are the Lamè coefficients of the middle surface associated with α_1 and α_2, and R_1 and R_2 are the principal radii of curvature of the middle surface of the shell. N_{ab} ($a, b = 1, 2$) are the shell membrane forces, $N_{3c} = N_{c3}$ ($c = 1, 2$) are the transverse shear forces, and M_{ab} are the bending and twisting moments. Surface loads on the shell are given by

$$\begin{aligned}
F_\beta^{(0)}&=A_1A_2[T_{3\beta}(h'/2)-T_{3\beta}(-h'/2)],\\
F_\beta^{(1)}&=\frac{h'}{2}A_1A_2[T_{3\beta}(h'/2)+T_{3\beta}(-h'/2)],
\end{aligned} \tag{4.8.5}$$

where $T_{3\beta}(h'/2)=0$ for the outer surface of the mass layer in our case. The shell constitutive relations are given by

$$N_\mu = h'\bar{c}_{\mu\nu}E_\nu, \quad \mu,\nu = 1,2,4,5,6,$$
$$M_\xi = \frac{h'^3}{12}\gamma_{\xi\eta}\kappa_\eta, \quad \xi,\eta = 1,2,6, \tag{4.8.6}$$

where the compact matrix indices have been used. The shell strains S_ν and curvatures κ_η are related to the shell middle surface displacements through

$$S_1 = S_{11} = \frac{1}{A_1}\left[\frac{\partial v_1}{\partial \alpha_1} + \frac{v_2}{A_2}\frac{\partial A_1}{\partial \alpha_2} + \frac{A_1 v_3}{R_1}\right],$$
$$S_2 = S_{22} = \frac{1}{A_2}\left[\frac{\partial v_2}{\partial \alpha_2} + \frac{v_1}{A_1}\frac{\partial A_2}{\partial \alpha_1} + \frac{A_2 v_3}{R_2}\right],$$
$$S_4 = 2S_{23} = w_2 - \frac{v_2}{R_2} + \frac{1}{A_2}\frac{\partial v_3}{\partial \alpha_2}, \tag{4.8.7}$$
$$S_5 = 2S_{31} = w_1 - \frac{v_1}{R_1} + \frac{1}{A_1}\frac{\partial v_3}{\partial \alpha_1},$$
$$S_6 = 2S_{12} = \frac{A_1}{A_2}\frac{\partial}{\partial \alpha_2}\left[\frac{v_1}{A_1}\right] + \frac{A_2}{A_1}\frac{\partial}{\partial \alpha_1}\left[\frac{v_2}{A_2}\right],$$

and

$$\kappa_1 = \kappa_{11} = \frac{1}{A_1}\left[\frac{\partial w_1}{\partial \alpha_1} + \frac{w_2}{A_2}\frac{\partial A_1}{\partial \alpha_2}\right],$$
$$\kappa_2 = \kappa_{22} = \frac{1}{A_2}\left[\frac{\partial w_2}{\partial \alpha_2} + \frac{w_1}{A_1}\frac{\partial A_2}{\partial \alpha_1}\right], \tag{4.8.8}$$
$$\kappa_6 = 2\kappa_{12} = \frac{A_1}{A_2}\frac{\partial}{\partial \alpha_2}\left[\frac{w_1}{A_1}\right] + \frac{A_2}{A_1}\frac{\partial}{\partial \alpha_1}\left[\frac{w_2}{A_2}\right].$$

The shell material constants are related to the three-dimensional material constants by

$$\bar{c}_{\mu\nu} = c'_{\mu\nu} - c'_{\mu 3}c'_{3\nu} / c'_{33},$$
$$\gamma_{\xi\eta} = c'_{\xi\eta} - c'_{\xi\lambda}c'^{-1}_{\lambda\delta}c'_{\delta\eta}, \quad \lambda,\delta = 3,4,5. \tag{4.8.9}$$

With successive substitutions from Eqs. (4.8.6) through (4.8.8), Eqs. (4.8.3) and (4.8.4) can be written in the following compact form:

$$
\begin{aligned}
&-\hat{L}_\beta(\mathbf{v},\mathbf{w})-T_{3\beta}(-h'/2)=\rho' h' \ddot{v}_\beta, \quad \beta=1,2,3, \\
&-\tilde{L}_\beta(\mathbf{v},\mathbf{w})+\frac{h'}{2}T_{3\beta}(-h'/2)=\frac{\rho' h'^3}{12}\ddot{w}_\beta, \quad \beta=1,2,
\end{aligned}
\tag{4.8.10}
$$

where $\hat{L}_\beta$ and $\tilde{L}_\beta$ are linear differential operators. The transformation coefficients between the local shell principal coordinates and the global Cartesian coordinates are denoted by $\delta_{i\beta}$. Multiplying Eq. $(4.8.10)_1$ by the transformation coefficients we have

$$
-\hat{L}_i(\mathbf{v},\mathbf{w})-T_{3i}(-h'/2)=\rho' h' \ddot{v}_i . \tag{4.8.11}
$$

The t_i in Eq. $(4.8.1)_3$ and the $-T_{3j}(-h'/2)$ in Eq. (4.8.11) are actions and reactions, equal in magnitude and opposite in direction ($-T_{3i}(-h'/2) = -t_i$). Substituting Eq. (4.8.11) into Eq. $(4.8.1)_3$, we have

$$
\begin{aligned}
&c_{jikl}u_{k,lj}=\rho\ddot{u}_i \quad \text{in} \quad V, \\
&u_i=0 \quad \text{on} \quad S_u, \\
&T_{ji}n_j=c_{jikl}u_{k,l}n_j=-\rho' h' \ddot{v}_i-L_i(\mathbf{v},\mathbf{w}) \quad \text{on} \quad S_T .
\end{aligned}
\tag{4.8.12}
$$

From Eq. (4.8.2) and the continuity of displacement between the crystal surface and the layer, on S we have

$$
u_1|_S=v_1-\frac{h'}{2}w_1, \quad u_2|_S=v_2-\frac{h'}{2}w_2, \quad u_3|_S=v_3 . \tag{4.8.13}
$$

In addition, we need to [20] eliminate the unknown $T_{3j}(-h'/2)$ from Eqs. $(4.8.10)_1$ and $(4.8.10)_2$ for $\beta = 1,2$, which results in

$$
-\tilde{L}_\beta(\mathbf{v},\mathbf{w})-\frac{h'}{2}\hat{L}_\beta(\mathbf{v},\mathbf{w})=\frac{\rho' h'^3}{12}\ddot{w}_\beta+\frac{h'}{2}\rho' h' \ddot{v}_\beta, \quad \beta=1,2. \tag{4.8.14}
$$

Equations (4.8.12) through (4.8.14) are the governing equations for **u**, **v** and **w** of a shell on a body.

4.8.2. Variational formulation

For time-harmonic motion with a frequency ω, denoting $\lambda=\omega^2$, factoring out the time-harmonic dependence from all fields and writing all fields by real functions, the variational formulation of the eigenvalue problem of an elastic body is given by the following functional (Rayleigh quotient) [10]:

$$\lambda = R = \frac{U}{T}, \tag{4.8.15}$$

where U is the potential or strain energy and T is the kinetic energy. Using three-dimensional equations for the crystal body and shell equations for the mass layer, we can write the Rayleigh quotient of the body including the mass layer as

$$\lambda = R(\mathbf{u}, \mathbf{v}, \mathbf{w})$$
$$= \frac{\int_V \frac{1}{2} T_{ij} S_{ij} dV + \int_{S_T} \frac{1}{2}(T_\nu S_\nu + M_\eta \kappa_\eta) dS}{\int_V \frac{1}{2} \rho u_i u_i dV + \int_{S_T} \left(\frac{1}{2} \rho' h' v_\beta v_\beta + \frac{1}{24} \rho' h'^3 w_\alpha w_\alpha \right) dS}, \tag{4.8.16}$$

where α = 1, 2. The last term in the numerator is the bending strain energy. The last term in the denominator is the kinetic energy due to the consideration of the rotatory inertia of the mass later.

4.8.3. Frequency shift

From Eq. (4.8.15), a perturbation of the frequency can be written as

$$\delta\lambda = \frac{T\delta U - U\delta T}{T^2} = \frac{\delta U - \lambda \delta T}{T}. \tag{4.8.17}$$

Denote the frequency, displacement, stress and strain components in the crystal body when the mass layer is not present by $\lambda^{(0)}$, $\mathbf{u}^{(0)}$, $\mathbf{T}^{(0)}$ and $\mathbf{S}^{(0)}$. They are governed by

$$\begin{aligned} &-c_{jikl} u_{k,lj}^{(0)} = \lambda^{(0)} \rho u_i^{(0)} \quad \text{in} \quad V, \\ &u_i^{(0)} = 0 \quad \text{on} \quad S_u, \\ &T_{ji}^{(0)} n_j = 0 \quad \text{on} \quad S_T, \end{aligned} \tag{4.8.18}$$

which is from Eq. (4.8.12) by dropping the mass layer. For a perturbation analysis, $\lambda^{(0)}$, $\mathbf{u}^{(0)}$, $\mathbf{T}^{(0)}$ and $\mathbf{S}^{(0)}$ are assumed known as usual. These fields are perturbed slightly when a thin mass layer is added. As an approximation, we use $\mathbf{u}^{(0)}$, $\mathbf{T}^{(0)}$ and $\mathbf{S}^{(0)}$ for **u, T** and **S** in the leading volume terms in the numerator and denominator of Eq. (4.8.16). When the crystal body is approximately carrying $\mathbf{u}^{(0)}$, $\mathbf{T}^{(0)}$ and $\mathbf{S}^{(0)}$ in V, the fields in the mass layer, $\mathbf{v}^{(0)}$ and $\mathbf{w}^{(0)}$, can be determined from

$$
\begin{aligned}
&u_1^{(0)}|_S = v_1^{(0)} - \frac{h'}{2}w_1^{(0)}, \quad u_2^{(0)}|_S = v_2^{(0)} - \frac{h'}{2}w_2^{(0)}, \quad u_3^{(0)}|_S = v_3^{(0)}, \\
&-\tilde{L}_\beta(\mathbf{v}^{(0)}, \mathbf{w}^{(0)}) - \frac{h'}{2}\hat{L}_\beta(\mathbf{v}^{(0)}, \mathbf{w}^{(0)}) \\
&\quad = -\lambda^{(0)}\frac{\rho' h'^3}{12}w_\beta^{(0)} - \lambda^{(0)}\frac{h'}{2}\rho' h' v_\beta^{(0)}, \quad \beta = 1,2,\ldots,
\end{aligned}
\tag{4.8.19}
$$

which are from Eqs. (4.8.13) and (4.8.14). When $\mathbf{u}^{(0)}$, $\mathbf{T}^{(0)}$ and $\mathbf{S}^{(0)}$ are known, Eq. (4.8.19) provides five equations for $\mathbf{v}^{(0)}$ and $\mathbf{w}^{(0)}$. Denoting

$$
\begin{aligned}
&T^{(0)} = \int_V \frac{1}{2}\rho u_i^{(0)} u_i^{(0)} dV, \quad U^{(0)} = \int_V \frac{1}{2} T_{ij}^{(0)} S_{ij}^{(0)} dV, \\
&\delta T = T' = \int_{S_T} \left(\frac{1}{2}\rho' h' v_\beta^{(0)} v_\beta^{(0)} + \frac{1}{24}\rho' h'^3 w_\alpha^{(0)} w_\alpha^{(0)}\right) dS, \\
&\delta U = U' = \int_{S_T} \frac{1}{2}(T_\nu^{(0)} S_\nu^{(0)} + M_\nu^{(0)} \kappa_\nu^{(0)}) dS, \\
&\lambda^{(0)} = \frac{U^{(0)}}{T^{(0)}},
\end{aligned}
\tag{4.8.20}
$$

we can write Eq. (4.8.17) as

$$
\begin{aligned}
\lambda - \lambda^{(0)} &\cong \frac{U' - \lambda^{(0)} T'}{T^{(0)}} \\
&= \left[\int_{S_T} \frac{1}{2}(T_\nu^{(0)} S_\nu^{(0)} + M_\nu^{(0)} \kappa_\nu^{(0)}) dS \right. \\
&\quad \left. -\lambda^{(0)} \int_{S_T} \left(\frac{1}{2}\rho' h' v_\beta^{(0)} v_\beta^{(0)} + \frac{1}{24}\rho' h'^3 w_\alpha^{(0)} w_\alpha^{(0)}\right) dS \right] \\
&\quad \Big/ \int_V \frac{1}{2}\rho u_i^{(0)} u_i^{(0)} dV,
\end{aligned}
\tag{4.8.21}
$$

where it has been assumed that T' and U' are small for a thin mass layer. Equation (4.8.21) is for the eigenvalue $\lambda = \omega^2$. For ω we write:

$$
\omega = \omega^{(0)} + \Delta\omega\,. \tag{4.8.22}
$$

Then,

$$
\begin{aligned}
\lambda = \omega^2 &= (\omega^{(0)} + \Delta\omega)^2 \cong (\omega^{(0)})^2 + 2\omega^{(0)}\Delta\omega \\
&= \lambda^{(0)} + 2\omega^{(0)}\Delta\omega.
\end{aligned}
\tag{4.8.23}
$$

Hence,

$$\frac{\Delta\omega}{\omega^{(0)}} \cong \frac{1}{2(\omega^{(0)})^2}(\lambda - \lambda^{(0)}) . \tag{4.8.24}$$

Substituting Eq. (4.8.21) into Eq. (4.8.24), we obtain

$$\begin{aligned}\frac{\omega-\omega^{(0)}}{\omega^{(0)}} &\cong \frac{1}{2(\omega^{(0)})^2}\times \\ &\left[\int_{S_T} \frac{1}{2}(T_\nu^{(0)}S_\nu^{(0)} + M_\nu^{(0)}\kappa_\nu^{(0)})dS \right. \\ &\left. -(\omega^{(0)})^2\int_{S_T}\left(\frac{1}{2}\rho' h' v_\beta^{(0)}v_\beta^{(0)} + \frac{1}{24}\rho' h'^3 w_\alpha^{(0)}w_\alpha^{(0)}\right)dS\right] \\ &\Big/ \int_V \frac{1}{2}\rho u_i^{(0)}u_i^{(0)}dV .\end{aligned} \tag{4.8.25}$$

4.9. Thickness-Shear Modes of a Plate with Thick Mass Layers

As an example of the application of Eq. (4.8.25), consider thickness-shear vibration of an AT-cut quartz plate (see Fig. 4.9.1). In Fig. 4.9.1 the mass layer normal is along the x_2 (or α_2) axis but the equations in the previous section were derived for a normal along α_3. Therefore some transformation is needed, which is straightforward because of the clear physical meaning of Eq. (4.8.25).

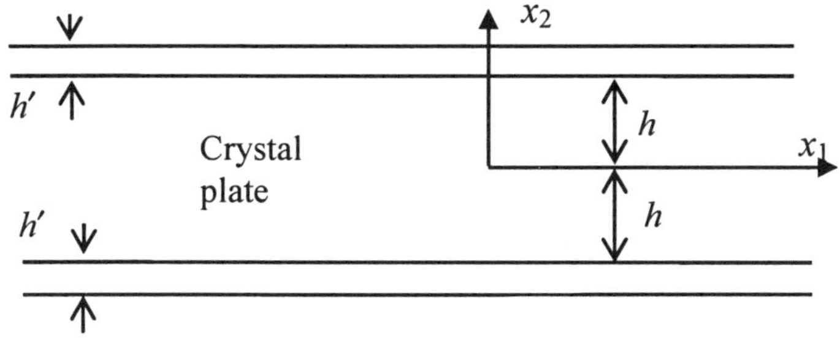

Fig. 4.9.1. An elastic plate with mass layers.

The unperturbed modes when the mass layers are not present are the same as Eq. (4.2.1):

$$\sqrt{\lambda^{(0)}} = \omega^{(0)} = \frac{\pi}{2h}\sqrt{\frac{c_{66}}{\rho}},$$
$$u_1^{(0)} = C\sin\frac{\pi}{2h}x_2, \quad u_2^{(0)} = u_3^{(0)} = 0, \tag{4.9.1}$$

where C is an arbitrary constant. For the corresponding fields in the mass layers, from Eqs. (4.8.19), (4.8.6) and (4.8.7) we obtain

$$C = v_1^{(0)} - \frac{h'}{2}w_1^{(0)},$$
$$-N_{21}^{(0)} = -\lambda^{(0)}\frac{\rho' h'^3}{12}w_1^{(0)} - \lambda^{(0)}\frac{h'}{2}\rho' h' v_1^{(0)}, \tag{4.9.2}$$
$$N_{21}^{(0)} = \frac{\pi^2}{12}h'\bar{c}_{66}E_6^{(0)}, \quad \bar{c}_{66} = c'_{66}, \quad S_6^{(0)} = w_1^{(0)},$$

where we have inserted a shear correction factor of $\pi^2/12$ [20]. From Eq. (4.9.2) we obtain

$$v_1^{(0)} = \frac{1-r}{1-4r}C \cong (1+3r)C,$$
$$w_1^{(0)} = \frac{6r}{1-4r}\frac{C}{h'} \cong 6r\frac{C}{h'}, \tag{4.9.3}$$

where

$$r = \frac{\rho' h'^2}{4\rho h^2}\frac{c_{66}}{c'_{66}} \tag{4.9.4}$$

is a small number. Eq. (4.9.3) shows that $v_1^{(0)}$ depends on r weakly, and $w_1^{(0)}$ depends on r linearly. Eq. (4.9.4) shows that the effect of the mass layer shear deformability is larger for heavier, thicker, and softer mass layers as expected. Then from Eq. (4.8.25) we obtain

$$\frac{\omega - \omega^{(0)}}{\omega^{(0)}} \cong \frac{1}{2(\omega^{(0)})^2}2$$
$$\frac{\frac{1}{2}\frac{\pi^2}{12}h'c'_{66}\left(\frac{6r}{h'}\right)^2 - (\omega^{(0)})^2\left[\frac{1}{2}\rho' h'(1+3r)^2 + \frac{1}{24}\rho' h'^3\left(\frac{6r}{h'}\right)^2\right]}{\frac{1}{2}\rho h}. \tag{4.9.5}$$

For the lowest order effect of the small r, we keep the linear terms only and obtain

$$\begin{aligned}\frac{\omega-\omega^{(0)}}{\omega^{(0)}} &\cong -\frac{\rho' h'}{\rho h}(1+3r) \\ &= -\frac{\rho' h'}{\rho h}\left(1+\frac{3}{4}\frac{\rho' h'^2}{\rho h^2}\frac{c_{66}}{c'_{66}}\right).\end{aligned} \tag{4.9.6}$$

If r is set to zero (or $c'_{66}\to\infty$, a shear-rigid mass layer), Eq. (4.9.6) reduces to the result in the second section of this chapter. The leading order of the mass layer shear deformability lowers the frequency because it is from the kinetic energy.

Some numerical results are shown in Fig. 4.9.2 for a few common metals for resonator electrodes as mass layers. For an AT-cut quartz plate ρ = 2649 kg/m^3 and c_{66} = 2.901 $\times 10^{10}$ N/m^2. The material data for the electrodes are

	Aluminum	Gold	Silver
ρ'	2695	19300	10490 kg/m^3
c'_{66}	2.5	2.85	2.7×10^{10} N/m^2

The figure shows that for thin electrodes (mass layers) the shear deformability is small because the curves have flat tangents for small h'.

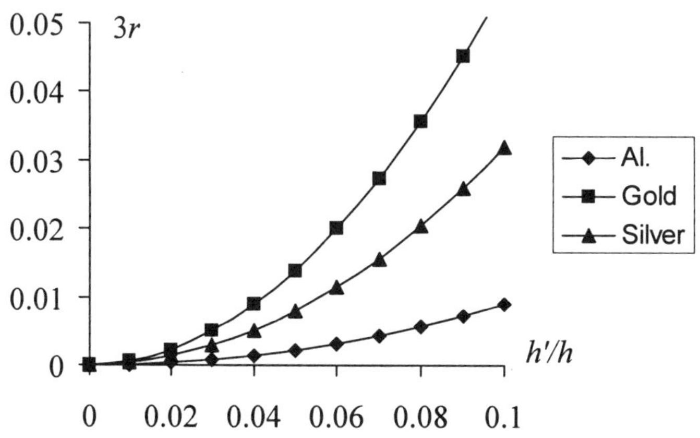

Fig. 4.9.2. Effects of mass layer shear deformability.

When the electrodes are relatively thick their shear deformability has an effect which may not be always negligible, especially for high frequency resonators that are thin and the electrodes that are relatively thick.

For this example, the exact solution is available [18]. The frequency equation for the anti-symmetric modes we care considering is given by the first factor of Eq. (2.7.13):

$$\cot \xi h = \sqrt{\frac{\rho' c'_{66}}{\rho c_{66}}} \tan \xi' h', \quad \xi' = \sqrt{\frac{\rho'}{c'_{66}}}\, \omega, \tag{4.9.7}$$

where piezoelectric coupling is neglected. For large c' or small ξ', we have

$$\tan \xi' h' \cong \xi' h' + \frac{1}{3}(\xi' h')^3. \tag{4.9.8}$$

Then Eq. $(4.9.7)_1$ can be approximately written as

$$\cot \xi h = R\xi h + \alpha(\xi h)^3, \tag{4.9.9}$$

where

$$R = \frac{\rho' h'}{\rho h}, \quad \alpha = \frac{1}{3}\frac{\rho c_{66}}{\rho' c'_{66}} R^3. \tag{4.9.10}$$

For small α we look for a perturbation solution to Eq. (4.9.9) in the following form:

$$\xi h = f^{(0)}(R) + \alpha f^{(1)}(R). \tag{4.9.11}$$

Substitution of Eq. (4.9.11) into Eq. (4.9.9) yields a series of perturbation problems. The zero- and first-order problems are

$$\begin{aligned} &\cot f^{(0)} = R f^{(0)}, \\ &-\frac{1}{\sin^2 f^{(0)}} f^{(1)} = R f^{(1)} + (f^{(0)})^3. \end{aligned} \tag{4.9.12}$$

For small R Eq. $(4.9.12)_1$ has an approximate solution

$$f^{(0)} \cong \frac{\pi}{2}(1 - R), \tag{4.9.13}$$

where higher-order terms of R can be found. This is not pursued because our main interest is the effect of the layer shear stiffness. Then from Eq. $(4.9.12)_2$ we find

$$f^{(1)} \cong -\frac{\pi^3}{8}. \tag{4.9.14}$$

Equations (4.9.11) through (4.9.14) imply that

$$\begin{aligned} \frac{\omega - \omega^{(0)}}{\omega^{(0)}} &\cong -\frac{\rho' h'}{\rho h}\left(1 + \frac{\pi^2}{12}\frac{\rho' h'^2}{\rho h^2}\frac{c_{66}}{c'_{66}}\right) \\ &= -\frac{\rho' h'}{\rho h}\left(1 + 0.822\frac{\rho' h'^2}{\rho h^2}\frac{c_{66}}{c'_{66}}\right). \end{aligned} \tag{4.9.15}$$

Comparison of Eq. (4.9.15) with Eq. (4.9.6) shows that they are close. The difference is less than 10%.

4.10. An Ill-Posed Problem in Elasticity for Mass Sensors

From the problems analyzed in the previous sections of this chapter on the frequency effects of a mass layer on an elastic body, it can be seen that both the inertia and stiffness of the mass layer can affect the frequencies of the elastic body. Most mass sensors are for measuring the density of the mass layer. Therefore, it is often desirable that the frequency shift depends only on the mass layer density but not on its stiffness. This can happen when, for a particular mode of the body, the surface of the body is in rigid-body motion without surface strains. Vanishing of surface strains over a traction-free surface normally implies an ill-posed boundary value problem in elasticity. In this section we examine body shapes for which the ill-posed free vibration eigenvalue problem of an elastic body without surface strains at its traction-free surface still has solutions.

4.10.1. Formulation of the problem

Consider an isotropic elastic body (see Fig. 4.10.1). Let the region occupied by the body be V and its traction-free boundary surface be S. The unit outward normal of S is $\mathbf{n}$.

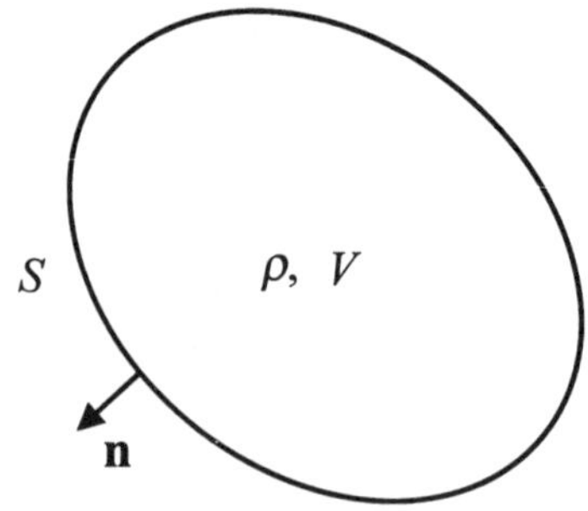

Fig. 4.10.1. An elastic body.

For free vibrations with a frequency ω, the governing equations and boundary conditions for modes with vanishing surface strains on the traction-free S are

$$\begin{aligned} &-T_{ji,j} = \rho\omega^2 u_i \quad \text{in} \quad V, \\ &T_{ji} = \lambda u_{k,k}\delta_{ji} + \mu(u_{i,j} + u_{j,i}) \quad \text{in} \quad V, \\ &T_{ji}n_j = 0 \quad \text{on} \quad S, \\ &S_{\alpha\beta} = 0 \quad \text{on} \quad S, \end{aligned} \tag{4.10.1}$$

where α, β = 1, 2 are the tangent plane indices associated with the principle directions on S. From the theory of elasticity, Eqs. $(4.10.1)_{1\text{-}3}$ constitute a well-posed eigenvalue problem. The presence of Eq. $(4.10.1)_4$ makes Eq. (4.10.1) ill-posed in general. However, this does not prevent, in special cases, e.g., for certain modes over some special domains V, Eq. $(4.10.1)_4$ from being automatically satisfied by solutions to Eq. $(4.10.1)_{1\text{-}3}$. We look for body shapes such that Eq. (4.10.1) still has a solution.

4.10.2. Known solutions

Consider an unbounded plate as shown in Fig. 4.10.2. When the plate is vibrating in pure thickness modes characterized by

$$u_j = u_j(x_2, t), \tag{4.10.2}$$

the two surfaces of the plate are in rigid-body motion without strains.

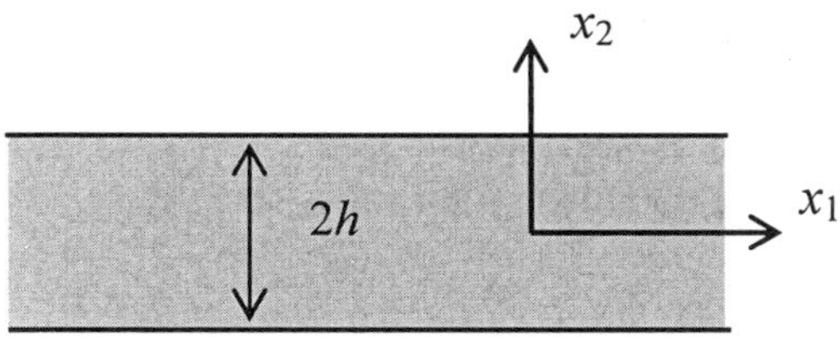

Fig. 4.10.2. An unbounded plate.

Next consider an infinite circular cylinder as shown in Fig. 4.10.3.

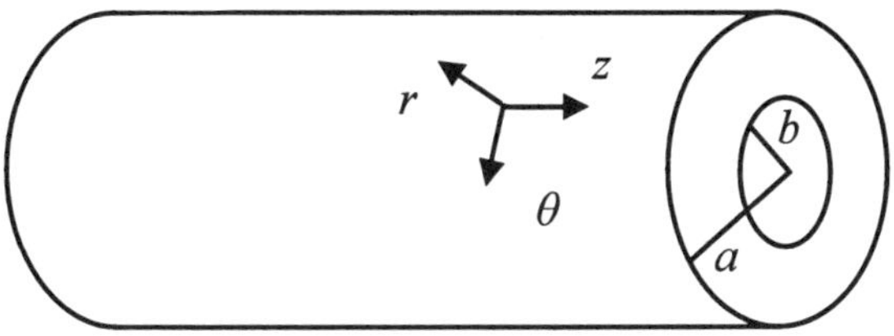

Fig. 4.10.3. A infinite circular cylinder.

Axial and tangential thickness-shear modes [10] described by

$$u_r = u_\theta = 0, \quad u_z = u_z(r,t), \tag{4.10.3}$$

and

$$u_r = 0, \quad u_\theta = u_\theta(r,t), \quad u_z = 0, \tag{4.10.4}$$

respectively do not have surface strains.

4.10.3. An open problem

The ill-posed problem in Eq. (4.10.1) is open within isotropic elasticity in the sense that it is unclear whether there exist body shapes and modes other than what are summarized above. The problem can be generalized to anisotropic elasticity and piezoelectricity.

4.11. Thickness-Shear Modes of a Circular Cylinder

In this section we identify two modes of a circular cylinder which do not have surface strains and analyze their mass sensitivity.

4.11.1. Thickness-shear modes in a circular cylinder

Consider an infinite circular cylinder. A cross section is shown in Fig. 4.11.1. The inner and outer surfaces are both traction-free. The mass density is ρ. The shear elastic constant is μ. A cylindrical coordinate system is defined by $x_1 = r\cos\theta$, $x_2 = r\sin\theta$, and $x_3 = z$. Two types of thickness-shear modes with no surface strains exist [10].

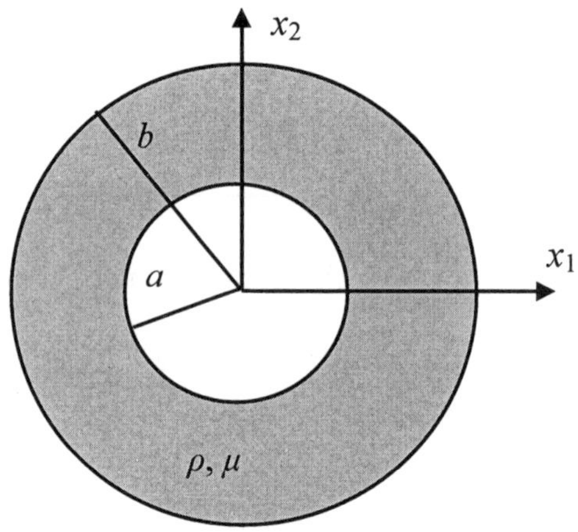

Fig. 4.11.1 Cross-section of a circular cylinder.

4.11.1.1. Axial thickness-shear

These modes are given by

$$\begin{aligned} &u_r = 0, \quad u_\theta = 0, \\ &u_z = C[Y_1(\xi b)J_0(\xi r) - J_1(\xi b)Y_0(\xi r)]\exp(i\omega t), \end{aligned} \tag{4.11.1}$$

where C is an arbitrary constant, J_0 and J_1 are the zero- and first-order Bessel functions of the first kind, Y_0 and Y_1 are the zero- and first-order Bessel functions of the second kind, and

$$\xi^2 = \frac{\rho\omega^2}{\mu}. \tag{4.11.2}$$

The frequency equation is

$$\begin{vmatrix} J_1(\xi a) & Y_1(\xi a) \\ J_1(\xi b) & Y_1(\xi b) \end{vmatrix} = 0. \tag{4.11.3}$$

4.11.1.2. Tangential thickness-shear

These modes are given by

$$\begin{aligned} &u_r = 0, \quad u_z = 0, \\ &u_\theta = C[Y_2(\xi b)J_1(\xi r) - J_2(\xi b)Y_1(\xi r)]\exp(i\omega t). \end{aligned} \tag{4.11.4}$$

The frequency equation is

$$\begin{vmatrix} J_2(\xi a) & Y_2(\xi a) \\ J_2(\xi b) & Y_2(\xi b) \end{vmatrix} = 0. \tag{4.11.5}$$

4.11.2. Mass sensitivity

Suppose a thin mass layer is added to the outer surface of the cylinder (see Fig. 4.11.2). Since there is no surface strain in the above thickness-shear modes, the mass layer does not deform. We only need to consider the inertial effect of the mass layer given by

$$\frac{\Delta\omega}{\omega} \cong -\frac{1}{2}\frac{\int_S \rho' h' u_k u_k dS}{\int_V \rho u_i u_i dV}, \tag{4.11.6}$$

but not the stiffness of the mass layer.

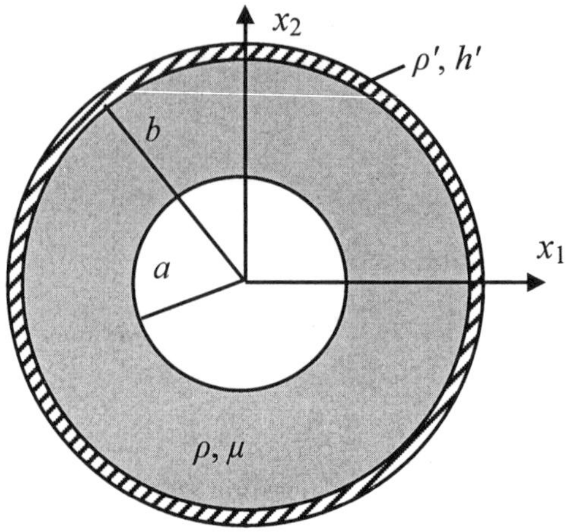

Fig. 4.11.2. A circular cylinder with a thin mass layer.

4.11.2.1. Axial thickness-shear

Substitution of Eq. (4.11.1) into Eq. (4.11.6) yields

$$\frac{\Delta\omega}{\omega} = -\frac{\rho' h'}{\rho(b-a)}\beta_1(a,b,\mu), \tag{4.11.7}$$

where

$$\beta_1(a,b,\mu) = \frac{b-a}{2}$$
$$\times \frac{[Y_1(\xi b)J_0(\xi b) - J_1(\xi b)Y_0(\xi b)]^2}{Y_1^2(\xi b)\int_a^b J_0^2(\xi r)dr + J_1^2(\xi b)\int_a^b Y_0^2(\xi r)dr - 2J_1(\xi b)Y_1(\xi b)\int_a^b J_0(\xi r)Y_0(\xi r)dr}. \tag{4.11.8}$$

4.11.2.2. Tangential thickness-shear

Similarly, substitution of Eq. (4.11.4) into Eq. (4.11.6) gives

$$\frac{\Delta\omega}{\omega} = -\frac{\rho' h'}{\rho(b-a)}\beta_2(a,b,\mu), \tag{4.11.9}$$

where

$$\beta_2(a,b,\mu)=\frac{b-a}{2}$$
$$\times\frac{[Y_2(\xi b)J_1(\xi b)-J_2(\xi b)Y_1(\xi b)]^2}{Y_2^2(\xi b)\int_a^b J_1^2(\xi r)dr+J_2^2(\xi b)\int_a^b Y_1^2(\xi r)dr-2J_2(\xi b)Y_2(\xi b)\int_a^b J_1(\xi r)Y_1(\xi r)dr}. \quad (4.11.10)$$

4.11.2.3. *Discussion*

We make the following observations from Eqs. (4.11.6) through (4.11.10):

(i) Only the inertia of the mass layer contributes to the frequency shift. Its stiffness does not;

(ii) The mass effect lowers the frequency as expected;

(iii) The frequency shift is linear in $\rho' h'$;

(iv) For a solid cylinder with a = 0, $Y_0(\xi r)$ in Eq. (4.11.1) and $Y_1(\xi r)$ in Eq. (4.11.4) have to be dropped because they are singular at r = 0. Then Eqs. (4.11.8) and (4.11.10) reduce to

$$\beta_1=\frac{b}{2}\frac{J_0^2(\xi b)}{\int_a^b J_0^2(\xi r)dr}, \quad \beta_2=\frac{b}{2}\frac{J_1^2(\xi b)}{\int_a^b J_1^2(\xi r)dr}. \quad (4.11.11)$$

4.12. Mass Sensitivity of Surface Waves

Surface waves over elastic and piezoelectric half-spaces have been used extensively to make surface acoustic wave (SAW) devices. In addition to the plane-strain Rayleigh waves, anti-plane (SH) surface waves can also propagate in certain piezoelectrics. These waves may rely on piezoelectric coupling and do not have an elastic counterpart. In this section, mass sensitivity of a surface wave in polarized ceramics called the Bleustein-Gulyaev wave is presented.

4.12.1. Governing equations

Consider a ceramic half-space with a thin mass layer as shown in Fig. 4.12.1. The ceramic is poled in the x_3 direction.

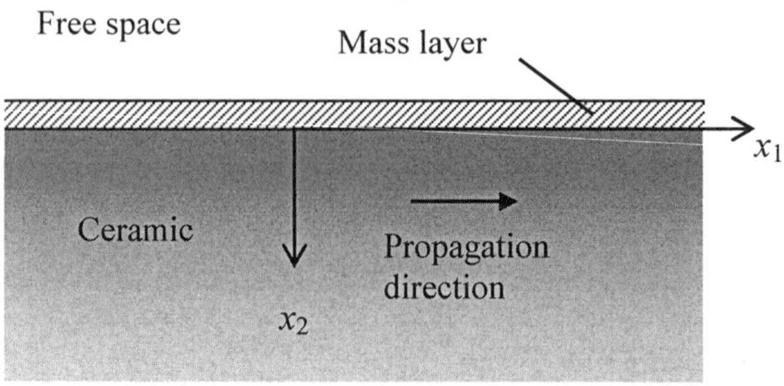

Fig. 4.12.1. A ceramic half-space with a thin mass layer.

Consider anti-plane motions of $u_3(x_1,x_2,t)$ and $\phi(x_1,x_2,t)$. The governing equations for the fields in the ceramic half-space are [27,10]

$$\bar{c}\nabla^2 u = \rho\ddot{u}, \quad \nabla^2\psi = 0, \quad \phi = \psi + \frac{e}{\varepsilon}u, \quad x_2 > 0, \tag{4.12.1}$$

where $u = u_3$, $c = c_{44}$, $e = e_{15}$, $\varepsilon = \varepsilon_{11}$, and $\bar{c} = c(1 + e^2 / c\varepsilon)$. Consider the possibility of solutions in the following form:

$$\begin{aligned}
&u = A\exp(-\xi_2 x_2)\cos(\xi_1 x_1 - \omega t),\\
&\psi = B\exp(-\xi_1 x_2)\cos(\xi_1 x_1 - \omega t),\\
&T_{23} = \bar{c}u_{,2} + e\psi_{,2}\\
&\quad = -[\bar{c}A\xi_2\exp(-\xi_2 x_2) + eB\xi_1\exp(-\xi_1 x_2)]\cos(\xi_1 x_1 - \omega t),\\
&\phi = [B\exp(-\xi_1 x_2) + \frac{e}{\varepsilon}A\exp(-\xi_2 x_2)]\cos(\xi_1 x_1 - \omega t),\\
&D_2 = -\varepsilon\psi_{,2} = \varepsilon\xi_1 B\exp(-\xi_1 x_2)\cos(\xi_1 x_1 - \omega t),
\end{aligned} \tag{4.12.2}$$

where A and B are undetermined constants, and ξ_2 should be positive for decaying behavior away from the surface. Eq. $(4.12.2)_2$ already satisfies Eq. $(4.12.1)_2$. For Eq. $(4.12.2)_1$ to satisfy Eq. $(4.12.1)_1$ we must have

$$\bar{c}(\xi_1^2 - \xi_2^2) = \rho\omega^2, \tag{4.12.3}$$

which determines

$$\xi_2^2 = \xi_1^2 - \frac{\rho\omega^2}{\bar{c}} = \xi_1^2\left(1 - \frac{v^2}{v_T^2}\right) > 0,$$
$$v^2 = \frac{\omega^2}{\xi_1^2}, \quad v_T^2 = \frac{\bar{c}}{\rho}. \tag{4.12.4}$$

4.12.2. A half-space with an electroded surface

First we consider the case when the surface of the half-space is electroded and the electrode is grounded. The corresponding boundary conditions are

$$T_{23} = \rho' h' \ddot{u}, \quad x_2 = 0,$$
$$\phi = 0, \quad x_2 = 0, \tag{4.12.5}$$

or, in terms of u and ψ,

$$T_{23} = \bar{c}u_{,2} + e\psi_{,2} = \rho' h' \ddot{u}, \quad x_2 = 0,$$
$$\phi = \psi + \frac{e}{\varepsilon}u = 0, \quad x_2 = 0. \tag{4.12.6}$$

Substituting Eq. (4.12.2) into Eq. (4.12.6):

$$-(\bar{c}A\xi_2 + eB\xi_1) = -\rho' h' \omega^2 A,$$
$$\frac{e}{\varepsilon}A + B = 0. \tag{4.12.7}$$

For non-trivial solutions,

$$\begin{vmatrix} \bar{c}\xi_2 - \rho' h' \omega^2 & e\xi_1 \\ e/\varepsilon & 1 \end{vmatrix} = \bar{c}\xi_2 - \rho' h' \omega^2 - \frac{e^2}{\varepsilon}\xi_1 = 0, \tag{4.12.8}$$

or

$$\xi_2 - \frac{\rho' h' \omega^2}{\bar{c}} = \bar{k}^2 \xi_1, \tag{4.12.9}$$

where

$$\bar{k}^2 = \frac{e^2}{\varepsilon \bar{c}}. \tag{4.12.10}$$

Substitution of Eq. (4.12.4) into Eq. (4.12.9) yields

$$\sqrt{\xi_1^2 - \frac{\rho\omega^2}{\bar{c}} - \frac{\rho' h' \omega^2}{\bar{c}}} = \bar{k}^2 \xi_1, \tag{4.12.11}$$

which determines the surface wave frequency or speed. In the special case when the mass layer is not present, i.e. $\rho' h' = 0$, Eq. (4.2.11) reduces to the well known Bleustein-Gulyaev wave with the following speed:

$$v_{B-G}^2 = \frac{\omega^2}{\xi_1^2} = \frac{\bar{c}}{\rho}(1 - \bar{k}^4) = v_T^2 (1 - \bar{k}^4) < v_T^2. \tag{4.12.12}$$

The difference between the frequencies of Eqs. (4.12.11) and (4.12.12) can be used to calculate $\rho' h'$. Note that Eq. (4.12.11) determines a dispersive wave and Eq. (4.12.12) determines a non-dispersive wave.

4.12.3. A half-space with an unelectroded surface

If the surface of the half-space is unelectroded, electric fields can also exist in the free space of $x_2 < 0$. Denoting the electric potential in the free space by $\hat{\phi}$, we have

$$\nabla^2 \hat{\phi} = 0, \quad x_2 < 0. \tag{4.12.13}$$

From Eq. (4.12.13), in the free space,

$$\begin{aligned} \hat{\phi} &= C \exp(\xi_1 x_2) \cos(\xi_1 x_1 - \omega t), \\ D_2 &= -\varepsilon_0 \hat{\phi}_{,2} = -\varepsilon_0 \xi_1 C \exp(\xi_1 x_2) \cos(\xi_1 x_1 - \omega t). \end{aligned} \tag{4.12.14}$$

The boundary and continuity conditions are

$$T_{23} = \rho' h' \ddot{u}, \quad \phi = \hat{\phi}, \quad D_2 = \hat{D}_2, \quad x_2 = 0, \tag{4.12.15}$$

or, in terms of u, ψ, and $\hat{\phi}$,

$$\begin{aligned} &T_{23} = \bar{c} u_{3,2} + e \psi_{,2} = \rho' h' \ddot{u}, \quad x_2 = 0, \\ &\phi = \psi + \frac{e}{\varepsilon} u = \hat{\phi}, \quad x_2 = 0, \\ &- \varepsilon \psi_{,2} = -\varepsilon_0 \hat{\phi}_{,2}, \quad x_2 = 0. \end{aligned} \tag{4.12.16}$$

Substituting Eqs. (4.12.2) and (4.12.14) into Eq. (4.12.16):

$$\begin{aligned}&\bar{c}(-A\xi_2)+e(-B\xi_1)=-\rho'h'\omega^2 A,\\ &-\varepsilon(-\xi_1 B)=-\varepsilon_0 C\xi_1,\\ &\frac{e}{\varepsilon}A+B=C.\end{aligned} \tag{4.12.17}$$

For non-trivial solutions,

$$\begin{aligned}&\begin{vmatrix}\bar{c}\xi_2-\rho'h'\omega^2 & e\xi_1 & 0\\ 0 & \varepsilon & \varepsilon_0\\ e/\varepsilon & 1 & -1\end{vmatrix}\\ &=\begin{vmatrix}\bar{c}\xi_2 & e\xi_1 & 0\\ 0 & \varepsilon & \varepsilon_0\\ e/\varepsilon & 1 & -1\end{vmatrix}+\begin{vmatrix}-\rho'h'\omega^2 & e\xi_1 & 0\\ 0 & \varepsilon & \varepsilon_0\\ 0 & 1 & -1\end{vmatrix}\\ &=-\bar{c}\xi_2\varepsilon+\frac{e^2}{\varepsilon}\varepsilon_0\xi_1-\bar{c}\xi_2\varepsilon_0-\rho'h'\omega^2\begin{vmatrix}\varepsilon & \varepsilon_0\\ 1 & -1\end{vmatrix}=0.\end{aligned} \tag{4.12.18}$$

When $\rho'h'=0$, Eq. (4.12.18) determines the following speed of the Bleustein-Gulyaev wave for an unelectroded half-space:

$$\xi_2=\bar{k}^2\xi_1\frac{1}{1+\varepsilon/\varepsilon_0}, \tag{4.12.19}$$

or

$$v_{B-G}^2=\frac{\omega^2}{\xi_1^2}=\frac{\bar{c}}{\rho}\left[1-\frac{\bar{k}^4}{(1+\varepsilon/\varepsilon_0)^2}\right]=v_T^2\left[1-\frac{\bar{k}^4}{(1+\varepsilon/\varepsilon_0)^2}\right]<v_T^2. \tag{4.12.20}$$

4.13. Thickness-Twist Waves in a Ceramic Plate

In this section we consider the inertial effect of identical mass layers on waves propagating in the x_1 direction in a ceramic plate poled in the x_3 direction (see Fig. 4.13.1). The major surfaces of the plate are traction-free and electroded, and the electrodes are grounded.

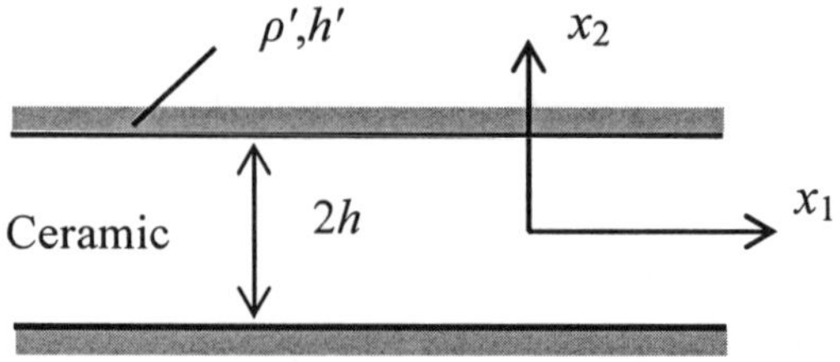

Fig. 4.13.1. A ceramic plate with identical mass layers.

4.13.1. Governing equations

We are interested in faces-shear and thickness-twist waves described by $u_3(x_1,x_2,t)$ and $\phi(x_1,x_2,t)$. With the notation in [27,10], from the first section of the third chapter, the governing equations are

$$\bar{c}\nabla^2 u=\rho\ddot{u},\quad \nabla^2\psi=0,\quad \phi=\psi+\frac{e}{\varepsilon}u,\quad -h<x_2<h, \tag{4.13.1}$$

where $u = u_3$, $c = c_{44}$, $e = e_{15}$, $\varepsilon = \varepsilon_{11}$, and $\bar{c}=c(1+e^2/c\varepsilon)$. The boundary conditions are

$$\begin{aligned} &T_{23}=\mp\rho'h'\ddot{u},\quad x_2=\pm h,\\ &\phi=0,\quad x_2=\pm h, \end{aligned} \tag{4.13.2}$$

or, in terms of u and ψ

$$\begin{aligned} &\bar{c}u_{,2}+e\psi_{,2}=\mp\rho'h'\ddot{u},\quad x_2=\pm h,\\ &\psi+\frac{e}{\varepsilon}u=0,\quad x_2=\pm h. \end{aligned} \tag{4.13.3}$$

The waves that we are interested in may be classified as symmetric and anti-symmetric. We discuss them separately below.

4.13.2. Anti-symmetric waves

For anti-symmetric waves we have

$$\begin{aligned} &u=A\sin\xi_2x_2\cos(\xi_1x_1-\omega t),\\ &\psi=B\sinh\xi_1x_2\cos(\xi_1x_1-\omega t), \end{aligned} \tag{4.13.4}$$

$$\xi_2^2 = \frac{\rho\omega^2}{\bar{c}} - \xi_1^2 = \xi_1^2\left(\frac{v^2}{v_T^2} - 1\right),$$
$$v^2 = \frac{\omega^2}{\xi_1^2}, \quad v_T^2 = \frac{\bar{c}}{\rho}. \tag{4.13.5}$$

where A and B are constants. Substitution of Eq. (4.13.4) into Eq. (4.13.3) yields

$$\bar{c}A\xi_2 \cos\xi_2 h + eB\xi_1 \cosh\xi_1 h = \rho'h'\omega^2 A\sin\xi_2 h,$$
$$\frac{e}{\varepsilon}A\sin\xi_2 h + B\sinh\xi_1 h = 0. \tag{4.13.6}$$

For non-trivial solutions we must have

$$\begin{vmatrix} \bar{c}\xi_2\cos\xi_2 h - \rho'h'\omega^2\sin\xi_2 h & e\xi_1\cosh\xi_1 h \\ \frac{e}{\varepsilon}\sin\xi_2 h & \sinh\xi_1 h \end{vmatrix}$$
$$= \begin{vmatrix} \bar{c}\xi_2\cos\xi_2 h & e\xi_1\cosh\xi_1 h \\ \frac{e}{\varepsilon}\sin\xi_2 h & \sinh\xi_1 h \end{vmatrix} + \begin{vmatrix} -\rho'h'\omega^2\sin\xi_2 h & e\xi_1\cosh\xi_1 h \\ 0 & \sinh\xi_1 h \end{vmatrix} \tag{4.13.7}$$
$$= \bar{c}\xi_2\cos\xi_2 h\sinh\xi_1 h - \frac{e^2}{\varepsilon}\xi_1\sin\xi_2 h\cosh\xi_1 h$$
$$- \rho'h'\omega^2\sin\xi_2 h\sinh\xi_1 h = 0,$$

or

$$\tanh\xi_1 h - \bar{k}^2\frac{\xi_1}{\xi_2}\tan\xi_2 h = \frac{\rho'h'\omega^2}{\bar{c}\xi_2}\tan\xi_2 h\tanh\xi_1 h\,. \tag{4.13.8}$$

With Eq. (4.13.5), Eq. (4.13.8) is an Eq. for ω, which determines the wave frequency or speed. When $\rho'h' = 0$, Eq. (4.13.8) reduces to

$$\tanh\xi_1 h - \bar{k}^2\frac{\xi_1}{\xi_2}\tan\xi_2 h = 0\,, \tag{4.13.9}$$

which is Eq. (3.1.15). The difference in frequencies determined from Eqs. (4.13.8) and (4.13.9) can be used to calculate $\rho'h'$. Asymptotic expressions for small $\rho'h'$ can be obtained from Eq. (4.13.8).

4.13.3. Symmetric waves

For symmetric waves we consider

$$\begin{aligned} u &= A\cos\xi_2 x_2 \cos(\xi_1 x_1 - \omega t), \\ \psi &= B\cosh\xi_1 x_2 \cos(\xi_1 x_1 - \omega t). \end{aligned} \tag{4.13.10}$$

We still have

$$\xi_2^2 = \frac{\rho\omega^2}{\bar{c}} - \xi_1^2 = \xi_1^2\left(\frac{v^2}{v_T^2} - 1\right). \tag{4.13.11}$$

Substitution of Eq. (4.13.11) into Eq. (4.13.3) leads to

$$\begin{aligned} &-\bar{c}A\xi_2 \sin\xi_2 h + eB\xi_1 \sinh\xi_1 h = \rho' h' \omega^2 A\cos\xi_2 h, \\ &\frac{e}{\varepsilon}A\cos\xi_2 h + B\cosh\xi_1 h = 0. \end{aligned} \tag{4.13.12}$$

For non-trivial solutions of A and/or B, we must have

$$\begin{aligned} &\begin{vmatrix} -\bar{c}\xi_2 \sin\xi_2 h - \rho' h' \omega^2 \cos\xi_2 h & e\xi_1 \sinh\xi_1 h \\ \frac{e}{\varepsilon}\cos\xi_2 h & \cosh\xi_1 h \end{vmatrix} \\ &= \begin{vmatrix} -\bar{c}\xi_2 \sin\xi_2 h & e\xi_1 \sinh\xi_1 h \\ \frac{e}{\varepsilon}\cos\xi_2 h & \cosh\xi_1 h \end{vmatrix} + \begin{vmatrix} -\rho' h' \omega^2 \cos\xi_2 h & e\xi_1 \sinh\xi_1 h \\ 0 & \cosh\xi_1 h \end{vmatrix} \\ &= -\bar{c}\xi_2 \sin\xi_2 h \cosh\xi_1 h - \frac{e^2}{\varepsilon}\xi_1 \cos\xi_2 h \sinh\xi_1 h \\ &\quad - \rho' h' \omega^2 \cos\xi_2 h \cosh\xi_1 h = 0, \end{aligned} \tag{4.13.13}$$

or

$$-\tan\xi_2 h - \bar{k}^2 \frac{\xi_1}{\xi_2}\tanh\xi_1 h = \frac{\rho' h' \omega^2}{\bar{c}\xi_2}. \tag{4.13.14}$$

When $\rho' h' = 0$, Eq. (4.13.14) reduces to

$$\tan\xi_2 h + \bar{k}^2 \frac{\xi_1}{\xi_2}\tanh\xi_1 h = 0, \tag{4.13.15}$$

which is Eq. (3.1.21).

4.14. Bechmann's Number for Thickness-Twist Waves

With the result from the previous section, in this section we make a digression to Bechmann's number for thickness-twist waves in a piezoelectric plate of 6 mm crystals. This requires the knowledge of dispersion relations for thickness-twist waves in an unbounded unelectroded plate, and an unbounded electroded plate.

4.14.1. Thickness-twist waves in an unbounded, unelectroded plate

Consider a piezoelectric plate of 6mm crystals with the six-fold axis along x_3 (see Fig. 4.14.1). First we consider the case of an unelectroded plate ($a = 0$).

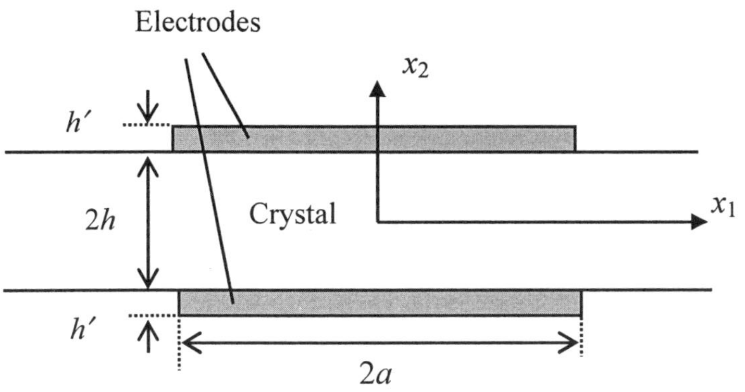

Fig. 4.14.1. A partially electroded piezoelectric plate of hexagonal crystals.

Thickness-twist waves are described by

$$\begin{aligned} &u_1 = u_2 = 0, \\ &u_3 = u(x_2)\exp[i(\xi_1 x_1 - \omega t)], \\ &\phi = \phi(x_2)\exp[i(\xi_1 x_1 - \omega t)]. \end{aligned} \tag{4.14.1}$$

A function ψ can be introduced through [27]

$$\phi = \psi + \frac{e}{\varepsilon} u, \tag{4.14.2}$$

where $e = e_{15}$ and $\varepsilon = \varepsilon_{11}$. The governing equations for u and ψ are [27]:

$$\bar{c}\nabla^2 u = \rho\ddot{u},$$
$$\nabla^2 \psi = 0, \tag{4.14.3}$$

where ∇^2 is the two-dimensional Laplacian, $\bar{c} = c + e^2/\varepsilon$, and $c = c_{44}$. At the plate surfaces we consider traction-free boundary conditions with

$$T_{23} = 0, \quad x_2 = \pm h,$$
$$D_2 = 0, \quad x_2 = \pm h. \tag{4.14.4}$$

Solution to Eqs. (4.14.3) and (4.14.4) are known [27]. The waves can be classified as anti-symmetric with u as an odd function of x_2, and symmetric with u as an even function of x_2. The dispersion relations for these waves can be obtained from Eq. (3.1.23) by setting $\varepsilon_0/\varepsilon_{11} = 0$:

$$\Omega^2 = Z^2 + n^2, \quad n = 0,1,2,3,4,\ldots, \tag{4.14.5}$$

where the dimensionless frequency and wave number are

$$\Omega^2 = \omega^2 \Big/ \left(\frac{\pi^2\bar{c}}{4\rho h^2}\right), \quad Z = \xi_1 \Big/ \left(\frac{\pi}{2h}\right). \tag{4.14.6}$$

When n is odd we have anti-symmetric waves. When n is even we have symmetric waves. In particular, $n = 0$ is a face-shear wave. This wave has no cutoff frequency and will not be considered in the following. For later determination of Bechmann's number we need the cutoff frequencies corresponding to $Z = 0$. These frequencies are simply $\Omega = n$.

4.14.2. Thickness-twist waves in an unbounded, electroded plate

To determine Bechmann's number we also need solutions for the same waves in a fully electroded plate ($a = \infty$ in Fig. 4.14.1). The electroded plate treated in [27] is for very thin electrodes of negligible mass. For our purpose we need to consider the inertia of the electrode mass with the following boundary conditions:

$$T_{23} = \mp\rho' h' \ddot{u}, \quad x_2 = \pm h,$$
$$\phi = 0, \quad x_2 = \pm h. \tag{4.14.7}$$

The solutions are given by Eq. (4.13.8) for anti-symmetric waves and Eq. (4.13.14) for symmetric waves:

$$\tanh \xi_1 h - \bar{k}^2 \frac{\xi_1}{\xi_2} \tan \xi_2 h = \frac{\rho' h' \omega^2}{\bar{c}\xi_2} \tan \xi_2 h \tanh \xi_1 h, \tag{4.14.8}$$

$$-\tan\xi_2 h - \bar{k}^2\frac{\xi_1}{\xi_2}\tanh\xi_1 h = \frac{\rho' h'\omega^2}{\bar{c}\xi_2}, \tag{4.14.9}$$

where

$$\xi_2^2 = \frac{\rho\omega^2}{\bar{c}} - \xi_1^2, \quad \bar{k}^2 = e^2/(\bar{c}\varepsilon). \tag{4.14.10}$$

With Eq. (4.14.10)$_1$, Eq. (4.14.8) and (4.14.9) are equations for ω versus ξ_1 . Consider the case when both $\rho' h'$ and $\bar{k}^2$ are small. We write the solution to Eqs. (4.14.8) and (4.14.9) as

$$\xi_2 h = \frac{n\pi}{2} + \Delta, \tag{4.14.11}$$

where Δ is small, $n = 1, 3, 5, \ldots$ for Eq. (4.14.8), and $n = 2, 4, 6, \ldots$ for Eq. (4.14.9). Substituting Eq. (4.14.11) into Eqs. (4.14.8) and (4.14.9), for small Δ, we obtain the following approximate dispersion relations for long waves with small Z:

$$\begin{aligned}&\Omega^2 = n^2(1-2R) - \frac{8}{\pi^2}\bar{k}^2 + Z^2 - 2(R+\bar{k}^2)Z^2,\\&n = 1,3,5,\ldots,\end{aligned} \tag{4.14.12}$$

$$\begin{aligned}&\Omega^2 = n^2(1-2R) + Z^2 - 2(R+\bar{k}^2)Z^2,\\&n = 2,4,6,\ldots,\end{aligned} \tag{4.14.13}$$

where the mass ratio $R = \rho' h'/(\rho h)$ is small. Eqs. (4.14.12) and (4.12.13) show that the electrode mass lowers the cutoff frequency as expected. It also affects the curvature near cutoff. In Eq. (4.14.12) the piezoelectric coupling lowers the cutoff frequency because shorted electrodes reduce the electric field in the plate and the related piezoelectric stiffening effect. The piezoelectric coupling also affects the curvature. In Eq. (4.14.13) the piezoelectric coupling does not lower the cutoff frequency because for symmetric waves values of the electric potential are the same at the two plate surfaces and shortening the electrodes does not affect the electric field.

4.14.3. Bechmann's number

Now we are ready to consider Bechmann's number for the partially electroded plate in Fig. 4.14.1. Bechmann's number is defined as the ratio, to the plate thickness, of the wavelength in the electroded portion at the cutoff frequency of the unelectroded portion of the plate [31].

We begin with anti-symmetric waves. Setting $\Omega^2 = n^2$ in Eq. (4.14.12), we obtain the corresponding wave number as

$$Z^2 = \frac{4}{\pi^2}(\xi_1 h)^2 \cong 2Rn^2 + \frac{8}{\pi^2}\bar{k}^2, \quad n = 1,3,5,\ldots. \tag{4.14.14}$$

Therefore, Bechmann's number for anti-symmetric thickness-twist waves is given by

$$B_{TT} = \frac{\pi}{\xi_1 h} = \frac{\pi}{\sqrt{\frac{\pi^2}{2}Rn^2 + 2\bar{k}^2}}, \quad n = 1,3,5,\ldots. \tag{4.14.15}$$

We make the following observations from Eq. (4.14.15):

(i) If $\bar{k}^2$ is set to zero, Eq. (4.14.15) reduces to

$$B_{TT} = \frac{\sqrt{2}}{n\sqrt{R}}, \tag{4.14.16}$$

which has the same dependence on n and R as the result in [31] for thickness-twist waves in a quartz plate from an elastic analysis. Only the coefficient is different. For a larger R, Bechmann's number is smaller, as expected. Eq. (4.14.16) also shows that higher-order modes imply a smaller Bechmann's number, which is less obvious.

(ii) Both R and $\bar{k}^2$ contribute to Bechmann's number. R is usually of the order of 0.01 or less for resonator electrodes. For thin film resonators R can be an order of magnitude larger, of the order of 0.1. For materials like ZnO and AlN, $\bar{k}^2$ is of the order of 0.1. In applications usually lower order modes with a moderate n are used. In this case the contribution from $\bar{k}^2$ is as large as the contribution from R. Therefore an accurate prediction of Bechmann's number requires a piezoelectric analysis. Even for quartz with $\bar{k}^2$ of the order of 0.01, a pure elastic analysis may be insufficient.

(iii) For symmetric waves, by setting $\Omega^2 = n^2$ in Eq. (4.14.13), we in fact obtain Eq. (4.14.16), as expected.

4.15. Thickness-Twist Waves in a Quartz Plate

Two cases of symmetric and asymmetric mass layers will be treated separately by different techniques.

4.15.1. Symmetric mass layers

Consider a plate of rotated Y-cut quartz with identical mass layers as shown in Fig. 4.15.1 [26].

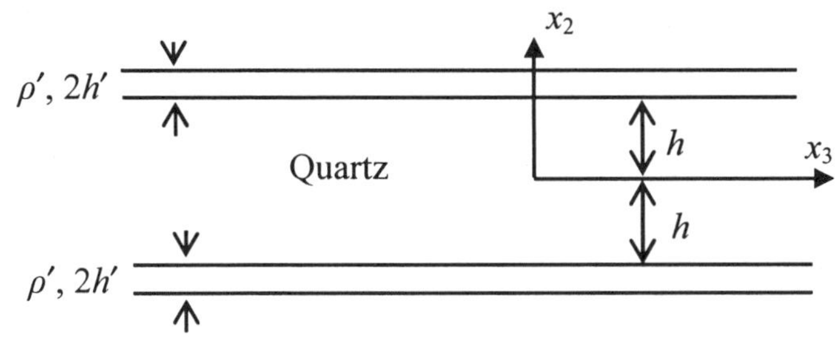

Fig. 4.15.1. A quartz plate with symmetric mass layers.

We neglect the weak piezoelectric coupling in quartz. Consider Thickness-twist modes described by

$$u_1 = U(y,z)\exp(i\omega t). \tag{4.15.1}$$

From Eqs. (3.1.28) through (3.1.30) we have

$$\begin{aligned} &U = C_1 \sin(\eta_1 y + \varsigma z) + C_2 \sin(\eta_2 y - \varsigma z), \\ &\rho\omega^2 = c_{66}\eta_1^2 + c_{55}\varsigma^2 + 2c_{56}\eta_1\varsigma \\ &\quad = c_{66}\eta_2^2 + c_{55}\varsigma^2 - 2c_{56}\eta_2\varsigma, \\ &c_{66}\eta_2 - c_{56}\varsigma = c_{66}\eta_1 + c_{56}\varsigma, \end{aligned} \tag{4.15.2}$$

where C_1 and C_2 are undetermined constants. The boundary conditions are

$$\begin{aligned} &T_{21} = c_{56}U_{,z} + c_{66}U_{,y} = \rho' 2h'\omega^2 U, \quad x_2 = h, \\ &T_{21} = c_{56}U_{,z} + c_{66}U_{,y} = -\rho' 2h'\omega^2 U, \quad x_2 = -h. \end{aligned} \tag{4.15.3}$$

Substituting Eq. (4.15.2)$_1$ into Eq. (4.15.3), with the use of Eq. (4.15.2)$_3$, we obtain

$$\frac{C_2}{C_1} = -\frac{k\cos\eta_1 h - \sin\eta_1 h}{k\cos\eta_2 h - \sin\eta_2 h} = \frac{k\sin\eta_1 h + \cos\eta_1 h}{k\sin\eta_2 h + \cos\eta_2 h}, \tag{4.15.4}$$

where

$$k = \frac{c_{66}\eta_1 + c_{56}\varsigma}{2\rho' h'\omega^2} = \frac{c_{66}\eta_2 - c_{56}\varsigma}{2\rho' h'\omega^2}. \tag{4.15.5}$$

Equations (4.15.5) and (4.15.2)$_{2,3}$ constitute the dispersion relation for thickness-twist waves in the plate with mass layers.

At cutoff ($\varsigma = 0$), the above equations reduce to

$$\begin{gathered}\eta_1 = \eta_2 = \eta, \quad \frac{\omega}{\omega_0} = \frac{2\eta h}{\pi}, \quad \omega_0^2 = \frac{\pi^2 c_{66}}{4h^2}, \\ \tan 2\eta h = \frac{2R\eta h}{R^2\eta^2 h^2 - 1}, \quad R = \frac{2\rho' h'}{\rho h}.\end{gathered} \tag{4.15.6}$$

The roots of technological interest are those when m is a small, odd integer. Letting

$$2\eta h = m\pi - \varepsilon, \quad 0 < \varepsilon << 1, \tag{4.15.7}$$

we can determine

$$2\eta h = m\pi(1 - R). \tag{4.15.8}$$

Hence, the cutoff frequencies are approximately

$$\frac{\omega}{\omega_0} = m(1 - R), \tag{4.15.9}$$

for small R.

4.15.2. Asymmetric mass layers

Next consider thickness-twist waves in a quartz plate with asymmetric mass layers (see Fig. 4.15.2).

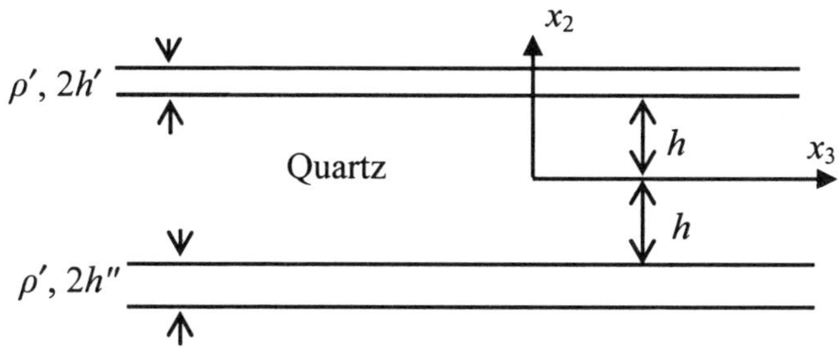

Fig. 4.15.2. A quartz plate with asymmetric mass layers.

From Eq. (3.1.25) the governing equation and the stress component relevant to boundary conditions are

$$\begin{aligned} &c_{66}u_{1,22} + 2c_{56}u_{1,23} + c_{55}u_{1,33} = \rho\ddot{u}_1, \\ &T_{21} = c_{56}u_{1,3} + c_{66}u_{1,2}. \end{aligned} \tag{4.15.10}$$

We study waves propagating in the x_3 direction:

$$u_1 = A\exp(i\eta y)\exp[i(\varsigma z - \omega t)], \tag{4.15.11}$$

where $y = x_2$ and $z = x_3$. Substitution of Eq. (4.15.11) into Eq. $(4.15.10)_1$ implies that

$$c_{66}\eta^2 + 2c_{56}\eta\varsigma + c_{55}\varsigma^2 - \rho\omega^2 = 0. \tag{4.15.12}$$

The two roots of Eq. (4.15.12) are

$$\begin{aligned} \eta_1 &= \frac{-c_{56}\varsigma + \sqrt{c_{56}^2\varsigma^2 + c_{66}(\rho\omega^2 - c_{55}\varsigma^2)}}{c_{66}}, \\ \eta_2 &= \frac{-c_{56}\varsigma - \sqrt{c_{56}^2\varsigma^2 + c_{66}(\rho\omega^2 - c_{55}\varsigma^2)}}{c_{66}}. \end{aligned} \tag{4.15.13}$$

Note that Eq. (4.15.13) implies that

$$c_{66}\eta_2 + c_{56}\varsigma = -(c_{66}\eta_1 + c_{56}\varsigma). \tag{4.15.14}$$

Then the general solution to Eq. $(4.15.10)_1$ in the form of Eq. (4.15.11), along with T_{21}, can be written as

$$\begin{aligned} u_1 &= [A_1\exp(i\eta_1 y) + A_2\exp(i\eta_2 y)]\exp[i(\varsigma z - \omega t)], \\ T_{21} &= [i(c_{56}\varsigma + c_{66}\eta_1)A_1\exp(i\eta_1 y) \\ &\quad + i(c_{56}\varsigma + c_{66}\eta_2)A_2\exp(i\eta_2 y)]\exp[i(\varsigma z - \omega t)], \end{aligned} \tag{4.15.15}$$

where A_1 and A_2 are undetermined constants. The boundary conditions to be satisfied are

$$\begin{aligned} -T_{21} &= \rho' 2h' \ddot{u}_1 = -\rho' 2h' \omega^2 u, \quad x_2 = h, \\ T_{21} &= \rho' 2h'' \ddot{u}_1 = -\rho' 2h'' \omega^2 u, \quad x_2 = -h. \end{aligned} \tag{4.15.16}$$

Substitution of Eq. (4.15.15) into Eq. (4.15.16) gives

$$\begin{aligned} & i(c_{56}\varsigma + c_{66}\eta_1)A_1 \exp(i\eta_1 h) + i(c_{56}\varsigma + c_{66}\eta_2)A_2 \exp(i\eta_2 h) \\ & \quad = \rho' 2h' \omega^2 [A_1 \exp(i\eta_1 h) + A_2 \exp(i\eta_2 h)], \\ & i(c_{56}\varsigma + c_{66}\eta_1)A_1 \exp(-i\eta_1 h) + i(c_{56}\varsigma + c_{66}\eta_2)A_2 \exp(-i\eta_2 h) \\ & \quad = -\rho' 2h'' \omega^2 [A_1 \exp(-i\eta_1 h) + A_2 \exp(-i\eta_2 h)]. \end{aligned} \tag{4.15.17}$$

For non-trivial solutions,

$$\begin{vmatrix} i(c_{56}\varsigma + c_{66}\eta_1)\exp(i\eta_1 h) - \rho' 2h' \omega^2 \exp(i\eta_1 h) & i(c_{56}\varsigma + c_{66}\eta_2)\exp(i\eta_2 h) - \rho' 2h' \omega^2 \exp(i\eta_2 h) \\ i(c_{56}\varsigma + c_{66}\eta_1)\exp(-i\eta_1 h) + \rho' 2h'' \omega^2 \exp(-i\eta_1 h) & i(c_{56}\varsigma + c_{66}\eta_2)\exp(-i\eta_2 h) + \rho' 2h'' \omega^2 \exp(-i\eta_2 h) \end{vmatrix} = 0. \tag{4.15.18}$$

With Eq. (4.15.14) we can write Eq. (4.15.18) as

$$\exp[i2(\eta_2 - \eta_1)h] = \frac{1 - k'k'' + i(k' + k'')}{1 - k'k'' - i(k' + k'')}, \tag{4.15.19}$$

where

$$k' = \frac{c_{66}\eta_1 + c_{56}\varsigma}{\rho' 2h' \omega^2}, \quad k'' = \frac{c_{66}\eta_1 + c_{56}\varsigma}{\rho' 2h'' \omega^2}. \tag{4.15.20}$$

Let θ be such that

$$\exp(i\theta) = \frac{1 - k'k'' + i(k' + k'')}{\sqrt{(1 - k'k'')^2 + (k' + k'')^2}}, \quad \tan\theta = \frac{k' + k''}{1 - k'k''}. \tag{4.15.21}$$

Then Eq. (4.15.19) can be written as

$$\exp[i2(\eta_2 - \eta_1)h] = \exp(i2\theta). \tag{4.15.22}$$

Hence,

$$\tan(\eta_1 - \eta_2)h = \frac{k' + k''}{1 - k'k''}. \tag{4.15.23}$$

Equations (4.15.12), (4.15.14) and (4.15.23) together determine the dispersion relations.

At cutoff ($\varsigma = 0$), the above equations reduce to

$$\begin{gathered}\eta_1 = -\eta_2 = \eta, \quad \frac{\omega}{\omega_0} = \frac{2\eta h}{\pi}, \quad \omega_0^2 = \frac{\pi^2 c_{66}}{4h^2 \rho}, \\ \tan 2\eta h = \frac{(R' + R'')\eta h}{R'R''\eta^2 h^2 - 1}, \quad R' = \frac{2\rho' h'}{\rho h}, \quad R'' = \frac{2\rho' h''}{\rho h}.\end{gathered} \tag{4.15.24}$$

When m is a small, odd integer, letting

$$2\eta h = m\pi - \varepsilon, \quad 0 < \varepsilon << 1, \tag{4.15.25}$$

we can determine

$$2\eta h = m\pi(1 - \overline{R}), \quad \overline{R} = (R' + R'')/2. \tag{4.15.26}$$

Hence, the cutoff frequencies are approximately

$$\frac{\omega}{\omega_0} = m(1 - \overline{R}). \tag{4.15.27}$$

Obviously, these results include those for symmetric mass layers as special cases.

Chapter 5
Fluid Sensors

A vibrating elastic body (resonator), put in contact with a fluid, changes its resonant frequencies. This effect has been used to make various fluid sensors for measuring fluid viscosity/density. In this chapter we discuss fluid sensors based on frequency shifts of resonators.

5.1. An Ill-Posed Problem in Elasticity for Fluid Sensors

Many fluid sensors operate with thickness-shear modes of a plate, which is a natural choice because in these modes the surface of the plate has no normal displacement so no compressional waves are generated in the fluid. The fluid produces a drag on the plate surface due to viscosity and the tangential motion of the plate surface, and thereby causes a frequency shift in the resonator. For fluid sensor applications, vibration modes of an elastic body without a normal displacement at its traction-free surface are of general interest. Vanishing normal displacement over a traction-free surface normally implies an ill-posed boundary value problem in elasticity. In this section we examine modes determined by the ill-posed free vibration eigenvalue problem of an elastic body without a normal displacement at its traction-free surface.

5.1.1. Formulation of the problem

Consider an isotropic elastic body. Let the region occupied by the body be V and its traction-free boundary surface be S. The unit outward normal of S is $\mathbf{n}$. For free vibrations with a frequency ω, the governing equations and boundary conditions for modes with vanishing normal displacement on S are [58]

$$
\begin{aligned}
&-T_{ji,j} = \rho\omega^2 u_i \quad \text{in} \quad V, \\
&T_{ji} = \lambda u_{k,k}\delta_{ji} + \mu(u_{i,j} + u_{j,i}) \quad \text{in} \quad V, \\
&T_{ji}n_j = 0 \quad \text{on} \quad S, \\
&u_j n_j = 0 \quad \text{on} \quad S.
\end{aligned} \tag{5.1.1}
$$

From the theory of elasticity, Eqs. $(5.1.1)_{1\text{-}3}$ constitute a well-posed eigenvalue problem. The presence of Eq. $(5.1.1)_4$ makes Eq. (5.1.1) ill-posed in general. However, this does not prevent that in special cases, e.g. for certain modes over some special domains V, Eq. $(5.1.1)_4$ may be automatically satisfied by solutions to Eq. $(5.1.1)_{1\text{-}3}$. We look for body shapes such that Eq. (5.1.1) has a solution.

5.1.2. Known solutions

A few known solutions are summarized below.

5.1.2.1. Anti-plane modes

Consider a cylindrical body with an arbitrary cross section in the (x_1, x_2) plane as shown in Fig. 5.1.1.

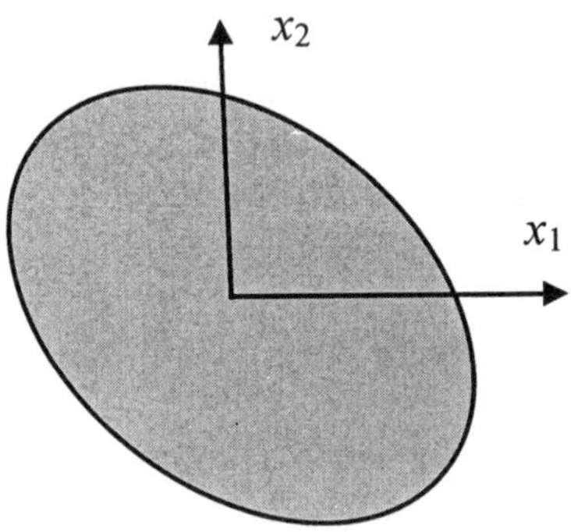

Fig. 5.1.1. Cross section of a cylindrical body.

Anti-plane (SH) motion of such a body is described by

$$u_1 = u_2 = 0, \quad u_3 = u_3(x_1, x_2, t)\,. \tag{5.1.2}$$

The motion is governed by

$$\mu\nabla^2 u_3 = \rho\ddot{u}_3, \tag{5.1.3}$$

where ∇^2 is the two-dimensional Laplacian $\nabla^2 = \partial_1^2 + \partial_2^2$. Anti-plane modes can exist in any two-dimensional region as shown in the figure.

In particular, consider an unbounded plate as shown in Fig. 5.1.2. The anti-plane modes are given by [59]:

$$u_3 = \begin{cases} \cos\dfrac{n\pi}{2h}x_2 \exp[i(\xi x_1 - \omega t)], & n = 0,2,4,\ldots, \\ \sin\dfrac{n\pi}{2h}x_2 \exp[i(\xi x_1 - \omega t)], & n = 1,3,5,\ldots, \end{cases}$$
$$\xi^2 = \frac{\omega^2}{c_2^2} - \left(\frac{n\pi}{2h}\right)^2, \quad c_2^2 = \frac{\mu}{\rho}. \tag{5.1.4}$$

These modes are called face-shear (n = 0) and thickness-twist (n = 1, 2, 3, …) waves.

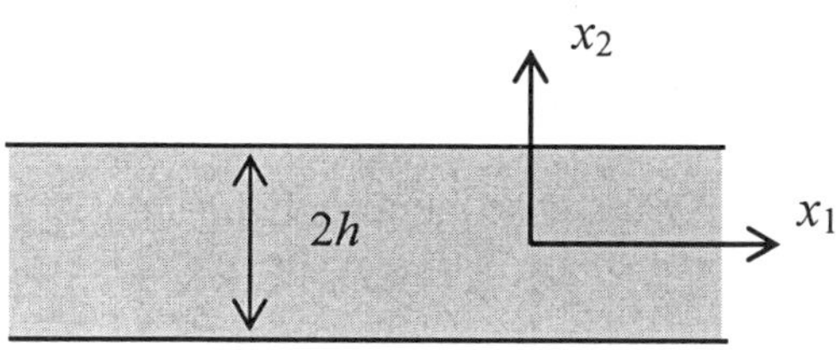

Fig. 5.1.2. An unbounded plate.

Anti-plane modes in an elastic wedge (see Fig. 5.1.3) are given by [41]:

$$u_3 = \begin{cases} J_\nu(\xi r)\cos\nu_n\theta \exp(-i\omega t), & n = 0,2,4\ldots, \\ J_\nu(\xi r)\sin\nu_n\theta \exp(-i\omega t), & n = 1,3,5,\ldots, \end{cases}$$
$$\xi = \frac{\omega}{c_2}, \quad \nu_n = \frac{n\pi}{2\alpha}. \tag{5.1.5}$$

For an infinite circular cylinder as shown in Fig. 5.1.4, anti-plane modes with $u_z(r,\theta,t)$ only are possible. For example, in the special case

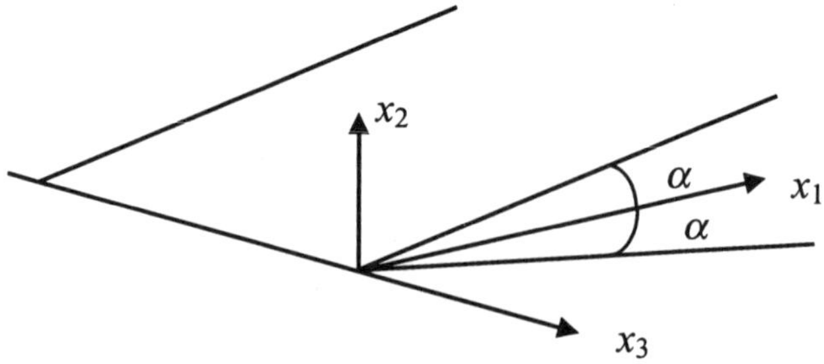

Fig. 5.1.3. An elastic wedge and coordinate system.

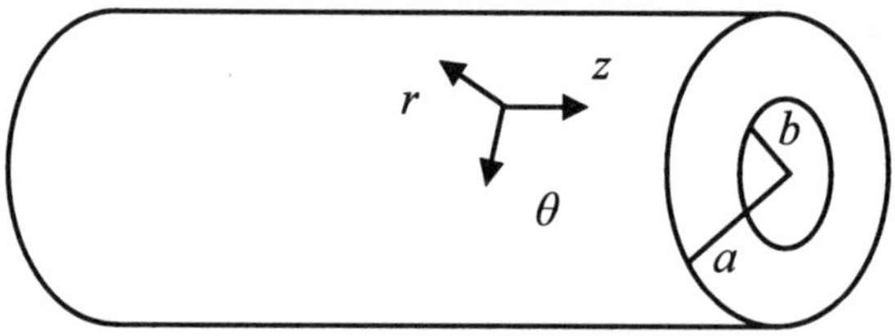

Fig. 5.1.4. A circular cylinder.

of a solid cylinder, axi-symmetric modes independent of θ are given by

$$u_z = J_0(\xi r),$$
$$\xi^2 = \frac{\omega^2}{c_2^2}, \quad J_0'(\xi a) = 0. \tag{5.1.6}$$

Anti-plane surface waves (Love wave [59]) can exist in an isotropic half-space carrying a surface layer of another material (see Fig. 5.1.5).

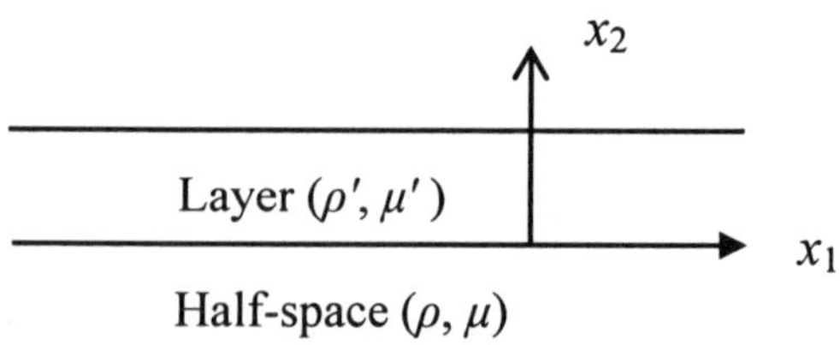

Fig. 5.1.5. A half-space with a layer of a different material.

5.1.2.2. Modes in a circular cylinder

For a circular cylinder there also exist torsional waves described by $u_\theta(r,t)$ only. For a solid cylinder [59],

$$\begin{aligned} &u_\theta = J_1(\xi r)\exp[i(\zeta z-\omega t)], \\ &\xi^2 = \frac{\omega^2}{c_2^2}-\zeta^2, \\ &\xi a J_0(\xi a)-2J_1(\xi a)=0. \end{aligned} \tag{5.1.7}$$

In the limit when $\xi \to 0$, Eq. (5.1.7) reduces to

$$\begin{aligned} &u_r = u_z = 0, \\ &u_\theta = r\exp[i(\xi z-\omega t)], \\ &\frac{\omega^2}{c_2^2} = \zeta^2. \end{aligned} \tag{5.1.8}$$

In addition, the tangential thickness-shear [10] described by

$$u_r = 0, \quad u_\theta = u_\theta(r,t), \quad u_z = 0 \tag{5.1.9}$$

is also without normal displacement at the surface.

5.1.2.3. Torsional modes in a sphere

For a sphere, in spherical coordinates (r, θ, φ), there exist the following torsional modes [60]:

$$\begin{aligned} &u_r = 0, \\ &u_\theta = \frac{m}{r\sin\theta} j_n(\beta r) p_n^m(\cos\theta)\begin{Bmatrix} \sin m\varphi \\ -\cos m\varphi \end{Bmatrix}\exp(-i\omega t), \\ &u_\varphi = \frac{1}{r} j_n(\beta r)\frac{d}{d\theta} p_n^m(\cos\theta)\begin{Bmatrix} \cos m\varphi \\ \sin m\varphi \end{Bmatrix}\exp(-i\omega t). \end{aligned} \tag{5.1.10}$$

5.1.3. Boundary integral equation formulation

The above ill-posed problem is open even within isotropic elasticity in the sense that it is unclear whether there exist body shapes and modes other than what are summarized in the above. Since the shape of the body is also an unknown, we use the boundary integral equation formulation to bring this unknown directly into the formulation. Consider the case of isotropic elasticity. The displacement **u** and surface traction **t** of an elastic body in a domain V with boundary S satisfy the following boundary integral equation [58]:

$$\begin{aligned}&\mathbf{C}(P_0)\mathbf{u}(P_0)+\int_S \hat{\mathbf{T}}(P,P_0,\omega)\cdot\mathbf{u}(P)d\Gamma(P)\\&\quad=\int_S \hat{\mathbf{U}}(P,P_0,\omega)\cdot\mathbf{t}(P)d\Gamma(P)+\int_V \hat{\mathbf{U}}(P,P_0,\omega)\cdot\mathbf{b}(P)d\Omega(P),\end{aligned} \tag{5.1.11}$$

where $C_{ij} = \delta_{ij}/2$ for a smooth boundary, $\mathbf{b}$ is the body force vector which will be set to zero, $\hat{\mathbf{U}}$ and $\hat{\mathbf{T}}$ are known second rank tensors and are related to the fundamental solution of the time-harmonic Navier operator in elasticity. They depend on the unknown frequency ω and can be found in a book on boundary element in elasticity. P is the source point and P_0 is the field point. Through the inner product with $\mathbf{n}(P_0)$ and Eq. $(5.1.1)_{3,4}$, from Eq. (5.1.11) we obtain

$$0=\int_S \mathbf{n}(P_0)\cdot\hat{\mathbf{T}}(P,P_0,\omega)\cdot\mathbf{u}(P)d\Gamma(P). \tag{5.1.12}$$

5.1.4. Generalization to piezoelectricity

Modes with vanishing normal displacement on a traction-free surface can also exist in certain piezoelectric crystals.

5.1.4.1. Polarized ceramics

Polarized ferroelectric ceramics are transversely isotropic. When the poling direction is along the x_3-axis, the following anti-plane motions are allowed [27]

$$\begin{aligned}&u_1=u_2=0,\quad u_3=u_3(x_1,x_2,t),\\&\phi=\phi(x_1,x_2,t).\end{aligned} \tag{5.1.13}$$

The governing equations are

$$\begin{aligned}&c\nabla^2u+e\nabla^2\phi=\rho\ddot{u},\\&e\nabla^2u-\varepsilon\nabla^2\phi=0,\end{aligned} \tag{5.1.14}$$

where $c = c_{44}$, $e = e_{15}$, and $\varepsilon = \varepsilon_{11}$ are the relevant elastic, piezoelectric, and dielectric constants. Quite a few anti-plane solutions have been obtained including surface waves, interface waves, Love type surface waves over a half-space carrying a layer of another material, surface waves on a circular cylinder, waves between two half-spaces with an air gap in between, axial motion of a circular cylinder, and plate waves.

These results are collected in [10]. Anti-plane wedge waves are given in [40]. For a circular ceramic cylinder poled tangentially, modes with $u_\theta(r,t)$ only can be found in [10].

5.1.4.2. Quartz

For rotated Y-cut quartz, anti-plane motions with only one displacement component, u_1, are particularly useful in device applications. Consider

$$\begin{aligned} &u_1 = u_1(x_2,x_3,t), \quad u_2 = u_3 = 0, \\ &\phi = \phi(x_2,x_3,t). \end{aligned} \tag{5.1.15}$$

Equation (5.1.15) yields the following non-vanishing components of strain, electric field, stress, and electric displacement:

$$\begin{aligned} &S_5 = u_{1,3}, \quad S_6 = u_{1,2}, \\ &E_2 = -\phi_{,2}, \quad E_3 = -\phi_{,3}, \\ &T_{31} = c_{55}u_{1,3} + c_{56}u_{1,2} + e_{25}\phi_{,2} + e_{35}\phi_{,3}, \\ &T_{21} = c_{56}u_{1,3} + c_{66}u_{1,2} + e_{26}\phi_{,2} + e_{36}\phi_{,3}, \\ &D_2 = e_{25}u_{1,3} + e_{26}u_{1,2} - \varepsilon_{22}\phi_{,2} - \varepsilon_{23}\phi_{,3}, \\ &D_3 = e_{35}u_{1,3} + e_{36}u_{1,2} - \varepsilon_{23}\phi_{,2} - \varepsilon_{33}\phi_{,3}. \end{aligned} \tag{5.1.16}$$

The equations left to be satisfied by u_1 and ϕ are

$$\begin{aligned} &c_{66}u_{1,22} + c_{55}u_{1,33} + 2c_{56}u_{1,23} \\ &\quad + e_{26}\phi_{,22} + e_{35}\phi_{,33} + (e_{25} + e_{36})\phi_{,23} = \rho\ddot{u}_1, \\ &e_{26}u_{1,22} + e_{35}u_{1,33} + (e_{25} + e_{36})u_{1,23} \\ &\quad - \varepsilon_{22}\phi_{,22} - \varepsilon_{33}\phi_{,33} - 2\varepsilon_{23}\phi_{,23} = 0. \end{aligned} \tag{5.1.17}$$

Pure thickness-shear modes of a rotated Y-cut quartz plate described by $u_1 = u_1(x_2,t)$ were studied in the second chapter. For a circular cylinder of monoclinic crystals, axial and tangential thickness-shear modes are given in [61].

5.2. Perturbation Analysis

Fluid sensors based on frequency shifts of resonators are for fluids whose density/viscosity is relatively low so that the fluid-induced frequency shifts are relatively small. For a fluid sensor we have a solid-fluid interaction problem over an unbounded, divided domain for the solid and fluid, respectively (interior-exterior problem). In this case a perturbation analysis may be considered [62]. Consider an anisotropic elastic body in an unbounded region of a viscous, incompressible fluid (see Fig. 5.2.1).

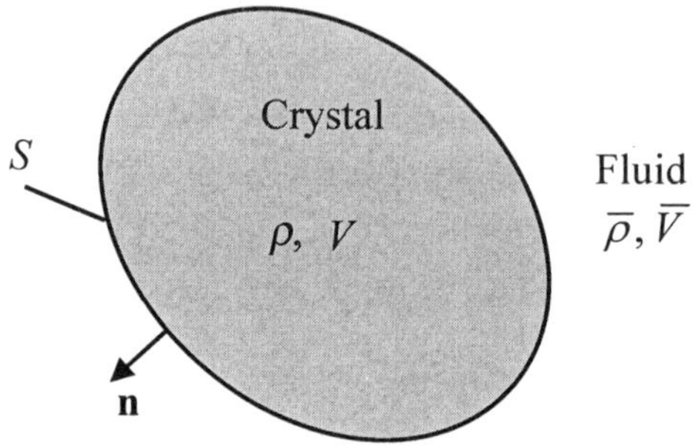

Fig. 5.2.1. A resonator in a fluid.

5.2.1. Governing equations

Let the interior region occupied by the crystal body be V and its boundary surface be S. The unit outward normal of S is $\mathbf{n}$. Let the exterior region occupied by the fluid be $\bar{V}$. For free vibrations, the governing equations and boundary conditions of the elastic body and the fluid are [63]

$$
\begin{aligned}
&T_{ji,j} = \rho \ddot{u}_i \quad \text{in} \quad V, \\
&T_{ji} = c_{jikl} u_{k,l} \quad \text{in} \quad V, \\
&\tau_{ji,j} - p_{,i} = \bar{\rho} \dot{v}_i \quad \text{in} \quad \bar{V}, \\
&v_{i,i} = 0 \quad \text{in} \quad \bar{V}, \\
&\tau_{ji} = \mu (v_{j,i} + v_{i,j}) \quad \text{in} \quad \bar{V}, \\
&T_{ji} n_j = \varepsilon (\tau_{ji} - p \delta_{ji}) n_j \quad \text{on} \quad S, \\
&\dot{u}_i = v_i \quad \text{on} \quad S,
\end{aligned}
\tag{5.2.1}
$$

where u_i is the displacement vector, T_{ij} is the stress tensor, c_{ijkl} are the elastic constants, and ρ is the mass density of the crystal. v_i is the velocity field, τ_{ij} is the viscous stress tensor, p is the pressure field, μ is the viscosity, and $\bar{\rho}$ is the mass density of the fluid. We consider small amplitude motions in which the material time derivative reduces to the partial derivative with respect to time, and Eq. (5.2.1) becomes linear. ε is a small parameter introduced to formally show the smallness of the effect of the low density/viscosity fluid on the crystal. For the unbounded fluid region, some boundary conditions at infinity are also needed. For a viscous fluid the fields die out far away under the disturbance from the crystal.

Consider time-harmonic motions. We use the complex notation. All fields are assumed to have an exp($i\omega t$) factor which is canceled from all terms. Then Eq. (5.2.1) becomes

$$\begin{aligned}
&-c_{jikl}u_{k,lj} = \rho\omega^2 u_i \quad \text{in} \quad V, \\
&\tau_{ji,j} - p_{,i} = \bar{\rho} i\omega v_i \quad \text{in} \quad \bar{V}, \\
&v_{i,i} = 0 \quad \text{in} \quad \bar{V}, \\
&\tau_{ji} = \mu(v_{j,i} + v_{i,j}) \quad \text{in} \quad \bar{V}, \\
&c_{jikl}u_{k,l}n_j = \varepsilon(\tau_{ji} - p\delta_{ji})n_j \quad \text{on} \quad S, \\
&i\omega u_i = v_i \quad \text{on} \quad S.
\end{aligned} \tag{5.2.2}$$

5.2.2. *Perturbation analysis*

For a perturbation solution of Eq. (5.2.2) we make the following expansions:

$$\begin{aligned}
&\omega \cong \omega^{(0)} + \varepsilon\,\omega^{(1)}, \quad \mathbf{u} \cong \mathbf{u}^{(0)} + \varepsilon\,\mathbf{u}^{(1)}, \\
&\mathbf{v} \cong \mathbf{v}^{(0)} + \varepsilon\,\mathbf{v}^{(1)}, \quad p \cong p^{(0)} + \varepsilon\, p^{(1)}, \\
&\tau_{ji} \cong \tau_{ji}^{(0)} + \varepsilon\tau_{ji}^{(1)}.
\end{aligned} \tag{5.2.3}$$

Substituting Eq. (5.2.3) into Eq. (5.2.2), collecting terms of equal powers of ε, we obtain a series of perturbation problems of successive orders. We are interested in the lowest order effect of the fluid. Therefore, we collect coefficients of terms with powers of ε^0 and ε^1 only. The zero-order problem consists of

$$\begin{aligned}
&-c_{jikl}u_{k,lj}^{(0)} = \rho(\omega^{(0)})^2 u_i^{(0)} \quad \text{in} \quad V, \\
&c_{jikl}u_{k,l}^{(0)}n_j = 0 \quad \text{on} \quad S,
\end{aligned} \tag{5.2.4}$$

and

$$\begin{aligned}
&\tau_{ji,j}^{(0)} - p_{,i}^{(0)} = \bar{\rho} i\omega^{(0)} v_i^{(0)} \quad \text{in} \quad \bar{V}, \\
&v_{i,i}^{(0)} = 0 \quad \text{in} \quad \bar{V}, \\
&\tau_{ji}^{(0)} = \mu(v_{j,i}^{(0)} + v_{i,j}^{(0)}) \quad \text{in} \quad \bar{V}, \\
&i\omega^{(0)} u_i^{(0)} = v_i^{(0)} \quad \text{on} \quad S.
\end{aligned} \tag{5.2.5}$$

Equation (5.2.4) shows that $\omega^{(0)}$ and $\mathbf{u}^{(0)}$ represent the frequency and modes when the fluid is not present, which can be written as real, nonnegative numbers and real functions, respectively. The zero-order problem is decoupled in the sense that the unperturbed frequency and modes of the crystal can be obtained from Eq. (5.2.4) by neglecting the fluid. Once $\omega^{(0)}$ and $\mathbf{u}^{(0)}$ of the crystal are determined from Eq. (5.2.4), the corresponding fluid motion is determined from Eq. (5.2.5). The interface continuity conditions in Eq. $(5.2.2)_{5,6}$ now become boundary conditions in Eqs. $(5.2.4)_2$ and $(5.2.5)_4$. The interior-exterior problem in Eq. (5.2.2) becomes an uncoupled interior problem in Eq. (5.2.4) and an exterior problem in Eq. (5.2.5). This is a simplification. The zero-order solution, $\omega^{(0)}$, $\mathbf{u}^{(0)}$, $v^{(0)}$, $p^{(0)}$ and $\tau_{ji}^{(0)}$, are assumed known as usual in a perturbation analysis.

For the first-order problem, for our purpose of determining the frequency shift $\varepsilon\omega^{(1)}$, only the following equations and boundary conditions for the crystal are needed:

$$\begin{aligned}
&-c_{jikl} u_{k,lj}^{(1)} = \rho(\omega^{(0)})^2 u_i^{(1)} + \rho 2\omega^{(0)}\omega^{(1)} u_i^{(0)} \quad \text{in} \quad V, \\
&c_{jikl} u_{k,l}^{(1)} n_j = (\tau_{ji}^{(0)} - p^{(0)}\delta_{ji}) n_j \quad \text{on} \quad S.
\end{aligned} \tag{5.2.6}$$

Comparison of Eq. (5.2.6) to Eq. (4.1.8) shows that mathematically $(\tau_{ji}^{(0)} - p^{(0)}\delta_{ji})n_j$ in Eq. $(5.2.6)_2$ plays the role of $\rho' h' \lambda^{(0)} u_i^{(0)}$ in Eq. $(4.1.8)_4$. Therefore the perturbation procedure is very similar and will not be repeated here. The result is

$$2\omega^{(0)}\omega^{(1)} \cong -\frac{\int_S (\tau_{ji}^{(0)} - p^{(0)}\delta_{ji}) n_j u_i^{(0)} dS}{\int_V \rho u_i^{(0)} u_i^{(0)} dV}, \tag{5.2.7}$$

or

$$\frac{\varepsilon\omega^{(1)}}{\omega^{(0)}} \cong -\frac{\varepsilon}{2(\omega^{(0)})^2} \frac{\int_S (\tau_{ji}^{(0)} - p^{(0)}\delta_{ji}) n_j u_i^{(0)} dS}{\int_V \rho u_i^{(0)} u_i^{(0)} dV}. \tag{5.2.8}$$

Finally, setting $\varepsilon = 1$, we obtain the first-order perturbation integral for the frequency shift as

$$\frac{\omega - \omega^{(0)}}{\omega^{(0)}} \cong -\frac{1}{2(\omega^{(0)})^2} \frac{\int_S (\tau_{ji}^{(0)} - p^{(0)}\delta_{ji}) n_j u_i^{(0)} dS}{\int_V \rho u_i^{(0)} u_i^{(0)} dV}. \tag{5.2.9}$$

5.3. Thickness-Shear Modes of a Plate

As an example of the applications of Eq. (5.2.9), consider the fundamental thickness-shear mode of an AT-cut quartz plate (see Fig. 5.3.1) [62].

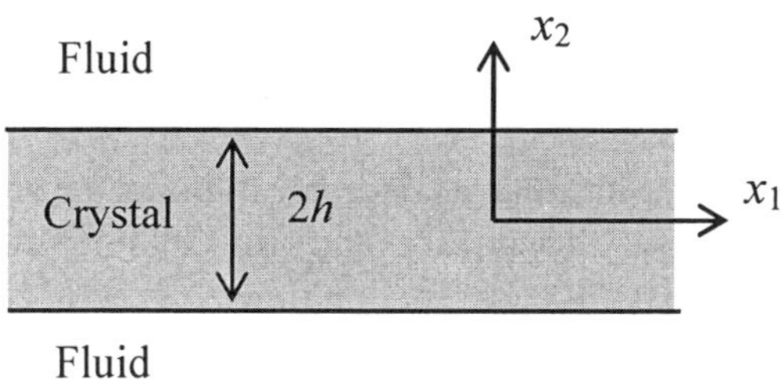

Fig. 5.3.1. An AT-cut quartz plate in a fluid.

The unperturbed frequency and mode when the fluid is not present is given by Eq. (4.2.1):

$$\omega^{(0)} = \frac{\pi}{2h}\sqrt{\frac{c_{66}}{\rho}}, \quad u_1^{(0)} = \sin\frac{\pi x_2}{2h}, \quad u_2^{(0)} = u_3^{(0)} = 0. \tag{5.3.1}$$

Then, from Eq. (5.2.5), the zero-order problem for the fluid motion is

$$\begin{aligned} &\tau_{21,2}^{(0)} = \bar{\rho} i\omega^{(0)} v_1^{(0)}, \quad x_2 > h, \\ &\tau_{21}^{(0)} = \mu v_{1,2}^{(0)}, \quad x_2 > h, \\ &i\omega^{(0)} = v_1^{(0)}, \quad x_2 = h, \end{aligned} \tag{5.3.2}$$

where, due to the anti-symmetry of the fundamental thickness-shear mode and the induced fluid motion, only the upper half-space of the fluid is considered. $p^{(0)}$ is a constant and is immaterial. The solution to Eq. (5.3.2) is

$$
\begin{aligned}
v_1^{(0)} &= i\omega^{(0)} \exp[-\xi(x_2 - h)], \\
\tau_{21}^{(0)} &= \mu i\omega^{(0)}(-\xi)\exp[-\xi(x_2 - h)], \\
\xi &= \sqrt{\frac{\bar{\rho} i\omega^{(0)}}{\mu}} = \sqrt{\frac{\bar{\rho}\omega^{(0)}}{\mu}}\frac{1+i}{\sqrt{2}}.
\end{aligned}
\tag{5.3.3}
$$

Substituting Eqs. (5.3.3) and (5.3.1) into Eq. (5.2.9), we obtain

$$
\frac{\omega - \omega^{(0)}}{\omega^{(0)}} \cong \frac{2}{\pi}\sqrt{\omega^{(0)}}\sqrt{\frac{\bar{\rho}\mu}{\rho c_{66}}}\frac{-1+i}{\sqrt{2}}.
\tag{5.3.4}
$$

Equation (5.3.4) shows that the fluid lowers the frequency and also causes a decay with respect to time (damped motion due to viscosity) as expected.

5.4. Torsional Modes of a Cylindrical Shell

Torsional modes of a circular cylindrical cylinder or shell have particles moving tangentially at the surface and are ideal for liquid sensing. We now study the propagation of torsional waves in a circular cylindrical elastic shell in contact with a viscous fluid [64] (see Fig. 5.4.1). Two cases will be considered: a shell filled with a fluid and a shell immersed in a fluid.

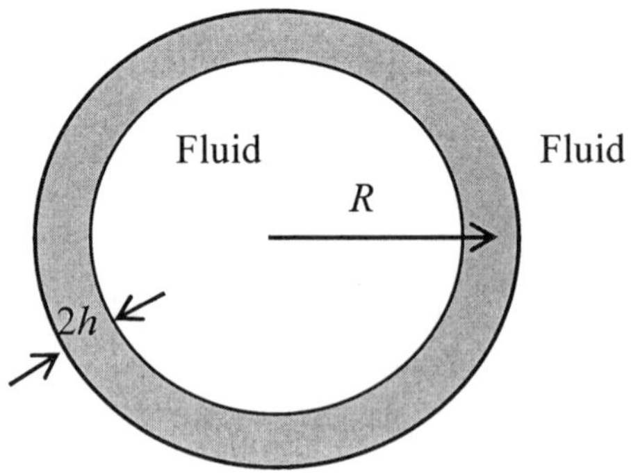

Fig. 5.4.1. An elastic shell in a viscous fluid.

5.4.1. Governing equations

Consider the following velocity field of the azimuthal motion of a linear, viscous fluid in cylindrical coordinates:

$$v_r = v_z = 0, \quad v_\theta = v(r,z,t). \tag{5.4.1}$$

The nontrivial stress components are [63]

$$T_{r\theta} = \mu r \frac{\partial}{\partial r}\left(\frac{v}{r}\right), \quad T_{rz} = \mu \frac{\partial v}{\partial z}, \tag{5.4.2}$$

where μ is the viscosity of the fluid. The relevant equation of motion is [63]

$$\rho' \frac{\partial v}{\partial t} = \mu\left(\frac{\partial^2 v}{\partial r^2} + \frac{1}{r}\frac{\partial v}{\partial r} - \frac{v}{r^2}\right) + \mu \frac{\partial^2 v}{\partial z^2}, \tag{5.4.3}$$

where ρ' is the fluid density. We look for a wave solution in the form

$$v = V(r)\sin kz \exp(i\omega t), \tag{5.4.4}$$

where k is the wave number and ω is the frequency. Substitution of Eq. (5.4.4) into Eq. (5.4.3) results in the following equation for V:

$$\frac{\partial^2 V}{\partial \xi^2} + \frac{1}{\xi}\frac{\partial V}{\partial \xi} - \left(1 + \frac{1}{\xi^2}\right)V = 0, \tag{5.4.5}$$

where

$$\xi = \lambda r, \quad \lambda^2 = k^2 + i\omega \frac{\rho'}{\mu}. \tag{5.4.6}$$

Equation (5.4.5) is the modified Bessel equation of order one. Its general solution is a linear combination of $I_1(\lambda r)$ and $K_1(\lambda r)$, the first order modified Bessel functions of the first and second kind. Hence

$$v = [C_1 I_1(\lambda r) + C_2 K_1(\lambda r)]\sin kz \exp(i\omega t), \tag{5.4.7}$$

where C_1 and C_2 are undetermined constants.

Consider torsional motions of a circular cylindrical shell with thickness $2h$ and middle surface radius R. The membrane theory [20] is sufficient. Let the middle surface displacement field be

$$
\begin{aligned}
u_1^{(0)} &= u_\theta^{(0)} = u(z,t), \\
u_2^{(0)} &= u_z^{(0)} = 0, \\
u_3^{(0)} &= u_r^{(0)} = 0.
\end{aligned} \tag{5.4.8}
$$

From [20] the equation that governs u is

$$
\rho 2h\ddot{u} = G2h\frac{\partial^2 u}{\partial z^2} + T_{r\theta}(R^+) - T_{r\theta}(R^-)\,, \tag{5.4.9}
$$

where ρ is the mass density and G is the shear modulus of the shell. $T_{r\theta}(R^+)$ and $T_{r\theta}(R^-)$ are the shear stress components at the outer and inner surfaces of the shell. They are due to the interaction with the fluid.

Substituting Eq. (5.4.2) into Eq. (5.4.9) we have

$$
\rho 2h\ddot{u} = G2h\frac{\partial^2 u}{\partial z^2} + \left[\mu r\frac{\partial}{\partial r}\left(\frac{v}{r}\right)\right]_{R^+} - \left[\mu r\frac{\partial}{\partial r}\left(\frac{v}{r}\right)\right]_{R^-}. \tag{5.4.10}
$$

Differentiating Eq. (5.4.10) with respect to time once, using the continuity of velocity (non-slip condition) between the shell and the fluid, we obtain

$$
\rho 2h\ddot{v}\,|_R = G2h\left.\frac{\partial^2 v}{\partial z^2}\right|_R + \left[\mu r\frac{\partial}{\partial r}\left(\frac{\dot{v}}{r}\right)\right]_{R^+} - \left[\mu r\frac{\partial}{\partial r}\left(\frac{\dot{v}}{r}\right)\right]_{R^-}. \tag{5.4.11}
$$

5.4.2. Interior problem

First consider a circular cylindrical shell filled with a fluid (interior problem). In this case, for the fluid region, since $K_1(0)$ is unbounded, we must have $C_2 = 0$ in Eq. (5.4.7). Substituting the remaining C_1 term into Eq. (5.4.11), we obtain

$$
\omega^2 = c_T^2 k^2 + i\frac{\mu\omega}{\rho 2hR}\left[\frac{\lambda R I_0(\lambda R)}{I_1(\lambda R)} - 2\right], \tag{5.4.12}
$$

where $c_T^2 = G/\rho$ is the torsional wave speed of the elastic shell when the fluid is not present. Equation (5.4.12) shows the effect of the fluid on the wave behavior, and hence the potential of measuring fluid properties through the wave frequency or speed. In the special case of a light fluid with a low density, from Eq. (5.4.6) we have

$$
\lambda \cong k + \frac{i\omega\rho'}{2\mu k}. \tag{5.4.13}
$$

Then,

$$I_0(\lambda R) \cong I_0(kR) + I_0'(kR)\frac{i\omega\rho' R}{2\mu k},$$
$$I_1(\lambda R) \cong I_1(kR) + I_1'(kR)\frac{i\omega\rho' R}{2\mu k}. \tag{5.4.14}$$

Substituting Eqs. (5.4.13) and (5.4.14) into Eq. (5.4.12), to the lowest order effect of the fluid mass density, we obtain

$$\omega^2\left(1+\frac{\beta}{k2h}\frac{\rho'}{\rho}\right) = c_T^2 k^2 + i\alpha\frac{\mu\omega}{\rho 2hR}, \tag{5.4.15}$$

where

$$\alpha = \frac{kRI_0(kR)}{I_1(kR)} - 2,$$
$$\beta = \frac{kR[I_1^2(kR) - I_0^2(kR)] + 2I_0(kR)I_1(kR)}{2I_1^2(kR)}. \tag{5.4.16}$$

In Eq. (5.4.15) the effects of the fluid inertia and viscosity become explicit. If we denote

$$\omega = \omega_0 + \Delta\omega, \quad \omega_0 = c_T k, \tag{5.4.17}$$

where ω_0 is the frequency when the fluid is not present, from Eq. (5.5.15) we obtain

$$\frac{\Delta\omega}{\omega_0} \cong -\frac{\beta}{4kh}\frac{\rho'}{\rho} + i\frac{\alpha\mu}{4\omega_0\rho hR}. \tag{5.4.18}$$

Equation (5.4.18) shows that ρ' causes frequency shifts and μ causes dissipation, as expected.

5.4.3. Exterior problem

Next consider a circular cylindrical shell immersed in an unbounded fluid (exterior problem). In this case, since $I_1(\infty)$ is unbounded, we must have $C_1 = 0$ in Eq. (5.4.7). Substituting the remaining C_2 term into Eq. (5.4.11), we obtain

$$\omega^2 = c_T^2 k^2 + i\frac{\mu\omega}{\rho 2hR}\left[\frac{\lambda R K_0(\lambda R)}{K_1(\lambda R)} + 2\right]. \tag{5.4.19}$$

Approximations similar to Eqs. (5.4.13) through (5.4.18) can also be made to Eq. (5.4.19).

5.5. Thickness-Shear Modes of a Circular Cylinder

In this section we analyze thickness-shear vibrations of a hollow circular cylinder interacting with a viscous fluid. A circular cylinder can vibrate in axial or tangential thickness-shear modes [10]. These modes are simpler than the torsional modes in the previous section. We use the three-dimensional equations of elasticity for the cylinder. We mainly study the tangential thickness-shear modes which can be considered as long torsional waves. Analysis of the axial thickness-shear modes is similar.

5.5.1. Governing equations

Consider a long cylinder whose cross-section is shown in Fig. 5.5.1.

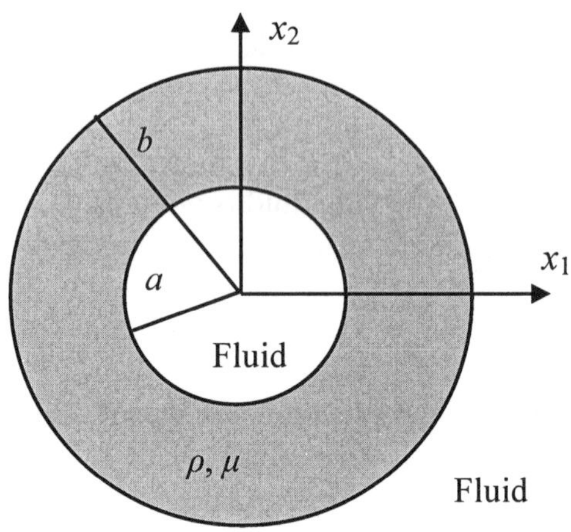

Fig. 5.5.1 Cross-section of a circular cylinder in a fluid.

The cylinder has mass density ρ and shear elastic constant G. It can be either immersed in or filled with a fluid of mass density ρ' and viscosity μ. A cylindrical coordinate system is defined by $x_1 = r\cos\theta$, $x_2 = r\sin\theta$, and $x_3 = z$.

5.5.1.1. The cylinder

Tangential thickness-shear modes in a circular cylinder are given by the following displacement and stress fields:

$$\begin{aligned} &u_r = 0, \quad u_z = 0, \\ &u_\theta = [C_1 J_1(\xi r) + C_2 Y_1(\xi r)]\exp(i\omega t), \\ &T_{\theta r} = -G\xi[C_1 J_2(\xi r) + C_2 Y_2(\xi r)]\exp(i\omega t), \end{aligned} \tag{5.5.1}$$

where C_1 and C_2 are arbitrary constants, J_1 and J_2 are the first- and second-order Bessel functions of the first kind, Y_1 and Y_2 are the first- and second-order Bessel functions of the second kind, and

$$\xi^2 = \frac{\rho\omega^2}{\mu}. \tag{5.5.2}$$

5.5.1.2. The fluid

For the fluid, from the previous section, we have the following fields of azimuthal motion in cylindrical coordinates:

$$\begin{aligned} &v_r = v_z = 0 \\ &v_\theta = [C_3 I_1(\lambda r) + C_4 K_1(\lambda r)]\exp(i\omega t), \\ &T_{\theta r} = \mu\lambda[C_3 I_2(\lambda r) - C_4 K_2(\lambda r)]\exp(i\omega t), \end{aligned} \tag{5.5.3}$$

where C_3 and C_4 are undetermined constants, I_1 and I_2 are the first- and second-order modified Bessel functions of the first kind, K_1 and K_2 are the first- and second-order modified Bessel functions of the second kind, and

$$\lambda^2 = i\omega\frac{\rho'}{\mu}. \tag{5.5.4}$$

5.5.2. Interior problem

First consider a circular cylinder filled with a fluid (interior problem). In this case, for the fluid region, since $K_1(0)$ is unbounded, we must have $C_4 = 0$ in Eq. (5.5.3). Substituting the remaining C_3 term and Eq. (5.5.1) into the following continuity and boundary conditions, we obtain

$$\begin{aligned} & i\omega[C_1 J_1(\xi a) + C_2 Y_1(\xi a)] = C_3 I_1(\lambda a), \\ & -G\xi[C_1 J_2(\xi a) + C_2 Y_2(\xi a)] = \mu\lambda C_3 I_2(\lambda a), \\ & -G\xi[C_1 J_2(\xi b) + C_2 Y_2(\xi b)] = 0. \end{aligned} \tag{5.5.5}$$

For non-trivial solutions, the determinant of the coefficient matrix has to vanish, which yields

$$\begin{vmatrix} i\omega J_1(\xi a) & i\omega Y_1(\xi a) & -I_1(\lambda a) \\ G\xi J_2(\xi a) & G\xi Y_2(\xi a) & \mu\lambda I_2(\xi a) \\ J_2(\xi b) & Y_2(\xi b) & 0 \end{vmatrix} = 0\,. \tag{5.5.6}$$

Equation (5.5.6) is an equation for ω. Complex roots are expected. Frequencies of a free cylinder not in contact with a fluid can be obtained from Eq. $(5.5.5)_{2,3}$ by setting $\lambda = 0$, which yields

$$\begin{vmatrix} J_2(\xi a) & Y_2(\xi a) \\ J_2(\xi b) & Y_2(\xi b) \end{vmatrix} = 0\,. \tag{5.5.7}$$

The difference between the corresponding roots of Eqs. (5.5.6) and (5.5.7) represents the effect of the fluid and can be used to calculate fluid properties.

5.5.3. Exterior problem

When the cylinder is immersed in an unbounded fluid (exterior problem), since $I_1(\infty)$ is unbounded, we must have $C_3 = 0$ in Eq. (5.5.3). Substituting the remaining C_4 term and Eq. (5.5.1) into the following continuity and boundary conditions, we obtain

$$\begin{aligned} & i\omega[C_1 J_1(\xi b) + C_2 Y_1(\xi b)] = C_4 K_1(\lambda b), \\ & -G\xi[C_1 J_2(\xi b) + C_2 Y_2(\xi b)] = -\mu\lambda C_4 K_2(\lambda b), \\ & -G\xi[C_1 J_2(\xi a) + C_2 Y_2(\xi a)] = 0. \end{aligned} \tag{5.5.8}$$

For non-trivial solutions,

$$\begin{vmatrix} i\omega J_1(\xi b) & i\omega Y_1(\xi b) & -K_1(\lambda b) \\ G\xi J_2(\xi b) & G\xi Y_2(\xi b) & -\mu\lambda K_2(\xi b) \\ J_2(\xi a) & Y_2(\xi a) & 0 \end{vmatrix} = 0\,. \tag{5.5.9}$$

In the special case of a solid cylinder, C_2 in Eq. (5.5.1) must vanish because Y_1 is unbounded at $r = 0$. From Eq. $(5.5.8)_{1,2}$, we obtain

$$
\begin{aligned}
i\omega C_1 J_1(\xi b) &= C_4 K_1(\lambda b), \\
-G\xi C_1 J_2(\xi b) &= -\mu\lambda C_4 K_2(\lambda b),
\end{aligned}
\tag{5.5.10}
$$

which implies that

$$
\begin{vmatrix} i\omega J_1(\xi b) & -K_1(\lambda b) \\ G\xi J_2(\xi b) & -\mu\lambda K_2(\lambda b) \end{vmatrix} = 0\,. \tag{5.5.11}
$$

Frequencies of a free solid cylinder not in contact with a fluid are given by

$$
J_2(\xi b) = 0\,. \tag{5.5.12}
$$

5.5.4. Axial thickness-shear

For a hollow cylinder with traction-free surfaces, axial thickness-shear modes are given by

$$
\begin{aligned}
&u_r = 0, \quad u_\theta = 0, \\
&u_z = C[Y_1(\xi b)J_0(\xi r) - J_1(\xi b)Y_0(\xi r)]\exp(i\omega t),
\end{aligned}
\tag{5.5.13}
$$

where C is an arbitrary constant, J_0 and J_1 are the zero- and first-order Bessel functions of the first kind, Y_0 and Y_1 are the zero- and first-order Bessel functions of the second kind, and

$$
\xi^2 = \frac{\rho\omega^2}{\mu}\,. \tag{5.5.14}
$$

The frequency equation is

$$
\begin{vmatrix} J_1(\xi a) & Y_1(\xi a) \\ J_1(\xi b) & Y_1(\xi b) \end{vmatrix} = 0\,. \tag{5.5.15}
$$

When the cylinder is put in contact with a fluid, the problem can be analyzed similar to that of the tangential thickness-shear modes. Axial thickness-shear modes have only tangential displacements on the cylinder surfaces and are ideal for fluid sensing.

5.6. Surface Wave Fluid Sensors

Surface waves can also be used to make fluid sensors. In this section we analyze the interaction between a surface wave and a viscous fluid.

5.6.1. Governing equations

Consider a ceramic half-space in contact with a half-space of a dielectric fluid (see Fig. 5.6.1). The ceramic is poled in the x_3 direction.

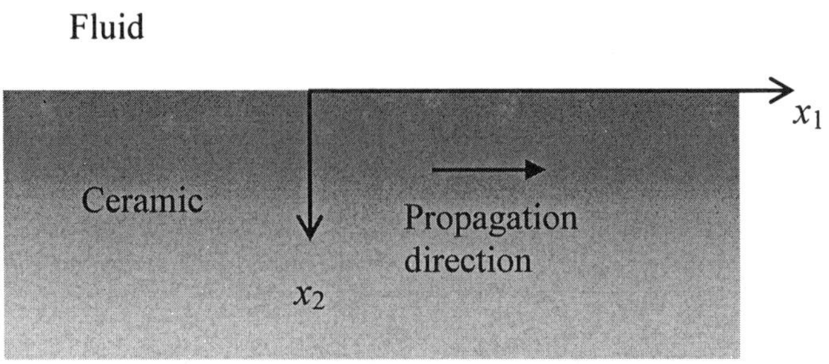

Fig. 5.6.1. A ceramic half-space in contact with a fluid.

5.6.1.1. The ceramic

Consider anti-plane motions of $u_3(x_1,x_2,t)$ and $\phi(x_1,x_2,t)$. The governing equations for the fields in the ceramic half-space are [27,10]

$$\bar{c}\nabla^2 u = \rho\ddot{u}, \quad \nabla^2\psi = 0, \quad \phi = \psi + \frac{e}{\varepsilon}u, \quad x_2 > 0, \tag{5.6.1}$$

where $u = u_3$, $c = c_{44}$, $e = e_{15}$, $\varepsilon = \varepsilon_{11}$, and $\bar{c} = c(1 + e^2/c\varepsilon)$. Consider the possibility of solutions in the following form:

$$\begin{aligned}
u &= A\exp(-\xi_2 x_2)\exp[i(\xi_1 x_1 - \omega t)],\\
\psi &= B\exp(-\xi_1 x_2)\exp[i(\xi_1 x_1 - \omega t)],\\
T_{23} &= \bar{c}u_{,2} + e\psi_{,2}\\
&= -[\bar{c}A\xi_2\exp(-\xi_2 x_2) + eB\xi_1\exp(-\xi_1 x_2)]\exp[i(\xi_1 x_1 - \omega t)],\\
\phi &= \left[B\exp(-\xi_1 x_2) + \frac{e}{\varepsilon}A\exp(-\xi_2 x_2)\right]\exp[i(\xi_1 x_1 - \omega t)],\\
D_2 &= -\varepsilon\psi_{,2} = \varepsilon\xi_1 B\exp(-\xi_1 x_2)\exp[i(\xi_1 x_1 - \omega t)],
\end{aligned} \tag{5.6.2}$$

where A and B are undetermined constants, $\xi_1 > 0$, and ξ_2 should have a positive real part for decaying behavior away from the surface.

Equation $(5.6.2)_2$ already satisfies Eq. $(5.6.1)_2$. For Eq. $(5.6.2)_1$ to satisfy Eq. $(5.6.1)_1$ we must have

$$\bar{c}(\xi_1^2 - \xi_2^2) = \rho\omega^2, \tag{5.6.3}$$

which determines

$$\begin{aligned} &\xi_2^2 = \xi_1^2 - \frac{\rho\omega^2}{\bar{c}} = \xi_1^2\left(1 - \frac{v^2}{v_T^2}\right), \\ &v^2 = \frac{\omega^2}{\xi_1^2}, \quad v_T^2 = \frac{\bar{c}}{\rho}. \end{aligned} \tag{5.6.4}$$

5.6.1.2. The fluid

Let $v_3(x_1, x_2, t)$ be the velocity field. The governing equation for v_3 and the relevant stress component for the interface continuity condition are [63]

$$\begin{aligned} &\mu(v_{3,11} + v_{3,22}) = \rho' \frac{\partial v_3}{\partial t}, \\ &T_{23} = \mu v_{3,2}. \end{aligned} \tag{5.6.5}$$

The following fields are relevant:

$$\begin{aligned} &v_3 = C\exp(\lambda x_2)\exp[i(\xi_1 x_1 - \omega t)], \\ &T_{23} = \mu\lambda C\exp(\lambda x_2)\exp[i(\xi_1 x_1 - \omega t)], \end{aligned} \tag{5.6.6}$$

where C is an arbitrary constant, λ has a positive real part and is determined by

$$\mu(\lambda^2 - \xi_1^2) = \rho' i\omega. \tag{5.6.7}$$

5.6.2. *A half-space with an electroded surface*

First consider the case when the surface of the half-space is electroded and the electrode is grounded. The corresponding boundary and continuity conditions are

$$\begin{aligned} &\dot{u}(x_2 = 0^+) = v_3(x_2 = 0^-), \\ &T_{23}(x_2 = 0^+) = T_{23}(x_2 = 0^-), \\ &\phi = 0, \quad x_2 = 0. \end{aligned} \tag{5.6.8}$$

Substituting Eqs. (5.6.2) and (5.6.6) into Eq. (5.6.8):

$$\begin{aligned} &-i\omega A = C, \\ &-(\bar{c}A\xi_2 + eB\xi_1) = \mu\lambda C, \\ &\frac{e}{\varepsilon}A + B = 0. \end{aligned} \tag{5.6.9}$$

For non-trivial solutions,

$$\begin{vmatrix} i\omega & 0 & 1 \\ \bar{c}\xi_2 & e\xi_1 & \mu\lambda \\ e/\varepsilon & 1 & 0 \end{vmatrix} = \begin{vmatrix} \bar{c}\xi_2 & e\xi_1 \\ e/\varepsilon & 1 \end{vmatrix} - \mu\lambda \begin{vmatrix} i\omega & 0 \\ e/\varepsilon & 1 \end{vmatrix} = 0, \tag{5.6.10}$$

or

$$\begin{vmatrix} \bar{c}\xi_2 & e\xi_1 \\ e/\varepsilon & 1 \end{vmatrix} = i\omega\mu\lambda, \tag{5.6.11}$$

which determines the surface wave frequency or speed. In the special case when the fluid is not present, i.e., $\mu = 0$, Eq. (5.6.11) reduces to the Bleustein-Gulyaev wave with the following speed:

$$v_{B-G}^2 = \frac{\omega^2}{\xi_1^2} = \frac{\bar{c}}{\rho}(1 - \bar{k}^4) = v_T^2(1 - \bar{k}^4) < v_T^2. \tag{5.6.12}$$

The difference between the frequencies of Eqs. (5.6.11) and (5.6.12) can be used to calculate μ or ρ'.

5.6.3. A half-space with an unelectroded surface

If the surface of the half-space is unelectroded, electric fields can also exist in the free space of $x_2 < 0$. Denoting the electric potential in the free space by $\hat{\phi}$, we have

$$\nabla^2 \hat{\phi} = 0, \quad x_2 < 0. \tag{5.6.13}$$

From Eq. (5.6.13), in the free space,

$$\begin{aligned} &\hat{\phi} = D\exp(\xi_1 x_2)\cos(\xi_1 x_1 - \omega t), \\ &D_2 = -\varepsilon_0 \hat{\phi}_{,2} = -\varepsilon_0 \xi_1 D\exp(\xi_1 x_2)\cos(\xi_1 x_1 - \omega t), \end{aligned} \tag{5.6.14}$$

where D is an undetermined constant. The boundary and continuity conditions are

$$
\begin{aligned}
& -i\omega u_3(x_2=0^+)=v_3(x_2=0^-), \\
& T_{23}(x_2=0^+)=T_{23}(x_2=0^-), \\
& \phi=\hat{\phi}, \quad D_2=\hat{D}_2, \quad x_2=0.
\end{aligned} \tag{5.6.15}
$$

Substituting Eqs. (5.6.2), (5.6.6) and (5.6.14) into Eq. (5.6.15):

$$
\begin{aligned}
& -i\omega A=C \\
& \bar{c}(-A\xi_2)+e(-B\xi_1)=\mu\lambda C, \\
& -\varepsilon(-\xi_1 B)=-\varepsilon_0 D\xi_1, \\
& \frac{e}{\varepsilon}A+B=D.
\end{aligned} \tag{5.6.16}
$$

For non-trivial solutions,

$$
\begin{aligned}
& \begin{vmatrix} i\omega & 0 & 1 & 0 \\ \bar{c}\xi_2 & e\xi_1 & \mu\lambda & 0 \\ 0 & \varepsilon & 0 & \varepsilon_0 \\ e/\varepsilon & 1 & 0 & -1 \end{vmatrix} \\
& = \begin{vmatrix} \bar{c}\xi_2 & e\xi_1 & 0 \\ 0 & \varepsilon & \varepsilon_0 \\ e/\varepsilon & 1 & -1 \end{vmatrix} + i\omega \begin{vmatrix} e\xi_1 & \mu\lambda & 0 \\ \varepsilon & 0 & \varepsilon_0 \\ 1 & 0 & -1 \end{vmatrix} \\
& = \begin{vmatrix} \bar{c}\xi_2 & e\xi_1 & 0 \\ 0 & \varepsilon & \varepsilon_0 \\ e/\varepsilon & 1 & -1 \end{vmatrix} + i\omega\mu\lambda(\varepsilon+\varepsilon_0)=0.
\end{aligned} \tag{5.6.17}
$$

When $\mu=0$, Eq. (5.6.17) determines the following speed of the Bleustein-Gulyaev wave for an unelectroded half-space:

$$
v_{B-G}^2=\frac{\omega^2}{\xi_1^2}=\frac{\bar{c}}{\rho}\left[1-\frac{\bar{k}^4}{(1+\varepsilon/\varepsilon_0)^2}\right]=v_T^2\left[1-\frac{\bar{k}^4}{(1+\varepsilon/\varepsilon_0)^2}\right]<v_T^2. \tag{5.6.18}
$$

5.7. Thickness-Twist Waves in a Ceramic Plate

In this section we consider waves propagating in the x_1 direction in a ceramic plate poled in the x_3 direction (see Fig. 5.7.1). The major surfaces of the plate are electroded, and the electrodes are grounded. The plate is immersed in a dielectric viscous fluid.

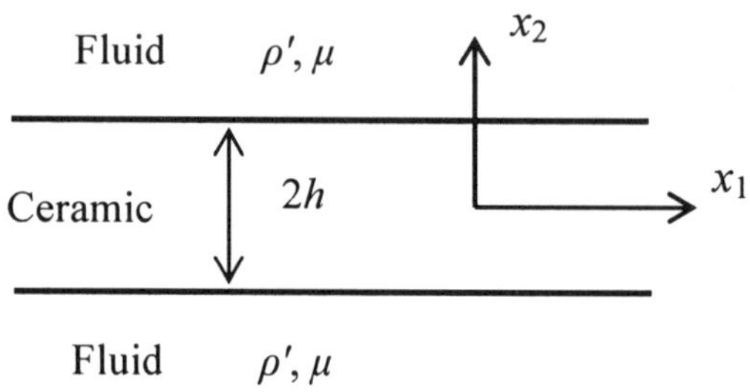

Fig. 5.7.1. A ceramic plate in a fluid.

5.7.1. Governing equations

5.7.1.1. The plate

Consider faces-shear and thickness-twist waves described by $u_3(x_1,x_2,t)$ and $\phi(x_1,x_2,t)$. With the notation in [27,10], from the first section of the third chapter, the governing equations are

$$\bar{c}\nabla^2 u = \rho\ddot{u}, \quad \nabla^2\psi = 0, \quad \phi = \psi + \frac{e}{\varepsilon}u, \quad -h < x_2 < h, \tag{5.7.1}$$

where $u = u_3$, $c = c_{44}$, $e = e_{15}$, $\varepsilon = \varepsilon_{11}$, and $\bar{c} = c(1 + e^2 / c\varepsilon)$. The stress component needed for the interface continuity conditions is

$$T_{23} = \bar{c}u_{,2} + e\psi_{,2} \tag{5.7.2}$$

5.7.1.2. The fluid

Let $v_3(x_1,x_2,t)$ be the velocity field. The governing equation for v_3 and the relevant stress component for the interface continuity condition are [63]

$$\mu(v_{3,11} + v_{3,22}) = \rho' \frac{\partial v_3}{\partial t}, \qquad (5.7.3)$$
$$T_{23} = \mu v_{3,2}.$$

For $x_2 > 0$, the following fields are relevant:

$$v_3 = C \exp[-\lambda(x_2 - h)] \exp[i(\xi_1 x_1 - \omega t)],$$
$$T_{23} = -\mu\lambda C \exp[-\lambda(x_2 - h)] \exp[i(\xi_1 x_1 - \omega t)], \qquad (5.7.4)$$

where C is an arbitrary constant, λ has a positive real part and is determined by

$$\mu(\lambda^2 - \xi_1^2) = \rho' i\omega. \qquad (5.7.5)$$

5.7.2. Anti-symmetric waves

For anti-symmetric waves we have

$$u = A \sin \xi_2 x_2 \exp[i(\xi_1 x_1 - \omega t)],$$
$$\psi = B \sinh \xi_1 x_2 \exp[i(\xi_1 x_1 - \omega t)], \qquad (5.7.6)$$
$$\xi_2^2 = \frac{\rho\omega^2}{\bar{c}} - \xi_1^2,$$

where A and B are constants. At $x_2 = h$ we require that

$$i\omega u(x_2 = h^-) = v_3(x_2 = h^+),$$
$$T_{23}(x_2 = h^-) = T_{23}(x_2 = h^+), \qquad (5.7.7)$$
$$\frac{e}{\varepsilon} u + \psi = 0.$$

Substitution of Eqs. (5.7.4) and (5.7.6) into Eq. (5.7.7) gives

$$i\omega A \sin \xi_2 h = C,$$
$$\bar{c} A \xi_2 \cos \xi_2 h + eB\xi_1 \cosh \xi_1 h = -\mu\lambda C, \qquad (5.7.8)$$
$$\frac{e}{\varepsilon} A \sin \xi_2 h + B \sinh \xi_1 h = 0.$$

For non-trivial solutions we must have

$$\begin{vmatrix} i\omega\sin\xi_2 h & 0 & -1 \\ \bar{c}\xi_2\cos\xi_2 h & e\xi_1\cosh\xi_1 h & \mu\lambda \\ \dfrac{e}{\varepsilon}\sin\xi_2 h & \sinh\xi_1 h & 0 \end{vmatrix}$$

$$= i\omega\sin\xi_2 h\begin{vmatrix} e\xi_1\cosh\xi_1 h & \mu\lambda \\ \sinh\xi_1 h & 0 \end{vmatrix} - \begin{vmatrix} \bar{c}\xi_2\cos\xi_2 h & e\xi_1\cosh\xi_1 h \\ \dfrac{e}{\varepsilon}\sin\xi_2 h & \sinh\xi_1 h \end{vmatrix} \quad (5.7.9)$$

$$= -i\omega\mu\lambda\sin\xi_2 h\sinh\xi_1 h$$

$$-\bar{c}\xi_2\cos\xi_2 h\sinh\xi_1 h + \frac{e^2}{\varepsilon}\xi_1\sin\xi_2 h\cosh\xi_1 h = 0,$$

or

$$\tanh\xi_1 h - \bar{k}^2\frac{\xi_1}{\xi_2}\tan\xi_2 h = -\frac{i\omega\mu\lambda}{\bar{c}\xi_2}\tan\xi_2 h\tanh\xi_1 h\,. \quad (5.7.10)$$

With Eq. (5.7.6)$_3$, Eq. (5.7.10) is an equation for ω, which determines the wave frequency or speed. When $\mu = 0$, Eq. (5.7.10) reduces to

$$\tanh\xi_1 h - \bar{k}^2\frac{\xi_1}{\xi_2}\tan\xi_2 h = 0\,, \quad (5.7.11)$$

which is Eq. (3.1.15). The difference in frequencies determined from Eqs. (5.7.10) and (5.7.11) can be used to calculate μ or ρ'.

5.7.3. Symmetric waves

For symmetric waves we consider

$$\begin{aligned} u &= A\cos\xi_2 x_2\cos(\xi_1 x_1 - \omega t), \\ \psi &= B\cosh\xi_1 x_2\cos(\xi_1 x_1 - \omega t). \end{aligned} \quad (5.7.12)$$

We still have

$$\xi_2^2 = \frac{\rho\omega^2}{\bar{c}} - \xi_1^2\,. \quad (5.7.13)$$

Substitution of Eqs. (5.7.4) and (5.7.12) into Eq. (5.7.7) leads to

$$\begin{aligned} & i\omega A\cos\xi_2 h = C, \\ & -\bar{c}A\xi_2\sin\xi_2 h + eB\xi_1\sinh\xi_1 h = -\mu\lambda C, \\ & \frac{e}{\varepsilon}A\cos\xi_2 h + B\cosh\xi_1 h = 0. \end{aligned} \tag{5.7.14}$$

For non-trivial solutions we must have

$$\begin{aligned} & \begin{vmatrix} i\omega\cos\xi_2 h & 0 & -1 \\ -\bar{c}\xi_2\sin\xi_2 h & e\xi_1\sinh\xi_1 h & \mu\lambda \\ \frac{e}{\varepsilon}\cos\xi_2 h & \cosh\xi_1 h & 0 \end{vmatrix} \\ & = i\omega\cos\xi_2 h\begin{vmatrix} e\xi_1\sinh\xi_1 h & \mu\lambda \\ \cosh\xi_1 h & 0 \end{vmatrix} - \begin{vmatrix} -\bar{c}\xi_2\sin\xi_2 h & e\xi_1\sinh\xi_1 h \\ \frac{e}{\varepsilon}\cos\xi_2 h & \cosh\xi_1 h \end{vmatrix} \\ & = -i\omega\mu\lambda\cos\xi_2 h\cosh\xi_1 h \\ & +\bar{c}\xi_2\sin\xi_2 h\cosh\xi_1 h + \frac{e^2}{\varepsilon}\xi_1\cos\xi_2 h\sinh\xi_1 h = 0, \end{aligned} \tag{5.7.15}$$

or

$$\tan\xi_2 h + \bar{k}^2\frac{\xi_1}{\xi_2}\tanh\xi_1 h = \frac{i\omega\mu\lambda}{\bar{c}\xi_2}. \tag{5.7.16}$$

When $\mu = 0$, Eq. (5.7.16) reduces to

$$\tan\xi_2 h + \bar{k}^2\frac{\xi_1}{\xi_2}\tanh\xi_1 h = 0, \tag{5.7.17}$$

which is Eq. (3.1.21).

Chapter 6
Gyroscopes — Frequency Effect

Two types of motion sensors, accelerometers and gyroscopes, are needed for detecting the complete motion of a moving object. Accelerometers sense linear motions, and gyroscopes sense angular motions. Traditional mechanical gyroscopes are based on the inertia of a rotating rigid body. New types of gyroscopes have also been developed, e.g. vibratory gyroscopes and optical gyroscopes. For vibratory gyroscopes, the excitation and detection of vibrations can be achieved in electrostatic, piezoelectric or other ways. Recently, using piezoelectric materials to make vibratory gyroscopes has been of increasing interest. Both angular rate induced frequency shifts [65–69] and charge (or current and voltage) [70,71] in vibrating elastic or piezoelectric structures have been be used to measure angular rates. More references can be found in a review article [72]. This chapter is on the frequency effect. The charge effect will be discussed in the next chapter.

6.1. High Frequency Vibrations of a Small Rotating Piezoelectric Body

The basic behaviors of a piezoelectric gyroscope can be described by the equations of a rotating piezoelectric body, which consist of the equations of linear piezoelectricity with rotation related Coriolis and centripetal accelerations. A Piezoelectric gyroscope is in small amplitude vibration in a reference frame rotating with it. The equilibrium state in the rotating reference frame is with initial deformations and stresses due to the centrifugal force. Therefore, an exact description of the motion of a piezoelectric gyroscope requires the equations for small, dynamic fields superposed on static initial fields due to the centrifugal force. The governing equations in the rotating frame can be written as

$$\begin{aligned} &T_{ji,j} - 2\rho\varepsilon_{ijk}\Omega_j\dot{u}_k - \rho(\Omega_i\Omega_j u_j - \Omega_j\Omega_j u_i) + O(\Omega^2) = \rho\ddot{u}_i, \\ &D_{i,i} = 0, \end{aligned} \tag{6.1.1}$$

and

$$
\begin{aligned}
&T_{ij} = c_{ijkl}S_{kl} - e_{kij}E_k, \quad D_i = e_{ijk}S_{jk} + \varepsilon_{ij}E_j, \\
&S_{ij} = (u_{i,j} + u_{j,i})/2, \quad E_i = -\phi_{,i},
\end{aligned}
\tag{6.1.2}
$$

where Ω_j is the angular velocity vector of the rotating body. The terms related to Ω_j represent the sum of the Coriolis and centrifugal forces. $O(\Omega^2)$ is due to the initial fields. The Coriolis force is linear in the angular velocity, which resembles damping but in fact the system is not dissipative because the Coriolis force is always perpendicular to the particle velocity and does not do any work during the motion.

Since piezoelectric gyroscopes are very small (of the order of 10 mm), their operating frequency ω_0 is very high, usually of the order of 10 kHz or higher. Piezoelectric gyroscopes are used to measure an angular rate Ω much smaller than its operating frequency ω_0. In this case the centrifugal force due to rotation, which is proportional to Ω^2, is much smaller compared to the Coriolis force which is proportional to $\omega_0\Omega$. Therefore, the effect of rotation on motions of piezoelectric gyroscopes is dominated by the Coriolis force. This is fundamentally different from the relatively well-studied subject of vibrations of a rotating elastic body in machine dynamics where large bodies with low resonant frequencies are in relatively fast rotations with the centrifugal force as one of the dominating forces. The high frequency vibration of a small rotating piezoelectric body with a relatively large Coriolis force presents a new class of modeling problems.

6.2. Propagation of Plane Waves

We first consider the propagation of plane waves in a rotating piezoelectric body of a material with general anisotropy, then we specialize the results to ceramics and quartz [73].

6.2.1. General solution

Let a piezoelectric body be rotating about a fixed axis at a constant angular rate Ω. Without loss of generality, we orient the x_3-axis of our Cartesian coordinate system along the rotation axis. Then the angular velocity vector has only one nontrivial component $(0,0,\Omega)$. Equation (6.1.1) takes the following simpler form:

$$T_{j1,j} = \rho(\ddot{u}_1 - 2\Omega\dot{u}_2 - \Omega^2 u_1),$$
$$T_{j2,j} = \rho(\ddot{u}_2 + 2\Omega\dot{u}_1 - \Omega^2 u_2), \tag{6.2.1}$$
$$T_{j3,j} = \rho\ddot{u}_3,$$

$$D_{i,i} = 0. \tag{6.2.2}$$

Consider the special case of plane waves propagating in the x_3 direction with $\partial_1 = \partial_2 = 0$. Then, from Eq. (6.2.2), with the related constitutive equation, we obtain

$$D_{3,3} = (e_{33}u_{3,3} + e_{34}u_{2,3} + e_{35}u_{1,3} - \varepsilon_{33}\phi_{,3})_{,3} = 0, \tag{6.2.3}$$

which implies

$$D_3 = e_{33}u_{3,3} + e_{34}u_{2,3} + e_{35}u_{1,3} - \varepsilon_{33}\phi_{,3} = -\varepsilon_{33}L_1, \tag{6.2.4}$$

where L_1 is an integration constant which may depend on the time t. Integrating Eq. (6.2.4) once more we have

$$\phi = \frac{e_{35}}{\varepsilon_{33}}u_1 + \frac{e_{34}}{\varepsilon_{33}}u_2 + \frac{e_{33}}{\varepsilon_{33}}u_3 + L_1 x_3 + L_2, \tag{6.2.5}$$

where L_2 is another integration constant. Substituting Eq. (6.2.5) into the relevant stress constitutive relations, we obtain

$$T_{31} = \bar{c}_{55}u_{1,3} + \bar{c}_{54}u_{2,3} + \bar{c}_{53}u_{3,3} + e_{35}L_1,$$
$$T_{32} = \bar{c}_{45}u_{1,3} + \bar{c}_{44}u_{2,3} + \bar{c}_{43}u_{3,3} + e_{34}L_1, \tag{6.2.6}$$
$$T_{33} = \bar{c}_{35}u_{1,3} + \bar{c}_{34}u_{2,3} + \bar{c}_{33}u_{3,3} + e_{33}L_1,$$

where

$$\bar{c}_{33} = c_{33} + e_{33}e_{33}/\varepsilon_{33}, \quad \bar{c}_{34} = c_{34} + e_{33}e_{34}/\varepsilon_{33},$$
$$\bar{c}_{35} = c_{35} + e_{33}e_{35}/\varepsilon_{33}, \quad \bar{c}_{44} = c_{44} + e_{34}e_{34}/\varepsilon_{33}, \tag{6.2.7}$$
$$\bar{c}_{45} = c_{45} + e_{34}e_{35}/\varepsilon_{33}, \quad \bar{c}_{55} = c_{55} + e_{35}e_{35}/\varepsilon_{33},$$

are the piezoelectrically stiffened elastic constants. Substitution of Eq. (6.2.6) into Eq. (6.2.1) yields the displacement equations of motion

$$\bar{c}_{55}u_{1,33} + \bar{c}_{54}u_{2,33} + \bar{c}_{53}u_{3,33} = \rho(\ddot{u}_1 - 2\Omega\dot{u}_2 - \Omega^2 u_1),$$
$$\bar{c}_{45}u_{1,33} + \bar{c}_{44}u_{2,33} + \bar{c}_{43}u_{3,33} = \rho(\ddot{u}_2 + 2\Omega\dot{u}_1 - \Omega^2 u_2), \tag{6.2.8}$$
$$\bar{c}_{35}u_{1,33} + \bar{c}_{34}u_{2,33} + \bar{c}_{33}u_{3,33} = \rho\ddot{u}_3.$$

Consider

$$u_i = A_i \exp[i(kx_3 + \omega t)], \tag{6.2.9}$$

where A_i are constants. Substitution of Eq. (6.2.9) into Eq. (6.2.8) results in the following homogeneous system of linear algebraic equations for A_i:

$$\begin{aligned}
&(\rho\omega^2 + \rho\Omega^2 - \bar{c}_{55}k^2)A_1 + (2i\rho\omega\Omega - \bar{c}_{45}k^2)A_2 - \bar{c}_{35}k^2 A_3 = 0,\\
&(-2i\rho\omega\Omega - \bar{c}_{45}k^2)A_1 + (\rho\omega^2 + \rho\Omega^2 - \bar{c}_{44}k^2)A_2 - \bar{c}_{34}k^2 A_3 = 0,\\
&-\bar{c}_{35}k^2 A_1 - \bar{c}_{34}k^2 A_2 + (\rho\omega^2 - \bar{c}_{33}k^2)A_3 = 0.
\end{aligned} \tag{6.2.10}$$

For non-trivial solutions the determinant of the coefficient matrix of the above system of linear equations must vanish, which leads to the following dispersion relation:

$$\begin{aligned}
&(\rho\omega^2 - \bar{c}_{33}k^2)(\rho\omega^2 - \bar{c}_{44}k^2)(\rho\omega^2 - \bar{c}_{55}k^2)\\
&\quad - 2\bar{c}_{34}\bar{c}_{35}\bar{c}_{45}k^6 - \bar{c}_{45}^2 k^4(\rho\omega^2 - \bar{c}_{33}k^2)\\
&\quad - \bar{c}_{35}^2 k^4(\rho\omega^2 - \bar{c}_{44}k^2) - \bar{c}_{34}^2 k^4(\rho\omega^2 - \bar{c}_{55}k^2)\\
&\quad = [(\rho\omega^2 - \bar{c}_{33}k^2)(2\rho\omega^2 + \bar{c}_{44}k^2 + \bar{c}_{55}k^2) + \bar{c}_{34}^2 k^4 + \bar{c}_{35}^2 k^4]\rho\Omega^2\\
&\quad -(\rho\omega^2 - \bar{c}_{33}k^2)\rho^2\Omega^4.
\end{aligned} \tag{6.2.11}$$

The terms on the right-hand side of Eq. (6.2.11) are due to the rotation of the body. When the body is not rotating, these terms are not present and Eq. (6.2.11) yields three real positive roots $\omega = \omega(k)$ or $k = k(\omega)$, which are linear and homogeneous functions; hence, the waves are not dispersive. The three roots represent a quasi-longitudinal wave for which $\bar{c}_{33}$ is the major elastic constant and two quasi-transverse waves for which $\bar{c}_{44}$ and $\bar{c}_{55}$ are important. When the body is rotating, for the quasi-longitudinal wave, the material particles move mainly in the x_3 direction and the associated Coriolis force is small. The two quasi-transverse waves are coupled by the Coriolis force, in addition to possible couplings by elastic constants. It can be expected that Eq. (6.2.11) will still have three real positive roots when Ω is small. However, because of the right-hand side terms in Eq. (6.2.11), the waves are clearly dispersive. The left-hand side of Eq. (6.2.11) is a polynomial of degree six, while the right-hand side is of degree four for ω and k. For very large ω, the right-hand side terms in Eq. (6.2.11) have little contribution, and the behavior of the dispersion relations is like those of a non-rotating body.

We can obtain an approximate expression of the dispersion relations when $\Omega << \omega$. Denoting the three wave speeds when the piezoelectric body is not rotating by v_1, v_2 and v_3, also denoting $v = \omega / k$, $v_{30}^2 = \bar{c}_{33} / \rho$, $v_{40}^2 = \bar{c}_{44} / \rho$, $v_{50}^2 = \bar{c}_{55} / \rho$ and $v_0^2 = (\bar{c}_{34} + \bar{c}_{535}) / \rho$, then Eq. (6.2.11) can be written as

$$
\begin{aligned}
&(v^2 - v_1^2)(v^2 - v_2^2)(v^2 - v_3^2) \\
&\quad \cong [(2v^2 + v_{40}^2 + v_{50}^2)(v^2 - v_{30}^2) + v_0^4] \frac{\Omega^2}{k^2},
\end{aligned}
\tag{6.2.12}
$$

where higher order terms of Ω are neglected. Since Ω is very small, the dispersion relations when the piezoelectric body is not rotating are perturbed by small quantities. For example, denoting $v^2 - v_1^2 = \Delta$ where Δ is a small quantity, substituting $v^2 = v_1^2 + \Delta$ into Eq. (6.2.12) and keeping the lowest order terms, we can determine Δ and obtain an approximation of the dispersion relation. Similarly, we can obtain approximations for the other two branches of the dispersion relations. The results are

$$
\begin{aligned}
\omega^2 &\approx v_1^2 k^2 + \Omega^2 \frac{(2v_1^2 + v_{40}^2 + v_{50}^2)(v_1^2 - v_{30}^2) + v_0^4}{(v_1^2 - v_2^2)(v_1^2 - v_3^2)}, \\
\omega^2 &\approx v_2^2 k^2 + \Omega^2 \frac{(2v_2^2 + v_{40}^2 + v_{50}^2)(v_2^2 - v_{30}^2) + v_0^4}{(v_2^2 - v_1^2)(v_2^2 - v_3^2)}, \\
\omega^2 &\approx v_3^2 k^2 + \Omega^2 \frac{(2v_3^2 + v_{40}^2 + v_{50}^2)(v_3^2 - v_{30}^2) + v_0^4}{(v_3^2 - v_1^2)(v_3^2 - v_2^2)},
\end{aligned}
\tag{6.2.13}
$$

which are clearly dispersive. Equation (6.2.13) represents hyperbolas with the dispersion curves of the non-rotating body as asymptotes.

For each specific root of Eq. (6.2.11), the normalized amplitude ratios can be obtained from Eq. (6.2.10) and are denoted by

$$
(A_1^{(i)} : A_2^{(i)} : A_3^{(i)}) = (\beta_1^{(i)} : \beta_2^{(i)} : \beta_3^{(i)}), \quad i = 1, 2, 3. \tag{6.2.14}
$$

Then, the following expressions are also solutions to Eq. (6.2.8):

$$u_j = \sum_{i=1}^{3} B^{(i)} \beta_j^{(i)} \exp[i(k^{(i)} x_3 + \omega t)],$$
$$\phi = \sum_{i=1}^{3} B^{(i)} \alpha^{(i)} \exp[i(k^{(i)} x_3 + \omega t)] + L_1 x_3 \exp(i\omega t) + L_2 \exp(i\omega t), \tag{6.2.15}$$

where $B^{(i)}$ are arbitrary constants, $k^{(i)}(\omega)$ is the ith of the three roots of Eq. (6.2.11), and

$$\alpha^{(i)} = \frac{e_{35}}{\varepsilon_{33}} \beta_1^{(i)} + \frac{e_{34}}{\varepsilon_{33}} \beta_2^{(i)} + \frac{e_{33}}{\varepsilon_{33}} \beta_3^{(i)} . \tag{6.2.16}$$

The integration constants in Eq. (6.2.15) differ from those in Eq. (6.2.5) in that the time dependence is explicitly written as harmonic. We note that there can also be waves with $\exp[i(-kx_3 + \omega t)]$ dependence.

For later applications to plates we want to consider stationary plane waves in the following form:

$$u_i = A_i \sin kx_3 \exp(i\omega t) . \tag{6.2.17}$$

Then, in a similar procedure, we can obtain the following solution to Eq. (6.2.8):

$$u_j = \sum_{i=1}^{3} B^{(i)} \beta_j^{(i)} \sin k^{(i)} x_3 \exp(i\omega t),$$
$$\phi = \sum_{i=1}^{3} B^{(i)} \alpha^{(i)} \sin k^{(i)} x_3 \exp(i\omega t) + L_1 x_3 \exp(i\omega t) + L_2 \exp(i\omega t). \tag{6.2.18}$$

The following stress and electric displacement components accompanying Eq. (6.2.18) will be useful later:

$$D_3 = -\varepsilon_{33} L_1 \exp(i\omega t),$$
$$T_{31} = \sum_{i=1}^{3} B^{(i)} \gamma_1^{(i)} k^{(i)} \cos k^{(i)} x_3 \exp(i\omega t) + e_{35} L_1 \exp(i\omega t),$$
$$T_{32} = \sum_{i=1}^{3} B^{(i)} \gamma_2^{(i)} k^{(i)} \cos k^{(i)} x_3 \exp(i\omega t) + e_{34} L_1 \exp(i\omega t), \tag{6.2.19}$$
$$T_{33} = \sum_{i=1}^{3} B^{(i)} \gamma_3^{(i)} k^{(i)} \cos k^{(i)} x_3 \exp(i\omega t) + e_{33} L_1 \exp(i\omega t),$$

where

$$\begin{aligned}\gamma_1^{(i)} &= \bar{c}_{55}\beta_1^{(i)} + \bar{c}_{45}\beta_2^{(i)} + \bar{c}_{35}\beta_3^{(i)},\\ \gamma_2^{(i)} &= \bar{c}_{45}\beta_1^{(i)} + \bar{c}_{44}\beta_2^{(i)} + \bar{c}_{34}\beta_3^{(i)},\\ \gamma_3^{(i)} &= \bar{c}_{35}\beta_1^{(i)} + \bar{c}_{34}\beta_2^{(i)} + \bar{c}_{33}\beta_3^{(i)}.\end{aligned} \tag{6.2.20}$$

Similar to Eq. (6.2.18), there exists another set of waves with

$$\begin{aligned}u_j &= \sum_{i=1}^{3} B^{(i)}\beta_j^{(i)} \cos k^{(i)} x_3 \exp(i\omega t),\\ \phi &= \sum_{i=1}^{3} B^{(i)}\alpha^{(i)} \cos k^{(i)} x_3 \exp(i\omega t) + L_1 x_3 \exp(i\omega t) + L_2 \exp(i\omega t),\end{aligned} \tag{6.2.21}$$

and

$$\begin{aligned}D_3 &= -\varepsilon_{33} L_1 \exp(i\omega t),\\ T_{31} &= \sum_{i=1}^{3} -B^{(i)}\gamma_1^{(i)} k^{(i)} \sin k^{(i)} x_3 \exp(i\omega t) + e_{35} L_1 \exp(i\omega t),\\ T_{32} &= \sum_{i=1}^{3} -B^{(i)}\gamma_2^{(i)} k^{(i)} \sin k^{(i)} x_3 \exp(i\omega t) + e_{34} L_1 \exp(i\omega t),\\ T_{33} &= \sum_{i=1}^{3} -B^{(i)}\gamma_3^{(i)} k^{(i)} \sin k^{(i)} x_3 \exp(i\omega t) + e_{33} L_1 \exp(i\omega t).\end{aligned} \tag{6.2.22}$$

6.2.2. Polarized ceramics

As an example, we consider a ceramic plate poled in the x_3 direction. Eqs. (6.2.5) through (6.2.8) reduce to the following two sets of uncoupled equations for longitudinal waves:

$$\begin{aligned}\bar{c}_{33} u_{3,33} &= \rho \ddot{u}_3,\\ T_{33} &= \bar{c}_{33} u_{3,3} + e_{33} L_1,\\ \phi &= \frac{e_{33}}{\varepsilon_{33}} u_3 + L_1 x_3 + L_2,\\ D_3 &= e_{33} u_{3,3} - \varepsilon_{33}\phi_{,3} = -\varepsilon_{33} L_1,\end{aligned} \tag{6.2.23}$$

and transverse waves:

$$\begin{aligned} c_{44}u_{1,33} &= \rho(\ddot{u}_1 - 2\Omega\dot{u}_2 - \Omega^2 u_1), \\ c_{44}u_{2,33} &= \rho(\ddot{u}_2 + 2\Omega\dot{u}_1 - \Omega^2 u_2), \\ T_{31} &= c_{44}u_{1,3}, \\ T_{32} &= c_{44}u_{2,3}. \end{aligned} \tag{6.2.24}$$

The dispersion relation in Eq. (6.2.11) now takes the following factored form

$$\begin{aligned} &(\rho\omega^2 - \bar{c}_{33}k^2)(\rho\omega^2 - c_{44}k^2)^2 \\ &\quad = (\rho\omega^2 - \bar{c}_{33}k^2)(2\rho\omega^2 - \rho\Omega^2 + 2c_{44}k^2)\rho\Omega^2. \end{aligned} \tag{6.2.25}$$

Equations (6.2.23) and (6.2.24) show that there exist two uncoupled waves. One is the longitudinal motion of u_3 and ϕ, the other is the transverse motion of u_1 and u_2. The common factor of both sides of Eq. (6.2.25) is the longitudinal branch of the dispersion relation

$$\omega = \sqrt{\frac{\bar{c}_{33}}{\rho}}\, k, \tag{6.2.26}$$

which can also be obtained from Eq. (6.2.23) directly. This branch is nondispersive. From Eq. (6.2.23) it can be seen that thickness-stretch is coupled to the electric field $E_3 = -\phi_{,3}$ but it is not affected by the rotation of the plate. The other two branches are for the transverse waves, which can be found from Eqs. (6.2.25) or (6.2.24) as

$$\omega = \sqrt{\frac{c_{44}}{\rho}}\, k \mp \Omega. \tag{6.2.27}$$

These two branches do not pass the origin of the (k,ω) plane and are therefore dispersive due to the rotation of the plate. The dispersion relations in Eqs. (6.2.26) and (6.2.27) can be written in the following dimensionless form:

$$\begin{aligned} &Y = \sqrt{\frac{\bar{c}_{33}}{c_{44}}}X, \quad Y = X - 1, \quad Y = X + 1, \\ &X = \sqrt{\frac{c_{44}}{\rho}}\frac{k}{\Omega}, \quad Y = \frac{\omega}{\Omega}. \end{aligned} \tag{6.2.28}$$

For PZT-5H, the dispersion relation in Eq. $(6.2.28)_1$ for longitudinal waves becomes $Y = 2.63X$. Dispersion relations from Eq. (6.2.28) are plotted in Fig. 6.2.1.

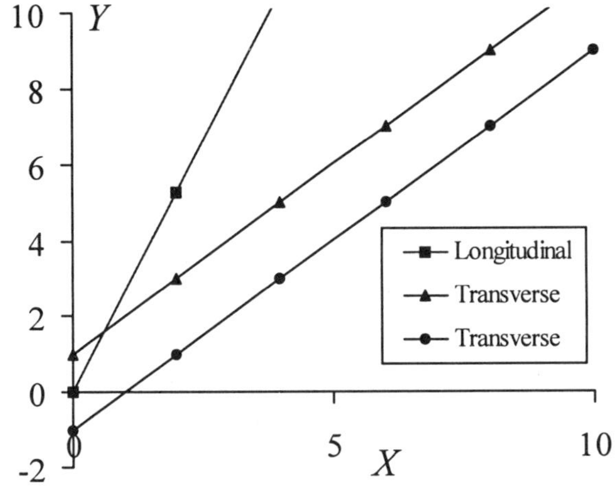

Fig. 6.2.1. Dispersion curves of plane waves in a rotating ceramic body from Eq. (6.2.28). $X = \sqrt{c_{44}/\rho}\, k/\Omega, \quad Y = \omega/\Omega$.

6.2.3. *Quartz*

Next we consider rotated Y-cut quartz as another example. In this case and in order to discuss the applications later, we orient the x_2-axis along the rotation axis. For plane waves propagating in the x_2 direction with $\partial_1 = \partial_3 = 0$, we have

$$\begin{aligned}
\bar{c}_{66}u_{1,22} &= \rho(\ddot{u}_1 + 2\Omega\dot{u}_3 - \Omega^2 u_1),\\
c_{22}u_{2,22} + c_{24}u_{3,22} &= \rho\ddot{u}_2,\\
c_{24}u_{2,22} + c_{44}u_{3,22} &= \rho(\ddot{u}_3 - 2\Omega\dot{u}_1 - \Omega^2 u_3),
\end{aligned} \qquad (6.2.29)$$

with

$$
\begin{aligned}
T_{21} &= \bar{c}_{66} u_{1,2} + e_{26} L_1, \\
T_{22} &= c_{22} u_{2,2} + c_{24} u_{3,2}, \\
T_{23} &= c_{24} u_{2,2} + c_{44} u_{3,2}, \\
D_2 &= -\varepsilon_{22} L_1, \quad \phi = \frac{e_{26}}{\varepsilon_{22}} u_1 + L_1 x_2 + L_2, \\
\bar{c}_{66} &= c_{66} + e_{26}^2 / \varepsilon_{22}.
\end{aligned}
\tag{6.2.30}
$$

For rotated Y-cut quartz there are two uncoupled groups of modes when the body is stationary. In the first group the shear motion u_1 is electrically coupled to ϕ. in the second group the shear motion u_3 is coupled to the longitudinal motion u_2, but they are not electrically coupled. Equation (6.2.29) shows that when the body is rotating, all these modes are coupled together because of the Coriolis force.

If a solution similar to Eq. (6.2.9) (depending on x_2 now, rather than x_3) is substituted into Eq. (6.2.29), we obtain

$$
\begin{aligned}
&(\rho\omega^2 + \rho\Omega^2 - \bar{c}_{66} k^2) A_1 - 2i\rho\omega\Omega A_3 = 0, \\
&(\rho\omega^2 - c_{22} k^2) A_2 - c_{24} k^2 A_3 = 0, \\
&2i\rho\omega\Omega A_1 - c_{24} k^2 A_2 + (\rho\omega^2 + \rho\Omega^2 - c_{44} k^2) A_3 = 0.
\end{aligned}
\tag{6.2.31}
$$

From Eq. (6.2.31) we obtain the dispersion relation as

$$
\begin{aligned}
&(\rho\omega^2 - \bar{c}_{66} k^2)[(\rho\omega^2 - c_{22} k^2)(\rho\omega^2 - c_{44} k^2) - c_{24}^2 k^4] \\
&\quad = (\rho\omega^2 - c_{22} k^2)(2\rho\omega^2 - \rho\Omega^2 + c_{44} k^2 + \bar{c}_{66} k^2)\rho\Omega^2.
\end{aligned}
\tag{6.2.32}
$$

For the large ω approximation, under small Ω, we have

$$
\begin{aligned}
\omega^2 &\approx v_1^2 k^2 + \Omega^2 \frac{(2v_1^2 + v_{40}^2 + v_{60}^2)(v_1^2 - v_{20}^2)}{(v_1^2 - v_2^2)(v_1^2 - v_3^2)}, \\
\omega^2 &\approx v_2^2 k^2 + \Omega^2 \frac{(2v_2^2 + v_{40}^2 + v_{60}^2)(v_2^2 - v_{20}^2)}{(v_2^2 - v_1^2)(v_2^2 - v_3^2)}, \\
\omega^2 &\approx v_3^2 k^2 + \Omega^2 \frac{(2v_3^2 + v_{40}^2 + v_{60}^2)(v_3^2 - v_{20}^2)}{(v_3^2 - v_1^2)(v_3^2 - v_2^2)},
\end{aligned}
\tag{6.2.33}
$$

where

$$
\begin{aligned}
v_1^2 &= \frac{\bar{c}_{66}}{\rho}, \\
v_2^2 &= \frac{1}{2\rho}\left[c_{22} + c_{44} + \sqrt{(c_{22} - c_{44})^2 + 4c_{24}^2}\,\right], \\
v_3^2 &= \frac{1}{2\rho}\left[c_{22} + c_{44} - \sqrt{(c_{22} - c_{44})^2 + 4c_{24}^2}\,\right], \\
v_{20}^2 &= \frac{c_{22}}{\rho}, \quad v_{40}^2 = \frac{c_{44}}{\rho}, \quad v_{60}^2 = \frac{\bar{c}_{66}}{\rho}.
\end{aligned}
\tag{6.2.34}
$$

Dispersion relations determined by Eq. (6.2.32) are obtained numerically and plotted as (a) in Fig. 6.2.2, along with their large ω approximation (b) from Eq. (6.2.33). For larger ω the approximation gets better.

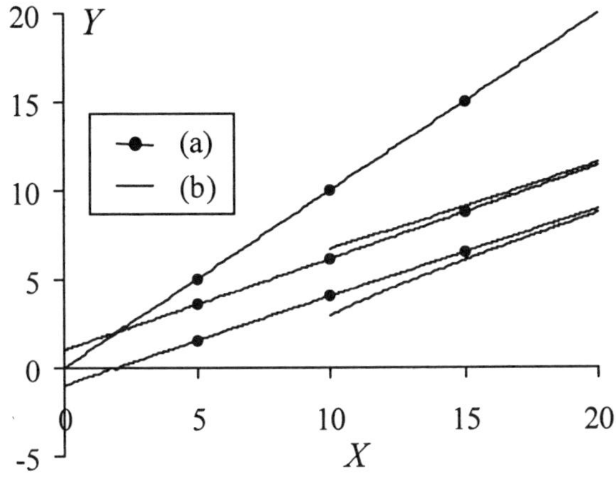

Fig. 6.2.2. Dispersion curves from Eq. (6.2.32) and their approximations from Eq. (6.2.33) for plane waves in a rotating quartz body. $X = \sqrt{c_{22}/\rho}\, k/\Omega, \quad Y = \omega/\Omega$.

Amplitude ratios can be obtained from Eq. (6.2.31) as

$$\begin{aligned}
&\beta_1^{(i)} = 2i\rho\omega\Omega/(\rho\omega^2 + \rho\Omega^2 - \bar{c}_{66}k^2), \\
&\beta_2^{(i)} = c_{24}k^2/(\rho\omega^2 - c_{22}k^2), \\
&\beta_3^{(i)} = 1.
\end{aligned} \tag{6.2.35}$$

In Eq. (6.2.35) the superscript i means that the ith of the three dispersion relations $k = k^{(i)}(\omega)$ determined from Eq. (6.2.32) is to be substituted into the right hand side of Eq. (6.2.35). These amplitude ratios are normalized with $\beta_3^{(i)}$ being equal to one. We then have the following plane stationary wave solution to Eq. (6.2.29):

$$\begin{aligned}
u_j &= \sum_{i=1}^{3} B^{(i)}\beta_j^{(i)} \sin k^{(i)}x_2 \exp(i\omega t), \\
\phi &= \sum_{i=1}^{3} B^{(i)}\alpha^{(i)} \sin k^{(i)}x_2 \exp(i\omega t) \\
&\quad + L_1 x_2 \exp(i\omega t) + L_2 \exp(i\omega t), \\
\alpha^{(i)} &= \frac{e_{26}}{\varepsilon_{22}}\beta_1^{(i)},
\end{aligned} \tag{6.2.36}$$

with

$$\begin{aligned}
D_2 &= -\varepsilon_{22}L_1 \exp(i\omega t), \\
T_{21} &= \sum_{i=1}^{3} B^{(i)}\gamma_1^{(i)}k^{(i)} \cos k^{(i)}x_2 \exp(i\omega t) + e_{26}L_1 \exp(i\omega t), \\
T_{22} &= \sum_{i=1}^{3} B^{(i)}\gamma_2^{(i)}k^{(i)} \cos k^{(i)}x_2 \exp(i\omega t), \\
T_{23} &= \sum_{i=1}^{3} B^{(i)}\gamma_3^{(i)}k^{(i)} \cos k^{(i)}x_2 \exp(i\omega t),
\end{aligned} \tag{6.2.37}$$

and

$$\begin{aligned}
&\gamma_1^{(i)} = \bar{c}_{66}\beta_1^{(i)}, \quad \gamma_2^{(i)} = c_{22}\beta_2^{(i)} + c_{24}\beta_3^{(i)}, \\
&\gamma_3^{(i)} = c_{24}\beta_2^{(i)} + c_{44}\beta_3^{(i)}.
\end{aligned} \tag{6.2.38}$$

Similarly, there also exists

$$u_j = \sum_{i=1}^{3} B^{(i)} \beta_j^{(i)} \cos k^{(i)} x_2 \exp(i\omega t),$$

$$\phi = \sum_{i=1}^{3} B^{(i)} \alpha^{(i)} \cos k^{(i)} x_2 \exp(i\omega t) \qquad (6.2.39)$$

$$+ L_1 x_2 \exp(i\omega t) + L_2 \exp(i\omega t),$$

with

$$D_2 = -\varepsilon_{22} L_1 \exp(i\omega t),$$

$$T_{21} = \sum_{i=1}^{3} - B^{(i)} \gamma_1^{(i)} k^{(i)} \sin k^{(i)} x_2 \exp(i\omega t)$$

$$+ e_{26} L_1 \exp(i\omega t), \qquad (6.2.40)$$

$$T_{22} = \sum_{i=1}^{3} - B^{(i)} \gamma_2^{(i)} k^{(i)} \sin k^{(i)} x_2 \exp(i\omega t),$$

$$T_{23} = \sum_{i=1}^{3} - B^{(i)} \gamma_3^{(i)} k^{(i)} \sin k^{(i)} x_2 \exp(i\omega t).$$

6.3. Thickness Vibrations of Plates

We now study the free vibration of a piezoelectric plate bounded by two major surfaces at $x_3 = \pm h$. The plate is rotating about its normal x_3 (see Fig. 6.3.1) [73,74]. The mechanical boundary conditions at the major surfaces are traction-free with $T_{3j}(\pm h) = 0$.

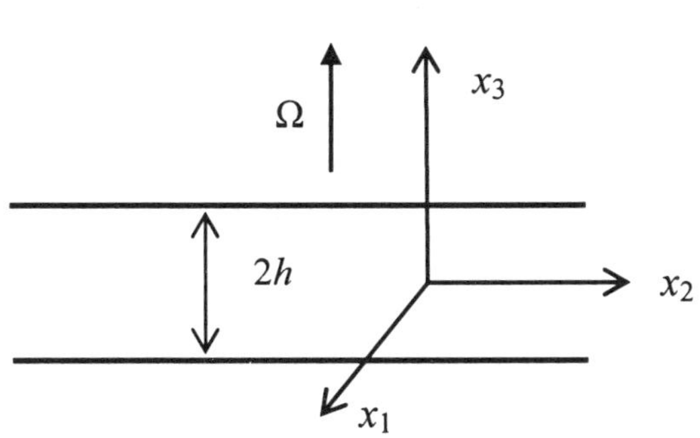

Fig. 6.3.1. A plate rotating about its normal.

6.3.1. General solution

First consider materials of general anisotropy.

6.3.1.1. Unelectroded plates or plates with open electrodes

For electrical boundary conditions we first study the case when the two major surfaces of the plate are unelectroded or with open electrodes. In both of these two cases the electrical boundary condition requires $D_3(\pm h)=0$ which implies $L_1=0$ from Eq. (6.2.4). The modes from Eq. (6.2.18) may be called anti-symmetric modes. It can be seen from Eq. (6.2.19) that when $L_1=0$ the traction-free boundary conditions require that

$$\sum_{i=1}^{3} B^{(i)}\gamma_1^{(i)}k^{(i)}\cos k^{(i)}h=0,$$

$$\sum_{i=1}^{3} B^{(i)}\gamma_2^{(i)}k^{(i)}\cos k^{(i)}h=0, \tag{6.3.1}$$

$$\sum_{i=1}^{3} B^{(i)}\gamma_3^{(i)}k^{(i)}\cos k^{(i)}h=0.$$

For non-trivial solutions of $B^{(i)}$ the determinant of the coefficient matrix must vanish which is equivalent to

$$\cos k^{(1)}h\cos k^{(2)}h\cos k^{(3)}h=0\,, \tag{6.3.2}$$

or

$$k_{(n)}^{(i)}h=\frac{n\pi}{2},\qquad n=1,3,5,\ldots, \tag{6.3.3}$$

which is the frequency equation for the anti-symmetric modes.

Similarly, for symmetric modes, from Eq. (6.2.22), we have

$$\sin k^{(1)}h\sin k^{(2)}h\sin k^{(3)}h=0,$$

$$k_{(n)}^{(i)}h=\frac{n\pi}{2},\qquad n=2,4,6,\ldots. \tag{6.3.4}$$

6.3.1.2. Plates with shorted electrodes

We now consider the case when the two major surfaces of the plate are electroded and shorted, with $\phi(h)=\phi(-h)$. For anti-symmetric modes, from Eqs. $(6.2.18)_2$ and $(6.2.19)_{2\text{-}4}$ we have

$$\begin{aligned}&\sum_{i=1}^{3} B^{(i)}\gamma_1^{(i)}k^{(i)}\cos k^{(i)}h + e_{35}L_1 = 0,\\&\sum_{i=1}^{3} B^{(i)}\gamma_2^{(i)}k^{(i)}\cos k^{(i)}h + e_{34}L_1 = 0,\\&\sum_{i=1}^{3} B^{(i)}\gamma_3^{(i)}k^{(i)}\cos k^{(i)}h + e_{33}L_1 = 0,\\&\sum_{i=1}^{3} B^{(i)}\alpha^{(i)}\sin k^{(i)}h + L_1 h = 0.\end{aligned} \tag{6.3.5}$$

For non-trivial solutions of $B^{(i)}$ and/or L_1 we must have

$$\begin{vmatrix}\gamma_1^{(1)}k^{(1)}\cos k^{(1)}h & \gamma_1^{(2)}k^{(2)}\cos k^{(2)}h & \gamma_1^{(3)}k^{(3)}\cos k^{(3)}h & e_{35}\\ \gamma_2^{(1)}k^{(1)}\cos k^{(1)}h & \gamma_2^{(2)}k^{(2)}\cos k^{(2)}h & \gamma_2^{(3)}k^{(3)}\cos k^{(3)}h & e_{34}\\ \gamma_3^{(1)}k^{(1)}\cos k^{(1)}h & \gamma_3^{(2)}k^{(2)}\cos k^{(2)}h & \gamma_3^{(3)}k^{(3)}\cos k^{(3)}h & e_{33}\\ \alpha^{(1)}\sin k^{(1)}h & \alpha^{(2)}\sin k^{(2)}h & \alpha^{(3)}\sin k^{(3)}h & h\end{vmatrix} = 0, \tag{6.3.6}$$

which is the frequency equation for the anti-symmetric modes.

For symmetric modes, from Eqs. (6.2.21) and (6.2.22), the frequency equation is found to be exactly the same as Eq. (6.3.4).

6.3.2. Ceramic plates

We now consider ceramic plates poled in the thickness direction x_3 as an example.

6.3.2.1. Thickness-shear modes

For thickness-shear vibrations, corresponding to the two dispersion relations in Eq. (6.2.27), the following two sets of anti-symmetric thickness-shear modes can be found from Eq. (6.2.24):

$$u_1 = \sin kx_3 \sin \omega t, \\ u_2 = -\sin kx_3 \cos \omega t, \tag{6.3.7}$$

and

$$u_1 = \sin kx_3 \sin \omega t, \\ u_2 = \sin kx_3 \cos \omega t, \tag{6.3.8}$$

with the frequency equation

$$\cos kh = 0, \qquad k_{(n)} h = \frac{n\pi}{2}, \qquad n = 1,3,5,\ldots. \tag{6.3.9}$$

The frequencies of the two modes given by Eqs. (6.3.7) and (6.3.8) are determined from the two dispersion relations in Eq. (6.2.27) respectively. Hence, the two modes have different frequencies for the same k

$$\omega_{(n)} = \omega_n \mp \Omega, \quad n = 1,3,5,\ldots, \\ \omega_n = \sqrt{\frac{c_{44}}{\rho}} \frac{\pi}{2h} n, \tag{6.3.10}$$

where ω_n are the thickness-shear resonant frequencies of a non-rotating plate. Frequencies predicted by Eq. (6.3.10) are plotted in Fig. 6.3.2 for small values of Ω and n = 1, where $\omega_{(1)}$ and Ω are normalized by the fundamental thickness-shear frequency ω_1 of a non-rotating plate. The figure shows that Ω causes the two shear frequencies to split. The Ω dependence of the shear mode frequencies can be used to detect angular rate. From the figure or Eq. (6.3.10) it is seen that the rotation effect on the thickness-shear frequencies is linear, which is desirable for make angular rate sensors.

For a non-rotating plate, the two thickness-shear modes in the x_1 and x_2 directions are not coupled and material particles move in straight lines. For a rotating plate, shears in both directions are coupled by the Coriolis force. Material particles move in planar circles with radii depending on x_3 (see Fig. 6.3.3). The middle plane of the plate is also the nodal plane of this shear mode, where material particles do not move. For the modes described by Eqs. (6.3.7) and (6.3.8), the material particles are moving in opposite directions, counterclockwise in one mode and clockwise in the other when viewed from the x_3 direction.

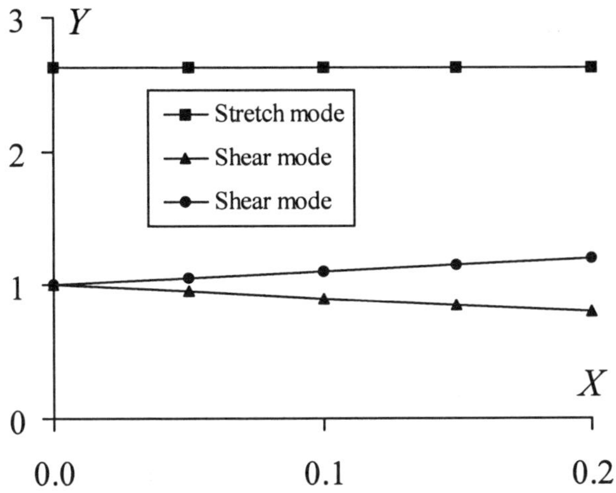

Fig. 6.3.2. Lowest thickness vibration frequencies of a rotating ceramic plate. $X = \Omega/\omega_1$, $Y = \omega/\omega_1$.

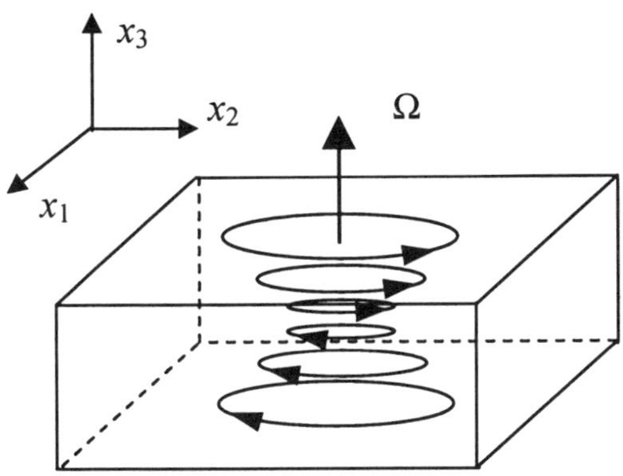

Fig. 6.3.3. Mode shape of a rotating ceramic plate in thickness-shear vibration.

Similarly to Eqs. (6.3.7) and (6.3.8), there exist two sets of symmetric modes

$$\begin{aligned} u_1 &= \cos kx_3 \sin \omega t, \\ u_2 &= -\cos kx_3 \cos \omega t, \end{aligned} \tag{6.3.11}$$

and

$$\begin{aligned} u_1 &= \cos kx_3 \sin \omega t, \\ u_2 &= \cos kx_3 \cos \omega t, \end{aligned} \tag{6.3.12}$$

with the frequency equation

$$\sin kh = 0, \qquad k_{(n)}h = \frac{n\pi}{2}, \qquad n = 2, 4, 6, \ldots. \tag{6.3.13}$$

The frequencies have the same expression as Eq. (6.3.10) except that n now assumes even numbers.

6.3.2.2. Thickness-stretch modes

For a ceramic plate, from Eq. (6.2.23), it can be seen that the thickness-stretch modes are electrically coupled but are not affected by the rotation of the plate. Material particles are in linear motion in the x_3 direction. For thoroughness we summarize the results below.

When the plate is unelectroded or with open electrodes, we have the following anti-symmetric and symmetric modes:

$$u_3 = \sin kx_3 \exp(i\omega t), \qquad k_{(n)}h = \frac{n\pi}{2}, \qquad n = 1, 3, 5, \ldots, \tag{6.3.14}$$

and

$$u_3 = \cos kx_3 \exp(i\omega t), \qquad k_{(n)}h = \frac{n\pi}{2}, \qquad n = 2, 4, 6, \ldots. \tag{6.3.15}$$

When the major surfaces of the plate are electroded and shorted, we have, for the anti-symmetric modes

$$u_3 = \sin kx_3 \exp(i\omega t), \quad \tan kh = \frac{e_{33}^2}{\varepsilon_{33}\bar{c}_{33}} kh, \tag{6.3.16}$$

and for the symmetric modes

$$u_3 = \cos kx_3 \exp(i\omega t), \qquad k_{(n)}h = \frac{n\pi}{2}, \qquad n = 2, 4, 6, \ldots. \tag{6.3.17}$$

The lowest thickness-stretch frequency is plotted in Fig. 6.3.2 and is unaffected by rotation, remaining a horizontal straight line.

6.3.3. Quartz plates

Next we consider rotated Y-cut quartz plates. For unelectroded plates or plates with open electrode, the frequency equations are exactly the same as Eqs. (6.3.2) and (6.3.4).

For a plate with shorted electrodes, the symmetric modes have a frequency equation the same as Eq. (6.3.4). For anti-symmetric modes we have the following frequency equation:

$$\begin{vmatrix} \gamma_1^{(1)}k^{(1)}\cos k^{(1)}h & \gamma_1^{(2)}k^{(2)}\cos k^{(2)}h & \gamma_1^{(3)}k^{(3)}\cos k^{(3)}h & e_{26} \\ \gamma_2^{(1)}k^{(1)}\cos k^{(1)}h & \gamma_2^{(2)}k^{(2)}\cos k^{(2)}h & \gamma_2^{(3)}k^{(3)}\cos k^{(3)}h & 0 \\ \gamma_3^{(1)}k^{(1)}\cos k^{(1)}h & \gamma_3^{(2)}k^{(2)}\cos k^{(2)}h & \gamma_3^{(3)}k^{(3)}\cos k^{(3)}h & 0 \\ \alpha^{(1)}\sin k^{(1)}h & \alpha^{(2)}\sin k^{(2)}h & \alpha^{(3)}\sin k^{(3)}h & h \end{vmatrix} = 0\,. \quad (6.3.18)$$

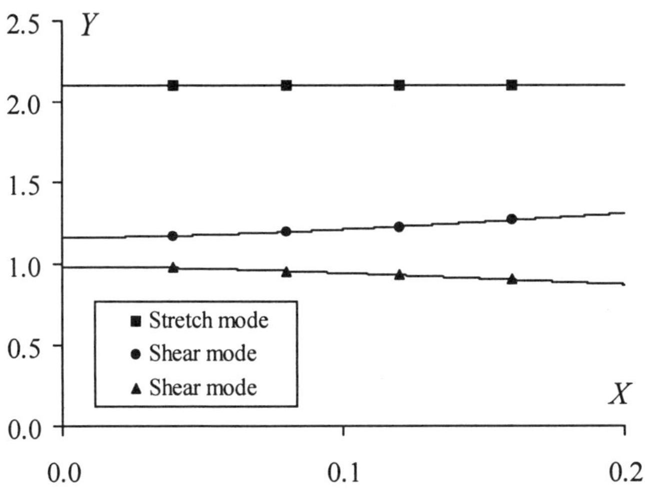

Fig. 6.3.4. Lowest thickness vibration frequencies of a rotating quartz plate. $X = \Omega/\omega_1$, $Y = \omega/\omega_1$.

Equation (6.3.18) is solved numerically with results plotted in Fig. 6.3.4 for the lowest thickness-shear and thickness-stretch frequencies. The frequencies are normalized by the fundamental thickness-shear frequency in the x_1 direction of a non-rotating plate

$$\omega_1 = \sqrt{\frac{\bar{c}_{66}\pi^2}{4\rho h^2}}, \quad \bar{c}_{66} = c_{66} + \frac{e_{26}^2}{\varepsilon_{22}}. \tag{6.3.19}$$

In the figure, the essentially thickness-stretch mode is affected little by rotation and its frequency remains an almost horizontal line. The other two are essentially thickness-shear modes which are clearly affected by the angular rate. There is a fundamental difference between Figs. 6.3.2 and 6.3.4. The dependence of the shear frequencies on the angular rate is linear in Fig. 6.3.2 but it has an essentially flat tangent near $X = 0$ in Fig. 6.3.4. The linear dependence in Fig. 6.3.2 is related to the transverse isotropy of polarized ceramics. The behavior in (6.3.4) is not suitable for making angular rate sensors.

6.4. Propagating Waves in a Rotating Piezoelectric Plate

In this section we analyze transversely varying thickness modes in a rotating piezoelectric plate [75]. Consider an unbounded piezoelectric plate of thickness $2h$ rotating at a constant angular rate Ω about its normal as shown in Fig. 6.4.1.

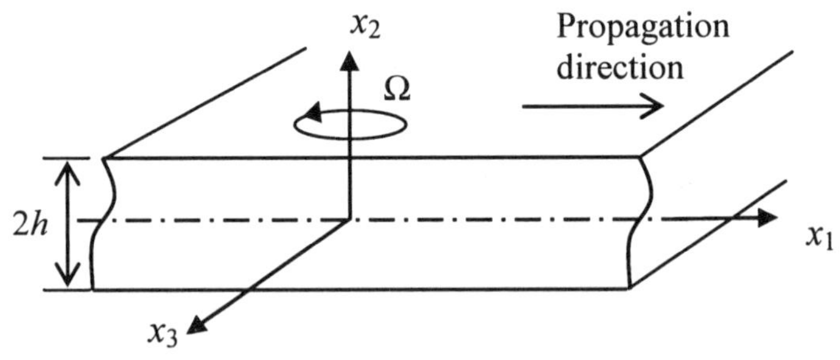

Fig. 6.4.1. A piezoelectric plate rotating about its normal.

6.4.1. General solution

Motions of the body are governed by Eqs. (6.1.1) and (6.1.2):

$$\begin{aligned} &c_{ijkl}u_{k,jl} + e_{kij}\phi_{,jk} = \rho[\ddot{u}_i + 2\varepsilon_{ijk}\Omega_j\dot{u}_k + (\Omega_i\Omega_j u_j - \Omega_j\Omega_j u_i)], \\ &e_{ijk}u_{j,ik} - \varepsilon_{ij}\phi_{,ij} = 0, \end{aligned} \tag{6.4.1}$$

where $(\Omega_1,\Omega_2,\Omega_3) = (0,\Omega,0)$. Over the major surfaces of the plate ($x_2 = \pm h$), the following traction-free mechanical boundary conditions are always assumed:

$$T_{2j}(\pm h) = [c_{2jkl}u_{k,l} + e_{k2j}\phi_{,k}]_{x_2=\pm h} = 0. \tag{6.4.2}$$

The electrical boundary conditions on the plate surfaces are a pair of shorted and grounded electrodes for which

$$\phi(h) = \phi(-h) = 0\ , \tag{6.4.3}$$

or a pair of unelectroded and charge-free surfaces for which

$$D_2(\pm h) = [e_{2jk}u_{j,k} - \varepsilon_{2j}\phi_{,j}]_{x_2=\pm h} = 0. \tag{6.4.4}$$

We consider straight-crested waves propagating in the x_1 direction with $\partial/\partial x_3 = 0$. Then Eq. (6.4.1) becomes

$$\begin{aligned} &c_{11}u_{1,11} + c_{12}u_{2,21} + c_{14}u_{3,21} + c_{15}u_{3,11} + c_{16}(u_{1,21} + u_{2,11}) \\ &\quad + c_{16}u_{1,12} + c_{26}u_{2,22} + c_{46}u_{3,22} + c_{56}u_{3,12} + c_{66}(u_{1,22} + u_{2,12}) \\ &\quad + e_{11}\phi_{,11} + e_{21}\phi_{,21} + e_{16}\phi_{,12} + e_{26}\phi_{,22} = \rho\ddot{u}_1 + 2\rho\Omega\dot{u}_3 - \rho\Omega^2 u_1, \\ &c_{16}u_{1,11} + c_{26}u_{2,21} + c_{46}u_{3,21} + c_{56}u_{3,11} + c_{66}(u_{1,21} + u_{2,11}) \\ &\quad + c_{12}u_{1,12} + c_{22}u_{2,22} + c_{24}u_{3,22} + c_{25}u_{3,12} + c_{26}(u_{1,22} + u_{2,12}) \\ &\quad + e_{16}\phi_{,11} + e_{26}\phi_{,21} + e_{12}\phi_{,12} + e_{22}\phi_{,22} = \rho\ddot{u}_2, \\ &c_{15}u_{1,11} + c_{25}u_{2,21} + c_{45}u_{3,21} + c_{55}u_{3,11} + c_{56}(u_{1,21} + u_{2,11}) \\ &\quad + c_{14}u_{1,12} + c_{24}u_{2,22} + c_{44}u_{3,22} + c_{45}u_{3,12} + c_{46}(u_{1,22} + u_{2,12}) \\ &\quad + e_{15}\phi_{,11} + e_{25}\phi_{,21} + e_{14}\phi_{,12} + e_{24}\phi_{,22} = \rho\ddot{u}_3 - 2\rho\Omega\dot{u}_1 - \Omega^2 u_3, \\ &e_{11}u_{1,11} + e_{12}u_{2,21} + e_{14}u_{3,21} + e_{15}u_{3,11} + e_{16}(u_{1,21} + u_{2,11}) \\ &\quad + e_{21}u_{1,12} + e_{22}u_{2,22} + e_{24}u_{3,22} + e_{25}u_{3,12} + e_{26}(u_{1,22} + u_{2,12}) \\ &\quad - \varepsilon_{11}\phi_{,11} - \varepsilon_{12}\phi_{,21} - \varepsilon_{12}\phi_{,12} - \varepsilon_{22}\phi_{,22} = 0. \end{aligned} \tag{6.4.5}$$

We seek solutions in the following form:

$$\begin{aligned} &u_j(\mathbf{x},t) = A_j \exp(k\eta x_2)\exp[i(kx_1 - \omega t)], \\ &\phi(\mathbf{x},t) = A_4 \exp(k\eta x_2)\exp[i(kx_1 - \omega t)], \end{aligned} \tag{6.4.6}$$

where k and ω are the wave number in the x_1 direction and the time frequency, respectively. η is related to the wave number in the x_2 direction. A_j (j = 1, 2, 3) and A_4 are complex constants, representing the wave amplitude. Substitution of Eq. (6.4.6) into Eq. (6.4.5) leads to the following four linear algebraic equations for A_j and A_4:

$$\begin{aligned}
&[\rho(\omega^2+\Omega^2)+k^2(\eta^2 c_{66}+i2\eta c_{16}-c_{11})]A_1 \\
&\quad +k^2(\eta^2 c_{26}+i\eta c_{66}+i\eta c_{12}-c_{16})A_2 \\
&\quad +[\rho i 2\Omega\omega+k^2(\eta^2 c_{46}+i\eta c_{56}+i\eta c_{14}-c_{15})]A_3 \\
&\quad +k^2(\eta^2 e_{26}+i\eta e_{21}+i\eta e_{16}-e_{11})A_4=0, \\
&k^2(\eta^2 c_{26}+i\eta c_{21}+i\eta c_{66}-c_{16})A_1 \\
&\quad +[\rho\omega^2+k^2(\eta^2 c_{22}+i2\eta c_{26}-c_{66})]A_2 \\
&\quad +k^2(\eta^2 c_{24}+i\eta c_{25}+i\eta c_{46}-c_{56})A_3 \\
&\quad +k^2(\eta^2 e_{22}+i\eta e_{26}+i\eta e_{12}-e_{16})A_4=0, \\
&[-\rho i 2\Omega\omega+k^2(\eta^2 c_{46}+i\eta c_{14}+i\eta c_{56}-c_{15})]A_1 \\
&\quad +k^2(\eta^2 c_{24}+i\eta c_{46}+i\eta c_{25}-c_{56})A_2 \\
&\quad +[\rho(\omega^2+\Omega^2)+k^2(\eta^2 c_{44}+i2\eta c_{45}-c_{55})]A_3 \\
&\quad +k^2(\eta^2 e_{24}+i\eta e_{25}+i\eta e_{14}-e_{15})A_4=0, \\
&(\eta^2 e_{26}+i\eta e_{21}+i\eta e_{16}-e_{11})A_1+(\eta^2 e_{22}+i\eta e_{26}+i\eta e_{12}-e_{16})A_2 \\
&\quad +(\eta^2 e_{24}+i\eta e_{25}+i\eta e_{14}-e_{15})A_3-(\eta^2\varepsilon_{22}+i2\eta\varepsilon_{12}-\varepsilon_{11})A_4=0.
\end{aligned} \tag{6.4.7}$$

For non-trivial solutions of A_j and/or A_4, the determinant of the coefficient matrix of the above equations must vanish. This leads to a polynomial equation of degree eight for η. We denote the eight roots of this equation by $\eta_{(m)}$, and the corresponding eigenvectors by ($\overline{A}_j^{(m)}, \overline{A}_4^{(m)}$), m = 1, 2, ..., 8. Thus, the general wave solution to Eq. (6.4.5) in the form of Eq. (6.4.6) can be written as

$$\begin{aligned}
u_i &= \sum_{m=1}^{8} C_{(m)}\overline{A}_i^{(m)}\exp(k\eta_{(m)}x_2)\exp[i(kx_1-\omega t)], \\
\phi &= \sum_{m=1}^{8} C_{(m)}\overline{A}_4^{(m)}\exp(k\eta_{(m)}x_2)\exp[i(kx_1-\omega t)],
\end{aligned} \tag{6.4.8}$$

where $C_{(m)}$ are eight undetermined constants. Substituting Eq. (6.4.8) into the boundary conditions in Eqs. (6.4.2) and (6.4.4) yields the following eight linear algebraic equations for $C_{(m)}$:

$$\begin{aligned}
&\sum_{m=1}^{8}(c_{12}i\overline{A}_1^{(m)} + c_{22}\eta_{(m)}\overline{A}_2^{(m)} + c_{24}\eta_{(m)}\overline{A}_3^{(m)} \\
&\quad + c_{25}i\overline{A}_3^{(m)} + c_{26}\eta_{(m)}\overline{A}_1^{(m)} + c_{26}i\overline{A}_2^{(m)} \\
&\quad + e_{12}i\overline{A}_4^{(m)} + e_{22}\eta_{(m)}\overline{A}_4^{(m)})\exp(\pm k\eta_{(m)}h)C_{(m)} = 0, \\
&\sum_{m=1}^{8}(c_{16}i\overline{A}_1^{(m)} + c_{26}\eta_{(m)}\overline{A}_2^{(m)} + c_{46}\eta_{(m)}\overline{A}_3^{(m)} \\
&\quad + c_{56}i\overline{A}_3^{(m)} + c_{66}\eta_{(m)}\overline{A}_1^{(m)} + c_{66}i\overline{A}_2^{(m)} \\
&\quad + e_{16}i\overline{A}_4^{(m)} + e_{26}\eta_{(m)}\overline{A}_4^{(m)})\exp(\pm k\eta_{(m)}h)C_{(m)} = 0, \\
&\sum_{m=1}^{8}(c_{14}i\overline{A}_1^{(m)} + c_{24}\eta_{(m)}\overline{A}_2^{(m)} + c_{44}\eta_{(m)}\overline{A}_3^{(m)} \\
&\quad + c_{45}i\overline{A}_3^{(m)} + c_{46}\eta_{(m)}\overline{A}_1^{(m)} + c_{46}i\overline{A}_2^{(m)} \\
&\quad + e_{14}i\overline{A}_4^{(m)} + e_{24}\eta_{(m)}\overline{A}_4^{(m)})\exp(\pm k\eta_{(m)}h)C_{(m)} = 0, \\
&\sum_{m=1}^{8}(e_{21}i\overline{A}_1^{(m)} + e_{22}\eta_{(m)}\overline{A}_2^{(m)} + e_{24}\eta_{(m)}\overline{A}_3^{(m)} \\
&\quad + e_{25}i\overline{A}_3^{(m)} + e_{26}\eta_{(m)}\overline{A}_1^{(m)} + e_{26}i\overline{A}_2^{(m)} \\
&\quad - \varepsilon_{12}i\overline{A}_4^{(m)} - \varepsilon_{22}\eta_{(m)}\overline{A}_4^{(m)})\exp(\pm k\eta_{(m)}h)C_{(m)} = 0.
\end{aligned} \tag{6.4.9}$$

If Eq. (6.4.3) is used instead of Eq. (6.4.4), Eq. $(6.4.9)_4$ should be replaced by

$$\sum_{j=1}^{8}\overline{A}_4^{(m)}\exp(\pm k\eta_{(m)}h)C_{(m)} = 0\,. \tag{6.4.10}$$

For non-trivial solutions of $C_{(m)}$, the determinant of the coefficient matrix of Eq. (6.4.9) has to vanish. This yields the frequency equation that determines the wave speed. The eigenvalue problem for ω represented by Eq. (6.4.9) is for an unbounded plate and, hence, is expected to have a continuous spectrum. Any value of ω is an eigenvalue with corresponding waves of certain wavelengths.

The eigenvalue problem in Eq. (6.4.9) is solved using an iterative numerical procedure. For a given real value of k, we choose a real trial value of ω. With this pair of tentative values of ω and k, we determine the eight roots $\eta_{(m)}$ of the polynomial equation of η, and the corresponding ($\overline{A}_j^{(m)}$, $\overline{A}_4^{(m)}$) from Eq. (6.4.7). Substituting the so-determined $\eta_{(m)}$ and ($\overline{A}_j^{(m)}$, $\overline{A}_4^{(m)}$) into the determinant of the coefficient matrix of Eq. (6.4.9) results in a formally complex determinant or, equivalently, two real determinants. Numerical results show that one of the two real determinants is always much smaller than the other, and is so small that it cannot be well captured within the machine precision and, thus, it is treated as zero numerically. If the above trial ω cannot make the remaining determinant vanish, we choose another trial value of ω to pair with the same k and repeat the above procedure until the remaining determinant vanishes. We are considering ideal materials without damping and, consequently, the waves propagate without attenuation. For lossy materials, a complex wave number is needed to satisfy the complex frequency equation because of wave attenuation. The so-determined ω together with the given k defines a point on the dispersion curve in the k-ω plane. Through this procedure, we obtain a dispersion relation in the form of $\omega = \omega(k)$. This dispersion relation is expected to have many branches, analogous to the waves propagating in a non-rotating plate (see Fig. 3.1.2).

6.4.2. Ceramic plates

As a numerical example, consider plates of polarized ceramics PZT-5H. The rotation-induced frequency shifts of the lowest thickness-shear and thickness-twist waves are plotted in Fig. 6.4.2 for a ceramic plate poled along the x_2 direction with shorted electrodes, and in Fig. 6.4.3 for the corresponding unelectroded plate, respectively.

The numerical results are plotted for $0.1 < kh/2\pi < 1$, because long waves with $\lambda = 2\pi/k >> h$ or $kh/2\pi << 1$ are usually used for devices working with thickness-shear modes. The frequency shift is defined by $\Delta\omega = \omega(\Omega) - \omega(0)$. In these figures, the relative frequency shifts $|\Delta\omega/\omega|$ are plotted with plus and minus signs indicating positive and negative frequency shifts, respectively. Figures 6.4.2 and 6.4.3 show that the

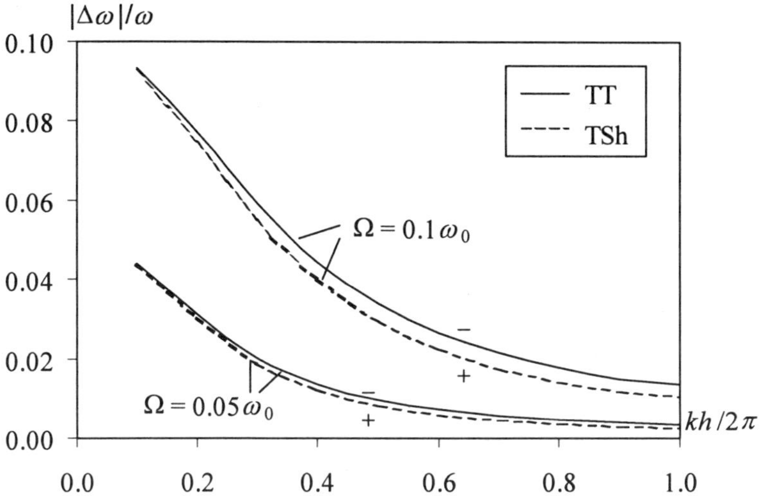

Fig. 6.4.2. Frequency shifts for the lowest thickness-shear and thickness-twist waves in an electroded plate poled along its normal direction.

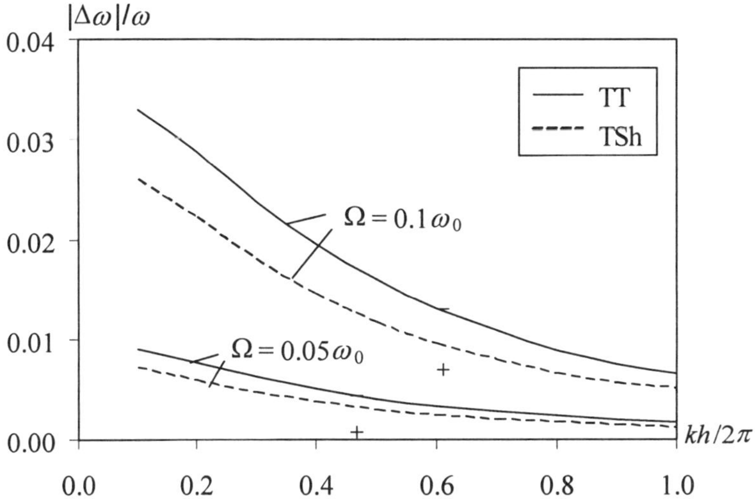

Fig. 6.4.3. Frequency shifts for the lowest thickness-shear and thickness-twist waves in an unelectroded plate poled along its normal direction.

frequency shifts are relatively large for small wave numbers corresponding to long waves, and the frequency shifts diminish as the wave number k becomes increasingly large. Figures 6.4.2 and 6.4.3 also show that the electroded plate has larger frequency shifts than the unelectroded plate. Numerical calculations show that this is also true for ceramic plates poled along the x_1-axis, or the x_3-axis, or along a direction bisecting the angle between the x_1- and x_3-axes. The corresponding numerical results are not shown because there are no characteristic differences from the two figures shown.

For unelectroded plates poled along the x_3- axis, the frequency shifts do not decrease monotonically with an increasing wave number. Instead, the frequency shifts of these modes initially rise with the wave number, reaching a maximum before they diminish as the wave number becomes increasingly large, as shown in Fig. 6.4.4.

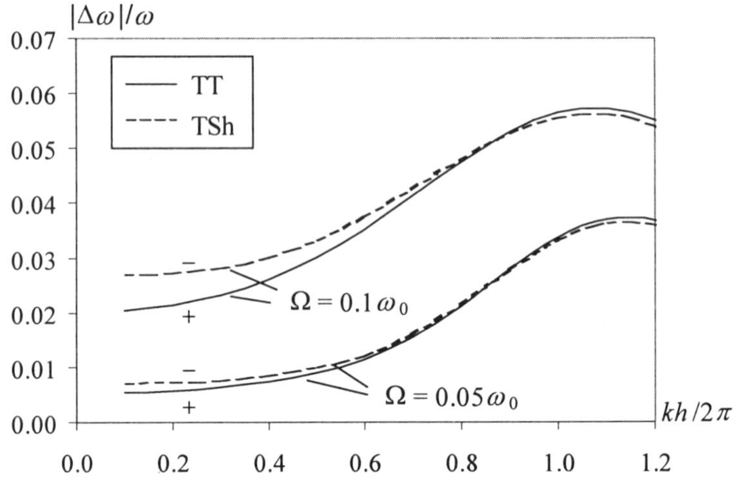

Fig. 6.4.4. Frequency shifts for the lowest thickness-shear and thickness-twist waves in an unelectroded plate poled long the x_3 direction.

Figures 6.4.5 and 6.4.6 show for electroded and unelectroded plates, respectively, the frequency shifts of the lowest thickness-shear and thickness-twist waves versus the rotation rate, for a fixed wave number $kh/2\pi = 0.1$, representing reasonably long waves. The frequency shifts

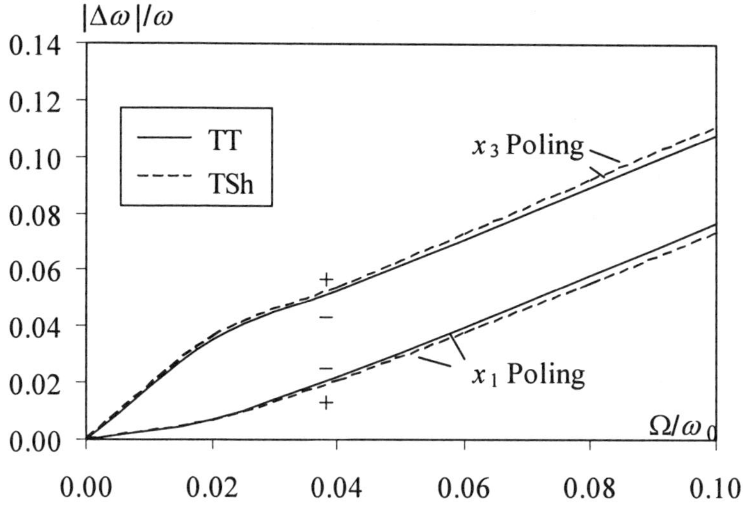

Fig. 6.4.5. Frequency shifts for the lowest thickness-shear and thickness-twist waves in an electroded ceramic plate ($kh/2\pi = 0.1$).

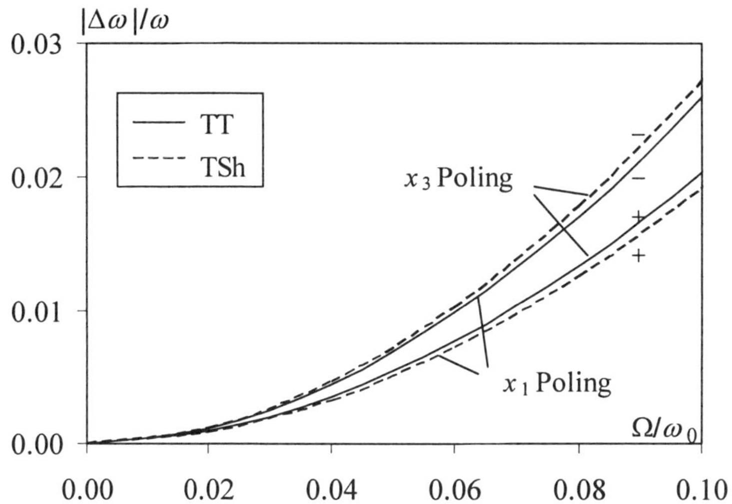

Fig. 6.4.6. Frequency shifts for the lowest thickness-shear and thickness-twist waves in an unelectroded ceramic plate ($kh/2\pi = 0.1$).

increase with the rotation rate monotonically in all the cases. The figures show that the frequency shifts are larger for plates poled along the x_3-axis than for plates poled along the x_1-axis. More importantly, the curves in Fig. 6.4.5 have finite initial slopes, indicating an initially linear relation between the rotation rate and the frequency shift, which is desirable for sensor applications. For the curves in Fig. 6.4.6, the tangents at the origin appear to be essentially flat, which is not useful for angular rate sensing.

6.5. Surface Waves over a Rotating Piezoelectric Half-Space

In this section, we study waves propagating along the surface of a rotating piezoelectric half-space (see Fig. 6.5.1) [76,77].

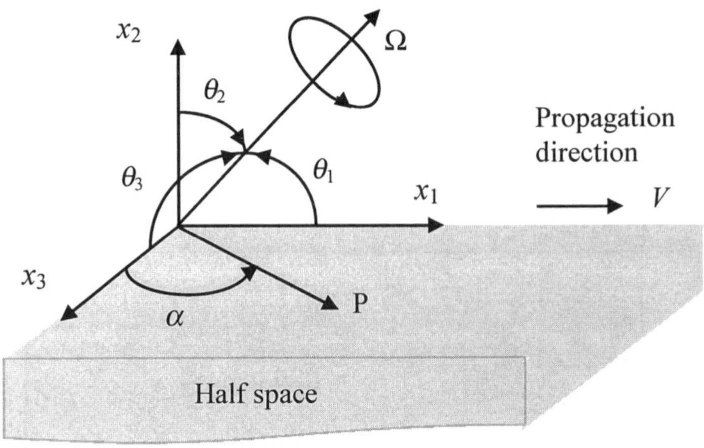

Fig. 6.5.1. A rotating piezoelectric half-space.

6.5.1. General solution

The half-space is rotating at a constant angular rate Ω about a fixed axis that is along a direction described by θ_i with respect to the coordinate axis x_i. We seek solutions in the following form:

$$\begin{aligned} u_j(\mathbf{x},t) &= A_j \exp(k\eta x_2)\exp[i(kx_1 - \omega t)], \\ \phi(\mathbf{x},t) &= A_4 \exp(k\eta x_2)\exp[i(kx_1 - \omega t)]. \end{aligned} \tag{6.5.1}$$

Substitution of Eq. (6.5.1) into Eqs. (6.1.1) and (6.1.2) results in the following four linear algebraic equations for A_j and A_4:

$$\begin{aligned}&[\rho(\omega^2+\Omega_2^2+\Omega_3^2)+k^2(\eta^2 c_{66}+i2\eta c_{16}-c_{11})]A_1\\&+[-\rho(i2\Omega_3\omega+\Omega_1\Omega_2)+k^2(\eta^2 c_{26}+i\eta c_{66}+i\eta c_{12}-c_{16})]A_2\\&+[\rho(i2\Omega_2\omega-\Omega_1\Omega_3)+k^2(\eta^2 c_{46}+i\eta c_{56}+i\eta c_{14}-c_{15})]A_3\\&+k^2(\eta^2 e_{26}+i\eta e_{21}+i\eta e_{16}-e_{11})A_4=0,\end{aligned}\tag{6.5.2}$$

$$\begin{aligned}&[\rho(i2\Omega_3\omega-\Omega_1\Omega_2)+k^2(\eta^2 c_{26}+i\eta c_{21}+i\eta c_{66}-c_{16})]A_1\\&+[\rho(\omega^2+\Omega_1^2+\Omega_3^2)+k^2(\eta^2 c_{22}+i2\eta c_{26}-c_{66})]A_2\\&+[-\rho(i2\Omega_1\omega+\Omega_2\Omega_3)+k^2(\eta^2 c_{24}+i\eta c_{25}+i\eta c_{46}-c_{56})]A_3\\&+k^2(\eta^2 e_{22}+i\eta e_{26}+i\eta e_{12}-e_{16})A_4=0,\end{aligned}\tag{6.5.3}$$

$$\begin{aligned}&[-\rho(i2\Omega_2\omega+\Omega_1\Omega_3)+k^2(\eta^2 c_{46}+i\eta c_{14}+i\eta c_{56}-c_{15})]A_1\\&+[\rho(i2\Omega_1\omega-\Omega_2\Omega_3)+k^2(\eta^2 c_{24}+i\eta c_{46}+i\eta c_{25}-c_{56})]A_2\\&+[\rho(\omega^2+\Omega_1^2+\Omega_2^2)+k^2(\eta^2 c_{44}+i2\eta c_{45}-c_{55})]A_3\\&+k^2(\eta^2 e_{24}+i\eta e_{25}+i\eta e_{14}-e_{15})A_4=0,\end{aligned}\tag{6.5.4}$$

$$\begin{aligned}&(\eta^2 e_{26}+i\eta e_{21}+i\eta e_{16}-e_{11})A_1+(\eta^2 e_{22}+i\eta e_{26}+i\eta e_{12}-e_{16})A_2\\&+(\eta^2 e_{24}+i\eta e_{25}+i\eta e_{14}-e_{15})A_3\\&-(\eta^2\varepsilon_{22}+i2\eta\varepsilon_{12}-\varepsilon_{11})A_4=0.\end{aligned}\tag{6.5.5}$$

Letting the determinant of the coefficient matrix vanish leads to a polynomial equation of degree eight for η. The roots for η may be complex in general. They represent combined sinusoidal and exponential behavior from the surface of the half-space. The four roots with positive real parts are chosen for decaying behavior from the surface, and are denoted by $\eta_{(m)}$, m = 1, 2, 3, 4. The corresponding eigenvectors are denoted by ($\overline{A}_j^{(m)}$, $\overline{A}_4^{(m)}$). We then construct a surface wave solution to Eqs. (6.1.1) and (6.1.2) as follows:

$$\begin{aligned}u_i&=\sum_{m=1}^{4}C_{(m)}\overline{A}_i^{(m)}\exp(k\eta_{(m)}x_2)\exp[i(kx_1-\omega t)],\\\phi&=\sum_{m=1}^{4}C_{(m)}\overline{A}_4^{(m)}\exp(k\eta_{(m)}x_2)\exp[i(kx_1-\omega t)],\end{aligned}\tag{6.5.6}$$

where $C_{(m)}$ are undetermined constants. Substituting Eq. (6.5.6) into the boundary conditions for a traction-free and unelectroded surface, we obtain the following four linear algebraic equations for $C_{(m)}$:

$$\sum_{m=1}^{4}(c_{12}i\overline{A}_1^{(m)} + c_{26}\eta_{(m)}\overline{A}_1^{(m)} + c_{26}i\overline{A}_2^{(m)} + c_{22}\eta_{(m)}\overline{A}_2^{(m)} + c_{24}\eta_{(m)}\overline{A}_3^{(m)} + c_{25}i\overline{A}_3^{(m)}$$
$$+ e_{12}i\overline{A}_4^{(m)} + e_{22}\eta_{(m)}\overline{A}_4^{(m)})C_{(m)} = 0,$$
$$\sum_{m=1}^{4}(c_{16}i\overline{A}_1^{(m)} + c_{66}\eta_{(m)}\overline{A}_1^{(m)} + c_{66}i\overline{A}_2^{(m)} + c_{26}\eta_{(m)}\overline{A}_2^{(m)} + c_{46}\eta_{(m)}\overline{A}_3^{(m)} + c_{56}i\overline{A}_3^{(m)}$$
$$+ e_{16}i\overline{A}_4^{(m)} + e_{26}\eta_{(m)}\overline{A}_4^{(m)})C_{(m)} = 0,$$
$$\sum_{m=1}^{4}(c_{14}i\overline{A}_1^{(m)} + c_{46}\eta_{(m)}\overline{A}_1^{(m)} + c_{46}i\overline{A}_2^{(m)} + c_{24}\eta_{(m)}\overline{A}_2^{(m)} + c_{44}\eta_{(m)}\overline{A}_3^{(m)} + c_{45}i\overline{A}_3^{(m)}$$
$$+ e_{14}i\overline{A}_4^{(m)} + e_{24}\eta_{(m)}\overline{A}_4^{(m)})C_{(m)} = 0,$$
$$\sum_{m=1}^{4}(e_{21}i\overline{A}_1^{(m)} + e_{26}\eta_{(m)}\overline{A}_1^{(m)} + e_{26}i\overline{A}_2^{(m)} + e_{22}\eta_{(m)}\overline{A}_2^{(m)} + e_{24}\eta_{(m)}\overline{A}_3^{(m)} + e_{25}i\overline{A}_3^{(m)}$$
$$- \varepsilon_{12}i\overline{A}_4^{(m)} - \varepsilon_{22}\eta_{(m)}\overline{A}_4^{(m)})C_{(m)} = 0. \quad (6.5.7)$$

For an electroded surface Eq. $(6.5.7)_4$ should be replaced by

$$\overline{A}_4^{(1)}C_{(1)} + \overline{A}_4^{(2)}C_{(2)} + \overline{A}_4^{(3)}C_{(3)} + \overline{A}_4^{(4)}C_{(4)} = 0\,. \quad (6.5.8)$$

For non-trivial solutions of $C_{(m)}$, the determinant of the coefficient matrix of Eq. (6.5.7) has to vanish, leading to the frequency equation that determines the surface wave speed.

6.5.2. *Ceramic half-space*

For numerical calculations, we consider polarized ceramics PZT-5H. When the poling direction is along the x_3 axis and the wave propagates along the x_1-axis, there exist two types of surface waves in the absence of rotation ($\Omega = 0$). One is the Rayleigh wave with u_1 and u_2, which is not electrically coupled. The other is the Bleustein-Gulyaev wave with u_3, which is coupled to ϕ and does not have an elastic counterpart. Numerical results on Rayleigh waves are presented below. The x_3-axis is the poling direction ($\alpha = 0^\circ$) except in Fig. 6.5.6 in which α varies from 0° to 90°.

Wave speed $V = \omega/k$ versus the angular rate is shown in Fig. 6.5.2 for a few directions of the rotation axis in the (x_2, x_3) plane, close to the x_3-axis. $V_1 = \sqrt{c_{11}/\rho}$. The change of wave speed appears to be linear when the angular rate is small. The three curves are close to each other, insensitive to small changes in the direction of the rotation axis.

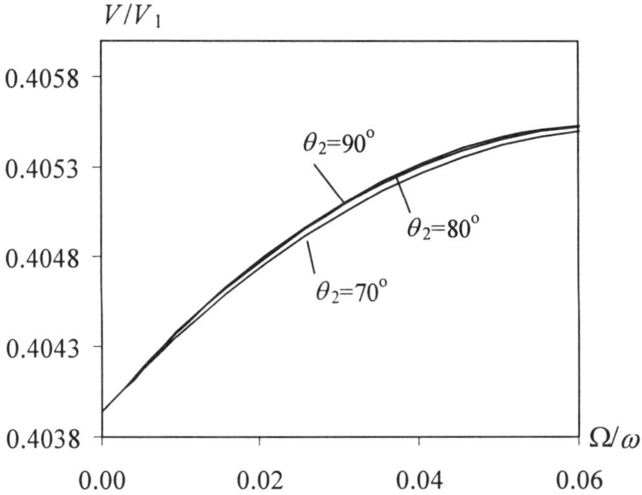

Fig. 6.5.2. Wave speed versus the angular rate ($\alpha = 0^{\circ}$, $\theta_1 = 90^{\circ}$).

Wave speed versus the angular rate for a few directions of the rotation axis in the (x_1, x_2) plane is shown in Fig. 6.5.3. In this case the curves seem to have a flat tangent at $\Omega = 0$, which implies little sensitivity for small angular rate.

Wave speed versus the direction of the rotation axis for different angular rates are shown in Fig. 6.5.4. The rotation axis starts from the x_1-axis, rotates to the x_2-axis through the (x_1, x_2) plane, then rotates to the x_3-axis through the (x_2, x_3) plane, and finally rotates back to the x_1-axis through the (x_3, x_1) plane. The curves correspond to equal increments of the angular rate. Near the x_3-axis, the sensitivity is large with essentially equal distance between the curves, indicating a linear dependence on the angular rate. The sensitivity near both the x_1- and x_2-axes is small. The curves are continuous but not smooth across

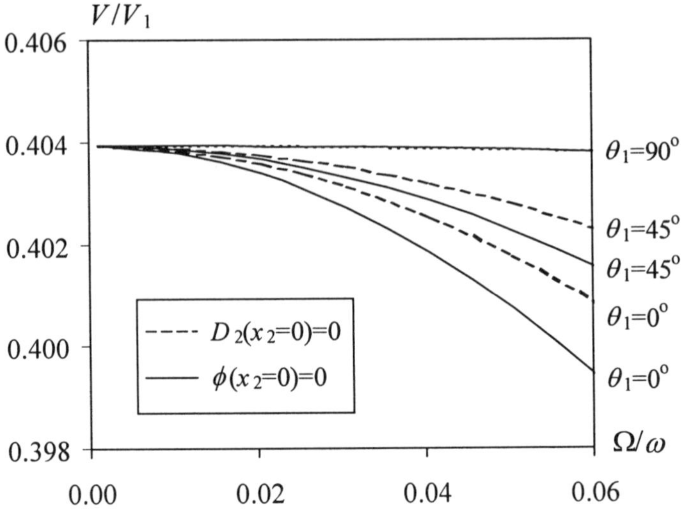

Fig. 6.5.3. Wave speed versus the angular rate ($\alpha = 0^{\circ}$, $\theta_3 = 90^{\circ}$).

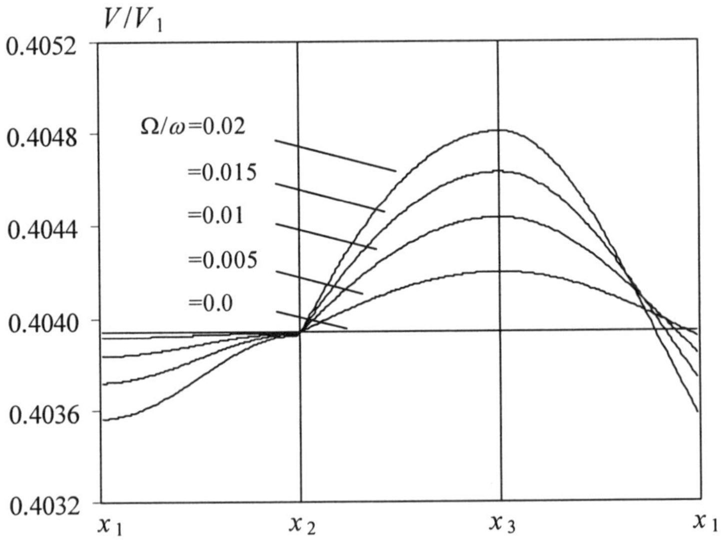

Fig. 6.5.4. Wave speed versus the direction of the rotation axis ($\alpha = 0^{\circ}$, $D_2(x_2 = 0) = 0$).

both the x_1- and x_2-axes, but they remain smooth at the x_3-axis. This is because the x_3-axis is an axis of rotational symmetry for polarized ceramics but the x_1- and x_2-axes are not.

To examine the sensitivity when the rotation axis is near the x_3-axis which has the highest sensitivity, consider the unit vector along the rotation axis as follows:

$$\mathbf{a} = (\cos\phi \mathbf{i}_1 + \sin\phi \mathbf{i}_2)\sin\theta_3 + \cos\theta_3 \mathbf{i}_3, \tag{6.5.9}$$

where $\mathbf{i}_1$, $\mathbf{i}_2$, and $\mathbf{i}_3$ are the unit vectors along the coordinate axes and ϕ is an angular coordinate. When θ_3 is fixed and ϕ varies from 0° to 360°, the unit vector **a** defines a cone whose axis of symmetry is along the x_3 axis. Results for different values of θ_3 and Ω are shown in Fig. 6.5.5.

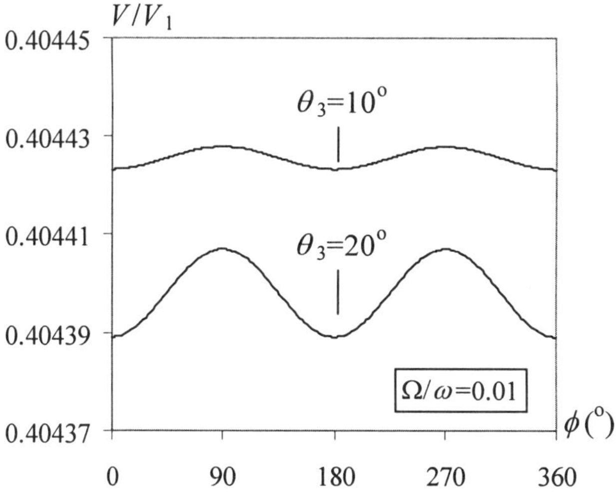

Fig. 6.5.5. Wave speed versus the direction of the rotation axis ($\alpha = 0°, \Omega = 0.01\omega$).

The relative change of wave speed versus the angular rate is shown in Fig. 6.5.6 for different values of the angle α that describes the poling direction. In the figure, $\Delta V = V(\Omega) - V(0)$, $V_0 = V(0)$, where V is considered as a function of the angular rate. In general the sensitivity is dependent on the poling direction. When $\alpha = 0°$, that is, the material is poled in the x_3 direction, the wave speed is most sensitive to the angular

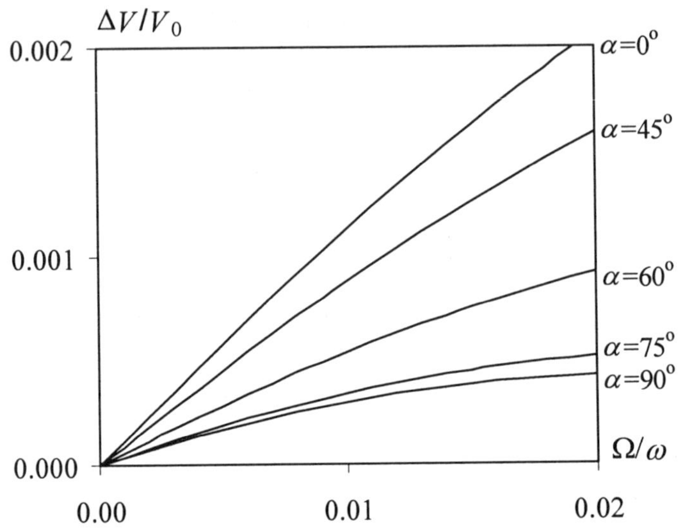

Fig. 6.5.6. Change of wave speed versus the poling direction ($\theta_3 = 0°$).

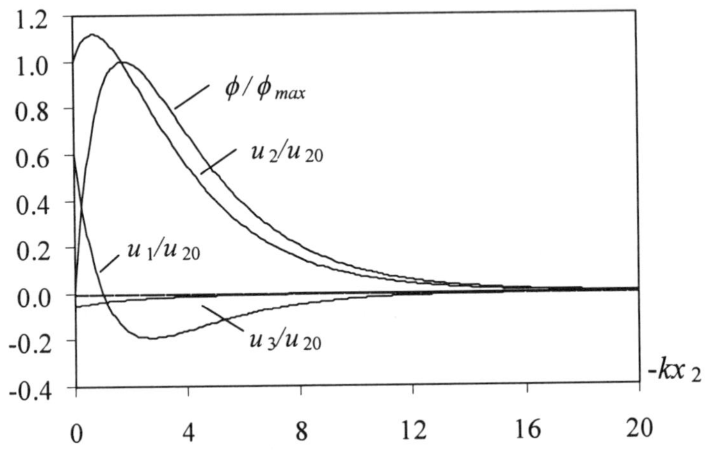

Fig. 6.5.7. Displacement and potential distribution near the surface ($\alpha = 0°$, $\theta_1 = 0°$, $\phi(x_2 = 0) = 0$).

rate. When $\alpha = 90^{\circ}$ or the material is poled in the x_1 direction, the wave speed is least sensitive.

Distributions of the displacement components and the electric potential according to the distance from the free surface are shown in Fig. 6.5.7 when the poling direction is along the x_3-axis and the rotation axis is along the x_1-axis. The displacement components are normalized by u_{20}, the amplitude of u_2 at the free surface. u_1 and u_2 correspond to the Rayleigh wave and the small u_3 is caused by rotation.

Calculations show that the wave speed corresponding to an unelectroded surface is always larger than the wave speed associated with an electroded surface. For an electroded surface, the electric field components E_1 and E_3 have to vanish on the electrode and, hence, they are small near the electrode. Therefore in a half-space with an electroded surface, the electric field is strongly limited by the electrode near the surface. Consequently, the related piezoelectric stiffening effect due to the electric field, which tends to raise the wave speed, is weakened near the surface. This has a strong effect on a surface wave whose energy is mostly confined within a thin layer near the surface. For materials of strong piezoelectric coupling, such as polarized ceramics, the electrical boundary conditions can noticeably influence the wave speed through the distribution of the electric field near the surface and the related piezoelectric stiffening effect.

Chapter 7
Gyroscopes — Charge Effect

In this chapter we analyze piezoelectric gyroscopes that detect angular rate by charge, current or voltage. Such a gyroscope makes use of two vibration modes of a piezoelectric body. The two modes have material particles moving in perpendicular directions so that they are coupled by the Coriolis force when the gyroscope is rotating. Furthermore, the frequencies of the two modes are close and tunable so that the gyroscope operates at the so-called double resonant condition with high sensitivity. When the gyroscope is excited into mechanical vibration by an applied alternating electric voltage in one of the two modes (the primary mode) and is attached to a body rotating with angular rate Ω, the Coriolis force excites the other mode (the secondary mode) through which Ω can be detected from electrical signals (voltage or current) accompanying the secondary mode. Double resonance occurs when the driving frequency, the frequency of the primary mode, and the frequency of the secondary mode are all about the same.

7.1. A Rectangular Beam

A rectangular elastic beam with piezoelectric films in flexural vibration was one of the earliest gyroscopes [70]. In this section we present an analysis on this gyroscope [78]. The structure of the beam gyroscope is shown in Fig. 7.1.1.

The elastic beam has a length L and a square cross section of equal sides $2a$. We consider a long and thin beam with $L >> a$. A cross section of the beam and the electrode configuration are also shown in Fig. 7.1.1. There are four identical piezoelectric films labeled by A, B, C, and D which are symmetrically attached to the four sides of the beam. The piezoelectric films are made from ferroelectric ceramics poled in the thickness directions as shown. The electrodes are shown by the thick lines in the cross section and they run through the entire length of the piezoelectric films. The beam is rotating about the x_1-axis with a constant

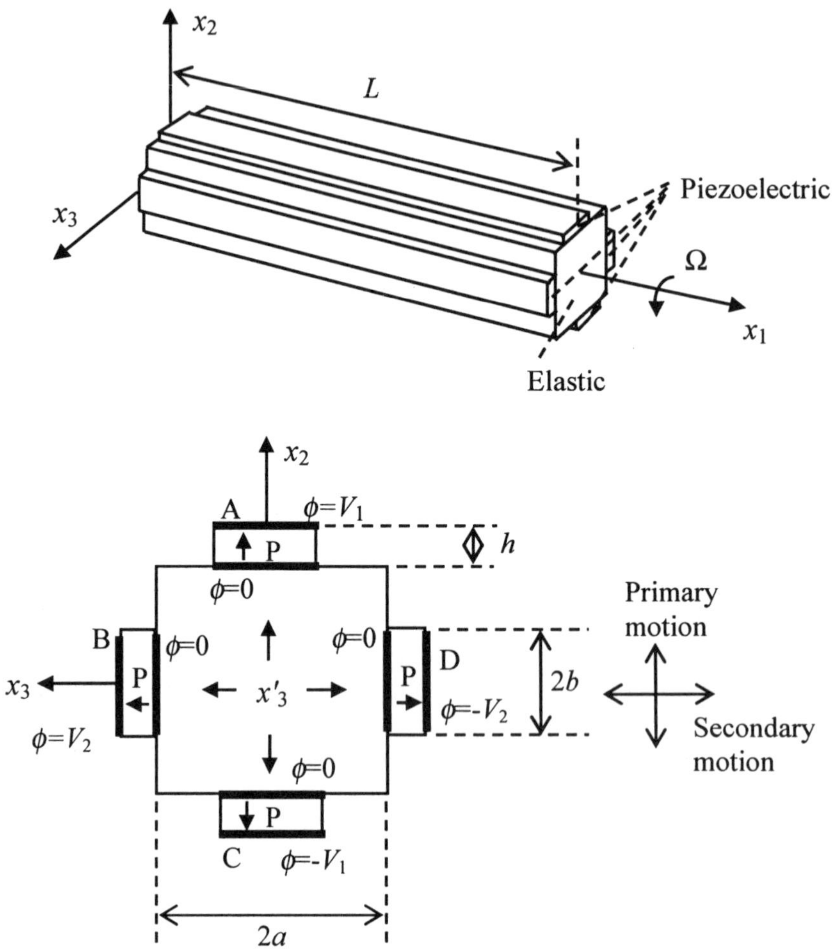

Fig. 7.1.1. A rectangular beam with piezoelectric films as a gyroscope.

angular rate Ω. The global coordinate system x_i is rotating with the beam. When $\pm V_1$ is applied across the piezoelectric films at A and C respectively, the film at A contracts and that at C expands or vice versa, thereby exciting the beam into flexural vibration in the x_2 direction (the primary motion). Since the beam is rotating about the x_1-axis, the Coriolis force which is perpendicular to the primary motion and the angular velocity vector causes a flexural motion in the x_3 direction (the secondary motion). In the secondary motion the piezoelectric film at B is compressed and that at D is stretched or vice versa, which produces an output voltage V_2. V_2 is related to Ω and can be used to detect it. If the beam is in flexural motion in the x_2 direction but is not rotating, there is no output V_2 at B and D.

7.1.1. Governing equations

Let the flexural motions of the beam in the x_2 and x_3 directions be $u_2(x_1,t)$ and $u_3(x_1,t)$. The axial normal strain for a beam in the classical theory of flexure is [20]

$$S_1 = -x_2 u_{2,11} - x_3 u_{3,11}. \tag{7.1.1}$$

$u_{2,11}$ and $u_{3,11}$ are bending curvatures in the x_2 and x_3 directions. Since there are four piezoelectric films poled in different directions, for each film we need a local coordinate system. The x'_3 axes of the four local coordinate systems are oriented along the poling directions of each film shown in Fig. 7.1.1. Then, with respect to the local coordinate systems, the electric field corresponding to the electrode configuration in Fig. 7.1.1 can be written as

$$E_3 = \begin{cases} -V_1/h & \text{at A}, \\ -V_2/h & \text{at B}, \\ V_1/h & \text{at C}, \\ V_2/h & \text{at D}. \end{cases} \tag{7.1.2}$$

For beams in flexure, the dominating stress component is the axial stress component T_1. For the piezoelectric films we are considering, the main electric field component is E_3 in the local coordinate systems. Then, for the piezoelectric films, under $T_2 = T_3 = T_4 = T_5 = T_6 = 0$, the constitutive relations relevant to our problem are

$$\begin{aligned} S_1 &= s_{11}T_1 + d_{31}E_3, \\ D_3 &= d_{31}T_1 + \varepsilon_{33}E_3. \end{aligned} \tag{7.1.3}$$

From Eq. (7.1.3) we solve for T_1 and D_3 in terms of S_1 and E_3, with the result

$$T_1 = s_{11}^{-1}(-x_2u_{2,11} - x_3u_{3,11}) - s_{11}^{-1}d_{31}E_3,$$
$$D_3 = s_{11}^{-1}d_{31}(-x_2u_{2,11} - x_3u_{3,11}) + \bar{\varepsilon}_{33}E_3, \qquad (7.1.4)$$

where Eq. (7.1.1) has been used, and

$$\bar{\varepsilon}_{33} = \varepsilon_{33}(1-k_{31}^2), \quad k_{31}^2 = d_{31}^2/(\varepsilon_{33}s_{11}). \qquad (7.1.5)$$

The constitutive relation for the elastic beam is simply

$$T_1 = ES_1 = E(-x_2u_{2,11} - x_3u_{3,11}), \qquad (7.1.6)$$

where E is the Young's modulus. The bending moments corresponding to flexures in the x_2 and x_3 directions are defined by the following integrals over the cross section A, which can be integrated with the expressions for T_1 in Eqs. $(7.1.4)_1$ and (7.1.6):

$$M_2 = \int_A x_2T_1dA = -Du_{2,11} + s_{11}^{-1}d_{31}V_1h^{-1}2G,$$
$$M_3 = \int_A x_3T_1dA = -Du_{3,11} + s_{11}^{-1}d_{31}V_2h^{-1}2G, \qquad (7.1.7)$$

where

$$D = E\frac{4a^4}{3} + 2s_{11}^{-1}\frac{2b}{3}[(a+h)^3 - a^3] + 2s_{11}^{-1}\frac{2h}{3}b^3,$$
$$G = 2bh\left(a + \frac{h}{2}\right). \qquad (7.1.8)$$

Equation (7.1.7) shows that bending in the x_2 and x_3 directions are coupled to V_1 and V_2 respectively. In terms of bending moments, the equations of motion for classical flexure take the form

$$M_{2,11} = m(\ddot{u}_2 - 2\Omega\dot{u}_3 - \Omega^2u_2),$$
$$M_{3,11} = m(\ddot{u}_3 + 2\Omega\dot{u}_2 - \Omega^2u_3), \qquad (7.1.9)$$

where we have included the Coriolis and centripetal accelerations and denoted

$$m = 4\rho a^2 + 4\rho'2bh. \qquad (7.1.10)$$

In Eq. (7.1.10), ρ is the mass density of the elastic beam, and ρ' is the mass density of the piezoelectric films. The charge on the outer electrode of the film at B is given by

$$Q_2 = -\int_0^L dx_1 \int_{-b}^{b} D_3(x_3 = a+h)dx_2$$
$$= s_{11}^{-1} d_{31}(a+h)2b[u_{3,1}(L,t) - u_{3,1}(0,t)] + \bar{\varepsilon}_{33} V_2 h^{-1} 2bL, \qquad (7.1.11)$$

where Eq. (7.1.4)$_2$ has been used. The current flowing out of this electrode is

$$I_2 = -\dot{Q}_2. \qquad (7.1.12)$$

The sensing current I_2 depends on the flexure in x_3 direction, but not on the flexure in the x_2 direction. In gyroscope applications the driving voltage V_1 is known and is time-harmonic. The sensing electrodes at B and D are connected to the grounded reference electrodes ($\phi = 0$) by an output circuit with impedance Z when the motion is time-harmonic. In the special cases when $Z = 0$ or ∞ we have short or open output circuit conditions with $V_2 = 0$ or $I_2 = 0$. In general, neither V_2 nor I_2 is known and we have the following circuit condition:

$$I_2 = V_2 / Z. \qquad (7.1.13)$$

Substituting Eq. (7.1.7) into Eq. (7.1.9), making use of the fact that V_1 and V_2 are at most functions of the time t but do not depend on x_1, we obtain

$$-Du_{2,1111} = m(\ddot{u}_2 - 2\Omega\dot{u}_3 - \Omega^2 u_2),$$
$$-Du_{3,1111} = m(\ddot{u}_3 + 2\Omega\dot{u}_2 - \Omega^2 u_3). \qquad (7.1.14)$$

For boundary conditions we consider a simply supported beam with

$$u_2(0,t) = u_3(0,t) = u_2(L,t) = u_3(L,t) = 0,$$
$$M_2(0,t) = M_3(0,t) = M_2(L,t) = M_3(L,t) = 0. \qquad (7.1.15)$$

Equation (7.1.14) is homogeneous, but the driving and sensing voltage are present in the moment conditions of Eqs. (7.1.15) through Eq. (7.1.7).

In summary, we need to solve the two fourth-order equations in Eq. (7.1.14) for time-harmonic motions driven by V_1, which will lead to eight integration constants. V_2 is another unknown constant that already exists in the expression for M_3. The expression for I_2 does not introduce any new constants. We have eight boundary conditions in Eq. (7.1.15) and a circuit condition given by Eq. (7.1.13).

7.1.2. Forced vibration solution

Since the driving voltage is time-harmonic, we use the complex notation

$$\{V_1, V_2, Q_2, I_2\} = \{\bar{V}_1, \bar{V}_2, \bar{Q}_2, \bar{I}_2\}\exp(i\omega t). \qquad (7.1.16)$$

For time-harmonic motions, Eq. (7.1.14) becomes linear ordinary differential equations with constant coefficients. The general solution can be readily found to be

$$\begin{aligned} u_2(x_1,t) &= \sum_{p=1}^{8} \gamma U^{(p)} \exp(k^{(p)}x_1)\exp(i\omega t), \\ u_3(x_1,t) &= \sum_{p=1}^{8} \beta^{(p)} U^{(p)} \exp(k^{(p)}x_1)\exp(i\omega t), \end{aligned} \qquad (7.1.17)$$

where $U^{(p)}$ are undetermined constants, $k^{(p)}$ for $p = 1, 2, \ldots, 8$ are the eight roots of the following equation for k:

$$k^4 = \frac{m}{D}(\omega \pm \Omega)^2, \qquad (7.1.18)$$

and

$$\gamma = 2im\omega\Omega, \quad \beta^{(p)} = D(k^{(p)})^4 - m(\omega^2 + \Omega^2). \qquad (7.1.19)$$

Substituting Eq. (7.1.17) into Eqs. (7.1.7), (7.1.11) and (7.1.12), the expressions for M_2, M_3 and I_2 can be obtained. Then, an application of the boundary condition in Eq. (7.1.15) and the circuit condition in Eq. (7.1.13) yields the following nine linear equations for the nine undetermined constants $U^{(p)}$ and $\bar{V}_2$:

$$\sum_{p=1}^{8} \gamma U^{(p)} = 0,$$

$$\sum_{p=1}^{8} \beta^{(p)} U^{(p)} = 0,$$

$$\sum_{p=1}^{8} \gamma \exp(k^{(p)}L) U^{(p)} = 0,$$

$$\sum_{p=1}^{8} \beta^{(p)} \exp(k^{(p)}L) U^{(p)} = 0,$$

$$\sum_{p=1}^{8} D\gamma(k^{(p)})^2 U^{(p)} = s_{11}^{-1} d_{31} \bar{V}_1 h^{-1} 2G,$$

$$\sum_{p=1}^{8} D\beta^{(p)}(k^{(p)})^2 U^{(p)} - s_{11}^{-1} d_{31} \bar{V}_2 h^{-1} 2G = 0,$$

$$\sum_{p=1}^{8} D\gamma(k^{(p)})^2 \exp(k^{(p)}L) U^{(p)} = s_{11}^{-1} d_{31} \bar{V}_1 h^{-1} 2G,$$

$$\sum_{p=1}^{8} D\beta^{(p)}(k^{(p)})^2 \exp(k^{(p)}L) U^{(p)} - s_{11}^{-1} d_{31} \bar{V}_2 h^{-1} 2G = 0,$$

$$\sum_{p=1}^{8} i\omega s_{11}^{-1} d_{31}(a+h) 2b\beta^{(p)} k^{(p)} [\exp(k^{(p)}L) - 1] U^{(p)} + \frac{1}{Z_2}\left(1 + \frac{Z_2}{Z}\right)\bar{V}_2 = 0, \tag{7.1.20}$$

where

$$Z_2 = \frac{1}{i\omega C_2}, \quad C_2 = \frac{\bar{\varepsilon}_{33} 2bL}{h}. \tag{7.1.21}$$

Equation (7.1.20) is driven by $\bar{V}_1$. In Eq. $(7.1.20)_9$, the impedance Z in general is a function of ω. The specific form of this function depends on the structure of the output circuit joining the sensing electrodes. When the output circuit is essentially capacitive, Z is pure imaginary and Z/Z_2 is a real number.

7.1.3. An example

As an example, we consider PZT-5H for the piezoelectric films. For the elastic beam, for comparison, we consider the following five materials with:

	E (GPa)	ρ (kg/m^3)
Aluminum alloy	70	2700
Titanium alloy	100	4500
Copper	110	8900
Tungsten	360	1900
Nylon	3.0	1000

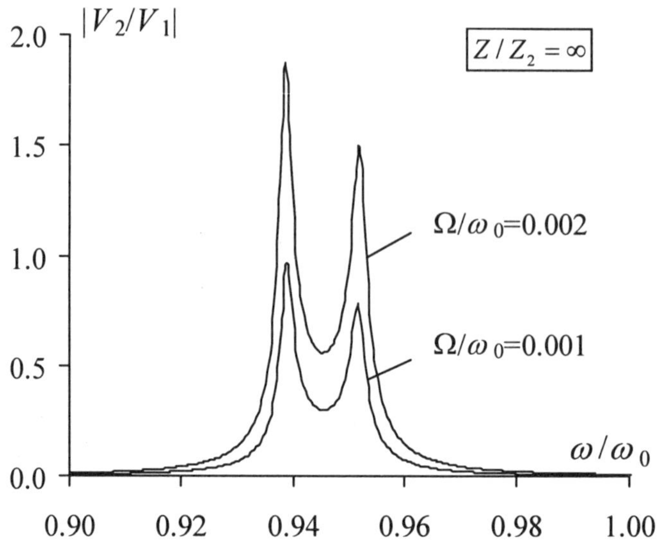

Fig. 7.1.2. Output voltage versus the driving frequency ω for different values of the rotation rate.

For geometric parameters, we choose L = 20 mm, a = 1 mm, and h = 0.1 mm, except in Fig. 7.1.7. b = 0.8a is fixed except in Figs. 7.1.5 and 7.1.6. The material of the elastic beam is fixed to be aluminum alloy except in Fig. 7.1.9. Damping is introduced by allowing the elastic constant of the piezoelectric films to be complex. The real elastic constant s_{11} in the above expressions are replaced by $s_{11}(1-iQ^{-1})$, where the value of the real number Q (the quality factor) for ceramics is usually in the order of 10^2 to 10^3. In the following, Q is fixed to be 10^2 except in Fig. 7.1.6. The above description of damping by a single number Q can be considered as a measure of the damping of the whole system when a proper value of Q is used.

Output voltage as a function of the driving frequency ω is plotted in Fig. 7.1.2 for the case of open sensing electrodes ($Z = \infty$) and two values of Ω. The rotation rate is normalized by

$$\omega_0 = \frac{\pi^2 a}{L^2}\sqrt{\frac{E}{3\rho}}, \tag{7.1.22}$$

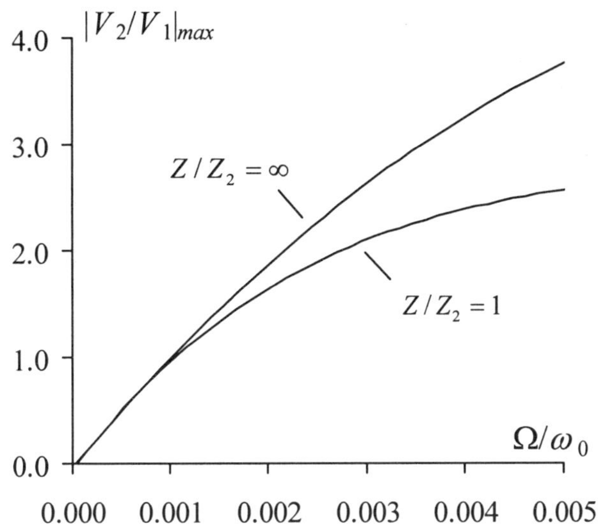

Fig. 7.1.3. Output voltage versus the rotation rate Ω.

which is the lowest flexural resonant frequency of the beam in Fig. 7.1.1 when the piezoelectric films are neglected. When the effects of the piezoelectric films are included, the resonant frequencies may become a little lower or higher, depending on the stiffness and the inertia of the film. In our case the density of the films is relatively high. Therefore there are two resonant frequencies with normalized values slightly lower than one. Near the two resonant frequencies the output voltage assumes maximum values. Figure 7.1.2 also shows that a larger rotation rate results in a larger output, which is as expected.

The dependence of the output voltage on the rotation rate Ω is shown in Fig. 7.1.3 for two values of the load Z. The figure shows that when Z is large the output voltage is high, as expected, while the output current is expected to be small. When Ω is much smaller than ω_0, the relation between the output voltage and Ω is essentially linear, which determines the range of the sensor for a linear response. These gyroscopes are useful for detecting angular rates relatively slow compared to the operating frequency. Since these gyroscopes are small with high resonant frequencies, the relatively slow rotation rate that the gyroscopes can detect linearly can still cover a variety of applications. Since the response is linear for small Ω, in the analysis of piezoelectric gyroscopes usually the centrifugal force which represents higher order effects of Ω can be neglected and the contribution to sensitivity is from the Coriolis force which is linear in Ω.

The variation of the output voltage according to the load Z is also of interest in practice and is given in Fig. 7.1.4 for two values of Ω. For small loads the sensing electrodes are almost shorted and the output voltage is low while the output current should be large. As the load increases, the voltage sensitivity increases and exhibits an essentially linear range. When the load is further increased the sensing electrodes are in fact nearly open with a saturated output voltage and a very small output current.

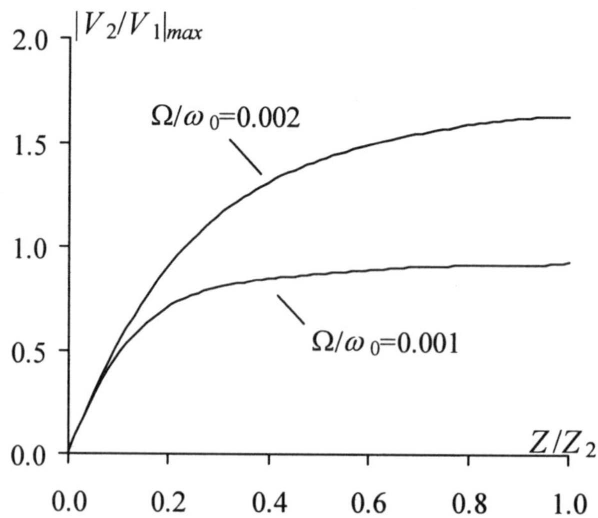

Fig. 7.1.4. Output voltage versus the load Z.

In Fig. 7.1.5 the effect of the ratio of b/a on the output voltage is shown. Intuitively it may seem that when b/a increases, i.e. when the piezoelectric films become wider, the output voltage should increase because wider piezoelectric films at A and C produce larger actuating forces so that the beam can be driven into stronger flexural motion in the x_2 direction (the primary motion). However, in the case shown in Fig. 7.1.5, the output voltage, in fact, decreases when the piezoelectric films become wider. The reason is that a stronger flexural motion in the

x_2 direction does not directly imply higher output voltage. As shown in Eq. (7.1.11), the output electric signal depends on the secondary motion, i.e. the flexure in the x_3 direction. Ideally, the resonant frequencies for flexures in the x_2 and the x_3 directions should be very close for the gyroscope to work at double-resonance. Since the piezoelectric films for driving at A and C and those for sensing at B and D are under different electric fields, their effective stiffness is different due to different piezoelectric stiffening by different electric fields. In such a case, increasing the width of the piezoelectric films makes the two resonant frequencies further apart, reducing the coupling between the two modes and thereby lowering the output voltage. This is evidenced by the larger distance between the two peaks of the curve corresponding to wider films shown in Fig. 7.1.5. Therefore, increasing the width of the piezoelectric films has combined effects and does not always lower or raise the output voltage in a simple manner. While wider piezoelectric films do not necessarily help on the output voltage, they allow wider electrodes and, hence, potentially larger sensing current.

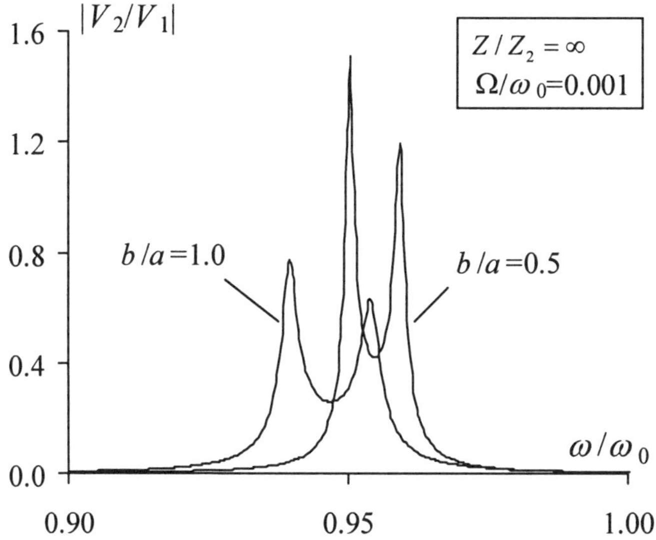

Fig. 7.1.5. Output voltage versus the driving frequency ω for different values of b/a.

Numerical tests show that if larger values of Q are used in the calculations, the peaks become narrower and higher. Although higher peaks suggest higher voltage sensitivity, narrower peaks require better control in tuning the device into resonance. In the extreme situation when Q is very large ($Q \cong \infty$), the resonance is almost singular and it becomes difficult to numerically capture the maxima of the peaks. The voltage sensitivity in this case is given in Fig. 7.1.6. The peaks are very high.

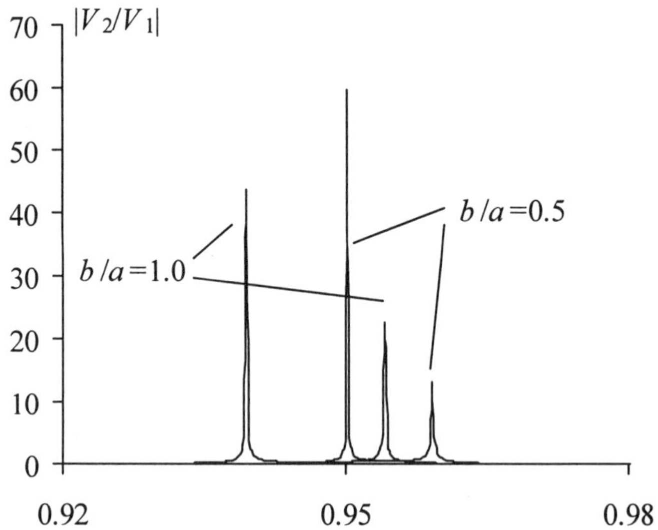

Fig. 7.1.6. Output voltage versus the driving frequency ω for different values of b/a ($Q \cong \infty$, $Z = \infty$, $\Omega/\omega_0 = 0.001$).

If the thickness of the piezoelectric films is varied, the results are as shown in Fig. 7.1.7. The dominating effect is that for thinner films the two resonant frequencies for flexures in the x_2 and x_3 directions are closer and the gyroscope is in better resonance and, hence, higher sensitivity.

When the piezoelectric constant d_{31} of the films is varied, the results are given in Fig. 7.1.8. Increasing d_{31} does not directly lead to higher sensitivity because stronger piezoelectric stiffening effect causes the two flexural resonant frequencies to become further apart and the resonance becomes poorer.

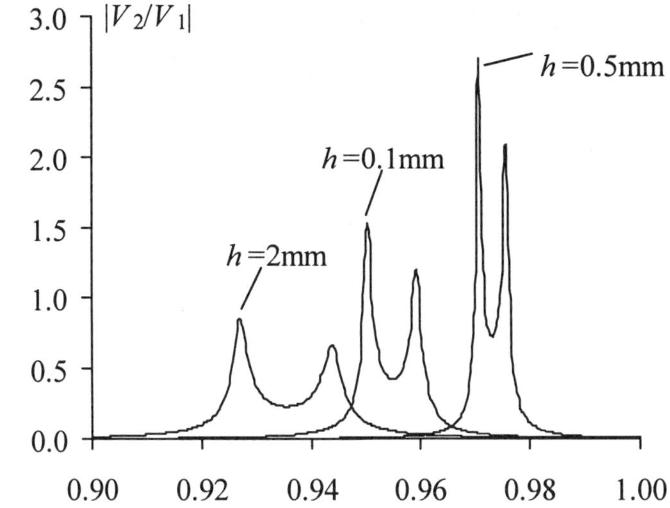

Fig. 7.1.7. Output voltage versus the driving frequency ω for different values of h ($Z = \infty$, $\Omega/\omega_0 = 0.001$).

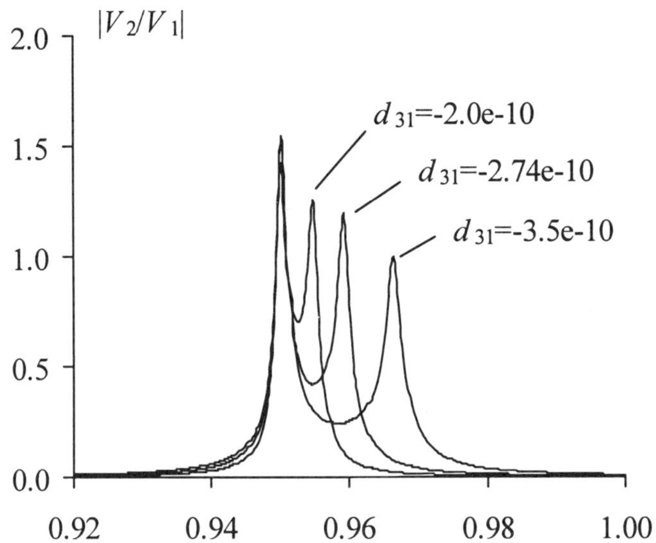

Fig. 7.1.8. Output voltage versus the driving frequency ω for different values of d_{31}.

Output voltage versus the rotation rate for different materials of the elastic beam is shown in Fig. 7.1.9. Among the materials considered are tungsten and nylon, the materials with the largest and smallest stiffness and the highest and lowest sensitivities. While stiffer beams are harder to drive which tends to lower the sensitivity, they have better resonance between the two flexural modes because for stiffer beams the mechanical effects of the piezoelectric films (which are partially responsible for the difference between the two resonant frequencies) are small. Again this is a combined effect and the behavior may be different for other materials.

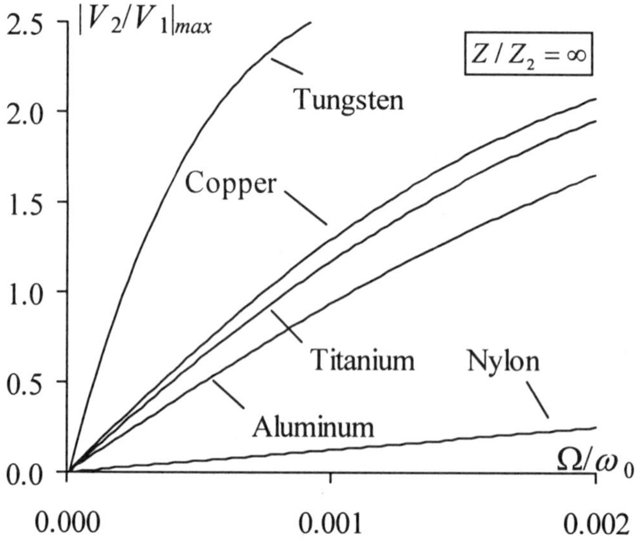

Fig. 7.1.9. Output voltage versus the rotation rate Ω for different materials of the elastic core.

Overall the condition that the two flexural frequencies be very close is the most important factor for high sensitivity. An elastic beam with a rectangular cross section is expected to produce better resonance by properly designing the width-height ratio of the cross section.

7.2. A Circular Tube

The beam gyroscope in the previous section has piezoelectric films on an elastic beam. In fact, piezoelectric materials alone can be used to make gyroscopes. Consider a long and thin circular tube of ceramics poled in the radial direction (see Fig. 7.2.1(a)) [79]. By long and thin we mean $L \gg R \gg h$, where L is the length, R is the mean radius, and h is the thickness of the tube. A cross section of the tube and the electrode configuration are shown in Fig. 1(b).

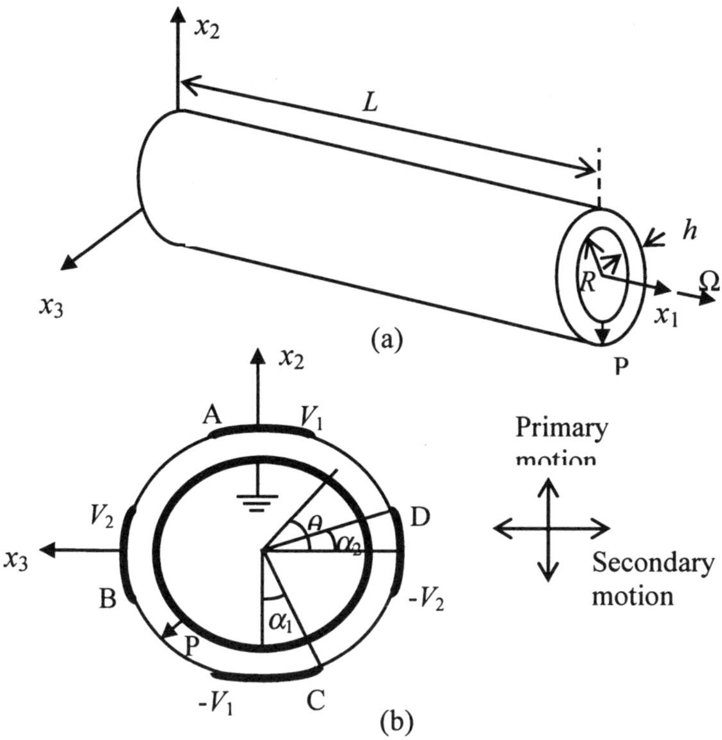

Fig. 7.2.1. A long and thin ceramic tube as a gyroscope.

The electrodes are shown by the thick lines and they run along the entire length of the tube. The inner surface of the tube is completely electroded and is grounded as a reference. The outer surface is electroded at four locations A, B, C, and D. When an alternating driving input voltage $\pm V_1$ is applied at electrodes A and C respectively, the material under A contracts and that under C expands or vice versa, thereby, exciting the tube into a flexural vibration in the x_2 direction (the primary motion). If the tube is rotating about the x_1-axis with a constant angular velocity Ω which is the quantity to be measured, the Coriolis force which is perpendicular to the directions of the excited primary motion (the x_2 direction) and the angular velocity vector (the x_1 direction) causes a flexural motion in the x_3 direction (the secondary motion). In the secondary motion the material under electrode B is compressed and that under electrode D is stretched or vice versa, which produces an output voltage V_2 at electrodes B and D. V_2 is related to Ω and can be used to detect it. When the tube is in flexural motion in the x_2 direction but is not rotating, there is no output V_2 at electrodes B and D. To avoid undesirable electrical couplings between the driving and sensing electrodes, α_1 and α_2 that describe the size of the electrodes have to be small when compared to $\pi/4$ with enough space between the driving and sensing electrodes.

7.2.1. Governing equations

Consider flexural motions of the tube in the x_2 and x_3 directions with flexural displacements $u_2(x_1,t)$ and $u_3(x_1,t)$. The electric fields can be written as

$$E_r = \begin{cases} V_2/h, & -\alpha_2 < \theta < \alpha_2, \\ -V_1/h, & \pi/2-\alpha_1 < \theta < \pi/2+\alpha_1, \\ -V_2/h, & \pi-\alpha_2 < \theta < \pi+\alpha_2, \\ V_1/h, & 3\pi/2-\alpha_1 < \theta < 3\pi/2+\alpha_1, \end{cases} \tag{7.2.1}$$

where r is the radial direction. The local axial, azimuthal, and radial directions are labeled as 1, 2 and 3. In a procedure similar to that in the previous section, the bending moments corresponding to flexure in the x_2 and x_3 directions have the following expressions:

$$\begin{aligned} M_2 &= \int_A x_2 T_1 dA = -EIu_{2,11} + s_{11}^{-1} d_{31} V_1 4R^2 \sin\alpha_1, \\ M_3 &= \int_A x_3 T_1 dA = -EIu_{3,11} + s_{11}^{-1} d_{31} V_2 4R^2 \sin\alpha_2, \end{aligned} \tag{7.2.2}$$

where we have denoted $EI = s_{11}^{-1}\pi R^3 h$. Equation (7.2.2) shows that bending in the x_2 and x_3 directions are coupled to V_1 and V_2 respectively. In terms of bending moments, the equations for classical flexure take the form

$$\begin{aligned} M_{2,11} &= m(\ddot{u}_2 - 2\Omega\dot{u}_3 - \Omega^2 u_2), \\ M_{3,11} &= m(\ddot{u}_3 + 2\Omega\dot{u}_2 - \Omega^2 u_3). \end{aligned} \quad (7.2.3)$$

In Eq. (7.2.3), $m = \rho 2\pi Rh$. The charge on the sensing electrode at B is given by

$$\begin{aligned} Q_2 &= -\int_0^L dx_1 \int_{\pi-\alpha_2}^{\pi+\alpha_2} D_r R d\theta \\ &= s_{11}^{-1} d_{31} 2R^2 \sin\alpha_2 [u_{3,1}(L,t) - u_{3,1}(0,t)] + \bar{\varepsilon}_{33} V_2 h^{-1} 2R\alpha_2 L. \end{aligned} \quad (7.2.4)$$

The current flowing out of this electrode is

$$I_2 = -\dot{Q}_2, \quad (7.2.5)$$

with the following circuit condition for time-harmonic motions:

$$I_2 = V_2 / Z. \quad (7.2.6)$$

The displacement equations of motion are

$$\begin{aligned} -EIu_{2,1111} &= m(\ddot{u}_2 - 2\Omega\dot{u}_3 - \Omega^2 u_2), \\ -EIu_{3,1111} &= m(\ddot{u}_3 + 2\Omega\dot{u}_2 - \Omega^2 u_3). \end{aligned} \quad (7.2.7)$$

Boundary conditions for a simply-supported beam are

$$\begin{aligned} &u_2(0,t) = u_3(0,t) = u_2(L,t) = u_3(L,t) = 0, \\ &M_2(0,t) = M_3(0,t) = M_2(L,t) = M_3(L,t) = 0. \end{aligned} \quad (7.2.8)$$

7.2.2. Forced vibration solution

With the complex notation,

$$\{V_1, V_2, Q_2, I_2\} = \{\bar{V}_1, \bar{V}_2, \bar{Q}_2, \bar{I}_2\} \exp(i\omega t). \quad (7.2.9)$$

The general solution can be readily found to be

$$\begin{aligned} u_2(x_1,t) &= \sum_{p=1}^{8} \gamma U^{(p)} \exp(k^{(p)} x_1) \exp(i\omega t), \\ u_3(x_1,t) &= \sum_{p=1}^{8} \beta^{(p)} U^{(p)} \exp(k^{(p)} x_1) \exp(i\omega t), \end{aligned} \quad (7.2.10)$$

where $U^{(p)}$ are undetermined constants, $k^{(p)}$ for p = 1, 2, ..., 8 are the eight roots of the following equation for k:

$$k^4 = \frac{m}{EI}(\omega \pm \Omega)^2, \tag{7.2.11}$$

and

$$\gamma = 2im\omega\Omega, \quad \beta^{(p)} = EI(k^{(p)})^4 - m(\omega^2 + \Omega^2). \tag{7.2.12}$$

The boundary conditions in Eq. (7.2.8) and the circuit condition yield the following nine linear equations for the nine undetermined constants $U^{(p)}$ and $\bar{V}_2$:

$$\begin{aligned}
&\sum_{p=1}^{8} \gamma U^{(p)} = 0, \\
&\sum_{p=1}^{8} \beta^{(p)} U^{(p)} = 0, \\
&\sum_{p=1}^{8} \gamma \exp(k^{(p)}L) U^{(p)} = 0, \\
&\sum_{p=1}^{8} \beta^{(p)} \exp(k^{(p)}L) U^{(p)} = 0, \\
&\sum_{p=1}^{8} EI\gamma (k^{(p)})^2 U^{(p)} = s_{11}^{-1} d_{31} \bar{V}_1 4R^2 \sin\alpha_1, \\
&\sum_{p=1}^{8} EI\beta^{(p)} (k^{(p)})^2 U^{(p)} - s_{11}^{-1} d_{31} \bar{V}_2 4R^2 \sin\alpha_2 = 0, \\
&\sum_{p=1}^{8} EI\gamma (k^{(p)})^2 \exp(k^{(p)}L) U^{(p)} = s_{11}^{-1} d_{31} \bar{V}_1 4R^2 \sin\alpha_1, \\
&\sum_{p=1}^{8} EI\beta^{(p)} (k^{(p)})^2 \exp(k^{(p)}L) U^{(p)} - s_{11}^{-1} d_{31} \bar{V}_2 4R^2 \sin\alpha_2 = 0, \\
&\sum_{p=1}^{8} i\omega s_{11}^{-1} d_{31} 2R^2 (\sin\alpha_2) \beta^{(p)} k^{(p)} [\exp(k^{(p)}L) - 1] U^{(p)} \\
&\qquad + \frac{1}{Z_2}\left(1 + \frac{Z_2}{Z}\right) \bar{V}_2 = 0,
\end{aligned} \tag{7.2.13}$$

where

$$Z_2 = \frac{1}{i\omega C_2}, \quad C_2 = \frac{\bar{\varepsilon}_{33} R 2\alpha_2 L}{h}. \tag{7.2.14}$$

7.2.3. An example

The basic behaviors of the tube gyroscope in this section are similar to Figs. 7.1.2 through 7.1.4 for the beam gyroscope in the previous section. We will only examine the effect of α_1 and α_2. As an example, consider PZT-5H with L = 20 mm, R = 1 mm, h = 0.1 mm, α_1 = 20°. $s_{pq}(1-iQ^{-1})$, $Q = 10^3$. Figure 7.2.2 shows the effect of α_2. The electric field in the electroded region causes piezoelectric stiffening and, hence, affects the resonant frequencies. The electric field under the driving electrodes is determined completely by the applied voltage V_1. The electric field under the output electrodes depends on the load Z. For the case of open output circuit shown ($Z = \infty$), the electric field under the output electrodes and the related stiffening effect is in the strongest possible case. In this case, when $|V_2| > |V_1|$, a smaller output electrode makes the two resonant frequencies closer. This leads to better resonance and higher sensitivity.

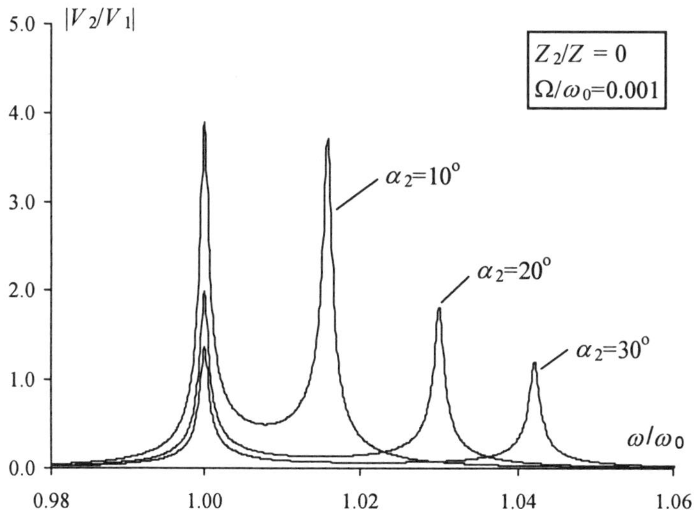

Fig. 7.2.2. The effect of electrode size α_2.

Calculations show that if the thickness of the tube is varied the normalized output voltage does not vary. This is as expected because the driving voltage and the output voltage are applied or picked up across the same dimension, i.e. the thickness of the tube.

7.3. A Beam Bimorph

An electric field in a piezoelectric material causes strain but not curvature. Therefore, in the previous two sections, different electric fields on opposite sides of a beam are needed to produce flexure. Another way of producing flexure in a beam is to use a bimorph. Consider the gyroscope shown in Fig. 7.3.1 [80].

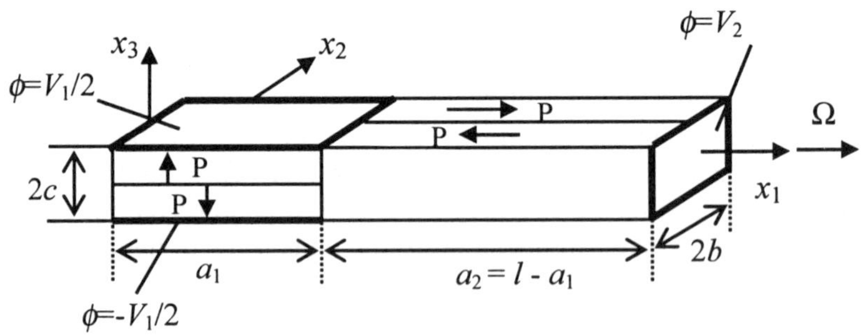

Fig. 7.3.1. A ceramic bimorph piezoelectric gyroscope.

The part in $0 < x_1 < a_1$ is the driving portion and $a_1 < x_1 < l$ is the receiving or sensing portion. The polarizations in the two layers are opposite. When a driving voltage V_1 is applied across the electrodes at $x_3 = \pm c$, one layer extends while the other contracts due to switched polarizations. This allows the beam to be driven into the lowest flexural mode in the x_3 direction with a proper driving frequency. If the beam is rotating about the x_1-axis with a constant angular rate Ω, the Coriolis force causes a flexural motion in the x_2 direction. This Coriolis force generated flexure produces a voltage V_2 in the receiving portion which can be picked up by the sensing electrodes. Flexure in the x_3 direction alone does not cause any output voltage.

7.3.1. Governing equations

To obtain one-dimensional equations for the elementary (classical) flexural motion of the gyroscope, we begin with the following approximations of the three-dimensional mechanical displacement and electric potential:

$$\begin{aligned}
&u_1(x_1,x_2,x_3,t) \cong -x_2 u_{2,1}^{(0,0)}(x_1,t) - x_3 u_{3,1}^{(0,0)}(x_1,t), \\
&u_2(x_1,x_2,x_3,t) \cong u_2^{(0,0)}(x_1,t), \\
&u_3(x_1,x_2,x_3,t) \cong u_3^{(0,0)}(x_1,t), \\
&\phi(x_1,x_2,x_3,t) \cong \phi^{(0,0)}(x_1,t) + x_2\phi^{(1,0)}(x_1,t) + x_3\phi^{(0,1)}(x_1,t),
\end{aligned} \tag{7.3.1}$$

where $u_2^{(0,0)}$ and $u_3^{(0,0)}$ are the flexural displacements in the x_2 and x_3 directions, $\phi^{(0,0)}$ will be shown later to be responsible for the axial electric field in the x_1 direction, and $\phi^{(1,0)}$ and $\phi^{(0,1)}$ are related to the lateral electric fields in the x_2 and x_3 directions. From [20] the relevant equations of flexure and electrostatics for the bimorph are

$$\begin{aligned}
T_{11,11}^{(1,0)} + F_{1,1}^{(1,0)} + F_2^{(0,0)} &= m\ddot{u}_2^{(0,0)}, \\
T_{11,11}^{(0,1)} + F_{1,1}^{(0,1)} + F_3^{(0,0)} &= m\ddot{u}_3^{(0,0)},
\end{aligned} \tag{7.3.2}$$

$$D_{1,1}^{(0,0)} + D^{(0,0)} = 0, \tag{7.3.3}$$

where $m = 4bc\rho$ is the mass per unit length of the beam. $T_{11}^{(1,0)}$ and $T_{11}^{(0,1)}$ are the bending moments in the x_2 and x_3 directions. $D_1^{(0,0)}$ is the one-dimensional electric displacement. $F_1^{(0,0)}$, $F_3^{(0,0)}$ and $D^{(0,0)}$ are related to surface force and charge. Equation (7.3.1) implies the following beam strains and electric fields relevant to our gyroscope problem:

$$S_1^{(1,0)} = -u_{2,11}^{(0,0)}, \quad S_1^{(0,1)} = -u_{3,11}^{(0,0)}, \tag{7.3.4}$$

$$E_1^{(0,0)} = -\phi_{,1}^{(0,0)}, \quad E_2^{(0,0)} = -\phi^{(1,0)}, \quad E_3^{(0,0)} = -\phi^{(0,1)}. \tag{7.3.5}$$

$S_1^{(1,0)}$ and $S_1^{(0,1)}$ are bending curvatures in the x_2 and x_3 directions. For long and thin beams, since $T_{22} = T_{23} = 0$ at $x_2 = \pm b$ and $T_{33} = T_{32} = 0$ at $x_3 = \pm c$, these stress components are also very small inside the beam and are approximately zero everywhere. In calculating the bending strains, the only stress component needs to be considered is T_{11}. The bending moment in the x_2 direction is

$$T_1^{(1,0)} = c_{11}^{(2,0)} S_1^{(1,0)} + c_{11}^{(1,1)} S_1^{(0,1)} - e_{k1}^{(1,0)} E_k^{(0,0)}, \tag{7.3.6}$$

where

$$c_{11}^{(2,0)} = \int_A \frac{x_2^2}{s_{11}} dA, \quad c_{11}^{(1,1)} = \int_A \frac{x_2 x_3}{s_{11}} dA, \quad e_{k1}^{(1,0)} = \int_A x_2 \frac{d_{k1}}{s_{11}} dA. \tag{7.3.7}$$

$c_{11}^{(m,n)}$ represent various cross-sectional properties related to the bending stiffness, etc. Similarly, the bending moment in the x_3 direction can be written as

$$T_1^{(0,1)} = c_{11}^{(1,1)} S_1^{(1,0)} + c_{11}^{(0,2)} S_1^{(0,1)} - e_{k1}^{(0,1)} E_k^{(0,0)}, \tag{7.3.8}$$

where

$$c_{11}^{(0,2)} = \int_A \frac{x_3^2}{s_{11}} dA, \quad e_{k1}^{(0,1)} = \int_A x_3 \frac{d_{k1}}{s_{11}} dA. \tag{7.3.9}$$

The electrical constitutive relation is found to be

$$D_i^{(0,0)} = e_{i1}^{(1,0)} S_1^{(1,0)} + e_{i1}^{(0,1)} S_1^{(0,1)} + \overline{\varepsilon}_{ik}^{(0,0)} E_k^{(0,0)}, \tag{7.3.10}$$

where

$$\overline{\varepsilon}_{ik}^{(0,0)} = \int_A \overline{\varepsilon}_{ik} dA, \quad \overline{\varepsilon}_{ik} = \varepsilon_{ik}^T - d_{i1} d_{k1} / s_{11}. \tag{7.3.11}$$

The transverse shearing forces in the beam are related to the bending moments by [20]

$$T_{12}^{(0,0)} = T_{11,1}^{(1,0)} + F_1^{(1,0)}, \quad T_{13}^{(0,0)} = T_{11,1}^{(0,1)} + F_1^{(0,1)}. \tag{7.3.12}$$

7.3.1.1. *Driving portion*

Specifically, for the driving portion in $0 < x_1 < a_1$, we have

$$\phi^{(0,0)} = \phi^{(1,0)} = 0, \quad \phi^{(0,1)} = \frac{V_1}{2c}, \tag{7.3.13}$$

$$E_1^{(0,0)} = 0, \quad E_2^{(0,0)} = 0, \quad E_3^{(0,0)} = -\frac{V_1}{2c}, \tag{7.3.14}$$

$$c_{11}^{(2,0)} = \frac{4b^3c}{3s_{11}}, \qquad c_{11}^{(1,1)} = 0, \qquad c_{11}^{(0,2)} = \frac{4bc^3}{3s_{11}},$$
$$e_{31}^{(0,1)} = \frac{2c^2bd_{31}}{s_{11}}, \quad e_{31}^{(1,0)} = 0, \tag{7.3.15}$$
$$\bar{\varepsilon}_{33}^{(0,0)} = 4bc\bar{\varepsilon}_{33}, \quad \bar{\varepsilon}_{33} = \varepsilon_{33}^T - d_{31}^2 / s_{11},$$

$$\begin{aligned} T_1^{(1,0)} &= c_{11}^{(2,0)} S_1^{(1,0)}, \\ T_1^{(0,1)} &= c_{11}^{(0,2)} S_1^{(0,1)} - e_{31}^{(0,1)} E_3^{(0,0)}, \end{aligned} \tag{7.3.16}$$

$$D_3^{(0,0)} = e_{31}^{(0,1)} S_1^{(0,1)} + \bar{\varepsilon}_{33}^{(0,0)} E_3^{(0,0)}. \tag{7.3.17}$$

The second term on the right hand side of Eq. $(7.3.16)_2$ can be interpreted as the moment of two equal and opposite forces with magnitude $2bcd_{31}E_3^{(0,0)} / s_{11}$ and at a distance c apart, which represents the moment due to the axial forces in each layer of the bimorph under the applied voltage, one in extension and the other is in compression. The charge and current on the driving electrode at $x_3 = c$ are

$$Q_1 = -\int_{-b}^{b}\int_{0}^{a_1} \frac{1}{4bc} D_3^{(0,0)} dx_1 dx_2, \quad I_1 = -\dot{Q}_1. \tag{7.3.18}$$

The equations of motion take the following form:

$$\begin{aligned} T_{1,11}^{(1,0)} &= m(\ddot{u}_2^{(0,0)} - 2\Omega\dot{u}_3^{(0,0)} - \Omega^2 u_2^{(0,0)}), \\ T_{11,11}^{(0,1)} &= m(\ddot{u}_3^{(0,0)} + 2\Omega\dot{u}_2^{(0,0)} - \Omega^2 u_3^{(0,0)}). \end{aligned} \tag{7.3.19}$$

7.3.1.2. *Sensing portion*

For the sensing portion in $a_1 < x_1 < l$, the lateral surfaces are unelectroded, hence, we have $D_2 = 0$ on $x_2 = \pm b$ and $D_3 = 0$ on $x_3 = \pm c$. Since the beam is thin, we have approximately $D_2 \cong 0$ and $D_3 \cong 0$ everywhere in this portion. This implies, from the constitutive relations of polarized ceramics, that $E_2 \cong 0$ and $E_3 \cong 0$ everywhere in the sensing portion. Therefore the major electric field in the receiving portion is the axial electric field E_1. We have

$$\phi^{(0,0)} = \phi^{(0,0)}(x_1, t), \quad \phi^{(1,0)} = \phi^{(0,1)} = 0, \tag{7.3.20}$$

$$E_1^{(0,0)} = -\phi_{,1}^{(0,0)}, \quad E_2^{(0,0)} = 0, \quad E_3^{(0,0)} = 0. \tag{7.3.21}$$

$$c_{11}^{(2,0)} = \frac{4b^3c}{3s_{33}} = \hat{c}_{11}^{(2,0)}, \quad c_{11}^{(1,1)} = 0, \quad c_{11}^{(0,2)} = \frac{4bc^3}{3s_{33}} = \hat{c}_{11}^{(0,2)},$$

$$e_{11}^{(0,1)} = 0, \quad e_{11}^{(1,0)} = \frac{2cb^2d_{33}}{s_{33}}, \tag{7.3.22}$$

$$\bar{\varepsilon}_{11}^{(0,0)} = 4bc\bar{\varepsilon}_{11}, \quad \bar{\varepsilon}_{11} = \varepsilon_{33}^T - d_{33}^2 / s_{33},$$

$$\begin{aligned} T_1^{(1,0)} &= \hat{c}_{11}^{(2,0)} S_1^{(1,0)} - e_{11}^{(1,0)} E_1^{(0,0)}, \\ T_1^{(0,1)} &= \hat{c}_{11}^{(0,2)} S_1^{(0,1)}, \end{aligned} \tag{7.3.23}$$

$$D_1^{(0,0)} = e_{11}^{(1,0)} S_1^{(1,0)} + \bar{\varepsilon}_{11}^{(0,0)} E_1^{(0,0)}. \tag{7.3.24}$$

$$D_{1,1}^{(0,0)} = 0, \tag{7.3.25}$$

$$D_1^{(0,0)} = e_{11}^{(1,0)} S_1^{(1,0)} + \bar{\varepsilon}_{11}^{(0,0)} E_1^{(0,0)} = -Q_2, \tag{7.3.26}$$

where Q_2 is an integration constant which may still be a function of the time t. Physically, Q_2 represents the electric charge on the sensing electrode at $x_1 = l$. The current flows out of this electrode is given by

$$I_2 = -\dot{Q}_2. \tag{7.3.27}$$

Integrating Eq. (7.3.26) once with respect to x_1, we obtain

$$e_{11}^{(1,0)} u_{2,1}^{(0,0)} + \bar{\varepsilon}_{11}^{(0,0)} \phi^{(0,0)} = Q_2 x_1 + L, \tag{7.3.28}$$

where L is another integration constant. Equation (7.3.28) allows us to obtain $\phi^{(0,0)}$ once $u_2^{(0,0)}$ is known. Solving Eq. (7.3.26) for $E_1^{(0,0)}$ and substituting the result into Eq. $(7.3.23)_1$ we obtain

$$T_1^{(1,0)} = \bar{c}_{11}^{(2,0)} S_1^{(1,0)} + \frac{e_{11}^{(1,0)}}{\bar{\varepsilon}_{11}^{(0,0)}} Q_2, \tag{7.3.29}$$

where

$$\bar{c}_{11}^{(2,0)} = \hat{c}_{11}^{(2,0)} + \frac{e_{11}^{(1,0)} e_{11}^{(1,0)}}{\bar{\varepsilon}_{11}^{(0,0)}} \tag{7.3.30}$$

is a piezoelectrically stiffened bending stiffness. We also have the following circuit condition for time-harmonic motions:

$$I_2 = V_2 / Z, \tag{7.3.31}$$

where

$$V_2 = \phi^{(0,0)}(l,t). \tag{7.3.32}$$

At the junction of the two portions we have

$$\phi^{(0,0)}(a_1^+,t)=0\,, \tag{7.3.33}$$

and the mechanical continuity conditions of the deflections $u_2^{(0,0)}$ and $u_3^{(0,0)}$, slopes $u_{2,1}^{(0,0)}$ and $u_{3,1}^{(0,0)}$, bending moments $T_1^{(1,0)}$ and $T_1^{(0,1)}$, as well as shear forces $T_{12}^{(0,0)}$ and $T_{13}^{(0,0)}$. Various combinations of mechanical boundary conditions can be prescribed at the two ends of the gyroscope.

7.3.1.3. Boundary value problem

With successive substitutions, we obtain, for the driving portion $0 < x_1 < a_1$, the following equations in terms of $u_2^{(0,0)}$ and $u_3^{(0,0)}$:

$$\begin{aligned} -c_{11}^{(2,0)}u_{2,1111}^{(0,0)} &= m(\ddot{u}_2^{(0,0)}-2\Omega\dot{u}_3^{(0,0)}-\Omega^2u_2^{(0,0)}), \\ -c_{11}^{(0,2)}u_{3,1111}^{(0,0)} &= m(\ddot{u}_3^{(0,0)}+2\Omega\dot{u}_2^{(0,0)}-\Omega^2u_3^{(0,0)}). \end{aligned} \tag{7.3.34}$$

Similarly, for the receiving portion $a_1 < x_1 < l$, we have

$$\begin{aligned} -\bar{c}_{11}^{(2,0)}u_{2,1111}^{(0,0)} &= m(\ddot{u}_2^{(0,0)}-2\Omega\dot{u}_3^{(0,0)}-\Omega^2u_2^{(0,0)}), \\ -\hat{c}_{11}^{(0,2)}u_{3,1111}^{(0,0)} &= m(\ddot{u}_3^{(0,0)}+2\Omega\dot{u}_2^{(0,0)}-\Omega^2u_3^{(0,0)}), \\ \bar{\varepsilon}_{11}^{(0,0)}\phi^{(0,0)} &= -e_{11}^{(1,0)}u_{2,1}^{(0,0)}+Q_2x_1+L. \end{aligned} \tag{7.3.35}$$

When the two ends of the beam are simply supported in both directions of x_2 and x_3, at the left end $x_1 = 0$ the boundary conditions are

$$\begin{aligned} &u_2^{(0,0)}=0, \quad u_3^{(0,0)}=0, \\ &T_1^{(1,0)}=-c_{11}^{(2,0)}u_{2,11}^{(0,0)}=0, \\ &T_1^{(0,1)}=-c_{11}^{(0,2)}u_{3,11}^{(0,0)}+e_{31}^{(0,1)}V_1/2c=0. \end{aligned} \tag{7.3.36}$$

At the right end $x_1 = l$, the boundary conditions are

$$\begin{aligned} &u_2^{(0,0)}=0, \quad u_3^{(0,0)}=0, \\ &T_1^{(1,0)}=-\bar{c}_{11}^{(2,0)}u_{2,11}^{(0,0)}+\frac{e_{11}^{(1,0)}}{\bar{\varepsilon}_{11}^{(0,0)}}Q_2=0, \\ &T_1^{(0,1)}=-\hat{c}_{11}^{(0,2)}u_{3,11}^{(0,0)}=0, \\ &\phi^{(0,0)}(l)=V_2, \quad -\dot{Q}_2=V_2/Z. \end{aligned} \tag{7.3.37}$$

At the junction of the two portions $x_1 = a_1$, the continuity conditions can be written as

$$\left\|u_2^{(0,0)}\right\|=0,\quad \left\|u_3^{(0,0)}\right\|=0,\quad \left\|u_{2,1}^{(0,0)}\right\|=0,\quad \left\|u_{3,1}^{(0,0)}\right\|=0,$$
$$\left\|T_1^{(1,0)}\right\|=0,\quad \left\|T_1^{(0,1)}\right\|=0,\quad \left\|T_{12}^{(0,0)}\right\|=0,\quad \left\|T_{13}^{(0,0)}\right\|=0, \tag{7.3.38}$$
$$\phi^{(0,0)}(a_1^+)=0,$$

where $\left\|u_2^{(0,0)}\right\| = u_2^{(0,0)}(a_1^+) - u_2^{(0,0)}(a_1^-)$ is the jump of $u_2^{(0,0)}$ across $x_1 = a_1$, and the others are similar. Effectively the beam is driven by a pair of concentrated moments at $x_1 = 0$ and $x_1 = a_1$.

7.3.2. Forced vibration solution

With the complex notation,

$$\{V_1,V_2,Q_2,L\}=\{\bar{V}_1,\bar{V}_2,\bar{Q}_2,\bar{L}\}\exp(i\omega t), \tag{7.3.39}$$

the general solution for the driving portion is found to be

$$u_2^{(0,0)}=\sum_{p=1}^{8}\alpha^{(p)}U^{(p)}\exp(k^{(p)}x_1)\exp(i\omega t),$$
$$u_3^{(0,0)}=\sum_{p=1}^{8}\beta U^{(p)}\exp(k^{(p)}x_1)\exp(i\omega t), \tag{7.3.40}$$

where $U^{(p)}$ are undetermined constants, $k^{(p)}$ are the eight roots of the equation

$$c_{11}^{(2,0)}c_{11}^{(0,2)}k^8 - m(\omega^2+\Omega^2)(c_{11}^{(2,0)}+c_{11}^{(0,2)})k^4 + m^2(\omega^2-\Omega^2)^2=0, \tag{7.3.41}$$

and

$$\alpha^{(p)}=\alpha(k^{(p)}),\quad \alpha(k)=c_{11}^{(0,2)}k^4-m(\omega^2+\Omega^2),$$
$$\beta=-2im\omega\Omega. \tag{7.3.42}$$

Similarly, the general solution for the sensing portion can be written as

$$u_2^{(0,0)} = \sum_{p=1}^{8} \bar{\alpha}^{(p)} \bar{U}^{(p)} \exp[\bar{k}^{(p)}(x_1 - l)] \exp(i\omega t),$$

$$u_3^{(0,0)} = \sum_{p=1}^{8} \beta \bar{U}^{(p)} \exp[\bar{k}^{(p)}(x_1 - l)] \exp(i\omega t), \tag{7.3.43}$$

$$\bar{\varepsilon}_{11}^{(0,0)} \phi^{(0,0)} = \sum_{p=1}^{8} -e_{11}^{(1,0)} \bar{k}^{(p)} \bar{\alpha}^{(p)} \bar{U}^{(p)} \exp[\bar{k}^{(p)}(x_1 - l)] \exp(i\omega t)$$

$$+ \bar{Q}_2 x_1 \exp(i\omega t) + \bar{L} \exp(i\omega t),$$

where $\bar{U}^{(p)}$ are undetermined constants, $\bar{k}^{(p)}$ are the eight roots of Eq. (7.3.41) when $c_{11}^{(2,0)}$ and $c_{11}^{(0,2)}$ are replaced by $\bar{c}_{11}^{(2,0)}$ and $\hat{c}_{11}^{(0,2)}$. $\bar{\alpha}^{(p)} = \alpha(\bar{k}^{(p)})$ with $c_{11}^{(0,2)}$ replaced by $\hat{c}_{11}^{(0,2)}$. Substitution of Eqs. (7.3.40) and (7.3.43) into Eqs. (7.3.36) through (7.3.38) gives 19 linear equations for the 19 undetermined constants $U^{(p)}$, $\bar{U}^{(p)}$, $\bar{V}_2$, $\bar{Q}_2$, and $\bar{L}$. These equations are driven by $\bar{V}_1$.

7.3.3. Free vibration solution

For free vibrations the driving voltage $V_1 = 0$ which means that the driving electrodes are shorted. The equations and boundary conditions become homogeneous and we have an eigenvalue problem. $u_2^{(0,0)}$, $u_3^{(0,0)}$, $\phi^{(0,0)}$ and ω are unknowns and non-trivial solutions are to be obtained. For gyroscopes, it is revealing to study the free vibration modes when the gyroscope is not rotating. For the beam we are considering, separate flexural modes in the x_2 and x_3 directions can be obtained. They are coupled by the Coriolis force when the beam is rotating.

Consider the sensing mode or flexure in the x_2 direction. When $\Omega = 0$, for the driving portion, we have

$$c_{11}^{(2,0)} k^4 - m\omega^2 = 0, \tag{7.3.44}$$

which has four roots of $k^{(1)} = \sqrt[4]{m\omega^2 / c_{11}^{(2,0)}} = k$, $k^{(2)} = ik^{(1)}$, $k^{(3)} = -k^{(1)}$, and $k^{(4)} = -k^{(2)}$. The solution for $u_2^{(0,0)}$ is

$$u_2^{(0,0)} = (U^{(1)} \sin kx_1 + U^{(2)} \sinh kx_1) \exp(i\omega t), \tag{7.3.45}$$

which already satisfies Eq. $(7.3.36)_{1,3}$.

Similarly, for the sensing portion, the solution can be written as

$$u_2^{(0,0)} = [\bar{U}^{(1)} \sin \bar{k}(x_1 - l) + \bar{U}^{(2)} \sinh \bar{k}(x_1 - l)] \exp(i\omega t), \tag{7.3.46}$$

where $\bar{k} = \sqrt[4]{m\omega^2 / \bar{c}_{11}^{(2,0)}}$, and Eq. $(7.3.37)_{1,3}$ have been satisfied. The continuity conditions of Eq. $(7.3.38)_{1,3,5,7}$ require that

$$\begin{bmatrix} \sin ka_1 & \sinh ka_1 & \sin \bar{k}a_2 & \sinh \bar{k}a_2 \\ k \cos ka_1 & k \cosh ka_1 & -\bar{k} \cos \bar{k}a_2 & -\bar{k} \cosh \bar{k}a_2 \\ c_{11}^{(2,0)} k^2 \sin ka_1 & -c_{11}^{(2,0)} k^2 \sinh ka_1 & \bar{c}_{11}^{(2,0)} \bar{k}^2 \sin \bar{k}a_2 & -\bar{c}_{11}^{(2,0)} \bar{k}^2 \sinh \bar{k}a_2 \\ c_{11}^{(2,0)} k^3 \cos ka_1 & -c_{11}^{(2,0)} k^3 \cosh ka_1 & -\bar{c}_{11}^{(2,0)} \bar{k}^3 \cos \bar{k}a_2 & \bar{c}_{11}^{(2,0)} \bar{k}^3 \cosh \bar{k}a_2 \end{bmatrix} \begin{bmatrix} U^{(1)} \\ U^{(2)} \\ \bar{U}^{(1)} \\ \bar{U}^{(2)} \end{bmatrix} = 0. \tag{7.3.47}$$

The determinant of the coefficient matrix has to vanish, which yields the frequency equation. Once the sensing mode is known, the corresponding electric potential distribution can be obtained from Eqs. $(7.3.35)_3$ and $(7.3.38)_9$ as

$$\begin{aligned} \bar{\varepsilon}_{11}^{(0,0)} \phi^{(0,0)} &= e_{11}^{(1,0)} \bar{U}^{(1)} \bar{k} [\cos \bar{k}a_2 - \cos \bar{k}(x_1 - l)] \exp(i\omega t) \\ &\quad + e_{11}^{(1,0)} \bar{U}^{(2)} \bar{k} [\cosh \bar{k}a_2 - \cosh \bar{k}(x_1 - l)] \exp(i\omega t). \end{aligned} \tag{7.3.48}$$

7.3.4. An example

As an example, we consider PZT-5H. For geometric parameters we choose $a_1 = a_2 = 10$ cm, $b = 1$ cm, and $c = 1$ cm.

7.3.4.1. Free vibration solution

The flexural deformation and electric potential distribution corresponding to the first flexural mode in the x_2 direction are plotted in Fig. 7.3.2, normalized by their maximum values. When the beam is deflected into this mode, in the sensing portion the potential keeps rising and produces a high output voltage.

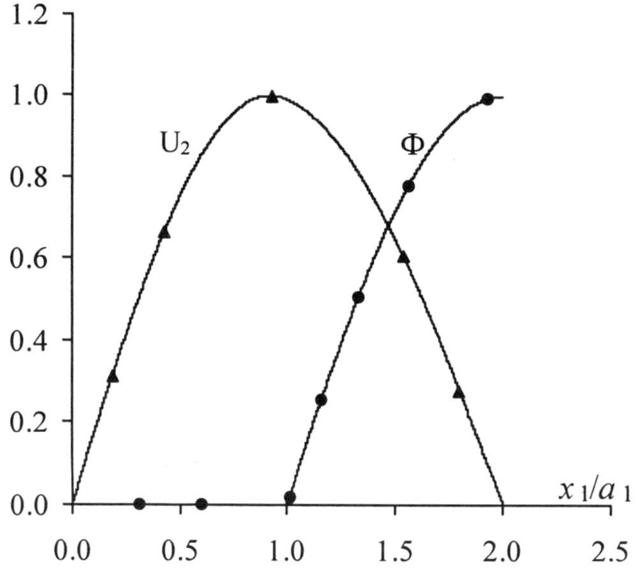

Fig. 7.3.2. Normalized deflection and potential distribution of the sensing mode showing voltage rise ($a_1 = a_2 = 10$ cm, $b = c = 1$ cm, $\Omega = 0$, $U_2 = u_2^{(0,0)} / u_{2\max}^{(0,0)}$, $\Phi = \phi^{(0,0)} / \phi_{\max}^{(0,0)}$).

7.3.4.2. *Forced vibration solution*

For forced vibrations we consider the case when the sensing electrodes are open with $Z = \infty$. In this case $\overline{Q_2} = 0$. We replace s_{pq} by $s_{pq}(1 - iQ^{-1})$, $Q = 10^2$. The basic behaviors are the same as Figs. 7.1.2 through 7.1.4. What is unique to the present gyroscope is the output voltage-aspect ratio relation shown in Fig. 7.3.3. An increasing behavior of the output voltage versus the aspect ratio is exhibited. This shows that when the sensing portion is relatively long compared to the thickness of the driving portion, high output voltage can be produced. This is in agreement with the behavior of the mode in Fig. 7.3.2.

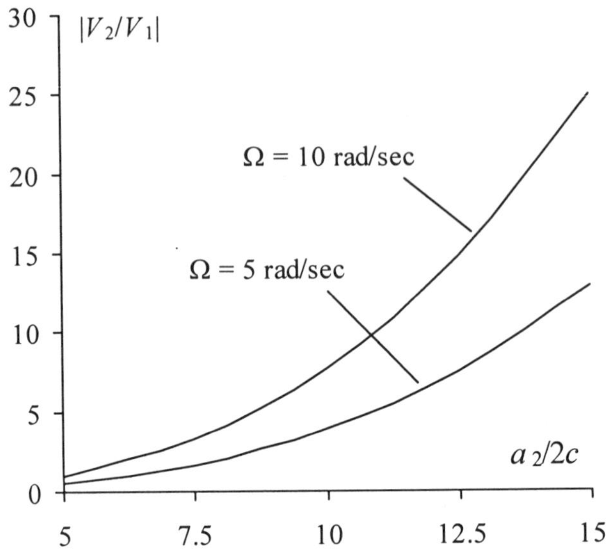

Fig. 7.3.3. Output voltage versus the aspect ratio $a_2/(2c)$.

7.4. An Inhomogeneous Shell

Consider the two modes of a circular cylindrical shell of radius R, length l, and thickness h shown in Fig. 7.4.1. The bottom of the shell is fixed and the top is free.

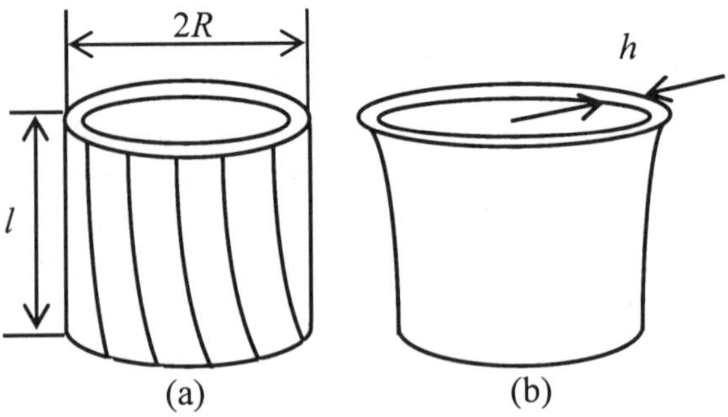

Fig. 7.4.1. Modes of a circular cylindrical shell. (a) Torsion. (b) Horn.

First, we note that material particles move in perpendicular directions in these two modes. Furthermore, the frequency of the torsional mode depends strongly on the length l, while the frequency of the horn mode is sensitive to the radius R. Therefore, by adjusting geometric parameters, the frequencies of these two modes can be tuned as close as desired. For simplicity consider a shell made of an isotropic elastic material. A finite element modal analysis by the commercial software ANSYS shows that the frequencies depend on the geometric parameters as shown in Fig. 7.3.2.

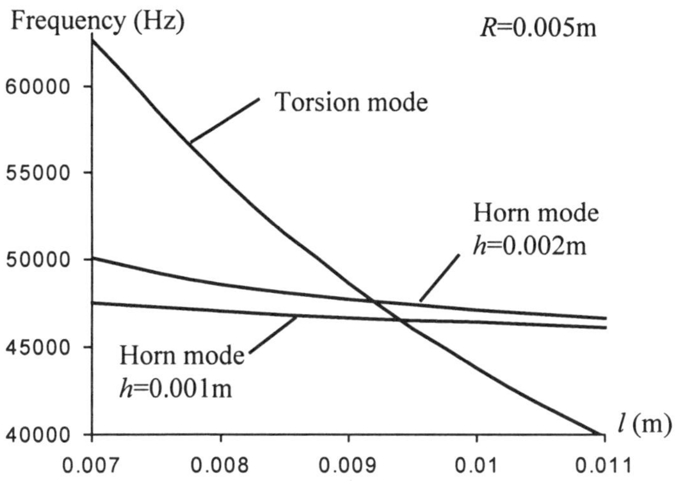

Fig. 7.4.2. Frequencies versus l ($E = 6.36\times10^{10}$ N/m^2, $\nu = 0.38$).

For certain values of the geometric parameters the frequencies of the two modes become the same (where the curves intersect). The above observations show that the torsional and horn modes can be used to make gyroscopes.

7.4.1. Structure

Consider the ceramic piezoelectric shell in Fig. 7.4.3 [81]. The shell is inhomogeneous. For the excitation of the horn mode and the detection of the torsional mode, the two portions of the shell are poled in different directions. In the driving portion $a_2 < z < l$, the ceramic is poled in the r (radial) direction. The driving portion is electroded at $r = R \pm h/2$ with an applied driving voltage V_1 which is time-harmonic. In the sensing portion $0 < z < a_2$, the ceramic is poled in the θ direction. One sensing electrode is placed at $z = 0$. The other sensing electrode is placed near $z = a_2$, or is shared with one of the driving electrodes. Under V_1, the shell can be driven into vibrations in the horn mode in Fig. 7.4.1(b) with u_r as the major displacement and a small u_z. When the shell is rotating about the z axis, the Coriolis force will drive the shell into the torsional motion in Fig. 7.4.1(a) with u_θ. Because of piezoelectric coupling, the torsional motion will generate a voltage between the sensing electrodes which is related to the Coriolis force and hence Ω. Geometric parameters of the shell are determined by that the radial mode and the torsional mode have almost the same resonant frequency.

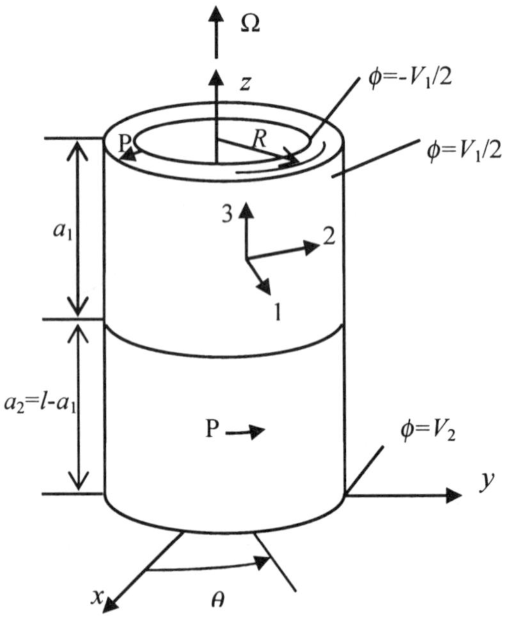

Fig. 7.4.3. A circular cylindrical ceramic shell.

7.4.2. Governing equations

For the driving portion $a_2 < z < l$ with ceramics poled in the radial (1) direction, the electric field is known

$$E_r = -V_1 / h, \quad E_\theta = E_z = 0 . \tag{7.4.1}$$

From the two-dimensional equations of piezoelectric shells [20], we have

$$
\begin{aligned}
& -Dh^{-1}u_{r,zzzz} - R^{-1}c_{33}^{p}u_r / R - R^{-1}c_{23}^{p}u_{z,z} - R^{-1}e_{13}^{p}V_1 / h \\
& \quad = \rho(\ddot{u}_r - 2\Omega\dot{u}_\theta - \Omega^2 u_r), \\
& c_{44}u_{\theta,zz} = \rho(\ddot{u}_\theta + 2\Omega\dot{u}_r - \Omega^2 u_\theta), \\
& c_{23}^{p}u_{r,z} / R + c_{33}^{p}u_{z,zz} = \rho\ddot{u}_z,
\end{aligned}
\tag{7.4.2}
$$

where

$$
\begin{aligned}
& c_{33}^{p} = \hat{c}_{11} - \hat{c}_{13}^{2} / \hat{c}_{33}, \quad c_{23}^{p} = \hat{c}_{12} - \hat{c}_{13}^{2} / \hat{c}_{33}, \quad c_{44} = \hat{c}_{66} \\
& e_{13}^{p} = \hat{e}_{31} - \hat{c}_{13}\hat{e}_{33} / \hat{c}_{33}, \quad \varepsilon_{11}^{p} = \hat{\varepsilon}_{33} + \hat{e}_{33}^{2} / \hat{c}_{33}, \quad D = h^3 c_{33}^{p} / 12.
\end{aligned}
\tag{7.4.3}
$$

$\hat{c}_{pq}$, $\hat{e}_{iq}$ and $\hat{\varepsilon}_{ij}$ are material constants when the poling direction is the three axis. The resultants needed for boundary conditions are

$$
\begin{aligned}
& N_z = hc_{23}^{p}u_r / R + hc_{33}^{p}u_{z,z} + he_{13}^{p}V_1 / h, \\
& N_{z\theta} = hc_{44}u_{\theta,z}, \quad M_z = Du_{r,zz}, \quad Q_z = Du_{r,zzz}, \\
& D_r^{(0)} = he_{13}^{p}(u_r / R + u_{z,z}) - h\varepsilon_{11}^{p}V_1 / h.
\end{aligned}
\tag{7.4.4}
$$

For the sensing portion $0 < z < a_2$, the electric field has the expression

$$
E_r = E_\theta = 0, \quad E_z = -\phi_{,z}. \tag{7.4.5}
$$

The equations of motion are

$$
\begin{aligned}
& -\overline{D}h^{-1}u_{r,zzzz} - R^{-1}\overline{c}_{22}^{p}u_r / R - R^{-1}\overline{c}_{23}^{p}u_{z,z} \\
& \quad = \rho(\ddot{u}_r - 2\Omega\dot{u}_\theta - \Omega^2 u_r), \\
& \overline{c}_{44}u_{\theta,zz} + \overline{e}_{34}\phi_{,zz} = \rho(\ddot{u}_\theta + 2\Omega\dot{u}_r - \Omega^2 u_\theta), \\
& \overline{c}_{23}^{p}u_{r,z} / R + \overline{c}_{33}^{p}u_{z,zz} = \rho\ddot{u}_z, \\
& \overline{e}_{34}u_{\theta,zz} - \overline{\varepsilon}_{33}^{p}\phi_{,zz} = 0,
\end{aligned}
\tag{7.4.6}
$$

where

$$
\begin{aligned}
& \overline{c}_{22}^{p} = \hat{c}_{33} - \hat{c}_{13}^{2} / \hat{c}_{11}, \quad \overline{c}_{23}^{p} = \hat{c}_{13} - \hat{c}_{12}\hat{c}_{13} / \hat{c}_{11}, \\
& \overline{c}_{33}^{p} = \hat{c}_{11} - \hat{c}_{12}^{2} / \hat{c}_{11}, \quad \overline{c}_{44} = \hat{c}_{44}, \\
& \overline{e}_{34} = \hat{e}_{15}, \quad \overline{\varepsilon}_{33}^{p} = \hat{\varepsilon}_{11}, \quad \overline{D} = h^3\overline{c}_{33}^{p} / 12.
\end{aligned}
\tag{7.4.7}
$$

The resultants needed for boundary conditions are

$$N_z = h\bar{c}_{23}^p u_r / R + h\bar{c}_{33}^p u_{z,z},$$
$$N_{z\theta} = h\bar{c}_{44} u_{\theta,z} + h\bar{e}_{34}\phi_{,z}, \quad M_z = \overline{D} u_{r,zz}, \quad Q_z = \overline{D} u_{r,zzz}, \tag{7.4.8}$$
$$D_z^{(0)} = h\bar{e}_{34} u_{\theta,z} - h\bar{\varepsilon}_{33}^p \phi_{,z}.$$

Integrating Eq. $(7.4.6)_4$ with respect to z twice, we obtain

$$\phi(z,t) = \frac{\bar{e}_{34}}{\bar{\varepsilon}_{33}^p} u_\theta(z,t) + \frac{Q_2(t)}{2\pi R h \bar{\varepsilon}_{33}^p} z + L(t), \tag{7.4.9}$$

where Q_2 and L are two integration constants (and may still depend on time). Physically, Q_2 represents the electric charge on the sensing electrode at $z = a_2$. The current flows out of this electrode is given by

$$I_2 = -\dot{Q}_2. \tag{7.4.10}$$

We also have the following circuit condition

$$I_2 = -V_2 / Z, \tag{7.4.11}$$

where

$$V_2 = \phi(0,t). \tag{7.4.12}$$

Substitution of Eq. (7.4.9) into Eqs. $(7.4.6)_2$ and $(7.4.8)_2$ leads to

$$\tilde{c}_{44} u_{\theta,zz} = \rho(\ddot{u}_\theta + \Omega \dot{u}_r - \Omega^2 u_\theta),$$
$$N_{z\theta} = h\tilde{c}_{44} u_{\theta,z} + \frac{\bar{e}_{34}}{2\pi R \bar{\varepsilon}_{33}^p} Q_2, \tag{7.4.13}$$

where

$$\tilde{c}_{44} = \bar{c}_{44} + \bar{e}_{34}^2 / \bar{\varepsilon}_{33}^p. \tag{7.4.14}$$

At the junction $z = a_2$ between the two portions of the shell, we have the following continuity conditions:

$$\|u_r\| = 0, \quad \|u_{r,z}\| = 0, \quad \|Q_z\| = 0, \quad \|M_z\| = 0,$$
$$\|u_\theta\| = 0, \quad \|N_{z\theta}\| = 0, \quad \|u_z\| = 0, \quad \|N_z\| = 0, \quad \phi(a_2,t) = 0, \tag{7.4.15}$$

where $\|u_r\| = u_r(a_2^+) - u_r(a_2^-)$ is the jump of u_r across $z = a_2$, and the rest are similar. At the upper end, the shell should be allowed to deform in the r and θ directions. Therefore, we consider the following conditions for a free end at $z = l$:

$$M_z = 0, \quad Q_z = 0, \quad N_{z\theta} = 0, \quad N_z = 0 . \tag{7.4.16}$$

At $z = 0$ we consider a fixed end with

$$u_r = 0, \quad u_{r,z} = 0, \quad u_\theta = 0, \quad u_z = 0 . \tag{7.4.17}$$

The circuit condition can be written as

$$\dot{Q}_2 = V_2 / Z . \tag{7.4.18}$$

We have one fourth-order and two second-order ordinary differential equations in each portion of the shell. This leads to sixteen integration constants. With V_2 and the two integration constants Q_2 and L that are already introduced, the total number of constants to be determined is nineteen. We have four boundary conditions at the upper end, four boundary conditions in at the lower end, nine conditions at the junction of the two portions, one circuit condition, and Eq. (7.4.12). The total number of conditions is also 19.

7.4.3. Forced vibration solution

Introduce the complex notation

$$\{V_1, V_2, Q_2, L\} = \{\overline{V}_1, \overline{V}_2, \overline{Q}_2, \overline{L}\} \exp(i\omega t) . \tag{7.4.19}$$

The general solution for the driving portion is found to be

$$\begin{Bmatrix} u_r \\ u_\theta \\ u_z \end{Bmatrix} = \sum_{p=1}^{8} \begin{Bmatrix} \alpha^{(p)} \\ \beta^{(p)} \\ \gamma^{(p)} \end{Bmatrix} U^{(p)} \exp[k^{(p)}(z - l)] \exp(i\omega t) + \begin{Bmatrix} \lambda \\ \mu \\ 0 \end{Bmatrix} \overline{V}_1 \exp(i\omega t) , \tag{7.4.20}$$

where the two terms on the right-hand side are the homogeneous and particular solutions respectively. $U^{(p)}$ are undetermined constants. $k^{(p)}$ are the eight roots of

$$\begin{vmatrix} -Dh^{-1}k^4 - R^{-2}c_{33}^p + \rho(\omega^2 + \Omega^2) & i2\rho\omega\Omega & -R^{-1}c_{23}^p k \\ -i2\rho\omega\Omega & c_{44}k^2 + \rho(\omega^2 + \Omega^2) & 0 \\ R^{-1}c_{23}^p k & 0 & c_{33}^p k^2 + \rho\omega^2 \end{vmatrix}$$

$$= 0, \tag{7.4.21}$$

$$\begin{aligned}
&\alpha^{(p)} = \alpha(k^{(p)}), \quad \alpha(k) = 1, \\
&\beta^{(p)} = \beta(k^{(p)}), \quad \beta(k) = i2\rho\omega\Omega/[c_{44}k^2 + \rho(\omega^2 + \Omega^2)], \\
&\gamma^{(p)} = \gamma(k^{(p)}), \quad \gamma(k) = -R^{-1}c_{23}^p k/(c_{33}^p k^2 + \rho\omega^2),
\end{aligned} \tag{7.4.22}$$

and

$$\begin{aligned}
&\lambda = (\omega^2 + \Omega^2)R^{-1}e_{13}^p/(h\Delta), \quad \mu = i2\omega\Omega R^{-1}e_{13}^p/(h\Delta), \\
&\Delta = [\rho(\omega^2 + \Omega^2) - R^{-2}c_{33}^p](\omega^2 + \Omega^2) - 4\rho\omega^2\Omega^2.
\end{aligned} \tag{7.4.23}$$

Similarly, the general solution for the sensing portion can be written as

$$\begin{Bmatrix} u_r \\ u_\theta \\ u_z \end{Bmatrix} = \sum_{p=1}^{8} \begin{Bmatrix} \bar{\alpha}^{(p)} \\ \bar{\beta}^{(p)} \\ \bar{\gamma}^{(p)} \end{Bmatrix} \bar{U}^{(p)} \exp(\bar{k}^{(p)} z)\exp(i\omega t), \tag{7.4.24}$$

where $\bar{U}^{(p)}$ are undetermined constants, $\bar{k}^{(p)}$ are the eight roots of

$$\begin{vmatrix} -\bar{D}h^{-1}\bar{k}^4 - R^{-2}\bar{c}_{22}^p + \rho(\omega^2 + \Omega^2) & i2\rho\omega\Omega & -R^{-1}\bar{c}_{23}^p\bar{k} \\ -i2\rho\omega\Omega & \tilde{c}_{44}\bar{k}^2 + \rho(\omega^2 + \Omega^2) & 0 \\ R^{-1}\bar{c}_{23}^p\bar{k} & 0 & \bar{c}_{33}^p\bar{k}^2 + \rho\omega^2 \end{vmatrix} = 0, \tag{7.4.25}$$

and

$$\begin{aligned}
&\bar{\alpha}^{(p)} = \bar{\alpha}(\bar{k}^{(p)}), \quad \bar{\alpha}(\bar{k}) = 1, \\
&\bar{\beta}^{(p)} = \bar{\beta}(\bar{k}^{(p)}), \quad \bar{\beta}(\bar{k}) = i2\rho\omega\Omega/[\tilde{c}_{44}\bar{k}^2 + \rho(\omega^2 + \Omega^2)], \\
&\bar{\gamma}^{(p)} = \bar{\gamma}(\bar{k}^{(p)}), \quad \bar{\gamma}(\bar{k}) = -R^{-1}\bar{c}_{23}^p\bar{k}/(\bar{c}_{33}^p\bar{k}^2 + \rho\omega^2).
\end{aligned} \tag{7.4.26}$$

The electric potential in the sensing portion is

$$\begin{aligned}
\phi(z,t) = &\sum_{p=1}^{8} \frac{\bar{e}_{34}}{\bar{\varepsilon}_{33}^p} \bar{\beta}^{(p)}\bar{U}^{(p)} \exp(\bar{k}^{(p)} z)\exp(i\omega t) \\
&+ \frac{Q_2(t)}{2\pi Rh\bar{\varepsilon}_{33}^p} z + L(t).
\end{aligned} \tag{7.4.27}$$

Substitution of the solutions into the boundary, continuity and circuit conditions yields the following 19 equations:

$$\sum_{p=1}^{8} \bar{\alpha}^{(p)}\bar{U}^{(p)} = 0, \quad \sum_{p=1}^{8} \bar{k}^{(p)}\bar{\alpha}^{(p)}\bar{U}^{(p)} = 0,$$
$$\sum_{p=1}^{8} \bar{\beta}^{(p)}\bar{U}^{(p)} = 0, \quad \sum_{p=1}^{8} \bar{\gamma}^{(p)}\bar{U}^{(p)} = 0, \tag{7.4.28}$$

$$\sum_{p=1}^{8} D(k^{(p)})^2\alpha^{(p)}U^{(p)} = 0, \quad \sum_{p=1}^{8} D(k^{(p)})^3\alpha^{(p)}U^{(p)} = 0,$$
$$\sum_{p=1}^{8} hc_{44}k^{(p)}\beta^{(p)}U^{(p)} = 0, \tag{7.4.29}$$
$$\sum_{p=1}^{8} (hc_{23}^{p}R^{-1}\alpha^{(p)} + hc_{33}^{p}k^{(p)}\gamma^{(p)})U^{(p)} = -(hc_{23}^{p}R^{-1}\lambda + e_{13}^{p})\bar{V}_1,$$

$$\sum_{p=1}^{8} \alpha^{(p)}U^{(p)}\exp(-k^{(p)}a_1) - \sum_{p=1}^{8} \bar{\alpha}^{(p)}\bar{U}^{(p)}\exp(\bar{k}^{(p)}a_2) = -\lambda\bar{V}_1,$$
$$\sum_{p=1}^{8} k^{(p)}\alpha^{(p)}U^{(p)}\exp(-k^{(p)}a_1) - \sum_{p=1}^{8} \bar{k}^{(p)}\bar{\alpha}^{(p)}\bar{U}^{(p)}\exp(\bar{k}^{(p)}a_2) = 0, \tag{7.4.30}$$

$$\sum_{p=1}^{8} D(k^{(p)})^2\alpha^{(p)}U^{(p)}\exp(-k^{(p)}a_1) - \sum_{p=1}^{8} \bar{D}(\bar{k}^{(p)})^2\bar{\alpha}^{(p)}\bar{U}^{(p)}\exp(\bar{k}^{(p)}a_2) = 0,$$
$$\sum_{p=1}^{8} D(k^{(p)})^3\alpha^{(p)}U^{(p)}\exp(-k^{(p)}a_1) - \sum_{p=1}^{8} \bar{D}(\bar{k}^{(p)})^3\bar{\alpha}^{(p)}\bar{U}^{(p)}\exp(\bar{k}^{(p)}a_2) = 0, \tag{7.4.31}$$

$$\sum_{p=1}^{8} \beta^{(p)} U^{(p)} \exp(-k^{(p)} a_1) - \sum_{p=1}^{8} \bar{\beta}^{(p)} \bar{U}^{(p)} \exp(\bar{k}^{(p)} a_2) = -\mu \bar{V}_1,$$

$$\sum_{p=1}^{8} h c_{44} k^{(p)} \beta^{(p)} U^{(p)} \exp(-k^{(p)} a_1) \tag{7.4.32}$$

$$-\sum_{p=1}^{8} h \tilde{c}_{44} \bar{k}^{(p)} \bar{\beta}^{(p)} \bar{U}^{(p)} \exp(\bar{k}^{(p)} a_2) + \frac{\bar{e}_{34}}{2\pi R \bar{\varepsilon}_{33}^{p}} \bar{Q}_2 = 0,$$

$$\sum_{p=1}^{8} \gamma^{(p)} U^{(p)} \exp(-k^{(p)} a_1) - \sum_{p=1}^{8} \bar{\gamma}^{(p)} \bar{U}^{(p)} \exp(\bar{k}^{(p)} a_2) = 0,$$

$$\sum_{p=1}^{8} (h c_{23}^{p} R^{-1} \alpha^{(p)} + h c_{33}^{p} k^{(p)} \gamma^{(p)}) U^{(p)} \exp(-k^{(p)} a_1) \tag{7.4.33}$$

$$-\sum_{p=1}^{8} (h \bar{c}_{23}^{p} R^{-1} \bar{\alpha}^{(p)} + h \bar{c}_{33}^{p} \bar{k}^{(p)} \bar{\gamma}^{(p)}) \bar{U}^{(p)} \exp(\bar{k}^{(p)} a_2)$$

$$= -(h c_{23}^{p} R^{-1} \lambda + e_{13}^{p}) \bar{V}_1,$$

$$\sum_{p=1}^{8} \frac{\bar{e}_{34}}{\bar{\varepsilon}_{33}^{p}} \bar{\beta}^{(p)} \bar{U}^{(p)} \exp(\bar{k}^{(p)} a_2) + \frac{\bar{Q}_2}{2\pi R h \bar{\varepsilon}_{33}^{p}} + \bar{L} = 0,$$

$$\sum_{p=1}^{8} \frac{\bar{e}_{34}}{\bar{\varepsilon}_{33}^{p}} \bar{\beta}^{(p)} \bar{U}^{(p)} + \frac{\bar{Q}_2}{2\pi R h \bar{\varepsilon}_{33}^{p}} + \bar{L} - \bar{V}_2 = 0, \tag{7.4.34}$$

$$i\omega \bar{Q}_2 - \bar{V}_2 / Z = 0.$$

7.4.4. Discussion

Results for a ceramic shell of PZT-5H can be found in [81]. In addition to the basic behaviors similar to Figs. 7.1.2 through 7.1.4, the ceramic shell gyroscope also has the behavior of high output voltage shown in Fig. 7.3.3. The voltage sensitivity increases for increasing *l/h* when it is driven across the small thickness dimension and the sensing signal is picked up from a larger dimension.

7.5. A Ceramic Ring

In this section we analyze vibrations of a rotating ceramic ring poled in the *z* direction as shown in Fig. 7.5.1 [82].

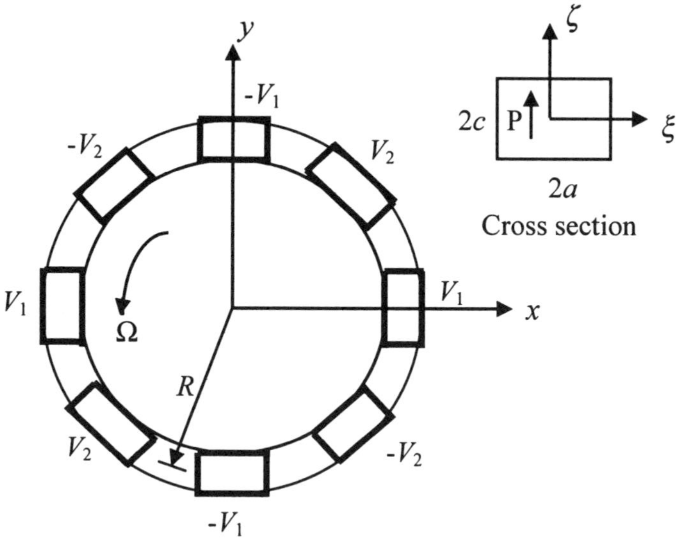

Fig. 7.5.1. A ceramic ring as a gyroscope.

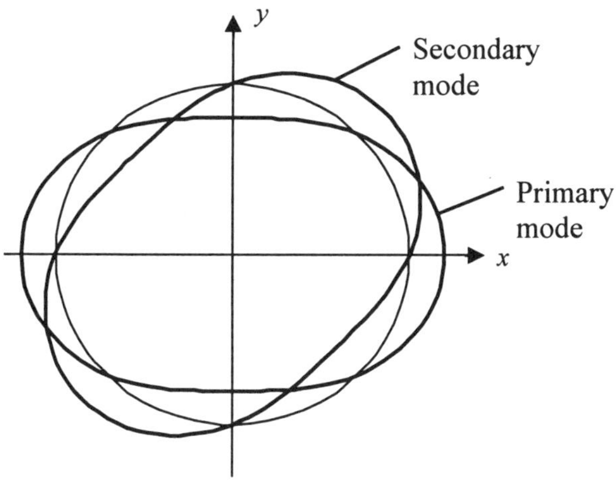

Fig. 7.5.2. Radial displacements of the primary and secondary modes of a non-rotating ring.

One possible arrangement of eight pairs of identical electrodes at $z = \pm c$ are shown in the areas bounded by the thick lines which can be used to excite and detect a particular pair of modes in Fig. 7.5.2. Equivalently, radially polarized ceramics can also be used which will require electrodes to be placed at $\xi = \pm a$.

7.5.1. Governing equations

In Fig. 7.5.1, V_1 is the input voltage across four pairs of driving electrodes and V_2 is the output voltage across the other four pairs of electrodes. We use one-dimensional equations for a piezoelectric ring [20]. The one-dimensional electric potentials are

$$\phi^{(0,0)} = 0, \quad \phi^{(1,0)} = 0, \quad \phi^{(0,1)} = V/2c, \tag{7.5.1}$$

where

$$\begin{aligned} V &\cong V_1(t)\cos(2\theta) + V_2(t)\sin(2\theta) \\ &= [\overline{V}_1 \cos(2\theta) + \overline{V}_2 \sin(2\theta)]\exp(i\omega t). \end{aligned} \tag{7.5.2}$$

$\overline{V}_1$ and $\overline{V}_2$ are complex constants. We have assumed harmonic time dependence and introduced the complex notation. Since the electrodes are equally spaced and that the electric potential is a constant on an electrode, the voltage distribution due to the electrode configuration in Fig. 7.5.1 is essentially piecewise constant. In Eq. (7.5.2), this piecewise constant voltage distribution is approximated by a trigonometric function. $\phi^{(0,1)}$ generates

$$E_z^{(0,0)} = -V/(2c), \quad E_\theta^{(0,1)} = -V_{,\theta}/(2cR). \tag{7.5.3}$$

Let the radial displacement be $u(\theta, t)$ and the tangential displacement be $v(\theta, t)$. The governing equations are [82,20]

$$\begin{aligned} &\frac{A\gamma^2}{s_{11}R^3}(v'''-u'''') - \frac{A}{s_{11}R}(v'+u) - \frac{Ad_{31}}{s_{11}}\frac{V}{2c} \\ &= \rho RA(\ddot{u} - 2\Omega\dot{v} - \Omega^2 u), \\ &\frac{A}{s_{11}R}(v''+u') + \frac{Ad_{31}}{s_{11}}\frac{V'}{2c} + \frac{A\gamma^2}{s_{11}R^3}(v''-u''') \\ &= \rho RA(\ddot{v} + 2\Omega\dot{u} - \Omega^2 v), \end{aligned} \tag{7.5.4}$$

where a prime indicates differentiation with respect to θ. The relevant constitutive relations take the following form:

$$N = \frac{A}{s_{11}R}(v_{,\theta} + u) + \frac{Ad_{31}}{s_{11}}\frac{V}{2c},$$
$$M = -\frac{A\gamma^2}{s_{11}R^2}(v_{,\theta} - u_{,\theta\theta}), \tag{7.5.5}$$
$$D_3^{(0,0)} = \frac{A\bar{e}_{31}}{R}(v_{,\theta} + u) - A\bar{\varepsilon}_{33}\frac{V}{2c},$$

where

$$\bar{e}_{31} = d_{31}/s_{11}, \quad \bar{\varepsilon}_{33} = \varepsilon_{33} - d_{31}d_{31}/s_{11},$$
$$A = 4ac, \quad \gamma^2 = a^2/3. \tag{7.5.6}$$

Equation (7.5.5) shows that $E_z^{(0,0)}$ is responsible for driving the ring into the primary motion through the axial force N and detecting the secondary motion electrically through $D_3^{(0,0)}$. The electric charges and the currents flowing out of the driving and sensing electrodes are given by

$$Q_1 = -4\int_{-\pi/8}^{\pi/8} \frac{D_3^{(0,0)}}{4ac} 2aRd\theta, \quad I_1 = -\dot{Q}_1,$$
$$Q_2 = -4\int_{\pi/8}^{3\pi/8} \frac{D_3^{(0,0)}}{4ac} 2aRd\theta, \quad I_2 = -\dot{Q}_2. \tag{7.5.7}$$

We have the following circuit condition when the motion is time-harmonic:

$$I_2 = V_2/Z. \tag{7.5.8}$$

For time-harmonic solutions let

$$u = [B_1\cos(2\theta) + B_2\sin(2\theta)]\exp(i\omega t),$$
$$v = [C_1\sin(2\theta) + C_2\cos(2\theta)]\exp(i\omega t), \tag{7.5.9}$$

where B_1, B_2, C_1 and C_2 are undetermined constants. The terms with B_1 and C_1 are related to the primary mode, and those with B_2 and C_2 are for the secondary mode. Substituting Eq. (7.5.9) into Eq. (7.5.4), requiring the coefficients of the terms with $\cos 2\theta$ and $\sin 2\theta$ to be equal separately, we obtain the following four equations:

$$\frac{A\gamma^2}{s_{11}R^3}(-8C_1-16B_1)-\frac{A}{s_{11}R}(2C_1+B_1)-\frac{Ad_{31}}{s_{11}}\frac{1}{2c}\overline{V}_1$$
$$=\rho RA(-\omega^2B_1-2i\Omega\omega C_2-\Omega^2B_1),$$
$$\frac{A\gamma^2}{s_{11}R^3}(8C_2-16B_2)-\frac{A}{s_{11}R}(-2C_2+B_2)-\frac{Ad_{31}}{s_{11}}\frac{1}{2c}\overline{V}_2$$
$$=\rho RA(-\omega^2B_2-2i\Omega\omega C_1-\Omega^2B_2),\tag{7.5.10}$$
$$\frac{A\gamma^2}{s_{11}R^3}(-4C_2+8B_2)+\frac{A}{s_{11}R}(-4C_2+2B_2)+\frac{Ad_{31}}{s_{11}}\frac{1}{2c}2\overline{V}_2$$
$$=\rho RA(-\omega^2C_2+2i\Omega\omega B_1-\Omega^2C_2),$$
$$\frac{A\gamma^2}{s_{11}R^3}(-4C_1-8B_1)+\frac{A}{s_{11}R}(-4C_1-2B_1)+\frac{Ad_{31}}{s_{11}}\frac{1}{2c}(-2\overline{V}_1)$$
$$=\rho RA(-\omega^2C_1+2i\Omega\omega B_2-\Omega^2C_1).$$

Equation (7.5.10) represents four equations for the five unknowns of B_1, B_2, C_1, C_2 and $\overline{V}_2$, with $\overline{V}_1$ as the known driving term. The additional equation comes from the circuit condition

$$i\omega 4\sqrt{2}a[\overline{e}_{31}(-2C_2+B_2)-R\overline{\varepsilon}_{33}\overline{V}_2/(2c)]=\overline{V}_2/Z\,.\tag{7.5.11}$$

Equations (7.5.10) and (7.5.11) show that the primary and secondary modes are coupled to $\overline{V}_1$ and $\overline{V}_2$ respectively, and the two modes are further coupled by rotation.

7.5.2. Free vibration solution

For free vibrations the driving voltage $\overline{V}_1$ vanishes. Then Eqs. (7.5.10) and (7.5.11) become homogeneous. Non-trivial solutions of B_1, B_2, C_1, C_2, $\overline{V}_2$ and ω are to be found. For piezoelectric gyroscopes, important information can be learned by studying free vibrations when the gyroscopes are not rotating. When the ring is not rotating we have $\Omega = 0$. Consider the case when the sensing electrodes are also shorted. In this case the circuit condition takes the special form of $\overline{V}_2 = 0$. Then Eq. (7.5.10) reduces to the following two sets of equations:

$$\begin{aligned}&\frac{A\gamma^2}{s_{11}R^3}(-8C_1-16B_1)-\frac{A}{s_{11}R}(2C_1+B_1)=\rho RA(-\omega^2B_1),\\&\frac{A\gamma^2}{s_{11}R^3}(-4C_1-8B_1)+\frac{A}{s_{11}R}(-4C_1-2B_1)=\rho RA(-\omega^2C_1),\end{aligned}\tag{7.5.12}$$

$$\begin{aligned}&\frac{A\gamma^2}{s_{11}R^3}(8C_2-16B_2)-\frac{A}{s_{11}R}(-2C_2+B_2)=\rho RA(-\omega^2B_2),\\&\frac{A\gamma^2}{s_{11}R^3}(-4C_2+8B_2)+\frac{A}{s_{11}R}(-4C_2+2B_2)=\rho RA(-\omega^2C_2).\end{aligned}\tag{7.5.13}$$

Equations (7.5.12) and (7.5.13) govern the primary and secondary modes respectively, which are not coupled without rotation. Clearly, Eq. (7.5.13) can be obtained from Eq. (7.5.12) under

$$B_2=-B_1,\quad C_2=C_1.\tag{7.5.14}$$

The two modes are in fact degenerate modes of the ring with the same frequency and one can be obtained from the other by a reorientation. The resonant frequency is obtained from the determinant of the coefficient matrix of either Eq. (7.5.12) or Eq. (7.5.13), which has to vanish for non-trivial solutions to exist. The two modes corresponding to B_1, C_1 and B_2, C_2 are shown in Fig. 7.5.2. Only the radial displacements of the two modes are shown in the figure. It can be verified that the material particles of these two modes move in perpendicular directions when $\theta = n\pi/4$, where $n = 0, 1, 2, \ldots, 7$. At these locations u or v of the primary mode and v or u of the secondary mode are large. Therefore when the ring is rotating the two modes can be well coupled by the Coriolis force, although they do not have perpendicular velocities at every point.

7.5.3. Forced vibration solution

The forced vibration behaviors of the ring gyroscope are similar to those shown in Figs. 7.1.2 through 7.1.4.

7.6. A Concentrated Mass and Ceramic Rods

In this section we study probably the simplest gyroscope [83,10]. Consider a concentrated mass M connected to two thin rods of polarized ceramics as shown in Fig. 7.6.1.

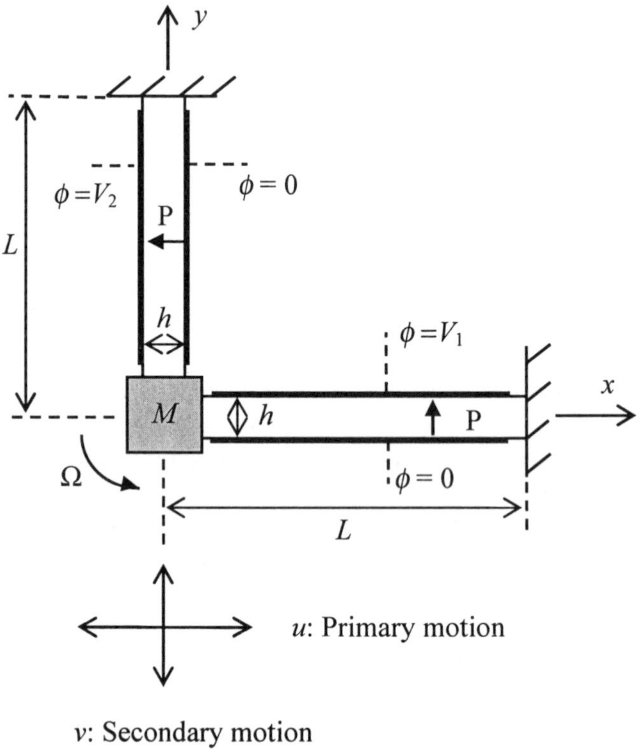

Fig. 7.6.1. A simple piezoelectric gyroscope.

The two rods are electroded at the side surfaces, with electrodes shown by the thick lines. Under a time-harmonic driving voltage $V_1(t)$, the rod along the x direction is driven into extensional vibrations. If the entire system is rotating about the normal of the (x, y) plane at an angular rate Ω, it results in a voltage output $V_2(t)$ across the width of the rod along the y direction. $V_2(t)$ is proportional to Ω when Ω is small, which can be used to detect Ω.

7.6.1. Governing equations

For long and thin rods ($L >> h$) the flexural rigidity is very small. The rods do not resist bending but can still provide extensional forces. When the mass of M is much larger than that of the rods, the inertial effect of the rods can be discounted. Then the mechanical behavior of the rods is effectively like two elastic springs with the addition of piezoelectric couplings. Let the displacements of M in the x and y directions be $u(t)$ and $v(t)$. For each rod we also associate a local coordinate system with the x_1 axis along the axis of the rod and the x_3 axis along the poling direction.

Consider the rod along the x direction first. Neglecting the dynamical effect in the rod due to inertia, the axial strain in the rod can be written as

$$S_1 = -u / L . \tag{7.6.1}$$

With respect to the local coordinate system, the electric field corresponding to the configuration of the driving electrodes can be written as

$$E_1 = E_2 = 0, \quad E_3 = -V_1 / h , \tag{7.6.2}$$

where the driving voltage V_1 is considered given and is time-harmonic. For thin rods in extension, the dominating stress component is the axial stress component T_1. All other stress components can be treated as zero. Under the above stress and electric field conditions, the constitutive relations take the form

$$\begin{aligned} S_1 &= s_{11} T_1 + d_{31} E_3 , \\ D_3 &= d_{31} T_1 + \varepsilon_{33} E_3 , \end{aligned} \tag{7.6.3}$$

where D_3 is the component of the electric displacement vector in the local coordinate system, and s_{11}, d_{31}, and ε_{33} are the relevant elastic, piezoelectric, and dielectric constants. From Eq. (7.6.3) we can solve for T_1 and D_3 in terms of S_1 and E_3, with the result

$$\begin{aligned} T_1 &= \frac{1}{s_{11}} S_1 - \frac{d_{31}}{s_{11}} E_3 = -\frac{1}{s_{11}} \frac{u}{L} + \frac{d_{31}}{s_{11}} \frac{V_1}{h} , \\ D_3 &= \frac{d_{31}}{s_{11}} S_1 + \bar{\varepsilon}_{33} E_3 = -\frac{d_{31}}{s_{11}} \frac{u}{L} - \bar{\varepsilon}_{33} \frac{V_1}{h} , \end{aligned} \tag{7.6.4}$$

where Eqs. (7.6.1) and (7.6.2) have been used, and

$$\bar{\varepsilon}_{33} = \varepsilon_{33} (1 - k_{31}^2), \quad k_{31}^2 = d_{31}^2 / (\varepsilon_{33} s_{11}) . \tag{7.6.5}$$

The axial force in the rod and the electric charge on the electrode at the upper surface of the rod are given by

$$F_1 = T_1 h = -Ku + \frac{d_{31}}{s_{11}} V_1,$$
$$Q_1 = -D_3 L = \frac{d_{31}}{s_{11}} u + C_0 V_1, \tag{7.6.6}$$

where

$$K = \frac{h}{s_{11} L}, \quad C_0 = \frac{\bar{\varepsilon}_{33} L}{h} \tag{7.6.7}$$

represent the elastic stiffness and the static capacitance of the rod. The electric current flowing out of the electrode is related to the charge by

$$I_1 = -\dot{Q}_1 = -\frac{d_{31}}{s_{11}} \dot{u} - C_0 \dot{V}_1. \tag{7.6.8}$$

Similarly, for the rod along the y direction, the axial force and electric current are given by

$$F_2 = -Kv + \frac{d_{31}}{s_{11}} V_2,$$
$$I_2 = -\frac{d_{31}}{s_{11}} \dot{v} - C_0 \dot{V}_2. \tag{7.6.9}$$

For time-harmonic motions, we have the following circuit condition:

$$I_2 = V_2 / Z, \tag{7.6.10}$$

where Z is the impedance of the output circuit.

The equations of motion are

$$F_1 = M(\ddot{u} - 2\Omega \dot{v} - \Omega^2 u),$$
$$F_2 = M(\ddot{v} + 2\Omega \dot{u} - \Omega^2 v). \tag{7.6.11}$$

7.6.2. Forced vibration solution

For time-harmonic motions we use the complex notation

$$(u, v, V_1, V_2, I_1, I_2) = (\bar{u}, \bar{v}, \bar{V}_1, \bar{V}_2, \bar{I}_1, \bar{I}_2) \exp(i\omega t). \tag{7.6.12}$$

The vibration of the system is governed by the following three linear equations for $\bar{u}$, $\bar{v}$, and $\bar{V}_2$, with $\bar{V}_1$ as the driving term:

$$[M(\omega^2+\Omega^2)-K]\bar{u}+2i\omega\Omega M\bar{v}=-\frac{d_{31}}{s_{11}}\bar{V}_1,$$

$$-2i\omega\Omega M\bar{u}+[M(\omega^2+\Omega^2)-K]\bar{v}+\frac{d_{31}}{s_{11}}\bar{V}_2=0, \qquad (7.6.13)$$

$$\frac{d_{31}}{s_{11}}\bar{v}+C_0\left(1+\frac{Z_0}{Z}\right)\bar{V}_2=0.$$

To include damping, s_{11} needs to be replaced by $s_{11}(1-iQ^{-1})$. The output voltage and current, as well as the driving current, are explicitly determined as

$$\frac{\bar{V}_2}{\bar{V}_1}=\frac{1}{\Delta}2i\omega\Omega\omega_0^2\frac{k_{31}^2}{1-k_{31}^2}\frac{Z}{Z+Z_0},$$

$$\frac{-\bar{I}_1}{\bar{V}_1/Z_0}=1-\frac{1}{\Delta}\frac{k_{31}^2}{1-k_{31}^2}\omega_0^2\left[\omega^2+\Omega^2-\omega_0^2\left(1+\frac{k_{31}^2}{1-k_{31}^2}\frac{Z}{Z+Z_0}\right)\right], \qquad (7.6.14)$$

$$\frac{\bar{I}_2}{\bar{V}_1/Z_0}=\frac{1}{\Delta}2i\omega\Omega\omega_0^2\frac{k_{31}^2}{1-k_{31}^2}\frac{Z_0}{Z+Z_0}.$$

Roles of various geometric and physical parameters can be directly seen from the above explicit solution. The basic behaviors of the gyroscope are as in Figs. 7.1.2 through 7.1.4.

7.7. A Ceramic Plate by Two-Dimensional Equations

Consider a rectangular ceramic plate poled in the thickness direction as shown in Fig. 7.7.1 [84].

The plate can vibrate at the fundamental thickness-shear modes in both the x_1 and the x_2 directions. For a ceramic plate poled in the thickness direction x_3, these thickness-shear modes can be excited or detected electrically by lateral electrodes on the sides at $x_1 = \pm a$ or $x_2 = \pm b$. Suppose a thickness-shear vibration u_1 in the x_1 direction is excited by a time-harmonic voltage $2V_1\exp(i\omega t)$ applied across a pair of electrodes at $x_1 = \pm a$. If the plate is rotating about its normal with an angular rate Ω, the Coriolis force F_2 will excite a thickness-shear motion

u_2 in the x_2 direction. This secondary thickness-shear will produce a voltage $2V_2\exp(i\omega t)$ across a pair of electrodes at $x_2 = \pm b$. V_2 can be shown to be proportional Ω and, therefore, can be used to detect Ω.

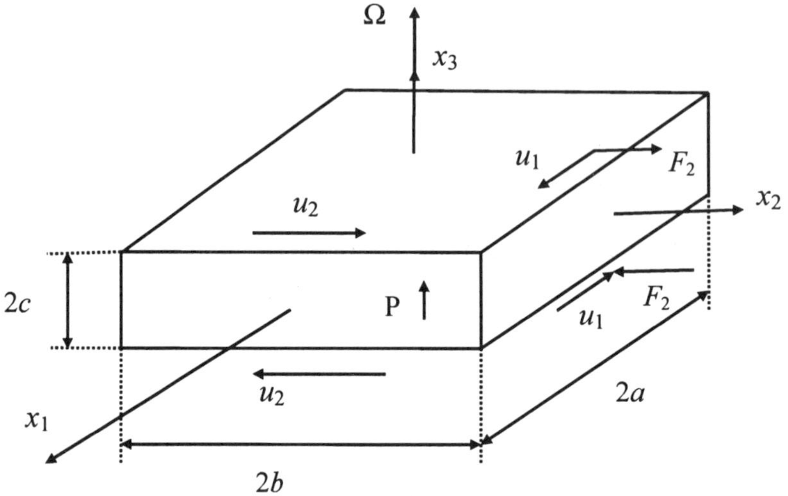

Fig. 7.7.1. A ceramic plate piezoelectric gyroscope.

7.7.1. Driving

We are going to use an approximate procedure to analyze the gyroscope, which can be considered as a perturbation procedure. For the thickness-shear motion excited by V_1, the main motion is $u_1(\mathbf{x},t) \cong x_3 u_1^{(1)}(x_1,x_2,t)$. We approximately have the following one-dimensional problem from the equation for piezoelectric plates [84,20]:

$$\begin{aligned}
&\bar{c}_{11}^{*}u_{1,11}^{(1)} - 3c^{-2}\kappa^2 c_{44}u_1^{(1)} - 3c^{-2}\kappa e_{15}\phi_{,1}^{(0)} = \rho\ddot{u}_1^{(1)}, \qquad |x_1| < a,\\
&\kappa e_{15}u_{1,1}^{(1)} - \varepsilon_{11}\phi_{,11}^{(0)} = 0, \qquad |x_1| < a,\\
&T_{13}^{(0)} = 2c(\kappa^2 c_{44}u_1^{(1)} + \kappa e_{15}\varphi_{,1}^{(0)}) = 0, \qquad x_1 = \pm a,\\
&\phi^{(0)} = \pm V_1 \sin\omega t, \qquad x_1 = \pm a.
\end{aligned} \tag{7.7.1}$$

As an approximation, we neglect the piezoelectric coupling term to $u_1^{(1)}$ in the electrostatic equation in Eq. (7.7.1)$_2$. This gives an approximate solution of the driving electric potential as

$$\phi^{(0)} = \frac{x_1}{a} V_1 \sin \omega t \cong \frac{8}{\pi^2} V_1 \sin \frac{\pi x_1}{2a} \sin \omega t \,. \tag{7.7.2}$$

In Eq. (7.7.2) we have approximated a linear function of x_1 over $[-a,a]$ by a sine function. It can be considered as a one-term approximation by Fourier series. This is sufficient for a qualitative study. Substituting Eq. (7.7.2) into Eq. (7.7.1)$_1$, we then obtain the following expression for $u_1^{(1)}$:

$$u_1^{(1)} = B_1 \cos \frac{\pi x_1}{2a} \sin \omega t \,, \tag{7.7.3}$$

where

$$B_1 = \frac{2\sqrt{3}}{\omega^2 - \omega_0^2 \left(1 + \dfrac{\overset{*}{c}_{11} c^2}{c_{44} a^2} \right)} \frac{e_{15} V_1 / a}{\rho c^2} \,, \quad \omega_0^2 = \frac{\pi^2 c_{44}}{4 \rho c^2} \,. \tag{7.7.4}$$

With $u_1^{(1)}$ known, we have the displacement field

$$u_1 = x_3 u_1^{(1)} = B_1 x_3 \cos \frac{\pi x_1}{2a} \sin \omega t \,, \tag{7.7.5}$$

which leads to a Coriolis force field

$$\mathbf{f} = -2\Omega \mathbf{i}_3 \times \dot{u}_1 \mathbf{i}_1 = -2\Omega \dot{u}_1 \mathbf{i}_2 \,, \tag{7.7.6}$$

or

$$\begin{aligned} f_2 &= -2\Omega \omega B_1 x_3 \cos \frac{\pi x_1}{2a} \cos \omega t \\ &\cong -\frac{4}{\pi} \Omega \omega B_1 x_3 \cos \omega t \cong -2\Omega \omega B_1 x_3 \cos \frac{\pi x_2}{2b} \cos \omega t, \end{aligned} \tag{7.7.7}$$

where we have approximated a cosine function over $[-a, a]$ by a constant, and similarly a constant over $[-b, b]$ by a cosine function. With F_2 we calculate the plate load

$$F_2^{(1)} = \int_{-c}^{c} x_3 \rho f_2 dx_3 = -\frac{2c^3}{3} \rho 2\Omega \omega B_1 \cos \frac{\pi x_2}{2b} \cos \omega t \,. \tag{7.7.8}$$

7.7.2. Sensing

Then the boundary value problem for the thickness-shear $u_2(\mathbf{x},t) \cong x_3 u_2^{(1)}(x_1, x_2, t)$ in the x_2 direction can be written as

$$\begin{aligned}
&\overset{*}{c}_{11} u^{(1)}_{2,22} - 3c^{-2}\kappa^2 c_{44} u_2^{(1)} \\
&\quad - 3c^{-2}\kappa e_{15}\phi^{(0)}_{,2} + \frac{3}{2c^3}F_2^{(1)} = \rho \ddot{u}_2^{(1)}, \quad |x_2| < b, \\
&\kappa e_{15} u^{(1)}_{2,2} - \varepsilon_{11}\phi^{(0)}_{,22} = 0, \quad |x_2| < b, \\
&T_{23}^{(0)} = 2c(\kappa^2 c_{44} u_2^{(1)} + \kappa e_{15}\phi^{(0)}_{,2}) = 0, \quad x_2 = \pm b \\
&\phi^{(0)} = \pm V_2 \sin\omega t, \quad x_2 = \pm b, \\
&D_2^{(0)} = 2c(\kappa e_{15} u_2^{(1)} - \varepsilon_{11}\phi^{(0)}_{,2}) = 0, \quad x_2 = \pm b.
\end{aligned} \tag{7.7.9}$$

In Eq. (7.7.9) where V_2 is unknown, we need some circuit condition joining the electrodes at $x_2 = \pm b$ to determine V_2. Consider the simple case of an open circuit in which $D_2^{(0)}$ or the charge, and hence the current on the electrodes, vanishes. As an approximation, we neglect the piezoelectric coupling term to $\phi^{(0)}$ in Eq. (7.7.9)$_1$. Then, under the Coriolis force $F_2^{(1)}$, we obtain

$$u_2^{(1)} = B_2 \cos\frac{\pi x_2}{2b}\cos\omega t, \tag{7.7.10}$$

where

$$B_2 = \frac{2\Omega\omega}{\omega^2 - \omega_0^2\left(1 + \dfrac{\overset{*}{c}_{11}c^2}{c_{44}b^2}\right)} B_1. \tag{7.7.11}$$

With $u_2^{(1)}$ known, from Eq. (7.7.9)$_2$, we obtain

$$\phi^{(0)} = A_2 \sin\frac{\pi x_2}{2b}\cos\omega t \cong \frac{\pi^2}{8} A_2 \frac{x_2}{b}\cos\omega t, \tag{7.7.12}$$

where

$$A_2 = \frac{b e_{15}}{\sqrt{3}\varepsilon_{11}} B_2. \tag{7.7.13}$$

The mechanical boundary conditions and the open circuit condition in Eq. (7.7.9) are satisfied. The electric boundary condition $(7.7.9)_4$ gives the output voltage

$$V_2 = \phi^{(0)}(x_2 = b)$$

$$= \left[\frac{\pi^2 b e_{15}}{8\sqrt{3}\varepsilon_{11}}\right]\left[\frac{2\Omega\omega}{\omega^2 - \omega_0^2\left(1+\dfrac{c_{11}^* c^2}{c_{44}b^2}\right)}\right]\left[\frac{2\sqrt{3}}{\omega^2 - \omega_0^2\left(1+\dfrac{c_{11}^* c^2}{c_{44}a^2}\right)}\frac{e_{15}V_1/a}{\rho c^2}\right]\cos\omega t, \tag{7.7.14}$$

or, for voltage sensitivity

$$\frac{V_2}{V_1\Omega} = \left[2k_{15}^2\frac{b}{a}\right]\left[\frac{\omega}{\omega^2 - \omega_0^2\left(1+\dfrac{c_{11}^* c^2}{c_{44}b^2}\right)}\right]\left[\frac{\omega_0^2}{\omega^2 - \omega_0^2\left(1+\dfrac{c_{11}^* c^2}{c_{44}a^2}\right)}\right]\cos\omega t. \tag{7.7.15}$$

7.7.3. Discussion

In Eq. (7.7.15), the three pairs of brackets on the right hand side represent, from right to left, respectively, the driving of the thickness-shear motion in the x_1 direction by the applied voltage V_1, the driving of the thickness-shear motion in the x_2 direction by the Coriolis force, and the sensing of the thickness-shear motion in the x_2 direction. The roles of various material constants and geometric parameters are clearly exhibited. At resonant frequencies, the above expressions become singular. Equation (7.7.15) shows that for high voltage sensitivity, the driving frequency and the two resonant frequencies of the thickness-shear motions in two directions must all be very close (double resonance). This implies that a must be very close to b and the plate is almost square. When that is the case, we can write

$$\omega_0^2\left(1+\frac{c_{11}^*c^2}{c_{44}a^2}\right)\cong(\omega+\Delta\omega_1)^2,$$
$$\omega_0^2\left(1+\frac{c_{11}^*c^2}{c_{44}b^2}\right)\cong(\omega+\Delta\omega_2)^2, \tag{7.7.16}$$

which implies that

$$\frac{V_2}{V_1\Omega}\cong k_{15}^2\frac{b}{2a}\frac{\omega}{(\Delta\omega_1)(\Delta\omega_2)}. \tag{7.7.17}$$

7.8. A Ceramic Plate by Zero-Dimensional Equations

In this section we analyzed the ceramic plate gyroscope in the previous section using zero-dimensional equations of a piezoelectric parallelepiped [85]. Consider the rectangular ceramic plate shown in Fig. 7.8.1.

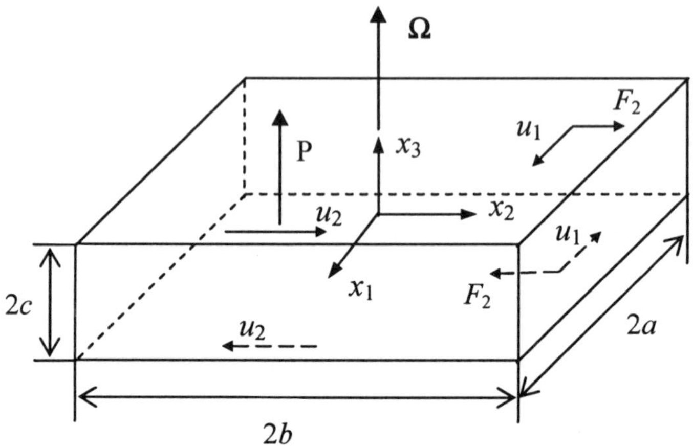

Fig. 7.8.1. A thickness-shear piezoelectric gyroscope.

A driving voltage V_1 is applied across the lateral electrodes at $x_1 = \pm a$ to excite the plate into thickness-shear motion u_1 in the x_1 direction. When the plate is rotating about the x_3-axis, the Coriolis force F_2 causes a thickness-shear motion u_2 in the x_2 direction. This shear in the x_2 direction generates a voltage V_2 between $x_2 = \pm b$, which can be used to detect the angular rate Ω.

7.8.1. Governing equations

In terms of the lowest order zero-dimensional equations [20], the equations for shear motions $u_1 \cong x_3 u_1^{(0,0,1)}$ and $u_2 \cong x_3 u_2^{(0,0,1)}$ are

$$\begin{aligned} -T_{31}^{(0,0,0)} &= \rho \frac{8abc^3}{3}(\ddot{u}_1^{(0,0,1)} - 2\Omega \dot{u}_2^{(0,0,1)} - \Omega^2 u_1^{(0,0,1)}), \\ -T_{32}^{(0,0,0)} &= \rho \frac{8abc^3}{3}(\ddot{u}_2^{(0,0,1)} + 2\Omega \dot{u}_1^{(0,0,1)} - \Omega^2 u_2^{(0,0,1)}). \end{aligned} \tag{7.8.1}$$

The relevant constitutive relations take the following form:

$$\begin{aligned} T_{32}^{(0,0,0)} &= 8abc(c_{44}\kappa^2 S_4^{(0,0,0)} - e_{15}\kappa E_2^{(0,0,0)}), \\ T_{31}^{(0,0,0)} &= 8abc(c_{44}\kappa^2 S_5^{(0,0,0)} - e_{15}\kappa E_1^{(0,0,0)}), \end{aligned} \tag{7.8.2}$$

$$\begin{aligned} D_1^{(0,0,0)} &= 8abc(e_{15}\kappa S_5^{(0,0,0)} + \varepsilon_{11} E_1^{(0,0,0)}), \\ D_2^{(0,0,0)} &= 8abc(e_{15}\kappa S_4^{(0,0,0)} + \varepsilon_{11} E_2^{(0,0,0)}). \end{aligned} \tag{7.8.3}$$

In Eqs. (7.8.2) and (7.8.3) we have introduced a thickness-shear correction factor κ. For a ceramic plate, we have $\kappa^2 = \pi^2/12$ when the plate is poled in the thickness direction. For convenience we denote

$$\begin{aligned} &u_1^{(0,0,1)} = U_1, \quad u_2^{(0,0,1)} = U_2, \\ &\phi^{(1,0,0)} = V_1/2a, \quad \phi^{(0,1,0)} = V_2/2b. \end{aligned} \tag{7.8.4}$$

Then

$$\begin{aligned} &S_4^{(0,0,0)} = U_2, \quad S_5^{(0,0,0)} = U_1, \\ &E_1^{(0,0,0)} = -V_1/2a, \quad E_2^{(0,0,0)} = -V_2/2b. \end{aligned} \tag{7.8.5}$$

With successive substitutions, we obtain

$$\begin{aligned} -c_{44}\kappa^2 U_1 - e_{15}\kappa V_1/2a &= \rho \frac{c^2}{3}(\ddot{U}_1 - 2\Omega \dot{U}_2 - \Omega^2 U_1), \\ -c_{44}\kappa^2 U_2 - e_{15}\kappa V_2/2b &= \rho \frac{c^2}{3}(\ddot{U}_2 + 2\Omega \dot{U}_1 - \Omega^2 U_2), \end{aligned} \tag{7.8.6}$$

$$D_1^{(0,0,0)} = 8abc(e_{15}\kappa U_1 - \varepsilon_{11}V_1 / 2a),$$
$$D_2^{(0,0,0)} = 8abc(e_{15}\kappa U_2 - \varepsilon_{11}V_2 / 2b). \tag{7.8.7}$$

The total electric charge on the electrodes at $x_1 = a$ or $x_2 = b$ and the electric currents flowing out of them are given by

$$Q_1 = -D_1^{(0,0,0)} /(2a), \quad Q_2 = -D_2^{(0,0,0)} /(2b),$$
$$I_1 = -\dot{Q}_1, \quad I_2 = -\dot{Q}_2. \tag{7.8.8}$$

Let the driving voltage be $V_1 = i\overline{V}_1 \exp(i\omega t)$. Introduce the complex notation

$$U_1 = i\overline{U}_1 \exp(i\omega t), \quad U_2 = \overline{U}_2 \exp(i\omega t),$$
$$I_2 = \overline{I}_2 \exp(i\omega t), \quad V_2 = \overline{V}_2 \exp(i\omega t), \tag{7.8.9}$$

we can write the circuit condition as

$$\overline{I}_2 = \overline{V}_2 / Z. \tag{7.8.10}$$

From Eqs. $(7.8.7)_2$, $(7.8.8)_{2,4}$ and (7.8.10), we obtain

$$4aci\omega(e_{15}\kappa\overline{U}_2 - \varepsilon_{11}\overline{V}_2 / 2b) = \overline{V}_2 / Z. \tag{7.8.11}$$

Then we have the following equations:

$$-c_{44}\kappa^2\overline{U}_1 - e_{15}\kappa\overline{V}_1 / 2a = \rho\frac{c^2}{3}(-\omega^2\overline{U}_1 - 2\Omega\omega\overline{U}_2 - \Omega^2\overline{U}_1),$$
$$-c_{44}\kappa^2\overline{U}_2 - e_{15}\kappa\overline{V}_2 / 2b = \rho\frac{c^2}{3}(-\omega^2\overline{U}_2 - 2\Omega\omega\overline{U}_1 - \Omega^2\overline{U}_2), \tag{7.8.12}$$
$$4ac\omega(e_{15}\kappa\overline{U}_2 - \varepsilon_{11}\overline{V}_2 / 2b) = \overline{V}_2 / iZ(\omega).$$

Equation (7.8.12) represents three linear algebraic equations for $\overline{U}_1$, $\overline{U}_2$ and $\overline{V}_2$ in which $\overline{V}_1$ is the inhomogeneous driving term.

7.8.2. Free vibration solution

For free vibrations we set $\overline{V}_1 = 0$, which physically means that the driving electrodes are shorted. Then Eq. (7.8.12) reduces to a system of homogeneous equations for $\overline{U}_1$, $\overline{U}_2$ and $\overline{V}_2$. For non-trivial solutions, the determinant of the coefficient matrix must vanish, which leads to the following frequency equation:

$$\begin{aligned}\Delta(\omega) &= (\omega^2+\Omega^2-\omega_0^2)^2 \\ &\quad -4\Omega^2\omega^2-\lambda(\omega)\omega_0^2(\omega^2+\Omega^2-\omega_0^2)=0,\end{aligned} \tag{7.8.13}$$

where we have denoted

$$\begin{aligned}\omega_0^2 &= \frac{3c_{44}\kappa^2}{\rho c^2}, \quad \lambda(\omega)=\frac{k_{15}^2}{1+Z_2(\omega)/Z(\omega)}, \\ k_{15}^2 &= \frac{e_{15}^2}{\varepsilon_{11}c_{44}}, \quad Z_2=\frac{1}{i\omega C_2}, \quad C_2=\frac{\varepsilon_{11}4ac}{2b}.\end{aligned} \tag{7.8.14}$$

ω_0 is the lowest thickness-shear resonant frequency of an elastic plate with shear constant c_{44}, mass density ρ, and thickness $2c$. C_2 is a static capacitance. Equation (7.8.13) can be written in the following form:

$$\omega^2=\left(1+\frac{1}{2}\lambda\right)\omega_0^2+\Omega^2\pm\frac{1}{2}\lambda\omega_0^2\sqrt{1+\frac{16+8\lambda}{\lambda^2}\frac{\Omega^2}{\omega_0^2}}\,. \tag{7.8.15}$$

Strictly speaking, Eq. (7.8.15) is not a frequency solution to Eq. (7.8.13) because, in general, λ is a function of ω. In the special case of shorted output electrodes we have $Z = 0$ and $\lambda = 0$. Then Eq. (7.8.15) reduces to

$$\omega=\omega_0\pm\Omega\,, \tag{7.8.16}$$

which is a frequency solution. When the output electrodes are open we have $Z = \infty$ and $\lambda = k_{15}^2$. In this case Eq. (7.8.15) also represents a frequency solution. Piezoelectric gyroscopes usually operate under the condition that $\Omega \ll \omega_0$. From Eq. (7.8.15) we can see that, for small Ω and open output electrodes, the effect of Ω on ω is quadratic. This is different from the special case of shorted output electrodes as shown by Eq. (7.8.16). If the load circuit is essentially capacitive with a capacitance C, we have $Z = 1/(i\omega C)$ and $Z_2/Z = C/C_2$ which is independent of ω. Then Eq. (7.8.15) represents a frequency solution.

7.8.3. Forced vibration solution

For forced vibration analysis, we use $c_{44}^* = c_{44}(1+iQ^{-1})$ for the shear elastic constant, where c_{44} and Q are real. From Eq. (7.8.12), we obtain the forced vibration solution for the output voltage as

$$\frac{\overline{V}_2}{\overline{V}_1} = -(k_{15}^*)^2 \frac{b}{a} \frac{2\Omega\omega(\omega_0^*)^2}{(1+Z_2/Z)\Delta(\omega)}, \tag{7.8.17}$$

where

$$(k_{15}^*)^2 = \frac{e_{15}^2}{\varepsilon_{11}c_{44}^*}, \quad (\omega_0^*)^2 = \frac{3c_{44}^*\kappa^2}{\rho c^2}. \tag{7.8.18}$$

The comparison of Eq. (7.8.17) with Eq. (7.7.17) is straightforward.

Chapter 8
Acceleration Sensitivity

Piezoelectric resonators are often used on moving objects, e.g. missiles and satellites. When these objects accelerate, inertial force due to the acceleration causes initial or biasing deformations in the resonator. The resonant frequencies of a resonator with the presence of acceleration induced biasing fields are slightly different from the frequencies without the biasing fields. This effect can be used to make acceleration sensors. For resonators used for control and telecommunication, acceleration induced frequency shifts need to be minimized. The requirements for military applications are driving the need to advance from the $|\Delta\omega/\omega_0| < 10^{-10}$ per g production technology available today to 10^{-12} per g or better in the future. To analyze acceleration sensitivity, the theory for small fields superposed on biasing fields is needed. Early results on resonator acceleration sensitivity were given in [86–89]. More references can be found in two review articles [90,91]. This chapter begins with the deformation of a crystal plate under acceleration, which is the biasing deformation needed in later sections on acceleration sensitivity. Since the acceleration effect is mechanical, piezoelectric coupling is neglected.

8.1. Deformation of a Quartz Plate under Normal Acceleration

Consider a rectangular plate of AT-cut quartz (see Fig. 8.1.1). The plate is simply supported and is under a normal acceleration.

8.1.1. Classical flexure

From the classical theory of flexure [88], the boundary value problem for the deflection $w_2^{(0)}$ is

$$\begin{aligned} &K_{AB,AB}^{(1)} - 2h\rho_0 a_2 = 0, \quad |X_1| < a, \quad |X_3| < b, \\ &w_2^{(0)} = 0, \quad K_{11}^{(1)} = 0, \quad X_1 = \pm a, \\ &w_2^{(0)} = 0, \quad K_{33}^{(1)} = 0, \quad X_3 = \pm b, \end{aligned} \tag{8.1.1}$$

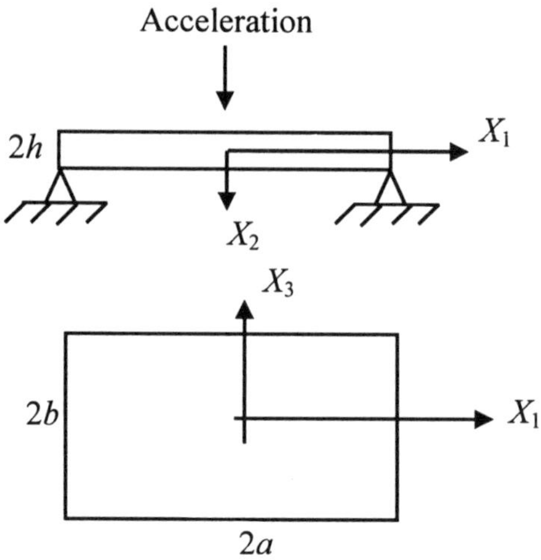

Fig. 8.1.1. A simply supported quartz plate.

where we have introduced the convention that *A*, *B*, *C*, *D* take the values 1 and 3 and skip 2. The bending and twisting moments are given by

$$K_{AB}^{(1)} = \int_{-h}^{h} X_2 K_{AB} dX_2 = \frac{2}{3} h^3 \gamma_{ABCD} S_{CD}^{(1)}, \tag{8.1.2}$$

where

$$S_{CD}^{(1)} = -w_{2,CD}^{(0)}, \tag{8.1.3}$$

and, with the compressed notation,

$$\begin{aligned} &\gamma_{RS} = c_{RS} - c_{RW} c_{WV}^{-1} c_{VS}, \\ &R,S = 1,3,5, \quad W,V = 2,4,6. \end{aligned} \tag{8.1.4}$$

To calculate acceleration sensitivity, the displacement gradient due to flexure is needed. Once $w_2^{(0)}$ is found, the displacement gradient directly due to flexure can be obtained from [88]

$$\begin{aligned} &w_{L,K} = w_{L,K}^S + \Omega_{KL}. \\ &w_{L,K}^S = \frac{1}{2}(w_{L,K} + w_{K,L}) = S_{KL} = X_2 S_{KL}^{(1)}, \quad S_W^{(1)} = -c_{WV}^{-1} c_{VS} S_S^{(1)}, \\ &\Omega_{KL} = \frac{1}{2}(w_{L,K} - w_{K,L}), \quad \Omega_{2A} = -w_{2,A}^{(0)} + X_2 S_{2A}^{(1)}, \quad \Omega_{13} = 0. \end{aligned} \tag{8.1.5}$$

8.1.2. Flexure induced in-plane extension and thickness contraction

When a plate is in flexure, its middle surface is usually stretched slightly. This is a second-order effect compared to the deflection. However, it has important implications on acceleration sensitivity. The in-plane extension of a plate $w_A(X_1, X_3)$ (A = 1, 3) is governed by the following boundary value problem:

$$\begin{aligned} &K_{AB,AB} = 0, \quad |X_1| < a, \quad |X_3| < b, \\ &w_1 = 0, \quad K_{13} = 0, \quad X_1 = \pm a, \\ &w_3 = 0, \quad K_{31} = 0, \quad X_3 = \pm b. \end{aligned} \tag{8.1.6}$$

The boundary conditions in Eq. (8.1.6) physically represent that at the edge of the plate there is no displacement perpendicular to the edge and there is no shear force tangential to the edge. The constitutive relations are

$$K_{AB} = \gamma_{ABCD} S_{CD}\,. \tag{8.1.7}$$

Flexure induced in-plane extension can be taken into account by considering certain quadratic terms in the expression of the finite strain tensor. The specific quadratic terms relevant are $w_{2,A}^{(0)} w_{2,B}^{(0)}$, where $w_2^{(0)}$ is determined by Eq. (8.1.1). Therefore,

$$\begin{aligned} S_1 &= S_{11} = w_{1,1} + (w_{1,1}w_{1,1} + w_{2,1}w_{2,1} + w_{3,1}w_{3,1})/2 \\ &\cong w_{1,1} + w_{2,1}^{(0)} w_{2,1}^{(0)}/2, \\ S_3 &= S_{33} = w_{3,3} + (w_{1,3}w_{1,3} + w_{2,3}w_{2,3} + w_{3,3}w_{3,3})/2 \\ &\cong w_{3,3} + w_{2,3}^{(0)} w_{2,3}^{(0)}/2, \\ S_5 &= 2S_{31} = w_{3,1} + w_{1,3} + w_{1,3}w_{1,1} + w_{2,3}w_{2,1} + w_{3,3}w_{3,1} \\ &\cong w_{3,1} + w_{1,3} + w_{2,3}^{(0)} w_{2,1}^{(0)}. \end{aligned} \tag{8.1.8}$$

With successive substitutions from Eqs. (8.1.7) and (8.1.8), Eq. $(8.1.6)_1$ can be written as two equations for w_1 and w_3

$$\begin{aligned} &\gamma_{11}w_{1,11} + 2\gamma_{15}w_{1,13} + \gamma_{55}w_{1,33} + \gamma_{15}w_{3,11} \\ &\quad + (\gamma_{13} + \gamma_{55})w_{3,13} + \gamma_{35}w_{3,33} + f_1 = 0, \\ &\gamma_{15}w_{1,11} + (\gamma_{13} + \gamma_{55})w_{1,13} + \gamma_{35}w_{1,33} + \gamma_{55}w_{3,11} \\ &\quad + 2\gamma_{35}w_{3,13} + \gamma_{33}w_{3,33} + f_3 = 0, \end{aligned} \tag{8.1.9}$$

where f_1 and f_3 represent

$$\begin{aligned} f_1 &= \gamma_{11} w_{2,1}^{(0)} w_{2,11}^{(0)} + 2\gamma_{15} w_{2,1}^{(0)} w_{2,13}^{(0)} + \gamma_{55} w_{2,1}^{(0)} w_{2,33}^{(0)} \\ &\quad + \gamma_{15} w_{2,3}^{(0)} w_{2,11}^{(0)} + (\gamma_{13} + \gamma_{55}) w_{2,3}^{(0)} w_{2,13}^{(0)} + \gamma_{35} w_{2,3}^{(0)} w_{2,33}^{(0)}, \\ f_3 &= \gamma_{15} w_{2,1}^{(0)} w_{2,11}^{(0)} + (\gamma_{13} + \gamma_{55}) w_{2,1}^{(0)} w_{2,13}^{(0)} + \gamma_{35} w_{2,1}^{(0)} w_{2,33}^{(0)} \\ &\quad + \gamma_{55} w_{2,3}^{(0)} w_{2,11}^{(0)} + 2\gamma_{35} w_{2,3}^{(0)} w_{2,13}^{(0)} + \gamma_{33} w_{2,3}^{(0)} w_{2,33}^{(0)}. \end{aligned} \tag{8.1.10}$$

For a plate of rotated Y-cut quartz, $\gamma_{15} = \gamma_{35} = 0$. Equations (8.1.9) and (8.1.10) reduce to

$$\begin{aligned} &\gamma_{11} w_{1,11} + \gamma_{55} w_{1,33} + (\gamma_{13} + \gamma_{55}) w_{3,13} + f_1 = 0, \\ &(\gamma_{13} + \gamma_{55}) w_{1,13} + \gamma_{55} w_{3,11} + \gamma_{33} w_{3,33} + f_3 = 0, \end{aligned} \tag{8.1.11}$$

and

$$\begin{aligned} f_1 &= \gamma_{11} w_{2,1}^{(0)} w_{2,11}^{(0)} + \gamma_{55} w_{2,1}^{(0)} w_{2,33}^{(0)} + (\gamma_{13} + \gamma_{55}) w_{2,3}^{(0)} w_{2,13}^{(0)}, \\ f_3 &= (\gamma_{13} + \gamma_{55}) w_{2,1}^{(0)} w_{2,13}^{(0)} + \gamma_{55} w_{2,3}^{(0)} w_{2,11}^{(0)} + \gamma_{33} w_{2,3}^{(0)} w_{2,33}^{(0)}, \end{aligned} \tag{8.1.12}$$

where

$$\begin{aligned} \gamma_{11} &= c_{11} - \frac{c_{44} c_{12}^2 + c_{22} c_{14}^2 - 2 c_{12} c_{14} c_{24}}{c_{22} c_{44} - c_{24}^2}, \\ \gamma_{13} &= c_{13} - \frac{c_{12} c_{44} c_{32} + c_{14} c_{22} c_{34} - c_{12} c_{24} c_{34} - c_{14} c_{24} c_{32}}{c_{22} c_{44} - c_{24}^2}, \\ \gamma_{33} &= c_{33} - \frac{c_{44} c_{32}^2 + c_{22} c_{34}^2 - 2 c_{32} c_{34} c_{24}}{c_{22} c_{44} - c_{24}^2}, \\ \gamma_{55} &= c_{55} - c_{56}^2 / c_{66}. \end{aligned} \tag{8.1.13}$$

With the solution for the extension w_A from Eq. (8.1.9), we can directly calculate the symmetric parts of $w_{A,B}$ due to extension (or indirectly due to flexure)

$$w_{1,1}^S, \quad w_{3,3}^S, \quad w_{1,3}^S = w_{3,1}^S. \tag{8.1.14}$$

We also have

$$\Omega_{13} = (w_{3,1} - w_{1,3}), \quad \Omega_{2A} = 0. \tag{8.1.15}$$

S_1, S_3 and S_5 can be directly calculated form the solution for extension. The other strain components, S_2, S_4 and S_6, can be calculated form

$$S_W = -c_{WV}^{-1} c_{VS} S_S \,. \tag{8.1.16}$$

Then, since

$$\begin{aligned}
S_2 &= S_{22} = w_{2,2} + (w_{1,2}w_{1,2} + w_{2,2}w_{2,2} + w_{3,2}w_{3,2})/2 \\
&\cong w_{2,2} + (w_{2,1}^{(0)}w_{2,1}^{(0)} + w_{2,3}^{(0)}w_{2,3}^{(0)})/2 = w_{2,2}^{S} + (w_{2,1}^{(0)}w_{2,1}^{(0)} + w_{2,3}^{(0)}w_{2,3}^{(0)})/2, \\
S_4 &= 2S_{23} = w_{2,3} + w_{3,2} + w_{1,2}w_{1,3} + w_{2,2}w_{2,3} + w_{3,2}w_{3,3} \\
&\cong w_{2,3} + w_{3,2} = 2w_{2,3}^{S} = 2w_{3,2}^{S}, \\
S_6 &= 2S_{12} = w_{1,2} + w_{2,1} + w_{1,1}w_{1,2} + w_{2,1}w_{2,2} + w_{3,1}w_{3,2} \\
&\cong w_{1,2} + w_{2,1} = 2w_{1,2}^{S} = 2w_{2,1}^{S},
\end{aligned} \tag{8.1.17}$$

we can calculate $w_{2,2}$, $w_{2,3} + w_{3,2}$, and $w_{2,1} + w_{1,2}$ from Eq. (8.1.17) and thus determine all the displacement gradient components due to extension through $w_{K,L} = w_{K,L}^{S} + \Omega_{LK}$.

8.1.3. One-dimensional deformation

First consider the case of cylindrical flexural deformation with $w_3 = 0$ and $\partial/\partial X_3 = 0$. In this case the plate equations become one-dimensional.

8.1.3.1. Classical flexure

Solution for the flexure of a rotated Y-cut quartz plate under a normal acceleration can be obtained either as a polynomial or a Fourier series as

$$\begin{aligned}
w_2^{(0)} &= \sum_m A_m \cos\alpha_m X_1, \\
\alpha_m &= \frac{m\pi}{2a}, \quad m = 1,3,5,\ldots, \\
A_m &= \frac{12\rho_0 a_2}{h^2 m\pi R_m} \sin\frac{m\pi}{2}, \quad R_m = -\gamma_{11}\alpha_m^4,
\end{aligned} \tag{8.1.18}$$

from which we can calculate the flexural strain

$$S_1^{(1)} = -w_{2,11}^{(0)} = \sum_m \alpha_m^2 A_m \cos\alpha_m X_1 \,. \tag{8.1.19}$$

From $w_3 = 0$ and $\partial / \partial X_3 = 0$ we have

$$S_3^{(1)} = S_4^{(1)} = S_5^{(1)} = 0, \quad \Omega_{13} = \Omega_{23} = 0 . \tag{8.1.20}$$

From the plate stress relaxation conditions $K_2^{(1)} = K_6^{(1)} = 0$, we obtain

$$S_2^{(1)} = -\frac{c_{21}}{c_{22}} S_1^{(1)} = -\frac{c_{21}}{c_{22}} \sum_m \alpha_m^2 A_m \cos\alpha_m X_1, \quad S_6^{(1)} = 0 . \tag{8.1.21}$$

Then, from Eq. (8.1.5)

$$\Omega_{21} = -w_{2,1}^{(0)} + X_2 S_{21}^{(1)} = -w_{2,1}^{(0)} = \sum_m \alpha_m A_m \sin\alpha_m X_1 . \tag{8.1.22}$$

The displacement gradient due to flexure is

$$w_{K,L} = X_2 S_{KL}^{(1)} + \Omega_{LK} . \tag{8.1.23}$$

8.1.3.2. Flexure induced extension

The in-plane extension is governed by

$$\begin{aligned} &K_{11,1} = 0, \quad -a < X_1 < a, \\ &w_1(X_1 = \pm a) = 0, \end{aligned} \tag{8.1.24}$$

where

$$\begin{aligned} &K_1 = \gamma_{11} S_1, \\ &S_1 \cong w_{1,1} + w_{2,1}^{(0)} w_{2,1}^{(0)} / 2. \end{aligned} \tag{8.1.25}$$

With Eq. (8.1.25), Eq. $(8.1.24)_1$ can be written as

$$\gamma_{11} w_{1,11} + f_1 = 0 , \tag{8.1.26}$$

where

$$f_1 = \gamma_{11} w_{2,1}^{(0)} w_{2,11}^{(0)} . \tag{8.1.27}$$

Then the following solution can be found for w_1:

$$w_1 = \sum_{m,n} C_{mn}^{(1)} \sin(\alpha_m + \alpha_n) X_1 + C_{mn}^{(2)} \sin(\alpha_m - \alpha_n) X_1 , \tag{8.1.28}$$

where

$$C_{mn}^{(1)} = \frac{\alpha_m \alpha_n^2}{2(\alpha_m + \alpha_n)^2} A_m A_n , \quad C_{mn}^{(2)} = \frac{\alpha_m \alpha_n^2}{2(\alpha_m - \alpha_n)^2} A_m A_n . \tag{8.1.29}$$

In Eqs. (8.1.28) and (8.1.29), it is understood that $C_{mn}^{(2)} = 0$ whenever $m = n$. Then the displacement gradient due to extension can be directly calculated:

$$\begin{aligned} w_{1,1}^S = w_{1,1} = \sum_{m,n} C_{mn}^{(1)}(\alpha_m + \alpha_n)\cos(\alpha_m + \alpha_n)X_1 \\ + C_{mn}^{(2)}(\alpha_m - \alpha_n)\cos(\alpha_m - \alpha_n)X_1. \end{aligned} \tag{8.1.30}$$

From $w_3 = 0$ and $\partial/\partial X_3 = 0$ we have

$$w_{3,3}^S = w_{3,1}^S = w_{3,2}^S = 0, \quad \Omega_{13} = \Omega_{23} = 0. \tag{8.1.31}$$

From Eq. $(8.1.25)_2$ the extensional strain can be obtained as

$$\begin{aligned} S_1 = \sum_{m,n} \left[C_{mn}^{(1)}(\alpha_m + \alpha_n) - \frac{1}{4}\alpha_m\alpha_n A_m A_n\right]\cos(\alpha_m + \alpha_n)X_1 \\ + \left[C_{mn}^{(2)}(\alpha_m - \alpha_n) + \frac{1}{4}\alpha_m\alpha_n A_m A_n\right]\cos(\alpha_m - \alpha_n)X_1. \end{aligned} \tag{8.1.32}$$

The other strain components can be obtained from the stress relaxation conditions $K_2 = K_6 = 0$ as

$$S_2 = w_{2,2} + \frac{1}{2}w_{2,1}^{(0)}w_{2,1}^{(0)} = -\frac{c_{21}}{c_{22}}S_1, \quad S_6 = 0, \tag{8.1.33}$$

which implies that

$$\begin{aligned} w_{2,2}^S = -\frac{c_{21}}{c_{22}}\sum_{m,n} \left[C_{mn}^{(1)}(\alpha_m + \alpha_n) - \frac{1}{4}\alpha_m\alpha_n A_m A_n\right]\cos(\alpha_m + \alpha_n)X_1 \\ -\frac{c_{21}}{c_{22}}\sum_{m,n} \left[C_{mn}^{(2)}(\alpha_m - \alpha_n) + \frac{1}{4}\alpha_m\alpha_n A_m A_n\right]\cos(\alpha_m - \alpha_n)X_1 \\ -\frac{1}{2}\sum_{m,n} \alpha_m\alpha_n A_m A_n \sin\alpha_m X_1 \sin\alpha_n X_1, \\ w_{1,2}^S = 0. \end{aligned} \tag{8.1.34}$$

We also have

$$\Omega_{12} = 0. \tag{8.1.35}$$

8.1.4. Two-dimensional deformation

8.1.4.1. Classical flexure

In this case the solution to Eq. (8.1.1) can be written as

$$w_2^{(0)} = \sum_{m,n} A_{mn} \cos\alpha_m X_1 \cos\kappa_n X_3, \tag{8.1.36}$$

where

$$\begin{aligned} &\alpha_m = m\pi/2a, \quad \kappa_n = n\pi/2b, \quad m,n \quad \text{odd}, \\ &A_{mn} = 48\rho_0 a_2 \sin\frac{m\pi}{2}\sin\frac{n\pi}{2} / h^2 mn\pi^2 R_{mn}, \\ &R_{mn} = -[\gamma_{11}\alpha_m^4 + \gamma_{33}\kappa_n^4 + (2\gamma_{13} + 4\gamma_{55})\alpha_m^2\kappa_n^2]. \end{aligned} \tag{8.1.37}$$

8.1.4.2. Flexure induced extension

Substituting Eq. (8.1.36) into Eq. (8.1.12), we obtain

$$\begin{aligned} f_1 = \sum_{m,n,p,q} & B_{mnpq}^{(1)} \sin\alpha_m X_1 \cos\kappa_n X_3 \cos\alpha_p X_1 \cos\kappa_q X_3 \\ & + B_{mnpq}^{(2)} \cos\alpha_m X_1 \sin\kappa_n X_3 \sin\alpha_p X_1 \sin\kappa_q X_3, \\ f_3 = \sum_{m,n,p,q} & B_{mnpq}^{(3)} \sin\alpha_m X_1 \cos\kappa_n X_3 \sin\alpha_p X_1 \sin\kappa_q X_3 \\ & + B_{mnpq}^{(4)} \cos\alpha_m X_1 \sin\kappa_n X_3 \cos\alpha_p X_1 \cos\kappa_q X_3, \\ m,n,p,q & \quad \text{odd}, \end{aligned} \tag{8.1.38}$$

where

$$\begin{aligned} B_{mnpq}^{(1)} &= (\gamma_{11}\alpha_p^2 + \gamma_{55}\kappa_q^2)\alpha_m A_{mn} A_{pq}, \\ B_{mnpq}^{(2)} &= -(\gamma_{13} + \gamma_{55})\kappa_n \alpha_p \kappa_q A_{mn} A_{pq}, \\ B_{mnpq}^{(3)} &= -(\gamma_{13} + \gamma_{55})\alpha_m \alpha_p \kappa_q A_{mn} A_{pq}, \\ B_{mnpq}^{(4)} &= (\gamma_{55}\alpha_p^2 + \gamma_{33}\kappa_q^2)\kappa_n A_{mn} A_{pq}. \end{aligned} \tag{8.1.39}$$

Equation (8.1.38) can be further written as

$$\begin{aligned} f_1 = \sum_{m,n,p,q} & G^{(1)}_{mnpq} \sin\alpha^{+}_{mp} X_1 \cos\kappa^{+}_{nq} X_3 \\ & + G^{(2)}_{mnpq} \sin\alpha^{+}_{mp} X_1 \cos\kappa^{-}_{nq} X_3 + G^{(3)}_{mnpq} \sin\alpha^{-}_{mp} X_1 \cos\kappa^{+}_{nq} X_3 \\ & + G^{(4)}_{mnpq} \sin\alpha^{-}_{mp} X_1 \cos\kappa^{-}_{nq} X_3, \\ f_3 = \sum_{m,n,p,q} & H^{(1)}_{mnpq} \cos\alpha^{+}_{mp} X_1 \sin\kappa^{+}_{nq} X_3 \\ & + H^{(2)}_{mnpq} \cos\alpha^{+}_{mp} X_1 \sin\kappa^{-}_{nq} X_3 + H^{(3)}_{mnpq} \cos\alpha^{-}_{mp} X_1 \sin\kappa^{+}_{nq} X_3 \\ & + H^{(4)}_{mnpq} \cos\alpha^{-}_{mp} X_1 \sin\kappa^{-}_{nq} X_3, \end{aligned} \tag{8.1.40}$$

where

$$\begin{aligned} G^{(1)}_{mnpq} &= G^{(4)}_{mnpq} = \frac{1}{4}(B^{(1)}_{mnpq} - B^{(2)}_{mnpq}), \\ G^{(2)}_{mnpq} &= G^{(3)}_{mnpq} = \frac{1}{4}(B^{(1)}_{mnpq} + B^{(2)}_{mnpq}), \\ H^{(1)}_{mnpq} &= H^{(4)}_{mnpq} = \frac{1}{4}(B^{(4)}_{mnpq} - B^{(3)}_{mnpq}), \\ H^{(2)}_{mnpq} &= H^{(3)}_{mnpq} = \frac{1}{4}(B^{(4)}_{mnpq} + B^{(3)}_{mnpq}), \\ \alpha^{+}_{mp} &= \alpha_m + \alpha_p, \quad \alpha^{-}_{mp} = \alpha_m - \alpha_p, \\ \kappa^{+}_{nq} &= \kappa_n + \kappa_q, \quad \kappa^{-}_{nq} = \kappa_n - \kappa_q. \end{aligned} \tag{8.1.41}$$

Next we solve Eq. (8.1.9) under the boundary conditions in Eq. (8.1.6), with f_1 and f_3 given by Eq. (8.1.40). Let

$$\begin{aligned} w_1 = \sum_{m,n,p,q} & C^{(1)}_{mnpq} \sin\alpha^{+}_{mp} X_1 \cos\kappa^{+}_{nq} X_3 \\ & + C^{(2)}_{mnpq} \sin\alpha^{+}_{mp} X_1 \cos\kappa^{-}_{nq} X_3 + C^{(3)}_{mnpq} \sin\alpha^{-}_{mp} X_1 \cos\kappa^{+}_{nq} X_3 \\ & + C^{(4)}_{mnpq} \sin\alpha^{-}_{mp} X_1 \cos\kappa^{-}_{nq} X_3, \\ w_3 = \sum_{m,n,p,q} & D^{(1)}_{mnpq} \cos\alpha^{+}_{mp} X_1 \sin\kappa^{+}_{nq} X_3 \\ & + D^{(2)}_{mnpq} \cos\alpha^{+}_{mp} X_1 \sin\kappa^{-}_{nq} X_3 + D^{(3)}_{mnpq} \cos\alpha^{-}_{mp} X_1 \sin\kappa^{+}_{nq} X_3 \\ & + D^{(4)}_{mnpq} \cos\alpha^{-}_{mp} X_1 \sin\kappa^{-}_{nq} X_3, \end{aligned} \tag{8.1.42}$$

where $C^{(1)}_{mnpq}$, $C^{(2)}_{mnpq}$, $C^{(3)}_{mnpq}$, $C^{(4)}_{mnpq}$, $D^{(1)}_{mnpq}$, $D^{(2)}_{mnpq}$, $D^{(3)}_{mnpq}$, and $D^{(4)}_{mnpq}$ are undetermined coefficients. Whenever $m = p$ and hence $\alpha^{-}_{mp} = 0$, it is understood that the corresponding $C^{(3)}_{mnpq} = C^{(4)}_{mnpq} = 0$. Similarly, when $\kappa^{-}_{nq} = 0$, we define the corresponding $D^{(2)}_{mnpq} = D^{(4)}_{mnpq} = 0$. Equation (8.1.42) automatically satisfies the boundary conditions in Eq. (8.1.6). Substituting Eq. (8.1.42) into Eq. (8.1.9), we obtain

$$\begin{aligned}
C^{(1)}_{mnpq} &= \frac{1}{\Delta_1}\{[\gamma_{55}(\alpha^{+}_{mp})^2 + \gamma_{33}(\kappa^{+}_{nq})^2]G^{(1)}_{mnpq} - \alpha^{+}_{mp}\kappa^{+}_{nq}(\gamma_{13}+\gamma_{55})H^{(1)}_{mnpq}\},\\
D^{(1)}_{mnpq} &= \frac{1}{\Delta_1}\{[\gamma_{11}(\alpha^{+}_{mp})^2 + \gamma_{55}(\kappa^{+}_{nq})^2]H^{(1)}_{mnpq} - \alpha^{+}_{mp}\kappa^{+}_{nq}(\gamma_{13}+\gamma_{55})G^{(1)}_{mnpq}\},\\
\Delta_1 &= [\gamma_{11}(\alpha^{+}_{mp})^2 + \gamma_{55}(\kappa^{+}_{nq})^2]\\
&\quad [\gamma_{55}(\alpha^{+}_{mp})^2 + \gamma_{33}(\kappa^{+}_{nq})^2] - (\alpha^{+}_{mp})^2(\kappa^{+}_{nq})^2(\gamma_{13}+\gamma_{55})^2,
\end{aligned} \tag{8.1.43}$$

$$\begin{aligned}
C^{(2)}_{mnpq} &= \frac{1}{\Delta_2}\{[\gamma_{55}(\alpha^{+}_{mp})^2 + \gamma_{33}(\kappa^{-}_{nq})^2]G^{(2)}_{mnpq} - \alpha^{+}_{mp}\kappa^{-}_{nq}(\gamma_{13}+\gamma_{55})H^{(2)}_{mnpq}\},\\
D^{(2)}_{mnpq} &= \frac{1}{\Delta_2}\{[\gamma_{11}(\alpha^{+}_{mp})^2 + \gamma_{55}(\kappa^{-}_{nq})^2]H^{(2)}_{mnpq} - \alpha^{+}_{mp}\kappa^{-}_{nq}(\gamma_{13}+\gamma_{55})G^{(2)}_{mnpq}\},\\
\Delta_2 &= [\gamma_{11}(\alpha^{+}_{mp})^2 + \gamma_{55}(\kappa^{-}_{nq})^2]\\
&\quad [\gamma_{55}(\alpha^{+}_{mp})^2 + \gamma_{33}(\kappa^{-}_{nq})^2] - (\alpha^{+}_{mp})^2(\kappa^{-}_{nq})^2(\gamma_{13}+\gamma_{55})^2,
\end{aligned} \tag{8.1.44}$$

$$\begin{aligned}
C^{(3)}_{mnpq} &= \frac{1}{\Delta_3}\{[\gamma_{55}(\alpha^{-}_{mp})^2 + \gamma_{33}(\kappa^{+}_{nq})^2]G^{(3)}_{mnpq} - \alpha^{-}_{mp}\kappa^{+}_{nq}(\gamma_{13}+\gamma_{55})H^{(3)}_{mnpq}\},\\
D^{(3)}_{mnpq} &= \frac{1}{\Delta_3}\{[\gamma_{11}(\alpha^{-}_{mp})^2 + \gamma_{55}(\kappa^{+}_{nq})^2]H^{(3)}_{mnpq} - \alpha^{-}_{mp}\kappa^{+}_{nq}(\gamma_{13}+\gamma_{55})G^{(3)}_{mnpq}\},\\
\Delta_3 &= [\gamma_{11}(\alpha^{-}_{mp})^2 + \gamma_{55}(\kappa^{+}_{nq})^2]\\
&\quad [\gamma_{55}(\alpha^{-}_{mp})^2 + \gamma_{33}(\kappa^{+}_{nq})^2] - (\alpha^{-}_{mp})^2(\kappa^{+}_{nq})^2(\gamma_{13}+\gamma_{55})^2,
\end{aligned} \tag{8.1.45}$$

$$C_{mnpq}^{(4)} = \frac{1}{\Delta_4}\{[\gamma_{55}(\alpha_{mp}^{-})^2 + \gamma_{33}(\kappa_{nq}^{-})^2]G_{mnpq}^{(4)} - \alpha_{mp}^{-}\kappa_{nq}^{-}(\gamma_{13}+\gamma_{55})H_{mnpq}^{(4)}\},$$

$$D_{mnpq}^{(4)} = \frac{1}{\Delta_4}\{[\gamma_{11}(\alpha_{mp}^{-})^2 + \gamma_{55}(\kappa_{nq}^{-})^2]H_{mnpq}^{(4)} - \alpha_{mp}^{-}\kappa_{nq}^{-}(\gamma_{13}+\gamma_{55})G_{mnpq}^{(4)}\},$$

$$\Delta_4 = [\gamma_{11}(\alpha_{mp}^{-})^2 + \gamma_{55}(\kappa_{nq}^{-})^2] \\ [\gamma_{55}(\alpha_{mp}^{-})^2 + \gamma_{33}(\kappa_{nq}^{-})^2] - (\alpha_{mp}^{-})^2(\kappa_{nq}^{-})^2(\gamma_{13}+\gamma_{55})^2. \tag{8.1.46}$$

8.2. First-Order Acceleration Sensitivity

Let a resonator be subject to a constant acceleration a_K. Typical response of the frequency shifts in the resonator under small accelerations in different directions is as shown in Fig. 8.2.1.

The first-order acceleration sensitivity of the resonator is a vector Γ_K which gives the frequency shift under a_K through

$$\frac{\Delta f}{f_0} = \Gamma_K a_K, \tag{8.2.1}$$

where $f = \omega/2\pi$, and $\Delta f = f - f_0$ is a frequency shift. Eq. (8.2.1) is valid for the case of small accelerations. For a plate resonator, an acceleration in an arbitrary direction can be broken into a component normal to the plate and another component in the plane of the plate. For the first-order acceleration sensitivity, the normal and in-plane acceleration sensitivities can be analyzed separately.

When the acceleration is small, the frequency shift can be calculated from the first-order perturbation integral in Eq. (1.3.19), which is repeated below for convenience:

$$\Delta\omega = \frac{1}{2\omega_0}\frac{\int_V \hat{c}_{L\gamma M\alpha} U_{\gamma,L} U_{\alpha,M} dV}{\int_V \rho_0 U_\alpha U_\alpha dV}, \tag{8.2.2}$$

where $\Delta\omega = \omega - \omega_0$. ω_0 and U_α are the frequency and mode when the biasing fields are not present.

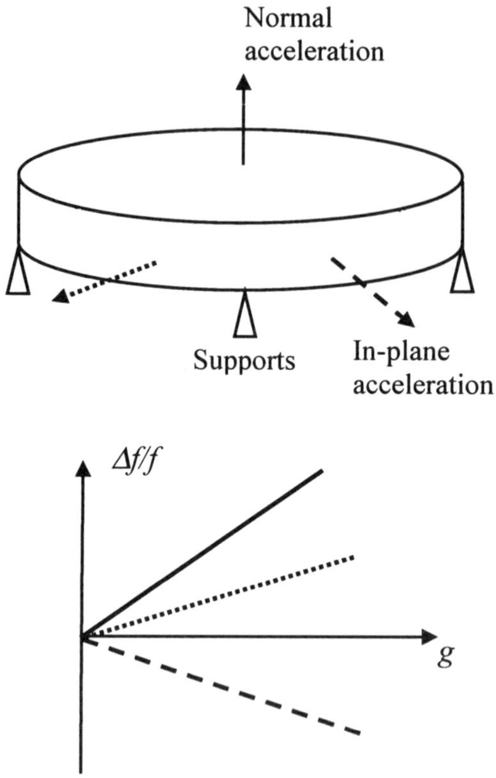

Fig. 8.2.1. Frequency shifts in a resonator under small accelerations.

As an example consider the normal acceleration sensitivity of a rotated Y-cut quartz thickness-shear resonator. The unperturbed fundamental mode is

$$\omega_0^2 = \frac{\pi^2 c_{66}}{4\rho_0 h^2}, \quad U_1 = \sin\frac{\pi}{2h}X_2, \quad U_2 = U_3 = 0\,. \tag{8.2.3}$$

For this particular mode, Eq. (8.2.2) becomes

$$\Delta\omega = \frac{1}{2\omega_0}\frac{\int_V \hat{c}_{2121}U_{1,2}U_{1,2}dV}{\int_V \rho_0 U_1 U_1 dV}, \tag{8.2.4}$$

where

$$\hat{c}_{2121} = (2c_{66} + c_{21} + c_{661})S_1^0 + (c_{22} + c_{662})S_2^0 + (c_{23} + c_{663})S_3^0 \\ + (c_{24} + c_{664})S_4^0 + (c_{46} + c_{25} + c_{665})S_5^0 + (2c_{26} + c_{666})S_6^0. \tag{8.2.5}$$

Since the biasing strains S_1^0, S_3^0, S_5^0 and S_2^0 due to flexure are odd functions of x_2 and the product of the shear mode is an even function, these biasing strains do not cause any frequency shift. S_4^0 and S_6^0 are even functions of x_2, but they are odd functions of x_3 and x_1, respectively. Since for a perfectly symmetric resonator the fundamental thickness-shear mode of interest is an even function of x_3 and x_1, it can be concluded that the first-order normal acceleration sensitivity vanishes.

8.3. An Estimate of Second-Order Acceleration Sensitivity and its Reduction

When second-order effects are included, Eq. (8.2.1) becomes $\Delta f / f_0 = \Gamma_K a_K + \Gamma_{KL} a_K a_L$. In this case, one immediate observation is that the normal and in-plane accelerations may interact. Reported results on second-order acceleration sensitivity are scattered. In this section we try to get an estimate on the second-order effect and study how to reduce it.

Consider a Y-cut quartz resonator under normal acceleration (see Fig. 8.3.1).

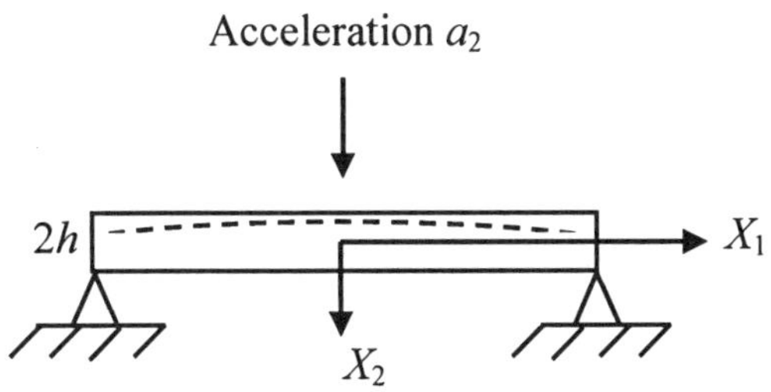

Fig. 8.3.1. A simply supported Y-cut quartz plate under normal acceleration.

The resonator operates with the fundamental thickness-shear mode in the X_1 direction. For our purpose it is sufficient to consider the case when there is no motion in the X_3 direction and no dependence on X_3. The normal acceleration causes a flexure $w(X_1)$ of the originally flat middle surface as shown by the dotted line in the figure, which is now curved. The first-order acceleration sensitivity vanishes according to the first-order perturbation integral.

It is obvious that the middle surface becomes a little longer during flexure, which is a second-order effect because the differential arc length of the deformed middle surface is given by

$$dl = [1+(w')^2]^{1/2}dX_1 \cong [1+(w')^2/2]dX_1, \tag{8.3.1}$$

which depends on the square of the slope of the small deflection and is neglected in the first-order analysis. This stretching strain S_1^0 of the middle surface also causes a contraction of the plate thickness which is symmetric about the middle surface and may produce a second-order frequency shift [92]. We give an estimate of this effect below [93]. The effect of the middle surface stretch on the fundamental thickness-shear mode can be obtained from Eq. (2.9.13) as:

$$\frac{\Delta\omega}{\omega} = \left(1+\frac{c_{661}}{2c_{66}} - \frac{c_{662}}{2c_{66}}\frac{c_{21}}{c_{22}}\right)S_1^0. \tag{8.3.2}$$

8.3.1. An estimate

The flexure $w(X_1)$ is given in Eq. (8.1.18). The first term in the series solution represents a good approximation of the deflection surface. Neglecting the rest of the terms in the series solution underestimates the center deflection of the plate by an error of 0.39%. Therefore, as an approximation, we take

$$\begin{aligned} &w = A\cos\alpha X_1, \\ &\alpha = \frac{\pi}{2a}, \quad A = \frac{12\rho_0 a_2}{h^2\pi R}, \quad R = -\gamma_{11}\alpha^4, \\ &\gamma_{11} = c_{11} - \frac{c_{12}^2c_{44} + c_{14}^2c_{22} - 2c_{12}c_{24}c_{14}}{c_{22}c_{44} - c_{24}^2} = s_{33}/(s_{11}s_{33} - s_{13}^2). \end{aligned} \tag{8.3.3}$$

The total elongation of the middle surface is

$$\Delta L = \int_{-a}^{a} \frac{1}{2}(w')^2 dX_1 = \frac{1}{2}A^2\alpha^2 a. \tag{8.3.4}$$

We use the average extensional strain for an estimate. Hence

$$S_1^0 \cong \frac{\Delta L}{2a} = \frac{1}{4} A^2 \alpha^2 . \tag{8.3.5}$$

With successive substitutions, we obtain

$$\begin{aligned} \frac{\Delta \omega}{\omega} &= \left(1 + \frac{c_{661}}{2c_{66}} - \frac{c_{662}}{2c_{66}} \frac{c_{21}}{c_{22}}\right) \frac{1}{4} A^2 \alpha^2 \\ &= 0.243 \left(1 + \frac{c_{661}}{2c_{66}} - \frac{c_{662}}{2c_{66}} \frac{c_{21}}{c_{22}}\right) \frac{\rho_0^2 a^6 g^2}{\gamma_{11}^2 h^4} \frac{a_2^2}{g^2} . \end{aligned} \tag{8.3.6}$$

Equation (8.3.6) is quadratic in a_2. This is qualitatively in agreement with the experimental fact that when the acceleration is not small the frequency shift depends on the acceleration nonlinearly [94]. The frequency shift depends strongly on the ratio of a/h. For thin plates with large a/h, the effect can be felt more.

As a numerical example, consider a Y-cut quartz resonator with a = 7 mm and h = 0.097001 mm. The produces an $\omega_0 = 10^6\,\text{Hz}$. When $a_2 = g$, from Eq. (8.3.6) and the elastic constants of quartz, we obtain

$$\frac{\Delta \omega}{\omega} = 0.773 \times 10^{-11} , \tag{8.3.7}$$

whose order of magnitude is consistent with the estimate in [92].

8.3.2. Reduction of normal acceleration sensitivity

Consider a crystal plate resonator as shown in Fig. 8.3.2. The plate has two overhang portions in $a < |X_1| < a+b$. The active portion of the resonator is the central portion of $|X_1| < a$.

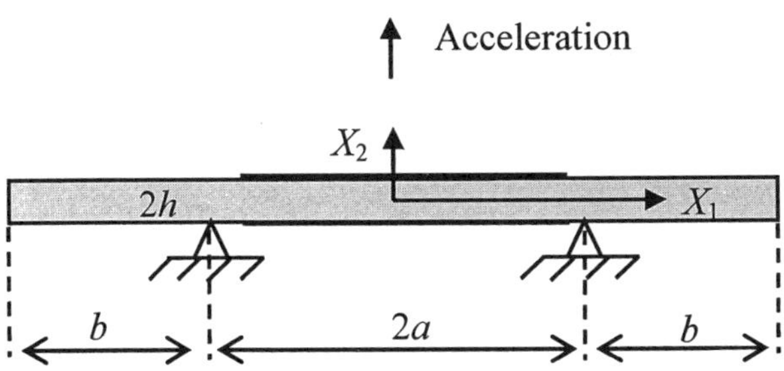

Fig. 8.3.2. A plate with overhangs under normal acceleration.

When the plate is under a normal acceleration, the inertial force in the overhang portions tends to reduce the deflection in the central portion due to the acceleration. This provides a possibility for design and optimization. By properly choosing the overhang length b, we can try to reduce or even remove the normal acceleration sensitivity.

First examine the effect of the overhangs on the deflection in the central portion of the plate under a normal acceleration. Due to symmetry consider half of the plate in $0 < X_1 < a+b$, with a unit thickness in the X_3 direction. The effect of the overhang portion $a < X_1 < a+b$ on the right half of the central portion $0 < X_1 < a$ can be represented by a shear force and a bending moment. Therefore we only need to analyze the portion in Fig. 8.3.3. In the figure, the acceleration a_2 is replaced by its inertial force.

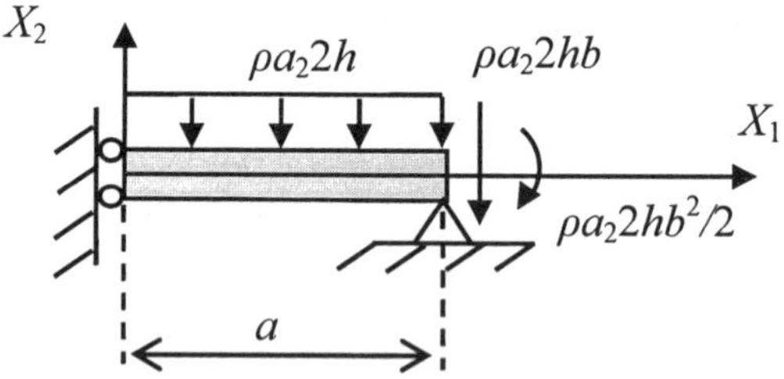

Fig. 8.3.3. Part of the plate in Fig. 8.3.2.

For this portion, the governing equation and boundary conditions for the deflection $w_2(X_1)$ are

$$\begin{aligned} &EIw_2'''' = -\rho a_2 2h, \quad 0 < X_1 < a, \\ &w_2'(0) = 0, \quad EIw_2'''(0) = 0, \\ &w_2(a) = 0, \quad EIw_2''(a) = -\rho a_2 hb^2, \end{aligned} \tag{8.3.8}$$

where a prime is a derivative with respect to X_1 and, for planar deformation independent of X_3 with $w_3 = 0$, we have the following effective Young's modulus E and the moment of inertia I:

$$E = c_{11} - c_{12}^2 / c_{22}, \quad I = \frac{2h^3}{3}. \tag{8.3.9}$$

The solution to Eq. (8.3.8) is

$$EIw_2 = \frac{1}{12}\rho a_2 h(a^2 - X_1^2)(X_1^2 + 6b^2 - 5a^2). \tag{8.3.10}$$

With Eq. (8.3.10), we can choose b to adjust the state of flexure in the central portion of the plate and the acceleration sensitivity. For example, with $b^2 = 5a^2/6$ we can make $w_2(0) = 0$. The total elongation of the middle surface of the plate is given by

$$\Delta L = \int_0^a \frac{1}{2}(w')^2 dX_1 = \frac{17\rho^2 a_2^2 a^7}{70E^2 h^4}\left(1 - \frac{42}{17}\frac{b^2}{a^2} + \frac{105}{68}\frac{b^4}{a^4}\right). \quad (8.3.11)$$

We use the average extensional strain for an estimate. Hence, $S_1^{(0)} \cong \Delta L / a$. The corresponding $S_2^{(0)}$ is given by stress relaxation. With successive substitutions, we obtain

$$\frac{\Delta\omega}{\omega} = 0.243\left(1 + \frac{c_{661}}{2c_{66}} - \frac{c_{662}}{2c_{66}}\frac{c_{21}}{c_{22}}\right)\frac{\rho^2 a^6 g^2}{E^2 h^4}\frac{a_2^2}{g^2} f(r),$$
$$f(r) = 1 - \frac{42}{17}r^2 + \frac{105}{68}r^4, \quad r = \frac{b}{a}. \quad (8.3.12)$$

$f(r)$ is plotted in Fig. 8.3.4. The figure shows that when r is approximately equal to 0.9, $f(r)$ assumes a minimum. The data show that this minimum is approximately 0.01. Therefore by properly choosing the overhang the normal acceleration sensitivity can be reduced by two orders of magnitude. Plates with overhangs can also be used for reducing the first-order normal acceleration sensitivity when it is not zero.

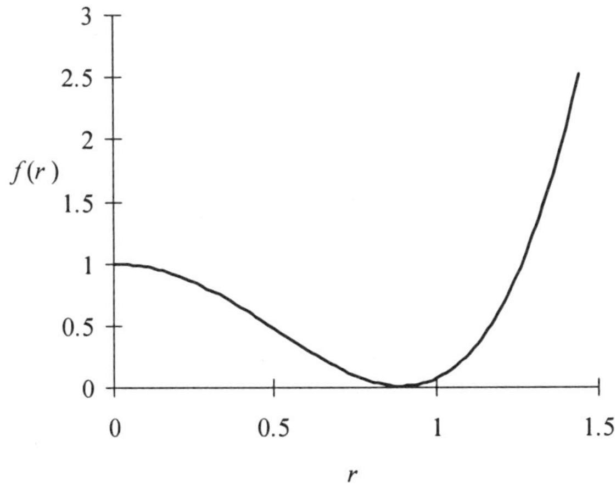

Fig. 8.3.4. Effects of overhang on acceleration sensitivity.

8.4. Second-Order Perturbation Analysis

In this section a systematic perturbation analysis is presented for second-order frequency shifts in crystal resonators.

8.4.1. Relatively large biasing deformations

For a complete description of the second-order effects of biasing fields on frequencies, a second-order perturbation analysis is needed [95]. Consider three states of a resonator (see Fig. 8.4.1). In the reference state a generic point is denoted by **X** with rectangular coordinates X_K. The mass density is ρ_0.

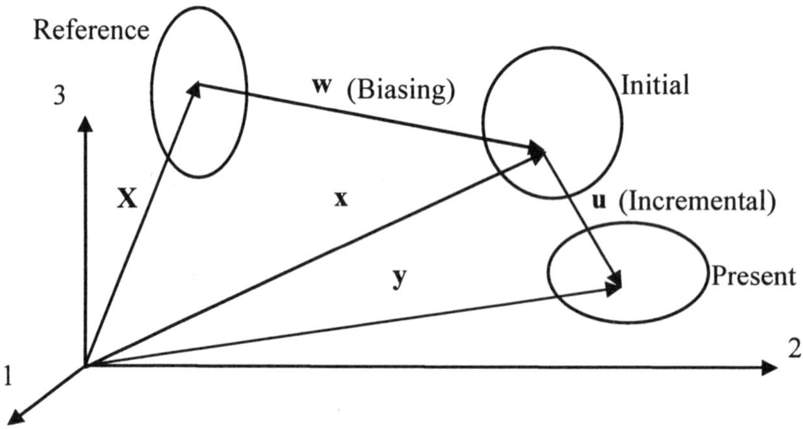

Fig. 8.4.1. Reference, initial, and present configurations of an elastic body.

Then the body deforms statically and finitely into the initial state. The position of the material point associated with **X** is given by $x_\alpha = x_\alpha(\mathbf{X})$. The initial displacement is given by $\mathbf{w} = \mathbf{x}\text{-}\mathbf{X}$. The initial deformations satisfy the following static equations of nonlinear elasticity:

$$K^0_{K\alpha,K} = 0, \quad K^0_{K\alpha} = x_{\alpha,L} T^0_{KL}, \quad T^0_{KL} = \rho_0 \left. \frac{\partial \psi}{\partial S_{KL}} \right|_{\mathbf{E}^0},$$

$$\rho_0 \psi (S_{KL}) = \frac{1}{2} c_{ABCD} S_{AB} S_{CD} + \frac{1}{6} c_{ABCDEF} S_{AB} S_{CD} S_{EF} \qquad (8.4.1)$$

$$+ \frac{1}{24} c_{ABCDEFGH} S_{AB} S_{CD} S_{EF} S_{GH},$$

$$S^0_{KL} = (x_{\alpha,K} x_{\alpha,L} - \delta_{KL})/2 = (w_{K,L} + w_{L,K} + w_{N,K} w_{N,L})/2.$$

We are interested in relatively large biasing deformations for which quadratic terms of the biasing displacement gradient $w_{K,L}$ need to be considered in solving the biasing deformation problem. For such a second-order analysis the inclusion of the third-order elastic constants in the strain energy density is sufficient. However, for a complete description of the second-order effects of the biasing deformation on the incremental motion the fourth-order elastic constants will be needed, which will become clear later. This is similar to the familiar situation that the third-order elastic constants are necessary for a complete description of the first-order effects of the biasing deformation on the incremental motion. We assume that the solution to the biasing deformation problem has been obtained and can be written in the following form:

$$\mathbf{w} = \varepsilon \overline{\mathbf{w}} + \varepsilon^2 \widetilde{\mathbf{w}} + O(\varepsilon^3), \qquad (8.4.2)$$

where ε is a small parameter in the biasing deformation problem, e.g., the intensity of the load or the acceleration. $O(\varepsilon^3)$ represents terms of ε^3 or higher orders.

8.4.2. Equations for incremental vibrations

To the deformed body at the initial configuration, time-dependent small deformations are applied. The final position of the material point associated with $\mathbf{X}$ is given by $y_i = y_i(\mathbf{X},t)$. The small, incremental displacement vector is denoted by $\mathbf{u} = \mathbf{y} - \mathbf{x}$. The equations of motion for the incremental fields are

$$K^1_{K\alpha,K} = \rho_0 \ddot{u}_\alpha,$$
$$K^1_{K\alpha} = G_{K\alpha L\gamma} u_{\gamma,L}, \tag{8.4.3}$$
$$G_{K\alpha L\gamma} = x_{\alpha,M} \left.\frac{\partial^2 \Sigma}{\partial S_{KM} \partial S_{LN}}\right|_{\mathbf{E}^0} x_{\gamma,N} + T^0_{KL}\delta_{\alpha\gamma} = G_{L\gamma K\alpha}.$$

Corresponding to Eq. (8.4.2), up to ε^2, we have the following expressions:

$$G_{K\alpha L\gamma} = c_{K\alpha L\gamma} + \varepsilon \bar{c}_{K\alpha L\gamma} + \varepsilon^2 \tilde{c}_{K\alpha L\gamma}, \tag{8.4.4}$$

$$\begin{aligned} \bar{c}_{K\alpha L\gamma} &= c_{K\alpha L\gamma AB} \bar{w}_{A,B} + c_{K\alpha LN} \bar{w}_{\gamma,N} \\ &\quad + c_{KML\gamma} \bar{w}_{\alpha,M} + c_{KLAB} \bar{w}_{A,B} \delta_{\alpha\gamma}, \end{aligned} \tag{8.4.5}$$

$$\begin{aligned} \tilde{c}_{K\alpha L\gamma} &= c_{K\alpha L\gamma AB} \tilde{w}_{A,B} + c_{K\alpha LN} \tilde{w}_{\gamma,N} \\ &+ c_{KML\gamma} \tilde{w}_{\alpha,M} + c_{KLAB} \tilde{w}_{A,B} \delta_{\alpha\gamma} \\ &+ \frac{1}{2} c_{K\alpha L\gamma AB} \bar{w}_{C,A} \bar{w}_{C,B} + \frac{1}{2} c_{K\alpha L\gamma ABCD} \bar{w}_{A,B} \bar{w}_{C,D} \\ &+ c_{K\alpha LNAB} \bar{w}_{\gamma,N} \bar{w}_{A,B} \\ &+ c_{KML\gamma AB} \bar{w}_{\alpha,M} \bar{w}_{A,B} + c_{KMLN} \bar{w}_{\alpha,M} \bar{w}_{\gamma,N} \\ &+ \frac{1}{2} c_{KLAB} \bar{w}_{C,A} \bar{w}_{C,B} \delta_{\alpha\gamma} + \frac{1}{2} c_{KLABCD} \bar{w}_{A,B} \bar{w}_{C,D} \delta_{\alpha\gamma}. \end{aligned} \tag{8.4.6}$$

The second-order biasing displacement $\tilde{\mathbf{w}}$ appears linearly in $\tilde{c}_{K\alpha L\gamma}$. The first-order biasing displacement $\bar{\mathbf{w}}$ appears quadratically in $\tilde{c}_{K\alpha L\gamma}$. The above equations show that the second-, third- and fourth-order elastic constants may all have contributions to the second-order effect of the biasing deformation. In particular, the fourth-order elastic constants are associated with $\bar{\mathbf{w}}$, but not $\tilde{\mathbf{w}}$.

For time-harmonic problems with a frequency ω, with successive substitutions, Eq. (8.4.3) can be written as an eigenvalue problem in the following form:

$$\begin{aligned} &[(c_{K\alpha L\gamma} + \varepsilon \bar{c}_{K\alpha L\gamma} + \varepsilon^2 \tilde{c}_{K\alpha L\gamma}) u_{\gamma,L}]_{,K} + \lambda \rho_0 u_\alpha = 0 \quad \text{in} \quad V, \\ &u_\alpha = 0 \quad \text{on} \quad S_1, \\ &N_K (c_{K\alpha L\gamma} + \varepsilon \bar{c}_{K\alpha L\gamma} + \varepsilon^2 \tilde{c}_{K\alpha L\gamma}) u_{\gamma,L} = 0 \quad \text{on} \quad S_2, \end{aligned} \tag{8.4.7}$$

where $\lambda = \omega^2$, $S_1 \cup S_2 = S$, and $S_1 \cap S_2 = 0$. The small parameter also appears in the boundary conditions. We look for nontrivial solutions to Eq. (8.4.7) that are normalized by $<\mathbf{u};\rho_0\mathbf{u}>$ = 1 with respect to the following inner product of two vectors:

$$\langle \mathbf{u};\mathbf{v}\rangle = \int_V u_\alpha v_\alpha dV \,. \tag{8.4.8}$$

8.4.3. Second-order perturbation analysis

The eignevaules are assumed to be simple, i.e. without degeneracy. Let

$$\mathbf{u} \cong \mathbf{U} + \varepsilon\overline{\mathbf{U}} + \varepsilon^2\widetilde{\mathbf{U}}, \quad \lambda \cong \mu + \varepsilon\overline{\mu} + \varepsilon^2\widetilde{\mu} \,. \tag{8.4.9}$$

Substituting Eq. (8.4.9) into Eq. (8.4.7), collecting the coefficients of terms of equal powers of ε, a series of perturbation problems are obtained. The zero-order problem is

$$\begin{aligned} &c_{K\alpha L\gamma}U_{\gamma,LK} + \mu\rho_0 U_\alpha = 0 \quad \text{in} \quad V, \\ &U_\alpha = 0 \quad \text{on} \quad S_1, \\ &N_K c_{K\alpha L\gamma}U_{\gamma,L} = 0 \quad \text{on} \quad S_2. \end{aligned} \tag{8.4.10}$$

The solution to the zero-order problem is assumed known. We denote the zero-order eigenvalues by $\mu^{(i)}$ (i = 1, 2, 3...) and the corresponding eigenvectors by $U_\alpha^{(i)}$ which form a complete set and are orthonormal, i.e. $\langle U_\alpha^{(i)};\rho_0 U_\alpha^{(j)}\rangle = \delta^{(ij)}$, where $\delta^{(ij)}$ is the Kronecker delta.

8.4.3.1. *First-order problem*

The first-order problem for the ith eigenvalue is

$$\begin{aligned} &(c_{K\alpha L\gamma}\overline{U}_{\gamma,L}^{(i)})_{,K} + (\overline{c}_{K\alpha L\gamma}U_{\gamma,L}^{(i)})_{,K} \\ &\quad + \mu^{(i)}\rho_0\overline{U}_\alpha^{(i)} + \overline{\mu}^{(i)}\rho_0 U_\alpha^{(i)} = 0 \quad \text{in} \quad V, \\ &\overline{U}_\alpha^{(i)} = 0 \quad \text{on} \quad S_1, \\ &N_K c_{K\alpha L\gamma}\overline{U}_{\gamma,L}^{(i)} + N_K\overline{c}_{K\alpha L\gamma}U_{\gamma,L}^{(i)} = 0 \quad \text{on} \quad S_2. \end{aligned} \tag{8.4.11}$$

We make the following generalized Fourier series expansion:

$$\overline{U}_\alpha^{(i)} = \sum_j a^{(ij)}U_\alpha^{(j)}, \tag{8.4.12}$$

where the expansion coefficients formally have the following expressions:

$$a^{(ij)} = \langle \overline{U}_\alpha^{(i)}; \rho_0 U_\alpha^{(j)} \rangle, \tag{8.4.13}$$

and are to be determined. Multiplying both sides of Eq. (8.4.11)$_1$ by $U_\alpha^{(j)}$ through the inner product in Eq. (8.4.8), we have

$$\begin{aligned} a^{(ij)}(\mu^{(j)} - \mu^{(i)}) &= \overline{\mu}^{(i)}\delta^{(ij)} - \int_V U_{\alpha,K}^{(j)} \overline{c}_{K\alpha L\gamma} U_{\gamma,L}^{(i)} dV \\ &= \overline{\mu}^{(i)}\delta^{(ij)} - \langle U_\alpha^{(i)}; \rho_0 U_\alpha^{(j)} \rangle, \end{aligned} \tag{8.4.14}$$

where we have introduced another inner product between two second-rank tensors. When $i = j$, Eq. (8.4.14) gives the first-order perturbation of eigenvalues

$$\overline{\mu}^{(i)} = \left\langle U_{\alpha,K}^{(i)}; \overline{c}_{K\alpha L\gamma} U_{\gamma,L}^{(i)} \right\rangle. \tag{8.4.15}$$

From Eq. (8.4.14), when $i \neq j$, we obtain

$$a^{(ij)} = \langle U_\alpha^{(i)}; \rho_0 U_\alpha^{(j)} \rangle / (\mu^{(i)} - \mu^{(j)}) = -a^{(ji)}. \tag{8.4.16}$$

The normalization condition

$$\langle U_\alpha^{(i)} + \varepsilon \overline{U}_\alpha^{(i)} + \varepsilon^2 \tilde{U}_\alpha^{(i)}; \rho_0 (U_\alpha^{(i)} + \varepsilon \overline{U}_\alpha^{(i)} + \varepsilon^2 \tilde{U}_\alpha^{(i)}) \rangle = 1 \tag{8.4.17}$$

implies that to the first-order $a^{(ii)} = 0$. Therefore the first-order perturbation of eigenvectors can be written as

$$\overline{U}_\alpha^{(i)} = \sum_{j \neq i} \frac{\langle U_{\alpha,K}^{(j)}; \overline{c}_{K\alpha L\gamma} U_{\gamma,L}^{(i)} \rangle}{\mu^{(i)} - \mu^{(j)}} U_\alpha^{(j)}. \tag{8.4.18}$$

8.4.3.2. *Second-order problem*

The second-order problem is

$$\begin{aligned} &(c_{K\alpha L\gamma} \tilde{U}_{\gamma,L}^{(i)})_{,K} + (\overline{c}_{K\alpha L\gamma} \overline{U}_{\gamma,L}^{(i)})_{,K} + (\tilde{c}_{K\alpha L\gamma} U_{\gamma,L}^{(i)})_{,K} \\ &\quad + \mu^{(i)} \rho_0 \tilde{U}_\alpha^{(i)} + \overline{\mu}^{(i)} \rho_0 \overline{U}_\alpha^{(i)} + \tilde{\mu}^{(i)} \rho_0 U_\alpha^{(i)} = 0 \quad \text{in} \quad V, \\ &\tilde{U}_\alpha^{(i)} = 0 \quad \text{on} \quad S_1, \\ &N_K c_{K\alpha L\gamma} \tilde{U}_{\gamma,L}^{(i)} + N_K \overline{c}_{K\alpha L\gamma} \overline{U}_{\gamma,L}^{(i)} + N_K \tilde{c}_{K\alpha L\gamma} U_{\gamma,L}^{(i)} = 0 \quad \text{on} \quad S_2. \end{aligned} \tag{8.4.19}$$

Multiplying both sides of Eq. (8.4.19)$_1$ by $U_\alpha^{(i)}$ through the inner product in Eq. (8.4.8), we obtain the second-order perturbation of eigenvalues as

$$\tilde{\mu}^{(i)} = \langle U_{\alpha,K}^{(i)}; \tilde{c}_{K\alpha L\gamma} U_{\gamma,L}^{(i)} \rangle + \langle U_{\alpha,K}^{(i)}; \overline{c}_{K\alpha L\gamma} \overline{U}_{\gamma,L}^{(i)} \rangle$$

$$= \langle U_{\alpha,K}^{(i)}; \tilde{c}_{K\alpha L\gamma} U_{\gamma,L}^{(i)} \rangle + \sum_{j \neq i} \frac{\langle U_{\alpha,K}^{(j)}; \overline{c}_{K\alpha L\gamma} U_{\gamma,L}^{(i)} \rangle^2}{\mu^{(i)} - \mu^{(j)}}. \tag{8.4.20}$$

The second-order perturbation of eigenvalues depends on $\tilde{c}_{K\alpha L\gamma}$ as expected. It also depends on $\overline{c}_{K\alpha L\gamma}$ quadratically.

The above expressions are for the eigenvalue $\lambda = \omega^2$. For ω we make the following expansion:

$$\omega \cong \omega_0 + \varepsilon\overline{\omega} + \varepsilon^2 \tilde{\omega}, \tag{8.4.21}$$

where ω_0 is the frequency when $\varepsilon = 0$, or when the biasing fields are not present. Then

$$\lambda = \omega^2 \cong (\omega_0 + \varepsilon\overline{\omega} + \varepsilon^2 \tilde{\omega})^2 \cong \mu + \varepsilon\overline{\mu} + \varepsilon^2 \tilde{\mu}. \tag{8.4.22}$$

From the terms linear in ε in Eq. (8.4.22), with the use of Eq. (8.4.15), we obtain

$$\varepsilon\overline{\omega}^{(i)} = \frac{\varepsilon\overline{\mu}^{(i)}}{2\omega_0^{(i)}} = \frac{1}{2\omega_0^{(i)}} \langle U_{\alpha,K}^{(i)}; \varepsilon\overline{c}_{K\alpha L\gamma} U_{\gamma,L}^{(i)} \rangle, \tag{8.4.23}$$

which is the first-order perturbation integral in Eq. (8.2.2) under $< U_\alpha^{(i)}; \rho_0 U_\alpha^{(i)} >= 1$. From the terms quadratic in ε in Eq. (8.4.22), we obtain the second-order frequency shift as

$$\varepsilon^2 \tilde{\omega} = \frac{\varepsilon^2 \tilde{\mu} - \varepsilon^2 \overline{\omega}^2}{2\omega_0}, \tag{8.4.24}$$

which, upon the use of Eqs. (8.4.20) and (8.4.23), takes the following form:

$$\varepsilon^2 \tilde{\omega}^{(i)} = \frac{1}{2\omega_0^{(i)}} \langle U_{\alpha,K}^{(i)}; \varepsilon^2 \tilde{c}_{K\alpha L\gamma} U_{\gamma,L}^{(i)} \rangle$$

$$+ \frac{1}{2\omega_0^{(i)}} \sum_{j \neq i} \frac{\langle U_{\alpha,K}^{(j)}; \varepsilon\overline{c}_{K\alpha L\gamma} U_{\gamma,L}^{(i)} \rangle^2}{(\omega_0^{(i)})^2 - (\omega_0^{(j)})^2} \tag{8.4.25}$$

$$- \frac{1}{2\omega_0^{(i)}} \left\{ \frac{1}{2\omega_0^{(i)}} \langle U_{\alpha,K}^{(i)}; \varepsilon\overline{c}_{K\alpha L\gamma} U_{\gamma,L}^{(i)} \rangle \right\}^2.$$

From Eq. (8.4.25), the second-order frequency shift can be calculated with the zero-order frequencies and modes, as well as $\tilde{c}_{K\alpha L\gamma}$ and $\bar{c}_{K\alpha L\gamma}$. The first term on the right-hand side of Eq. (8.4.25) is due to $\tilde{c}_{K\alpha L\gamma}$ as expected. The second term is quadratic in $\bar{c}_{K\alpha L\gamma}$. This term is involved with all of the zero-order frequencies and modes. If $\omega_0^{(i)}$ is an isolated frequency, i.e., $\omega_0^{(i)} - \omega_0^{(j)}$ is relatively large for all j, then this term is small. The third term is due to the fact that our perturbation analysis is performed on $\lambda = \omega^2$, not directly on ω.

8.5. Second-Order Normal Acceleration Sensitivity

Consider a rotated Y-cut plate under normal acceleration as shown in Fig. 8.5.1.

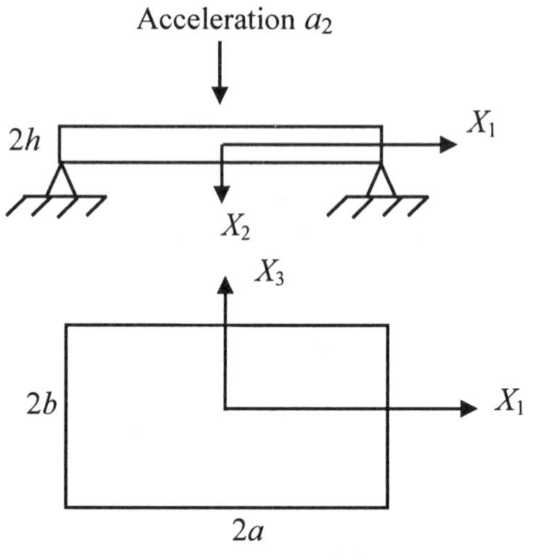

Fig. 8.5.1. A simply supported quartz plate.

The biasing deformation is due to the normal acceleration a_2. In this case a_2 or a dimensionless number proportional to a_2 can be chosen to be the small parameter ε. The deflection of the plate changes its sign when the acceleration is reversed, but the in-plane extension does not. Therefore, the deflection is an odd function of a_2 or ε, with the leading term linear in ε and the next term cubic in ε if a Taylor expansion in ε is made. By the same reasoning, the deflection induced in-plane extension is an even function of ε, with the leading term quadratic in ε. Therefore, to the second-order in ε, the deflection is classical. The deflection from the classical theory and the flexure induced in-plane extension which is of order ε^2 were determined in the first section of this chapter. The first order acceleration sensitivity is zero as shown in the second section of this chapter.

8.5.1. Second-order acceleration sensitivity

In order to use Eq. (8.4.25) to calculate the second-order frequency shift, we need all of the unperturbed modes and frequencies. We assume that the particular frequency of interest is isolated. Then the second term on the right hand side of Eq. (8.4.25) can be neglected and we only need to consider the particular frequency and mode of interest, which in our case is the fundamental thickness-shear mode in the X_1 direction. If the plate is very thin, then, for thickness-shear modes which are near constant in the central portion of the plate, the edge effect can be neglected and we approximately have

$$\omega_0^2 = \frac{\pi^2 c_{66}}{4\rho_0 h^2}, \quad u_1 = \sqrt{\frac{3}{8\rho_0 abh^3}} X_2, \quad u_2 = u_3 = 0, \tag{8.5.1}$$

which is from the two-dimensional plate equations, normalized by $\langle \mathbf{u}, \rho_0 \mathbf{u} \rangle = 1$ over the volume of a finite plate of length $2a$, width $2b$, and thickness $2h$. The exact fundamental thickness-shear mode from the three-dimensional equations has a sinusoidal dependence on X_2 while the one in Eq. (8.5.1) has a linear dependence. This difference is not important to the problem we are interested in, i.e. some estimate of the second-order frequency shifts.

For an isolated mode with a vanishing first-order frequency shift, the second-order frequency shift in Eq. (8.4.25) simplifies to

$$\varepsilon^2 \tilde{\omega} = \frac{1}{2\omega_0} \langle U_{\alpha,K}; \varepsilon^2 \tilde{c}_{K\alpha L\gamma} U_{\gamma,L} \rangle, \tag{8.5.2}$$

where the superscript (i) has been dropped. In particular, for the mode in Eq. (8.5.1), Eq. (8.5.2) reduces to

$$\varepsilon^2\tilde{\omega} = \frac{1}{2\omega_0}\frac{3}{8\rho_{0}\ {}^{3}}\int \varepsilon^{2}\tilde{\cdot}_{2121} dV > . \tag{8.5.3}$$

We break the second-order frequency shift into two parts due to the in-plane extension and flexure as follows:

$$\begin{aligned}
\varepsilon^2\tilde{\omega} &= \varepsilon^2\tilde{\omega}^{ext} + \varepsilon^2\tilde{\omega}^{flex}, \\
\varepsilon^2\tilde{\omega}^{ext} &= \frac{1}{2\omega_0}\langle U_{\alpha,K}; \varepsilon^2\tilde{c}^{ext}_{K\alpha L\gamma} U_{\gamma,L}\rangle, \\
\varepsilon^2\tilde{\omega}^{flex} &= \frac{1}{2\omega_0}\langle U_{\alpha,K}; \varepsilon^2\tilde{c}^{flex}_{K\alpha L\gamma} U_{\gamma,L}\rangle,
\end{aligned} \tag{8.5.4}$$

$$\begin{aligned}
\tilde{c}^{ext}_{K\alpha L\gamma} &= c_{K\alpha L\gamma AB}\tilde{w}_{A,B} \\
&\quad + c_{K\alpha LN}\tilde{w}_{\gamma,N} + c_{KML\gamma}\tilde{w}_{\alpha,M} + c_{KLAB}\tilde{w}_{A,B}\delta_{\alpha\gamma},
\end{aligned} \tag{8.5.5}$$

$$\begin{aligned}
\tilde{c}^{flex}_{K\alpha L\gamma} &= \frac{1}{2}c_{K\alpha L\gamma AB}\overline{w}_{C,A}\overline{w}_{C,B} + \frac{1}{2}c_{K\alpha L\gamma ABCD}\overline{w}_{A,B}\overline{w}_{C,D} \\
&\quad + c_{K\alpha LNAB}\overline{w}_{\gamma,N}\overline{w}_{A,B} + c_{KML\gamma AB}\overline{w}_{\alpha,M}\overline{w}_{A,B} + c_{KMLN}\overline{w}_{\alpha,M}\overline{w}_{\gamma,N} \\
&\quad + \frac{1}{2}c_{KLAB}\overline{w}_{C,A}\overline{w}_{C,B}\delta_{\alpha\gamma} + \frac{1}{2}c_{KLABCD}\overline{w}_{A,B}\overline{w}_{C,D}\delta_{\alpha\gamma}.
\end{aligned} \tag{8.5.6}$$

The fourth-order elastic constants appear in $\varepsilon^2\tilde{\omega}^{flex}$ only.

8.5.2. One-dimensional deformation

For the one-dimensional case of $w_3 = 0$ and $\partial/\partial X_3 = 0$, from the solution in the first section of this chapter, we obtain

$$\begin{aligned}
\frac{\varepsilon^2\tilde{\omega}^{ext}}{\omega_0} &= \frac{1}{2\omega_0^2}\frac{3}{4\rho_0 a h^3}\int_{-h}^{h}\int_{-a}^{a}\varepsilon^2\tilde{c}^{ext}_{2121}dX_1 dX_2 \\
&\quad - \frac{3}{8\omega_0^2\rho_0 h^2}(c_{22} + c_{662})\left(1 + \frac{c_{21}}{c_{22}}\right)\sum_{m=1}^{\infty}\alpha_m^2 A_m^2,
\end{aligned} \tag{8.5.7}$$

and, similarly,

$$\frac{\varepsilon^2 \tilde{\omega}^{flex}}{\omega_0} = -\frac{3}{8\omega_0^2 \rho_0 h^3} \sum_m \alpha_m^2 A_m^2$$
$$\left\{ \frac{\alpha_m^2 h^3}{3} \left[c_{21} + 2c_{66} + c_{211} + 5c_{661} + c_{6611} \right.\right.$$
$$\left. + \frac{c_{21}^2}{c_{22}^2}(c_{22} + c_{222} + c_{662} + c_{6622}) - \frac{c_{21}}{c_{22}}(2c_{212} + 4c_{662} + 2c_{6612}) \right] \tag{8.5.8}$$
$$\left. + h(c_{21} + c_{22} + c_{661} + c_{662}) \right\}.$$

The above frequency shift depends on the acceleration nonlinearly through A_m which is proportional to the acceleration.

As a numerical example, we consider the same Y-cut quartz resonator with a = 7 mm, h = 0.097001 mm and $\omega_0 = 10^6$ Hz as in the third section of this chapter. When $a_2 = g$, from Eq. (8.5.7) and the elastic constants of quartz, we obtain part of the second-order acceleration sensitivity as

$$\frac{\varepsilon^2 \tilde{\omega}^{ext}}{\omega_0} = -8.63 \times 10^{-11}. \tag{8.5.9}$$

Since the first term in the series solution represents a good approximation of the deflection surface, we keep only one term in the summation in Eq. (8.5.7) and obtain the following expression which shows the dependence on the parameters explicitly:

$$\frac{\varepsilon^2 \tilde{\omega}^{ext}}{\omega_0} = -\frac{13824 a^6 \rho_0^2 g^2}{\pi^{10} h^4 \gamma_{1111}^2} \left(\frac{a_2}{g}\right)^2 \left(\frac{c_{22}}{c_{66}} + \frac{c_{662}}{c_{66}}\right)\left(1 + \frac{c_{21}}{c_{22}}\right). \tag{8.5.10}$$

For the same resonator, when $a_2 = g$, from Eq. (8.5.10) we find

$$\frac{\varepsilon^2 \tilde{\omega}^{ext}}{\omega_0} = -8.62 \times 10^{-11}, \tag{8.5.11}$$

which is very close to Eq. (8.5.9).

The other part of the second-order frequency shift requires knowledge of three fourth-order constants which are not available. As an estimate we set the unknown fourth-order constants to zero and obtain from Eq. (8.5.8) the following:

$$\frac{\varepsilon^2 \tilde{\omega}^{flex}}{\omega_0} = 2.09 \times 10^{-9}. \tag{8.5.12}$$

If the leading term in Eq. (8.5.8) is used alone, the result is about the same.

8.5.3. Two-dimensional deformation

For the two-dimensional case, from Eqs. (8.5.5) and (8.5.6) and the biasing fields in the first section of this chapter, we obtain the following expression for the frequency shift:

$$\begin{aligned}\frac{\varepsilon^2 \tilde{\omega}^{ext}}{\omega_0} &= -\frac{3}{16\omega_0^2 \rho_0 h^2} \\ &\sum_{m,n} A_{mn}^2 \{(c_{22} + c_{662})[(1 + \nu_{21})\alpha_m^2 + (1 + \nu_{23})\kappa_n^2] \\ &+ 2(c_{23} + c_{664})(\nu_{41}\alpha_m^2 + \nu_{43}\kappa_n^2)\},\end{aligned} \tag{8.5.13}$$

and

$$\begin{aligned}\frac{\varepsilon^2 \tilde{\omega}^{flex}}{\omega_0} &= -\frac{3}{16\omega_0^2 \rho_0 h^3} \sum_{m,n} A_{mn}^2 \\ &\left\{ \frac{\alpha_m^4 h^3}{3} [c_{21} + 2c_{66} + c_{211} + 5c_{661} + c_{6611} \right. \\ &+ \nu_{21}^2 (c_{22} + c_{222} + c_{662} + c_{6622}) - \nu_{21}(2c_{212} + 4c_{662} + 2c_{6612})] \\ &+ \frac{\kappa_n^4 h^3}{3} [c_{22} + c_{23} + c_{222} + c_{233} + c_{662} + c_{663} + c_{6622} + c_{6633} \\ &+ \nu_{23}^2 (c_{22} + c_{222} + c_{662} + c_{6622}) - \nu_{23}(2c_{223} + 2c_{6623})] \\ &+ \frac{\alpha_m^2 \kappa_n^2 h^3}{3} [c_{21} + c_{23} + 2c_{44} + 2c_{213} + 8c_{645} + c_{661} + 5c_{663} + 2c_{6613} \\ &- \nu_{21}(2c_{223} + 2c_{6623}) - \nu_{23}(2c_{212} + 4c_{662} + 2c_{6612}) \\ &+ 2\nu_{21}\nu_{23}(c_{22} + c_{222} + c_{662} + c_{6622})] \\ &\left. + \alpha_m^2 h(c_{21} + 3c_{22} + c_{661} + c_{662}) + \kappa_n^2 h(c_{22} + c_{23} + c_{662} + c_{663}) \right\},\end{aligned} \tag{8.5.14}$$

where

$$\nu_{21}=\frac{c_{12}c_{44}-c_{14}c_{24}}{c_{22}c_{44}-c_{24}^2},\quad \nu_{23}=\frac{c_{23}c_{44}-c_{24}c_{34}}{c_{22}c_{44}-c_{24}^2},$$
$$\nu_{41}=\frac{c_{14}c_{22}-c_{12}c_{24}}{c_{22}c_{44}-c_{24}^2},\quad \nu_{43}=\frac{c_{34}c_{22}-c_{23}c_{24}}{c_{22}c_{44}-c_{24}^2},\quad \nu_{65}=\frac{c_{56}}{c_{66}}. \tag{8.5.15}$$

For the same square Y-cut quartz resonator, under $a_2 = g$, Eq. (8.4.13) yields

$$\frac{\varepsilon^2\tilde{\omega}^{ext}}{\omega_0}=-2.25\times10^{-10}. \tag{8.5.16}$$

If the leading term $m = n = 1$ in Eq. (8.5.13) is used alone in the calculation (which causes an underestimate of the plate center deflection by 2.6%), the acceleration sensitivity is

$$\frac{\varepsilon^2\tilde{\omega}^{ext}}{\omega_0}=-2.24\times10^{-10}. \tag{8.5.17}$$

Equation (8.5.14) requires knowledge of six fourth-order constants. As an estimate we set the unknown fourth-order constants to zero and obtain the following:

$$\frac{\varepsilon^2\tilde{\omega}^{flex}}{\omega_0}=3.84\times10^{-10}. \tag{8.5.18}$$

If the leading term of Eq. (8.5.14) is used alone, the result is 3.83×10^{-10}.

8.6. Effects of Middle Surface Curvature

As pointed out in the last section of the third chapter, the middle surface of a plano-convex resonator is not flat. This causes coupling between flexure and extension. The implication is that the first-order normal acceleration sensitivity of a perfectly symmetric plano-convex resonator may not vanish. Consider a resonator with a slightly curved middle surface as shown in Fig. 8.6.1.

8.6.1. Biasing deformation

When the resonator is under a normal acceleration a_2, for the biasing deformation due to the acceleration, the classical theory for coupled

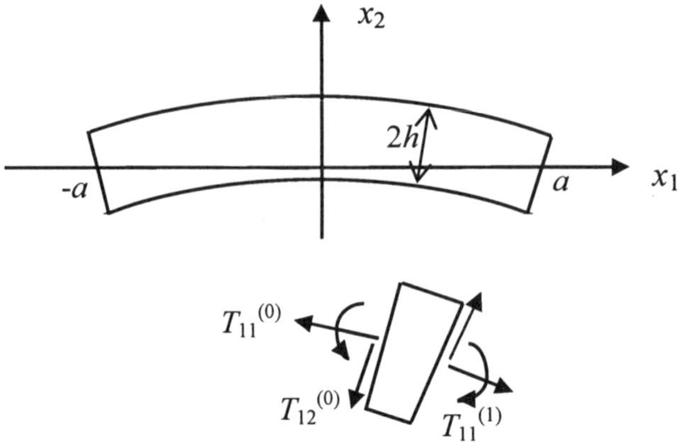

Fig. 8.6.1. A concave-convex resonator.

extension and flexure without shear deformation is sufficient. This can be obtained from the equations in the last section of the third chapter by eliminating thickness-shear as follows. First, we neglect the effect of the rotatory inertia in Eq. $(3.12.2)_3$ and obtain

$$T_{21}^{(0)} = T_{11,1}^{(1)}. \tag{8.6.1}$$

Substitution of Eq. (8.6.1) into Eq. $(3.12.2)_2$ yields

$$-k_1 T_{11}^{(0)} + T_{11,11}^{(1)} + F_2^{(0)} = 2h\rho_0 \ddot{w}_2^{(0)}, \tag{8.6.2}$$

where we have used **w** for the biasing displacement. We also set the shell shear strain in Eq. $(3.12.4)_2$ to zero and obtain

$$w_1^{(1)} = -w_{2,1}^{(0)} + k_1 w_1^{(0)}. \tag{8.6.3}$$

Under Eqs. (8.6.1) and (8.6.3), Eq. (3.12.4) becomes

$$\begin{aligned} T_{11}^{(0)} &= 2h\bar{c}_{11}(w_{1,1}^{(0)} + k_1 w_2^{(0)}), \\ T_{12}^{(0)} &= -\frac{2h^3}{3}\bar{c}_{11}(w_{2,111}^{(0)} - k_1 w_{1,11}^{(0)}), \\ T_{11}^{(1)} &= -\frac{2h^3}{3}\bar{c}_{11}(w_{2,11}^{(0)} - k_1 w_{1,1}^{(0)}). \end{aligned} \tag{8.6.4}$$

Substituting Eq. (8.6.4) into Eq. $(3.12.2)_1$ and Eq. (8.6.2) we have the following two equations for $w_1^{(0)}$ and $w_2^{(0)}$:

$$\begin{aligned} &2h\bar{c}_{11}(w_{1,11}^{(0)} + k_1 w_{2,1}^{(0)}) = 0, \\ &-k_1 2h\bar{c}_{11}(w_{1,1}^{(0)} + k_1 w_2^{(0)}) \\ &\quad -\frac{2h^3}{3}\bar{c}_{11}(w_{2,1111}^{(0)} - k_1 w_{1,111}^{(0)}) + \rho_0 2ha_2 = 0, \end{aligned} \tag{8.6.5}$$

where, for a shallow shell under a constant normal acceleration a_2, we have set

$$F_2^{(0)} = \rho_0 2ha_2 . \tag{8.6.6}$$

The boundary conditions are

$$\begin{aligned} &T_{11}^{(0)}(X_1 = \pm a) = 0, \\ &w_2^{(0)}(X_1 = \pm a) = 0, \\ &T_{11}^{(1)}(X_1 = \pm a) = 0, \end{aligned} \tag{8.6.7}$$

which can be satisfied by

$$\begin{aligned} &w_2^{(0)} = \sum_{m\ \text{odd}} A_m \cos\alpha_m X_1, \quad w_1^{(0)} = \sum_{m\ \text{odd}} B_m \sin\alpha_m X_1, \\ &\alpha_m = \frac{m\pi}{2a}, \quad m = 1,3,5.... \end{aligned} \tag{8.6.8}$$

Substituting Eq. (8.6.8) into Eq. (8.6.5), to the first order of the small curvature k_1, we have

$$A_m = \frac{12\rho_0 a_2}{h^2 m\pi \bar{c}_{11}\alpha_m^4}\sin\frac{m\pi}{2}, \quad B_m = -\frac{k_1}{\alpha_m}A_m . \tag{8.6.9}$$

We need $w_{K,L}$ to calculate a frequency shift. We begin with

$$\begin{aligned} &w_{K,L} = S_{LK} + \Omega_{LK}, \\ &S_{LK} = (w_{K,L} + w_{L,K})/2 = S_{KL}, \\ &\Omega_{LK} = (w_{K,L} - w_{L,K})/2 = -\Omega_{KL}. \end{aligned} \tag{8.6.10}$$

Since $w_3 = 0$ and $\partial/\partial X_3 = 0$, we immediately have

$$S_{31} = S_{32} = S_{33} = 0, \quad \Omega_{31} = \Omega_{32} = 0 . \tag{8.6.11}$$

The biasing displacement field is given by

$$w_1 = w_1^{(0)} + Y(-w_{2,1}^{(0)} + k_1 w_1^{(0)}), \quad w_2 = w_2^{(0)}, \tag{8.6.12}$$

from which we have

$$\begin{aligned} S_{11} &= w_{1,1}^{(0)} + Y(-w_{2,11}^{(0)} + k_1 w_{1,1}^{(0)}), \\ S_{21} &= k_1 w_1^{(0)}, \\ \Omega_{21} &= -2w_{2,1}^{(0)} + k_1 w_1^{(0)}. \end{aligned} \tag{8.6.13}$$

From the following stress relaxation condition of a thin shell:

$$c_{21}S_{11} + c_{22}S_{22} = 0, \tag{8.6.14}$$

we obtain the thickness strain as

$$S_{22} = -\frac{c_{12}}{c_{22}} S_{11} = -\frac{c_{12}}{c_{22}} [w_{1,1}^{(0)} + Y(-w_{2,11}^{(0)} + k_1 w_{1,1}^{(0)})]. \tag{8.6.15}$$

8.6.2. Unperturbed modes

For the unperturbed modes we are interested in the mode with $n = 1$ in Eq. (3.12.12)

$$u_1 = Yu_1^{(1)} = Y\cos\frac{\pi}{2a}X_1 \exp(i\omega t), \tag{8.6.16}$$

with frequency

$$\frac{\omega^2}{\omega_0^2} = 1 + \frac{h^2}{a^2}\left(\frac{\bar{c}_{11}}{c_{66}} + \kappa^2 + \frac{2k_1^2h^2}{3}\frac{\bar{c}_{11}}{c_{66}}\right). \tag{8.6.17}$$

In Eq. (8.6.17) the second-order effect of k_1 can be neglected.

8.6.3. Frequency shift

Since the flexure in the form of Eq. (8.6.8) produces no first-order frequency shift, we consider the effect of the biasing extensional deformation only. To the first order of k_1, we have the following components of the biasing deformation gradient:

$$w_{1,1} = w_{1,1}^{(0)}, \quad w_{2,2} = -\frac{c_{12}}{c_{22}} w_{1,1}^{(0)}. \tag{8.6.18}$$

The non-zero components of $\hat{c}_{K\alpha L\gamma}$ are

$$\hat{c}_{1111} = \left[3c_{11} + c_{111} - \frac{c_{12}}{c_{22}}(c_{12} + c_{112})\right] w_{1,1},$$
$$\hat{c}_{2121} = \left[2c_{66} + c_{12} + c_{166} - \frac{c_{12}}{c_{22}}(c_{22} + c_{266}) w_{1,1}\right]. \tag{8.6.19}$$

Then

$$\frac{\Delta\omega}{\omega} = -k_1 \sum_{m \text{ odd}} \frac{A_m \sin\frac{m\pi}{2}}{m\pi(m^2-4)} \times \frac{1}{2\omega_0^2 \rho_0 a^2 h^2}$$
$$\left\{\left[3c_{11} + c_{111} - \frac{c_{12}}{c_{22}}(c_{12} + c_{112})\right]\pi^2 h^2 (m^2 - 2) \right. \tag{8.6.20}$$
$$\left. - \left[2c_{66} + c_{12} + c_{166} - \frac{c_{12}}{c_{22}}(c_{22} + c_{266})\right] 24a^2 \right\}.$$

The series converges rapidly. If we keep the leading term of the series only, we have

$$\frac{\Delta\omega}{\omega} = -128\rho_0 k_1 a_2 a^2$$
$$\times \frac{1}{\pi^8 \bar{c}_{11} c_{66}\left[1 + \left(\frac{\bar{c}_{11}}{c_{66}} + \kappa^2 + \frac{2k_1^2 h^2}{3}\frac{\bar{c}_{11}}{c_{66}}\right)/(a/h)^2\right]} \tag{8.6.21}$$
$$\times \left\{\pi^2 \left[3c_{11} + c_{111} - \frac{c_{12}}{c_{22}}(c_{12} + c_{112})\right]\right.$$
$$\left. + 24(a/h)^2 \left[2c_{66} + c_{12} + c_{166} - \frac{c_{12}}{c_{22}}(c_{22} + c_{266})\right]\right\}.$$

Equation (8.6.21) shows a first-order effect linear in k_1 and a_2.

As a numerical example, consider an AT-cut quartz resonator: under $a_2 = g$. From Eq. (8.6.21) we find

$$\frac{\Delta\omega}{\omega} = 9.56 \times 10^{-8}. \tag{8.6.22}$$

For a Y-cut langasite resonator, when $a_2 = g$,

$$\frac{\Delta\omega}{\omega} = 1.26\times10^{-8}. \qquad (8.6.23)$$

8.7. Vibration Sensitivity

When a resonator is under a constant acceleration $\mathbf{a}$, the acceleration sensitivity $\mathbf{\Gamma}$ defines the acceleration perturbed frequency by $\omega(\mathbf{a}) = \omega_0(1+\mathbf{\Gamma}\cdot\mathbf{a})$, where ω_0 is the unperturbed frequency. In certain applications, the acceleration is also time-dependent (vibration). In the case of a time-harmonic acceleration with a frequency ω_v, the acceleration can be written as $\mathbf{a}\cos\omega_v t$, where $\mathbf{a}$ is a constant vector. In this case, it is more appropriate to consider the vibration sensitivity. When the frequency of the vibration is much lower than the operating frequency of the resonator, i.e. $\omega_v << \omega_0$, it has been widely assumed that the acceleration sensitivity $\mathbf{\Gamma}$ calculated from a constant acceleration

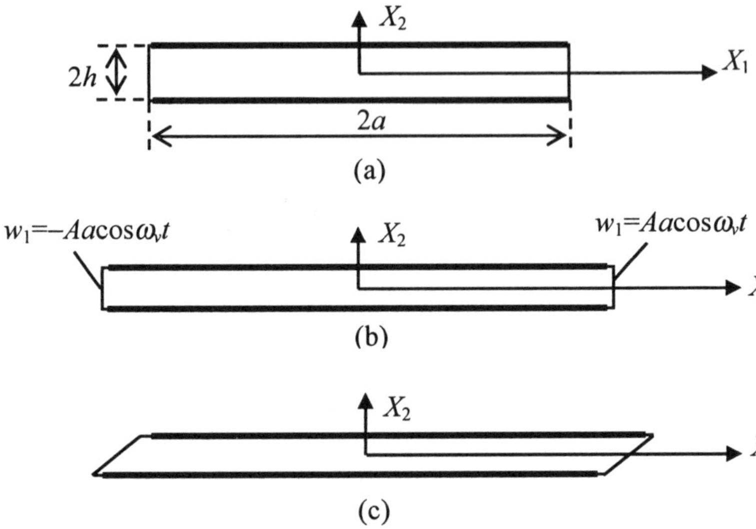

Fig. 8.7.1. (a) The reference configuration, (b) the biasing extensional deformation, and (c) the incremental thickness-shear deformation of a thin crystal plate.

analysis is still applicable. The frequency of a resonator under a time-dependent acceleration is then approximately given by $\omega(\mathbf{a}) = \omega_0(1 + \mathbf{\Gamma} \cdot \mathbf{a}\cos\omega_v t)$. This approach may be called a quasistatic analysis. Mathematically, since the vibration induced biasing fields are time-dependent, the equations for the incremental motion have time-dependent coefficients. In this section, the problem of resonator vibration sensitivity is analyzed from the equations for small fields superposed on time-dependent biasing fields [96].

Consider a thin crystal plate as shown in Fig. 8.7.1(a), with $a >> h$.

8.7.1. Governing equations

Consider motions independent of X_3, with $w_3 = u_3 = 0$. The two major surfaces of the plate are traction-free. The two ends of the plate at $X_1 = \pm a$ move with prescribed extensional displacements $\pm Aa\cos\omega_v t$, where A is a dimensionless number. The biasing fields are infinitesimal. Hence A is a very small number ($A << 1$). The plate is driven into extensional motions under the prescribed end displacements (see Fig. 8.7.1(b)). Since the plate is thin, $T^0_2 = 0$ is taken to be approximately true throughout. The equation for extensional motions takes the following form:

$$\bar{c}_{11} w_{1,11} = \rho_0 \ddot{w}_1, \quad -a < X_1 < a, \tag{8.7.1}$$

with boundary conditions

$$w_1(X_1 = \pm a) = \pm Aa\cos\omega_v t, \tag{8.7.2}$$

where $w_1 = w_1(X_1,t)$ and $\bar{c}_{11} = c_{11} - c_{12}^2 / c_{22}$. Once the extensional displacement is found, the related thickness expansion or contraction due to Poisson's effect can be obtained from the plate stress relaxation condition as

$$w_{2,2} = -\frac{c_{21}}{c_{22}} w_{1,1}. \tag{8.7.3}$$

Consider low frequency forcing with ω_v much lower than the resonant frequency of the first extensional mode of the plate, i.e. $\omega_v << \omega_{(1)} = \dfrac{\pi}{2a}\sqrt{\dfrac{\bar{c}_{11}}{\rho_0}}$. Then the inertial term $\rho_0 \ddot{w}_1$ in Eq. (8.7.1) may be neglected. The solution to Eqs. (8.7.1) and (8.7.2) is

$$w_1 = AX_1 \cos \omega_v t, \quad w_2 = -\frac{c_{21}}{c_{22}} AX_2 \cos \omega_v t\,, \tag{8.7.4}$$

which produces the following spatially homogeneous, time-dependent biasing strains:

$$S_1^0 = A\cos \omega_v t, \quad S_2^0 = -\frac{c_{21}}{c_{22}} A\cos \omega_v t\,. \tag{8.7.5}$$

Since the biasing deformations are uniform and the plate is assumed to be thin, for the incremental motion the edge effects are neglected and motions are assumed to be independent of X_1. Consider thickness-shear vibration in the X_1 direction (see Fig. 8.7.1(c)),

$$u_1(X_2,t) = u(t)\sin\frac{n\pi}{2b}X_2, \quad n = 1,3,5\ldots. \tag{8.7.6}$$

From the ninth section of the second chapter, the governing equation for $u(t)$ is

$$\ddot{u} + \omega_0^2(1 + \varepsilon \cos \omega_v t)u = 0\,, \tag{8.7.7}$$

where

$$\omega_0^2 = \frac{n^2\pi^2 c_{66}}{4b^2\rho_0}, \quad \varepsilon = \left(2 + \frac{c_{661}}{c_{66}} - \frac{c_{662}c_{21}}{c_{66}c_{22}}\right)A\,. \tag{8.7.8}$$

Equation (8.7.7) is the well-known Mathieu equation.

8.7.2. First-order solution

Consider the case when the frequency of the biasing fields is much lower than the operating frequency of the resonator, i.e., $\omega_v << \omega_0$. We seek solutions to Eq. (8.7.7) in the following form:

$$u(t) = \alpha(t)\cos[\omega_0 t + \beta(t)]\,, \tag{8.7.9}$$

where $\alpha(t)$ and $\beta(t)$ are dimensionless functions. They are slowly varying in the following sense

$$\dot{\alpha} << \omega_0, \quad \dot{\beta} << \omega_0\,. \tag{8.7.10}$$

Differentiating Eq. (8.7.9) with respect to time twice, substituting the expressions for u and $\ddot{u}$ into Eq. (8.7.7), setting the sum of the coefficients of the terms with $\sin(\omega_0 t+\beta)$ and the sum of the coefficients of the terms with $\cos(\omega_0 t+\beta)$ to zero separately, we obtain the following two equations for $\alpha(t)$ and $\beta(t)$:

$$\begin{aligned} &\alpha\ddot{\beta}+2\dot{\alpha}\dot{\beta}+2\dot{\alpha}\omega_0=0, \\ &\ddot{\alpha}-2\omega_0\alpha\dot{\beta}-\alpha\dot{\beta}\dot{\beta}+\varepsilon\alpha\omega_0^2\cos\omega_v t=0. \end{aligned} \tag{8.7.11}$$

Under Eq. (8.7.10), we neglect terms associated with the second derivatives of $\alpha(t)$ and $\beta(t)$ or the product of their first derivatives in Eq. (8.7.11) and obtain the following approximate system of first-order equations:

$$\begin{aligned} &2\dot{\alpha}\omega_0=0, \\ &-2\omega_0\alpha\dot{\beta}+\varepsilon\alpha\omega_0^2\cos\omega_v t=0. \end{aligned} \tag{8.7.12}$$

Equation (8.7.12) shows that to the first order, α is a constant and the corresponding β is found to be

$$\beta(t)=\frac{\omega_0}{2\omega_v}\varepsilon\sin\omega_v t+\gamma\,, \tag{8.7.13}$$

where γ is an integration constant. Equation (8.7.13) shows that $\beta(t)$ is indeed slowly varying in the sense of Eq. (8.7.10) due to the smallness of ε. Within the first-order approximation

$$u(t)=\alpha\cos\left(\omega_0 t+\frac{\omega_0}{2\omega_v}\varepsilon\sin\omega_v t+\gamma\right), \tag{8.7.14}$$

which is a frequency modulated signal. The modulated frequency may be defined as

$$\omega(t)=\frac{d}{dt}[\omega_0 t+\beta(t)]=\omega_0\left(1+\frac{\varepsilon}{2}\cos\omega_v t\right). \tag{8.7.15}$$

8.7.3. Second-order solution

For a more refined solution that shows the time dependence of α and a more refined expression for β, some second-order terms in Eq. (8.7.11) need to be kept. The first-order solution shows that although both $\alpha(t)$ and $\beta(t)$ are slowly varying, $\alpha(t)$ is even slower than $\beta(t)$. Therefore we drop $\ddot{\alpha}$ but keep $\dot{\beta}\dot{\beta}$ (which was dropped in the first-order approximation) in Eq. $(8.7.11)_2$ and obtain

$$-2\omega_0\dot{\beta}-\dot{\beta}\dot{\beta}+\varepsilon\omega_0^2\cos\omega_v t=0\,. \tag{8.7.16}$$

From Eq. (8.7.16) we obtain

$$\dot{\beta} = -\omega_0 + \omega_0 \sqrt{1+\varepsilon \cos \omega_v t} \,. \tag{8.7.17}$$

Equation (8.7.17) implies the following approximate expression of the frequency:

$$\begin{aligned} \omega &= \omega_0 + \dot{\beta} = \omega_0 \sqrt{1+\varepsilon \cos \omega_v t} \\ &= \omega_0 \left(1 + \frac{\varepsilon}{2} \cos \omega_v t - \frac{\varepsilon^2}{8} \cos^2 \omega_v t + \cdots \right), \end{aligned} \tag{8.7.18}$$

which is more refined than Eq. (8.7.15). Multiplying Eq. $(8.7.11)_1$ by $\alpha(t)$ gives

$$\alpha^2 \ddot{\beta} + 2\alpha \dot{\alpha} \dot{\beta} + 2\alpha \dot{\alpha} \omega_0 = 0 \,, \tag{8.7.19}$$

which can be written as

$$\frac{d}{dt}(\alpha^2 \dot{\beta} + \alpha^2 \omega_0) = 0 \,. \tag{8.7.20}$$

Integration of Eq. (8.7.20) yields

$$\alpha^2 \dot{\beta} + \alpha^2 \omega_0 = \lambda^2 \omega_0 \,, \tag{8.7.21}$$

where λ is an integration constant. Solving Eq. (8.7.21) for $\alpha(t)$, we obtain

$$\alpha^2 = \frac{\lambda^2}{1 + \dot{\beta} / \omega_0} \,. \tag{8.7.22}$$

To the lowest order, we have, approximately,

$$\alpha = \lambda \left(1 - \frac{\dot{\beta}}{2\omega_0} \right) = \frac{\lambda}{2}(3 - \sqrt{1+\varepsilon \cos \omega_v t}) \,, \tag{8.7.23}$$

which is indeed much slower than $\beta(t)$. Equations (8.7.18) and (8.7.23) show that both the frequency and the amplitude of $u(t)$ are modulated.

8.7.4. An example

As a numerical example, consider a Y-cut quartz resonator. For geometric parameters we choose $a = 14.25$ mm and $b = 0.9696$ mm. The biasing deformation is specified by $A = 1$. Then $\varepsilon = 0.4221$. The above parameters imply that for the extensional biasing deformation, $\omega_{(1)}/2\pi = 10^5$ Hz. For the thickness-shear motion without biasing deformation, the fundamental thickness-shear frequency (n = 1 in Eq. (8.7.8)) is $\omega_0/2\pi$ = 10^6 Hz. We choose $\omega_v/2\pi$ = 10^5 Hz. We are considering a very small effect. The above parameters are chosen to greatly exaggerate this small effect. Therefore, some of the parameters do not quite satisfy the assumptions that $A \ll 1$, $\varepsilon \ll 1$, $\omega_v \ll \omega_{(1)}$ and $\omega_v \ll \omega_0$. The zero-, first-, and approximate second-order solutions are summarized below:

$$u(t) = \cos\omega_0 t,$$

$$u(t) = \cos\left(\omega_0 t + \frac{\omega_0}{2\omega_v}\varepsilon\sin\omega_v t\right), \tag{8.7.24}$$

$$u(t) = \frac{1}{2}(3 - \sqrt{1 + \varepsilon\cos\omega_v t})\cos\left(\omega_0 t + \frac{\omega_0}{2\omega_v}\varepsilon\sin\omega_v t\right).$$

where we have set λ = 1 and γ = 0. Eq. $(8.7.24)_1$ is the zero-order solution when the biasing field is not present, which is a pure sinusoidal signal. Equation $(8.7.24)_2$ shows the first-order effect of the biasing field. This is a frequency modulated signal as shown in Fig. 8.7.2. This is the same as the result from the quasistatic analysis.

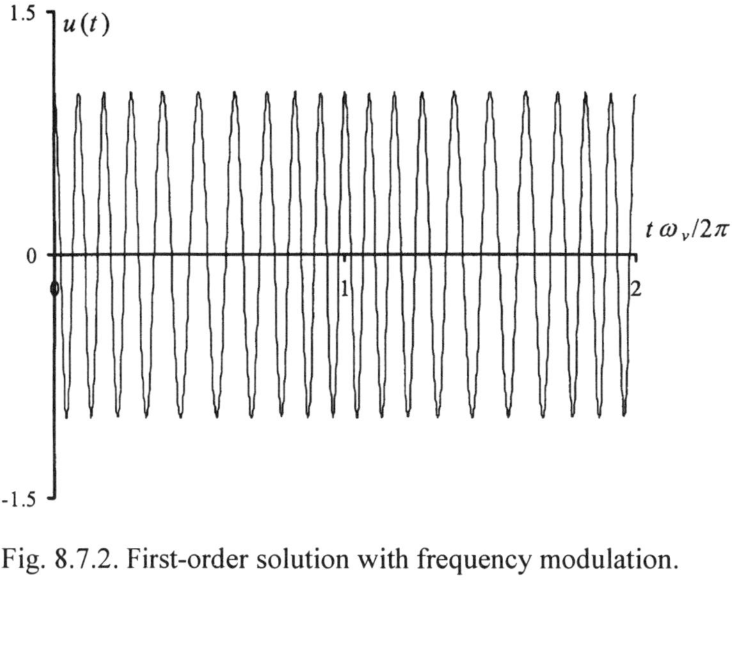

Fig. 8.7.2. First-order solution with frequency modulation.

Equation $(8.7.24)_3$ includes some second-order effects on the amplitude modulation and is shown in Fig. 8.7.3.

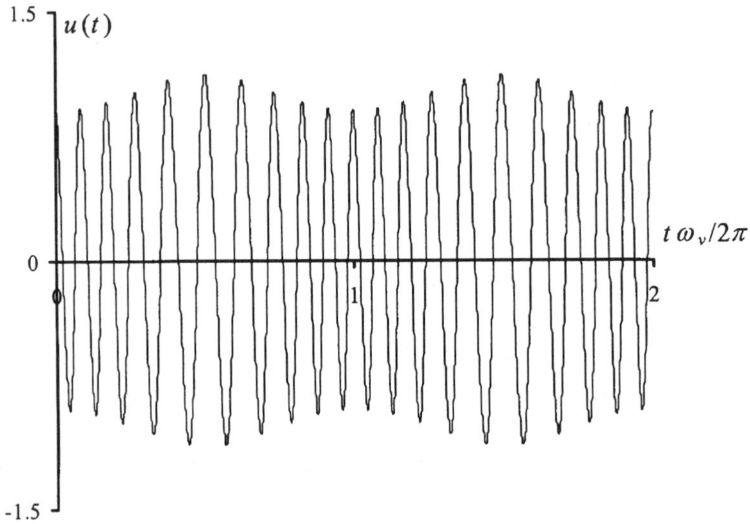

Fig. 8.7.3. Second-order solution with frequency and amplitude modulation.

Chapter 9
Pressure Sensors

Piezoelectric pressure sensors can be made by measuring either pressure induced charge or frequency shifts. In this chapter we analyzed a few pressure sensors based on frequency shifts in resonators. References on this topic can be found in a few review articles [97–99].

9.1. A Rectangular Plate in a Circular Cylindrical Shell

Consider a rotated Y-cut quartz thickness-shear resonator sealed in a metallic circular cylindrical shell as shown in Fig. 9.1.1 [100]. The structure is long in the X_3 direction and only a cross-section is shown. A uniform external pressure p is applied on the outer surface of the shell. The crystal plate experiences compression in the X_1 direction due to the pressure. This compression causes biasing deformations and frequency shifts in the crystal plate, which can be used to measure the pressure. The deformation is with $\partial/\partial X_3 = 0$ and $u_3 = 0$.

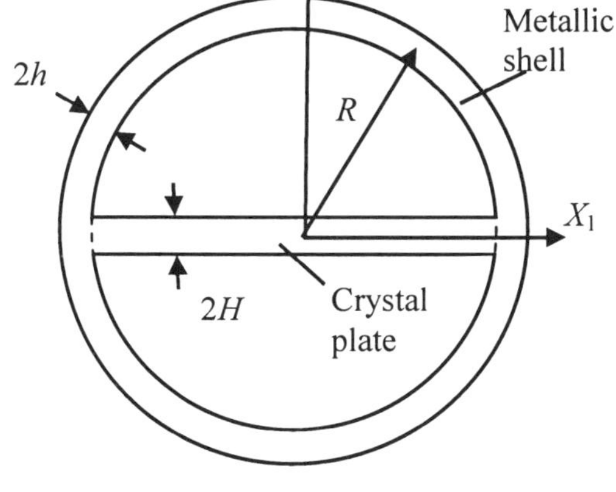

Fig. 9.1.1. Cross-section of a crystal plate in a cylindrical shell as a pressure sensor.

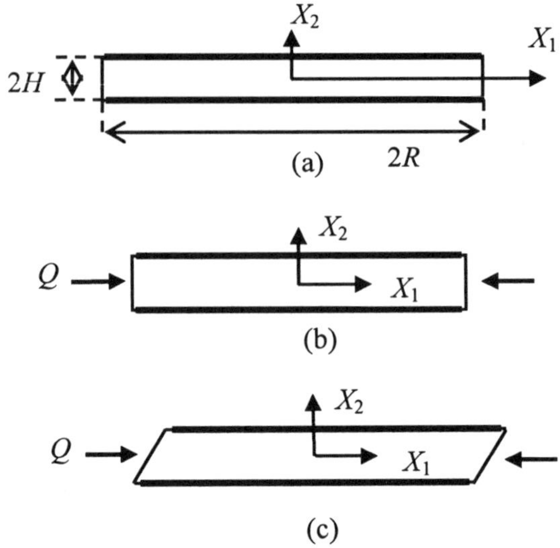

Fig. 9.1.2. (a) Reference configuration, (b) biasing extensional deformation, and (c) incremental thickness-shear deformation of a thin quartz plate.

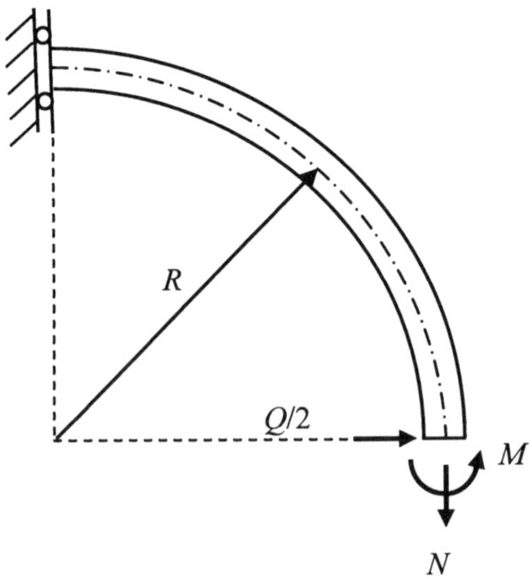

Fig. 9.1.3. Internal forces in the shell.

The reference configuration of the crystal plate when there is no pressure is shown in Fig. 9.1.2(a). The plate is assumed to be thin with $R >> H$. The pressure induces a compressional force Q on the crystal plate through the shell (see Fig. 9.1.2(b)). Q acts as biasing fields in the crystal plate. Under Q the plate vibrates at the fundamental thickness-shear mode (see Fig. 9.1.2(c)) with a frequency slightly different from the case when $Q = 0$.

First we need to determine Q. Due to symmetry consider a quarter of the shell as shown in Fig. 9.1.3.

9.1.1. Analysis

The shell is assumed to be thin with thickness $2h << R$. Under p, the shell undergoes coupled extension and flexure. Consider thin shells whose shear deformation can be neglected. The problem for determining Q is statically indeterminate. The strain energy per unit length along the X_3 direction is

$$\begin{aligned} U = &\frac{R}{4Eh}\int_0^{\pi/2}\left[Q\sin\frac{\theta}{2} + N\cos\theta - p(R+h)(1-\cos\theta)\right]^2 d\theta \\ &+\frac{3R}{4Eh^3}\int_0^{\pi/2}\left[M + QR\sin\frac{\theta}{2} - NR(1-\cos\theta)\right. \\ &\left. - pR(R+h)(1-\cos\theta)\right]^2 d\theta, \end{aligned} \tag{9.1.1}$$

where E should be understood as the Young's modulus modified for plane-strain. According to the theory of elastic structures, the partial derivatives of U with respect to M, N or Q yield the corresponding displacements. From the symmetry of the structure we are considering, we can write the following equilibrium and compatibility conditions:

$$\begin{aligned} &N = -pR, \quad \frac{\partial U}{\partial M} = 0, \\ &\frac{\partial U}{\partial (Q/2)} = -\frac{Q}{\bar{c}_{11}2H}(R-h), \end{aligned} \tag{9.1.2}$$

where $\bar{c}_{11} = c_{11} - c_{12}^2/c_{22}$. The right-hand side of Eq. (9.1.2)$_3$ is half of the shortening of the quartz plate under Q. c_{11} and c_{12} are the elastic constants of the quartz plate. The solution to Eq. (9.1.2) is

$$\frac{Q}{2H} = \frac{1}{2}\frac{\gamma_1}{\gamma_2} p,$$
$$\gamma_1 = 4\bar{c}_{11} hR[Eh\pi(h+2R) - 3E'R^2(\pi - 4)], \tag{9.1.3}$$
$$\gamma_2 = 3\bar{c}_{11} E'HR^3(\pi^2 - 8) + Eh^2\pi[8E'h(R-h) + \bar{c}_{11} HR\pi].$$

For the biasing fields in the crystal plate under Q, from the equations for planar deformation we have the following extensional strain:

$$S_1^0 = w_{1,1} = -\frac{Q}{\bar{c}_{11} 2H}. \tag{9.1.4}$$

The related thickness expansion due to Poisson's effect can be obtained from the plate stress relaxation condition $K_{22}^0 = 0$ as

$$S_2^0 = w_{2,2} = -\frac{c_{21}}{c_{22}} w_{1,1}. \tag{9.1.5}$$

Since the biasing deformations are uniform and the plate is thin with $R >> H$, for the incremental motion we neglect edge effects and consider motions independent of X_1. For thickness-shear vibration in the X_1 direction with

$$u_1(X_2, t) = u(t)\sin\frac{n\pi}{2H} X_2, \quad n = 1, 3, 5, \ldots, \tag{9.1.6}$$

from Eq. (2.9.10), the governing equation for $u(t)$ is

$$\ddot{u} + \omega_0^2(1 + cS_1^0)u = 0, \tag{9.1.7}$$

where

$$c = 2 + \frac{c_{661}}{c_{66}} - \frac{c_{662} c_{21}}{c_{66} c_{22}}, \quad \omega_0^2 = \frac{n^2\pi^2 c_{66}}{4H^2 \rho_0}. \tag{9.1.8}$$

From Eq. (2.9.13), the frequency shift is given by

$$\frac{\omega - \omega_0}{\omega_0} = \frac{1}{2} cS_1^0 = -\frac{c}{4}\frac{\gamma_1}{\gamma_2}\frac{p}{\bar{c}_{11}}. \tag{9.1.9}$$

The following observations can be made from Eq. (9.1.9):

(i) The frequency shift is linear in p;

(ii) The third-order nonlinear elastic constants are relevant to the sensitivity through c.

9.1.2. Numerical results

A few numerical examples are calculated below. For the shell we consider Al, Cu, and W with the Young's modulus E of 70, 110 and $360 \times 10^9 \text{N/m}^2$, respectively. We ignore the difference between the Young's modulus and its plane-strain modification as an approximation.

Frequency shifts versus pressure in a Y-cut quartz plate for different shell materials are shown in Fig. 9.1.4, where $h/R = H/R = 1/10$ are fixed. Similar results are plotted in Fig. 9.1.5 for a plate of Y-cut langasite. The frequency shifts vary linearly with p. The slopes of the straight lines are related to sensitivity. Larger slopes represent higher sensitivity. Clearly, for softer shell materials the sensitivity is slightly higher.

The dependence of frequency shifts upon pressure in a Y-cut quartz plate for different H/h is shown in Fig. 9.1.6, where $H/R = 1/10$ is fixed and the outer shell material is fixed to be Al. The figure shows that smaller H/h results in higher sensitivity.

The above analysis is based on the assumption of small bias or low applied pressure. When subjected to a large pressure, the response will be nonlinear and the equations for finite biasing fields (nonlinear) will be required to obtain an accurate prediction of frequency shifts.

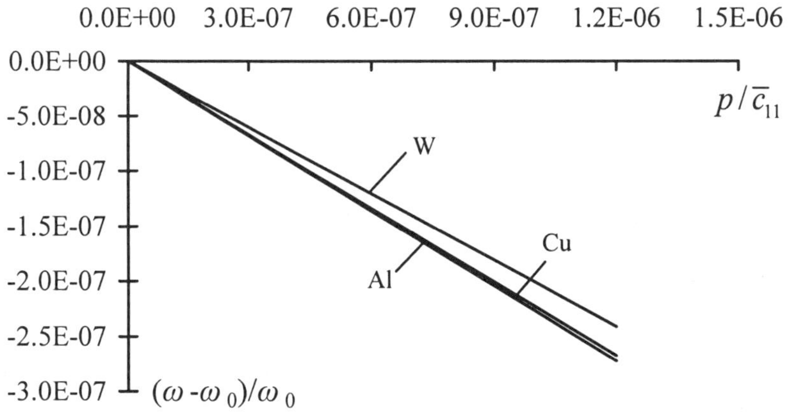

Fig. 9.1.4. Frequency shifts in a quartz resonator for different shell materials.

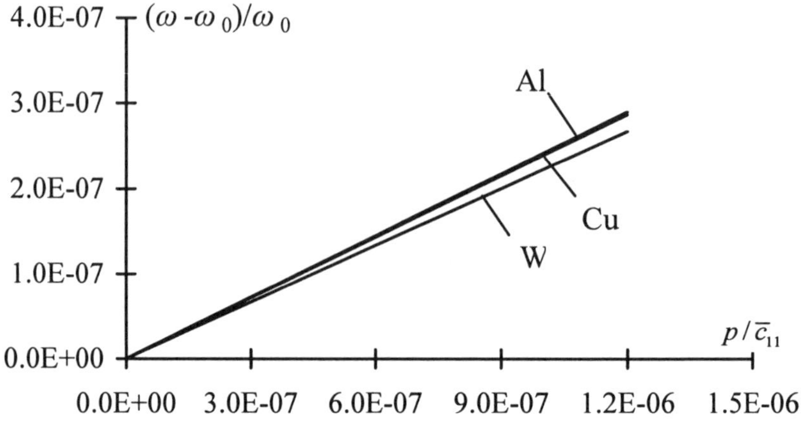

Fig. 9.1.5. Frequency shifts in a langasite resonator for different shell materials.

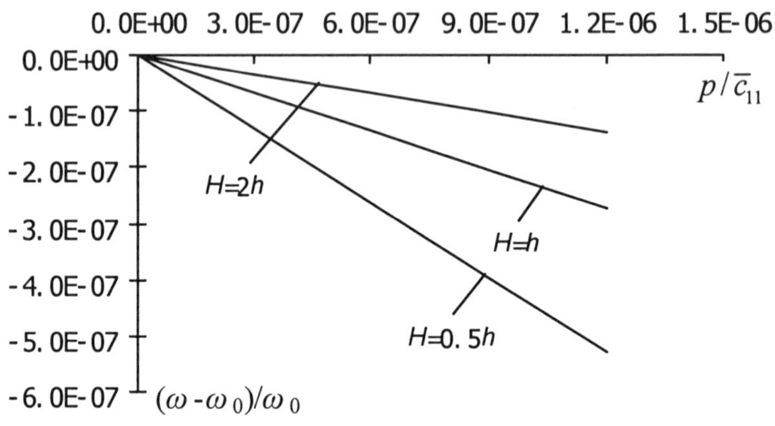

Fig. 9.1.6. Frequency shifts in a quartz resonator for different geometric parameters.

9.2. A Circular Plate in a Circular Cylindrical Shell

Consider a circular plate quartz thickness-shear resonator sealed in a circular cylindrical shell as shown in Fig. 9.2.1 [101]. The structure is long in the X_2 direction. A uniform external pressure p is applied.

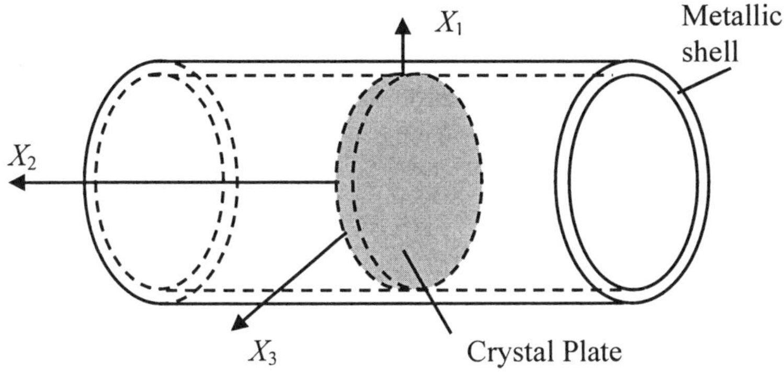

Fig. 9.2.1. A crystal plate in a cylindrical shell as a pressure sensor.

9.2.1. Analysis

The pressure induces a compressional force Q on the crystal plate through the shell (see Fig. 9.2.2). Q acts as biasing fields in the crystal plate. The plate vibrates at the fundamental thickness-shear mode. Q causes a frequency shift in this mode. First we need to determine Q. Due to symmetry consider half of the shell.

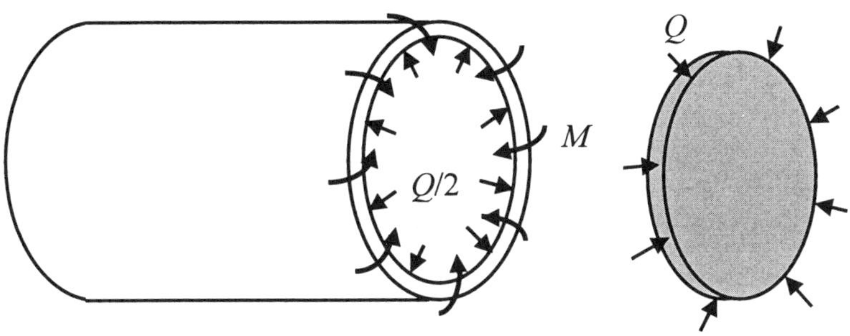

Fig. 9.2.2. Free body diagrams of the crystal plate and the shell.

Under p, Q and a bending moment M develop as shown in the figure. The problem for determining Q is statically indeterminate. For AT- or SC-cut quartz plates the anisotropy in the (X_3, X_1) plane is weak, ranging from 5.5–23.1% as compared to the isotropic part. Therefore, as an approximation, only when determining Q, we treat the crystal plate as isotropic. Using the theory of elastic shells, we can determine the deflection at $X_2 = 0$ of the cylindrical shell under the joint action of p, $Q/2$ and M as follows

$$w(0) = \frac{Q}{8\alpha^3 D} - \frac{pR^2}{Eh}, \tag{9.2.1}$$

where

$$\alpha^4 = \frac{Eh}{4R^2 D}, \quad D = \frac{Eh^3}{12(1-\nu^2)}. \tag{9.2.2}$$

E and ν are the Young's modulus and the Poisson's ratio, respectively, of the shell. In obtaining $w(0)$ in the above, we have eliminated M through the compatibility condition

$$\left. \frac{\partial w}{\partial X_2} \right|_{X_2=0} = 0. \tag{9.2.3}$$

Then the radial contraction of the crystal plate should be equal to $w(0)$, which results in

$$\begin{aligned} &\frac{Q}{2H} = \frac{p}{\gamma}, \\ &\gamma = \frac{1}{(1+\bar{c}_{33}/\bar{c}_{11})/2 + \bar{c}_{13}/\bar{c}_{11}} \frac{E}{\bar{c}_{11}} \frac{h}{R} + \sqrt[4]{3(1-\nu^2)} \frac{H}{\sqrt{Rh}}, \end{aligned} \tag{9.2.4}$$

where

$$\begin{aligned} &\bar{c}_{11} = c_{11} - c_{12}^2/c_{22}, \quad \bar{c}_{33} = c_{33} - c_{13}^2/c_{22}, \\ &\bar{c}_{13} = c_{13} - c_{12}c_{13}/c_{22}. \end{aligned} \tag{9.2.5}$$

For the biasing fields in the crystal plate under Q, the stresses are approximately planar and isotropic:

$$T_{11}^0 = T_{33}^0 = -Q/2H, \quad \text{all other} \quad T_{IJ}^0 = 0. \tag{9.2.6}$$

The biasing strains are determined by inverting the following anisotropic plane-stress constitutive relations:

$$
\begin{aligned}
T_{11}^0 &= c_{11}S_{11}^0 + c_{12}S_{22}^0 + c_{13}S_{33}^0 + c_{14}S_{23}^0 = -Q/2H, \\
T_{22}^0 &= c_{12}S_{11}^0 + c_{22}S_{22}^0 + c_{23}S_{33}^0 + c_{24}S_{23}^0 = 0, \\
T_{33}^0 &= c_{13}S_{11}^0 + c_{23}S_{22}^0 + c_{33}S_{33}^0 + c_{34}S_{23}^0 = -Q/2H, \\
T_{23}^0 &= c_{14}S_{11}^0 + c_{24}S_{22}^0 + c_{34}S_{33}^0 + c_{44}S_{23}^0 = 0, \\
T_{31}^0 &= c_{55}S_{31}^0 + c_{56}S_{12}^0 = 0, \\
T_{12}^0 &= c_{56}S_{31}^0 + c_{66}S_{12}^0 = 0,
\end{aligned}
\tag{9.2.7}
$$

with the following result:

$$
\begin{aligned}
S_{11}^0 &= -(Q/2H)(G_{11} + G_{13}), \\
S_{22}^0 &= -(Q/2H)(G_{21} + G_{23}), \\
S_{33}^0 &= -(Q/2H)(G_{31} + G_{33}), \\
S_{23}^0 &= -(Q/2H)(G_{41} + G_{43}), \\
S_{31}^0 &= S_{12}^0 = 0,
\end{aligned}
\tag{9.2.8}
$$

where

$$
\left(G_{ij}\right) = \begin{pmatrix} c_{11} & c_{12} & c_{13} & c_{14} \\ c_{12} & c_{22} & c_{23} & c_{24} \\ c_{13} & c_{23} & c_{33} & c_{34} \\ c_{14} & c_{24} & c_{34} & c_{44} \end{pmatrix}^{-1} . \tag{9.2.9}
$$

For the incremental motion we consider thickness-shear in the X_1 direction. When the biasing fields are not present, the odd modes of practical interest are given by:

$$
\begin{aligned}
U_1 &= \sin\frac{n\pi X_2}{2H}\exp(i\omega_0 t), \quad n = 1,3,5,\ldots, \\
U_2 &= U_3 = 0, \quad \omega_0^2 = \frac{n^2\pi^2 c_{66}}{4\rho_0 H^2}.
\end{aligned}
\tag{9.2.10}
$$

We use the perturbation integral in Eq. (1.3.19) which is repeated below for convenience:

$$
\Delta\omega = \omega - \omega_0 = \frac{1}{2\omega_0}\frac{\int_V \hat{c}_{L\gamma M\alpha} U_{\gamma,L} U_{\alpha,M}\, dV}{\int_V \rho_0 U_\alpha U_\alpha\, dV} . \tag{9.2.11}
$$

Substituting the biasing deformation and the unperturbed mode into Eq. (9.2.11), we obtain

$$\Delta\omega = \frac{k^2 \hat{c}_{2121}}{2\omega_0 \rho_0}, \tag{9.2.12}$$

where

$$k^2 = \left(\frac{n\pi}{2H}\right)^2,$$
$$\hat{c}_{2121} = (2c_{66} + c_{21} + c_{661})S_1^0 + (c_{22} + c_{662})S_2^0 + (c_{23} + c_{663})S_3^0 + (c_{24} + c_{664})S_4^0 + (c_{46} + c_{25} + c_{665})S_5^0 + (2c_{26} + c_{666})S_6^0. \tag{9.2.13}$$

9.2.2. Numerical results

As numerical examples, for the shell we consider Al, Cu, and W with the Young's modulus E of 70, 110 and 360 $\times 10^9$ N/m^2, respectively. Frequency shifts versus pressure in a Y-cut quartz plate for different shell materials are shown in Fig. 9.2.3, where $h/R = H/R = 1/10$ is fixed.

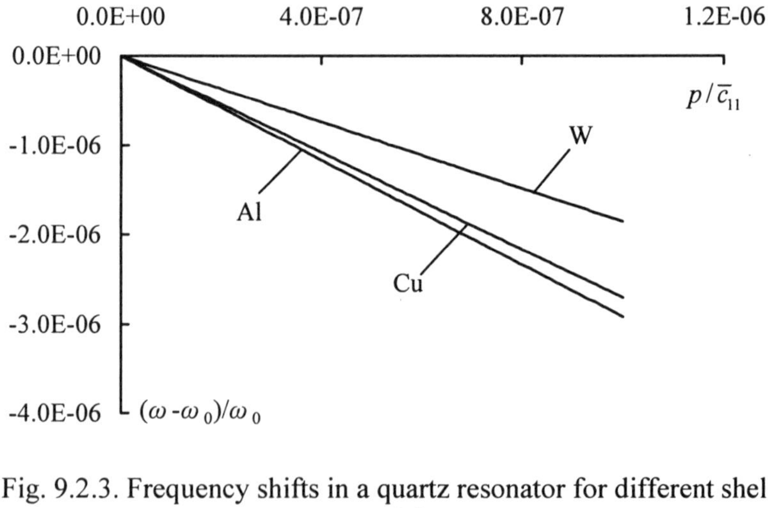

Fig. 9.2.3. Frequency shifts in a quartz resonator for different shell materials.

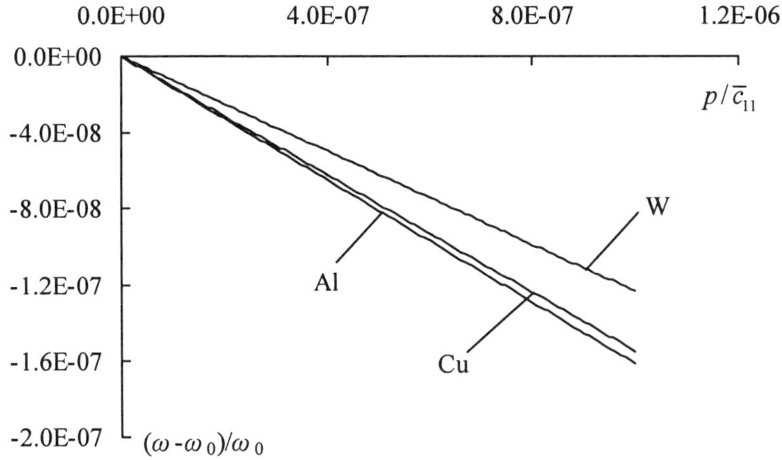

Fig. 9.2.4. Frequency shifts in a langasite resonator for different shell materials.

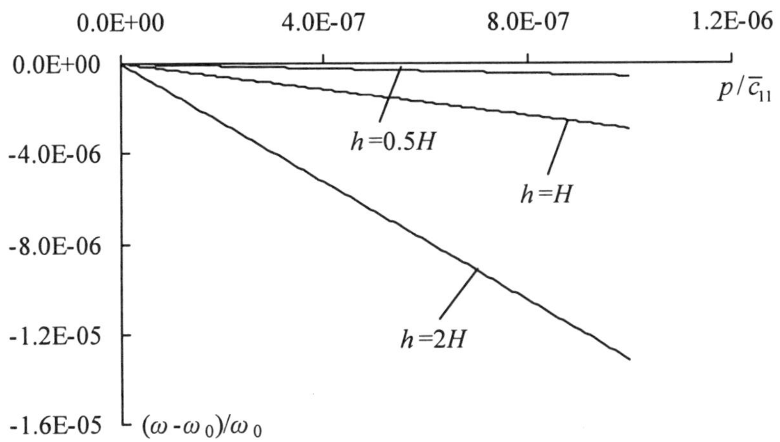

Fig. 9.2.5. Frequency shifts in a quartz resonator for different geometric parameters.

Similar results are plotted in Fig. 9.2.4 for a plate of Y-cut langasite. The frequency shifts vary linearly with p. The slopes of the straight lines are related to sensitivity. Clearly, for softer shell materials the sensitivity is higher.

The dependence of frequency shifts upon pressure in a Y-cut quartz plate for different H/h is shown in Fig. 9.2.5, where $H/R = 1/10$ is fixed and the outer shell material is Al. The figure shows that smaller H/h results in higher sensitivity.

9.3. A Rectangular Plate in a Shallow Shell

Consider the reference state of the structure shown in Fig. 9.3.1 [102].

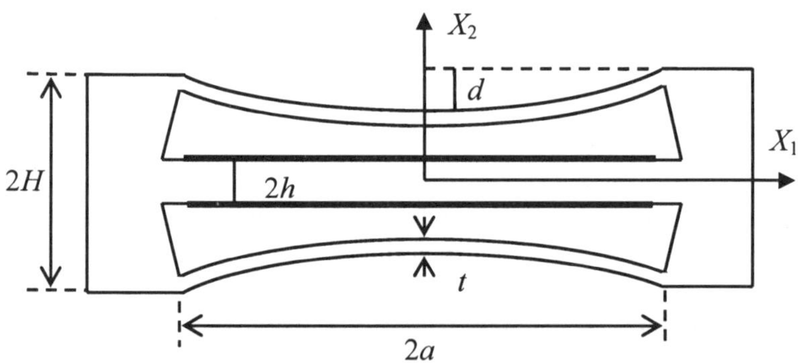

Fig. 9.3.1. A crystal plate in a shallow shell as a pressure sensor.

9.3.1. Analysis

The shell is a cylindrical structure which is long in the X_3 direction. Figure 9.3.1 shows a cross-section with a unit length in the X_3 direction. A quartz thickness-shear resonator of thickness $2h$ between two electrodes represented by the thick lines is sealed in the shell. The shallow shell is part of a circular cylindrical shell. The shell is shallow in the sense that $a >> d$. The shell is also assumed to be very thin with $t << a$. The radius of the corresponding circular cylindrical shell (see Fig. 9.3.2) can be found as

$$R = \frac{a^2 + d^2}{2d} >> a. \tag{9.3.1}$$

When the structure in Fig. 9.3.1 is subject to a surrounding pressure p, extensional stresses develop in the shell. As an approximation, we neglect bending moments and shear forces everywhere in the shell including the edges, and treat it as a membrane that does resist bending.

The free body diagram of the lower piece of the shell when under pressure p is shown in Fig. 9.3.2. The problem is statically determinate. The membrane extensional stress N/t induced by p is simply the hoop stress for a circular cylindrical shell under a uniform pressure

$$\frac{N}{t}=\frac{pR}{t},$$
$$N=pR=p\frac{a^2+d^2}{2d}\cong\frac{pa^2}{2d}, \tag{9.3.2}$$

which is large when $a >> d$. The membrane stress in Eq. (9.3.2) compresses the crystal plate axially and generates a pressure-induced initial stress in the resonator. Under the axial stress induced by p, the resonator vibrates in a thickness-shear mode at a resonant frequency that is slightly perturbed from that of a stress-free resonator.

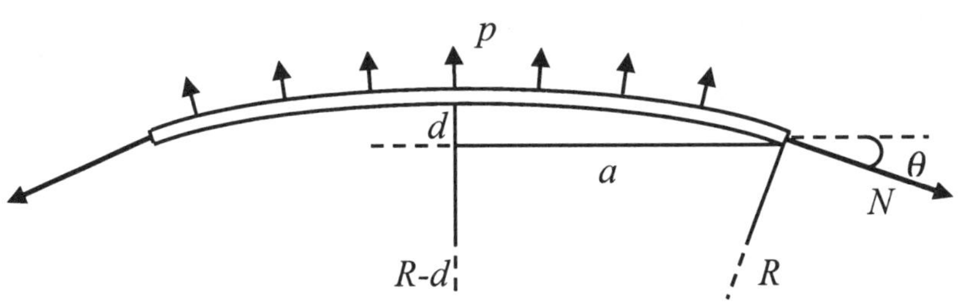

Fig. 9.3.2. Free body diagram of the shell.

The crystal plate is under the axial compressional force Q (see Fig. 9.1.2) which depends on the pressure p as follows

$$Q=2N\cos\theta+p2H\cong 2(N+pH), \tag{9.3.3}$$

where the approximation is valid for a shallow shell with a small θ. The two terms on the right-hand side of Eq. (9.3.3) are due to the pressure p on the shells and the pressure that acts on the side walls of the structure with height $2H$. Q produces the following strain in the crystal plate:

$$S_1^0=w_{1,1}=-\frac{Q}{\bar{c}_{11}2H}, \tag{9.3.4}$$

and the associated thickness expansion due to Poisson's effect. Then from Eq. (2.9.13) the frequency shift due to pressure is

$$\frac{\omega-\omega_0}{\omega_0}=\frac{cS_1^0}{2}=-\frac{c}{2}\frac{p}{\bar{c}_{11}}\left(\frac{a^2}{2dh}+\frac{H}{h}\right). \tag{9.3.5}$$

The frequency shift depends on a^2/dh, which is a large number for a shallow shell. Thus, a shallow shell magnifies the pressure and increases the sensitivity. A thicker resonator or a more rigid resonator with a large h or a large c_{11} results in lower sensitivity. Under the assumption of the membrane state of thin shells, shell thickness does not have an effect on the sensitivity.

9.3.2. *Numerical results*

Consider the fundamental thickness-shear mode. Relative frequency shift versus pressure is shown in Fig. 9.3.3 for different values of a/d, while a/h = 20 and H/h = 5 are fixed. Clearly, for larger values of a/d, i.e. shallower shells, the sensitivity is higher.

The above analysis is based on the membrane theory of shells. When the shells become thicker, the effect of bending moment and transverse shear force cannot be ignored, especially near the edge of the shell

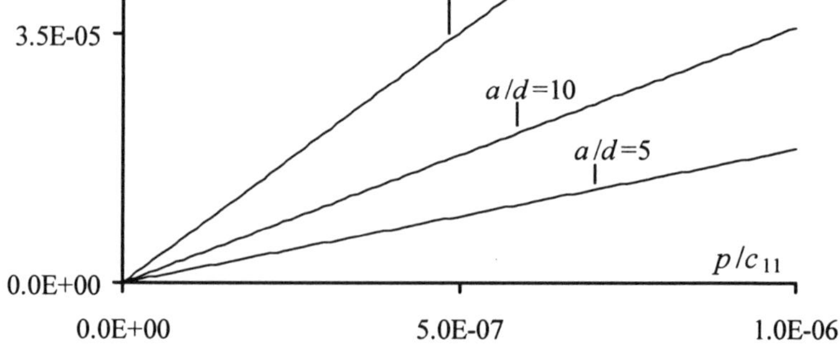

Fig. 9.3.3. Frequency shift versus pressure for different values of a/d.

9.4. A Bimorph

It is well known that for a thin plate the flexural stiffness is much smaller than the extensional stiffness. Therefore pressure-induced flexure is more favorable for pressure sensing. A fundamental obstacle for using pressure induced flexure in thickness-shear pressure sensors is that for a rotated Y-cut quartz thickness-shear resonator the frequency shift under pressure induced flexure is zero mainly because of the anti-symmetry of the flexural deformation (see the second section of the eighth chapter for normal acceleration sensitivity, a similar problem). To overcome this obstacle some asymmetry needs to be introduced into the structure. In this section we study a dual plate structure (see Fig. 9.4.1) for a thickness-shear mode pressure sensor [103].

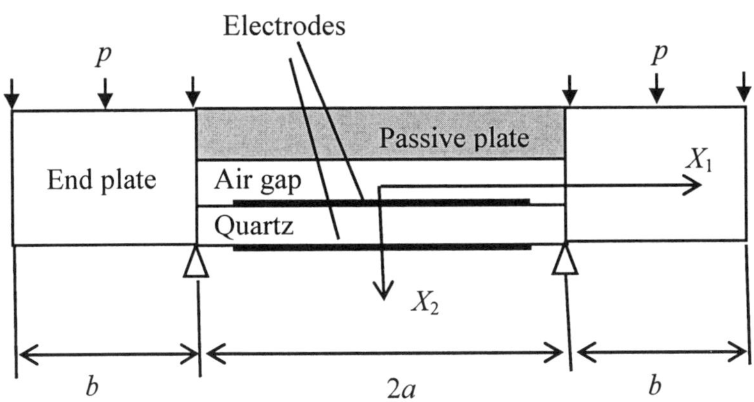

Fig. 9.4.1. A dual plate thickness-shear resonator as a pressure sensor.

9.4.1. *Analysis*

The plates in the structure are assumed thin and the air gap is of negligible thickness. When a uniform pressure p is applied to the end plates, it causes pure bending in the central dual plate portion. When the thickness of the crystal and passive plates are properly designed, the crystal plate has compressive strain only and the passive plate has extensional strain only. Then a net frequency shift proportional to the applied pressure can be obtained from the crystal plate. Since thin plates bend easily, a high sensitivity is expected.

We determine the initial fields using the classical plate theories of flexure and extension. The central part of the plate as a whole is subjected to a constant bending moment $M = pb^2/2$, while the crystal plate and the passive plate are in combined bending and compression (or extension) individually. The free body diagrams of the crystal and passive plates are shown in Fig. 9.4.2.

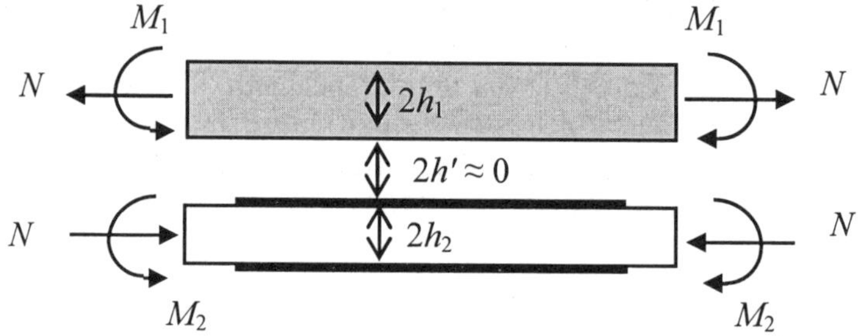

Fig. 9.4.2. Free body diagrams of the plates.

For equilibrium we must have

$$M_1 + M_2 + N(h_1 + h_2) = pb^2/2. \tag{9.4.1}$$

Consider the case when the end plates are much more rigid than the crystal and the passive plates. Therefore rotations at the ends of the crystal and the passive plates must be the same

$$\theta_1 = \frac{M_1}{E_1 I_1}\frac{a}{2} = \frac{M_2}{E_2 I_2}\frac{a}{2} = \theta_2, \tag{9.4.2}$$

where E is the effective Young's modulus and $I = 2h^3/3$ is the moment of inertia of the crystal or the passive plate. For rotated Y-cut quartz in planar deformation,

$$E_2 = \gamma_{11} = c_{11} - c_{12}^2/c_{22}. \tag{9.4.3}$$

When the end plates are relatively rigid, the central portion of the structure bends like a whole plate. The extensional strains of the crystal and the passive plates at the air gap must be the same. Therefore

$$\frac{N}{E_1 2h_1} - \frac{M_1}{E_1 I_1}\frac{a}{2} = \frac{M_2}{E_2 I_2}\frac{a}{2} - \frac{N}{E_2 2h_2}. \tag{9.4.4}$$

Equations (9.4.1), (9.4.2) and (9.4.4) are three equations for M_1, M_2 and N. We want to choose E and h of the crystal and the passive plates by design such that the narrow air gap is also the neutral surface without extension or compression. This requires that:

$$\frac{N}{E_1 2h_1} - \frac{M_1}{E_1 I_1}\frac{a}{2} = 0, \quad \frac{M_2}{E_2 I_2}\frac{a}{2} - \frac{N}{E_2 2h_2} = 0. \tag{9.4.5}$$

Equation (9.4.5) is a stronger restriction than Eq. (9.4.4). Therefore we solve Eqs. (9.4.1), (9.4.2) and (9.4.5) for M_1, M_2 and N. Since there are four equations and three unknowns, E and h of the crystal and the passive plates can be varied to make sure that these equations are all satisfied. In addition, we expect a constraint among the E and h of the plates. Substituting Eq. (9.4.5) into Eq. (9.4.2), we obtain

$$\frac{1}{E_1 h_1} = \frac{1}{E_2 h_2}. \tag{9.4.6}$$

From Eqs. (9.4.5) and (9.4.1) we obtain

$$\begin{gathered} M_1 = \frac{h_1^2}{a^2}\lambda pab, \quad M_2 = \frac{h_2^2}{a^2}\lambda pab, \quad N = \frac{3}{2}\lambda pb, \\ \lambda = \frac{b/a}{2(h_1/a)^2 + 2(h_2/a)^2 + 3(h_1/a + h_2/a)}. \end{gathered} \tag{9.4.7}$$

The flexure $w_2(X_1)$ of the crystal plate is given by

$$\frac{2}{3}h_2^3 E_2 w_{2,11} = M_2. \tag{9.4.8}$$

This leads to

$$\frac{2}{3}h_2^3 E_2 w_2 = M_2(X_1^2 - a^2)/2. \tag{9.4.9}$$

The extensional displacement is given by

$$w_1(X_1, X_2) = -w_{2,1}(X_1)(X_2 - h_2) - \frac{N}{E_2 2h_2}X_1. \tag{9.4.10}$$

The first term on the right-hand side of Eq. (9.4.10) does not cause any first-order frequency shift when substituted into the perturbation integral. Therefore in the following analysis we will only consider the second term in Eq. (9.4.10) due to N. The biasing extensional strain in the crystal plate under N is

$$S_1^0 = w_{1,1} = -\frac{N}{E_2 2h_2}. \tag{9.4.11}$$

The related thickness contraction or expansion due to Poisson's effect can be obtained from the plate stress relaxation condition $K_{22}^0 = 0$ as

$$S_2^0 = w_{2,2} = -\frac{c_{21}}{c_{22}} w_{1,1}. \tag{9.4.12}$$

For odd thickness-shear modes

$$u_1(X_2,t) = u(t)\sin\frac{n\pi}{2h_2}(X_2 - h_2), \quad n = 1,3,5,\ldots, \tag{9.4.13}$$

which have $X_2 = h_2$ as a nodal plane. From Eq. (2.9.13) the frequency shift is

$$\frac{\omega - \omega_0}{\omega_0} = \frac{1}{2}cS_1^0, \tag{9.4.14}$$

where

$$\omega_0^2 = \frac{n^2\pi^2 c_{66}}{4h_2^2 \rho_0}. \tag{9.4.15}$$

9.4.2. *Numerical results*

As numerical examples, for the passive plate we consider Al, Cu, and W with the Young's modulus E of 70, 110 and $360\times10^9\,\mathrm{N/m^2}$, respectively. Figure 9.4.3 shows frequency shifts of a quartz resonator for different passive plate materials. When the passive plate is stiff (W), the thickness of the passive plate as determined by Eq. (9.4.6) is small. This implies a smaller bending stiffness of the central portion of the sensor and hence higher sensitivity. Similar results for a langasite sensor are shown in Fig. 9.4.4. The sign of the frequency shifts of a langasite sensor is opposite to that of a quartz sensor. This depends on the signs of certain elastic constants.

Figure 9.4.5 shows the effect of the length/thickness ratio of the crystal plate. Thinner crystal plates lead to smaller bending stiffness and higher sensitivity. Figure 9.4.6 shows that longer end plates imply higher sensitivity. These are as expected.

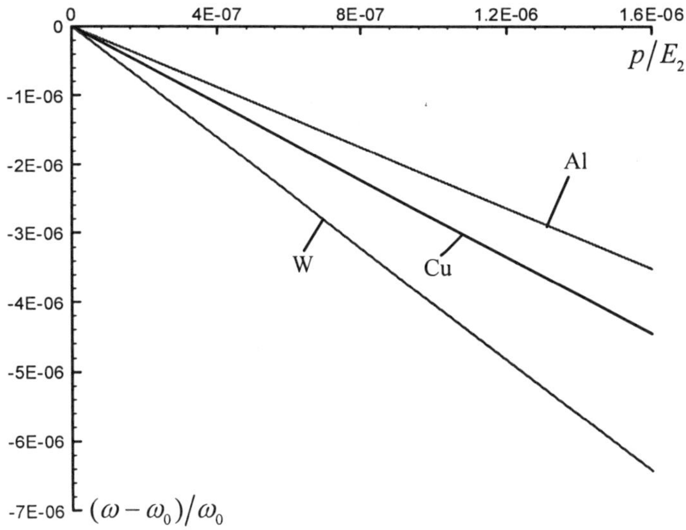

Fig. 9.4.3. Frequency shifts of a Y-cut quartz resonator for different passive plate materials ($b = a = 10\ h_2$).

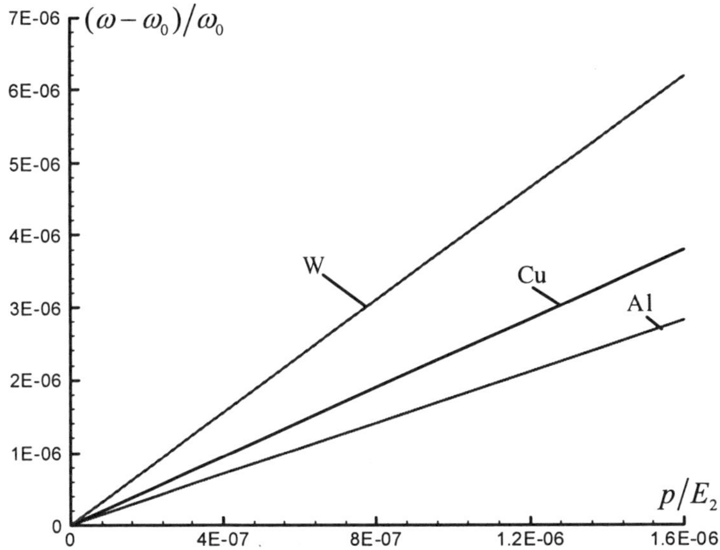

Fig. 9.4.4. Frequency shifts of a Y-cut langasite resonator for different passive plate materials ($b = a = 10\ h_2$).

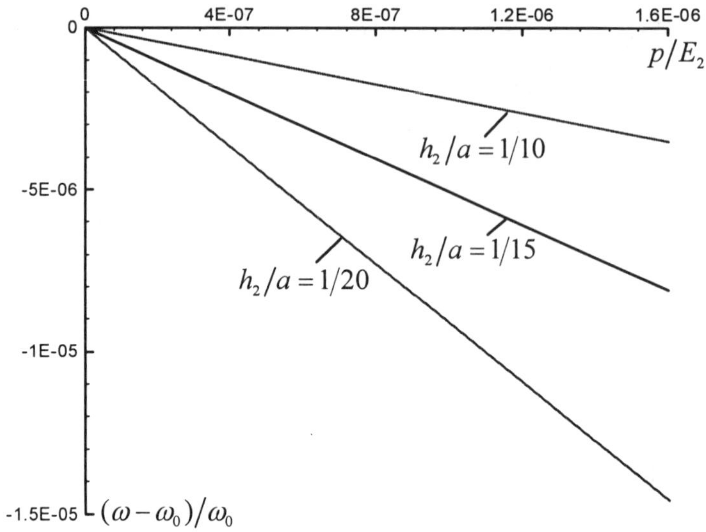

Fig. 9.4.5. Frequency shifts of a Y-cut quartz resonator of different aspect ratios ($b = a$, Al for the passive plate).

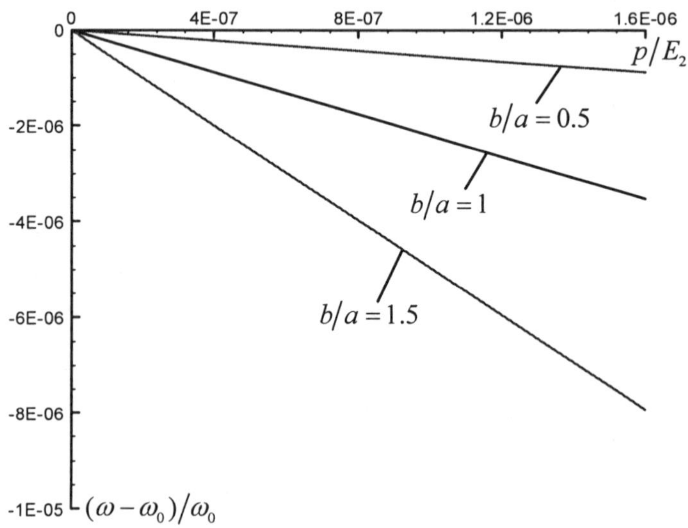

Fig. 9.4.6. Frequency shifts of a Y-cut quartz resonator with different end plates ($a = 10\ h_2$, Al for the passive plate).

The most important result is presented in Fig. 9.4.7 in which the passive plate is fixed to be Cu. Compared to the rectangular quartz plate under compression analyzed in the first section of this chapter and the circular quartz plate under compression analyzed in the second section of this chapter, the present dual-plate sensor has a much higher sensitivity. In plotting Fig. 9.4.7, $a/h_2 = R/h_2 = 20$ has been used for all three structures.

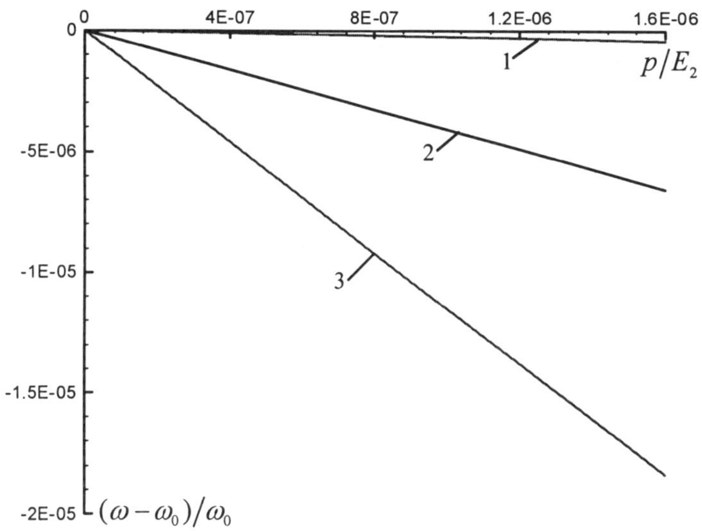

Fig. 9.4.7. Frequency shifts of a Y-cut quartz resonator with different pressure transfer structures ($b = a = 20\ h_2$, Al for the passive plate). Case 1: rectangular plate under compression. Case 2: circular plate under compression. Case 3: dual-plate in bending.

9.5. Surface Wave Pressure Sensors Based on Extension

Pressure sensors can also be made using surface acoustic waves (SAW). In this section we examine surface waves propagating in plates with pressure induced extensional deformations [104]. In the next section we will examine surface waves propagating in plates with pressure induced flexure. Consider a homogeneous material body occupying a half-space V with $X_2 > 0$ in a Cartesian coordinate system X_K (see Fig. 9.5.1)

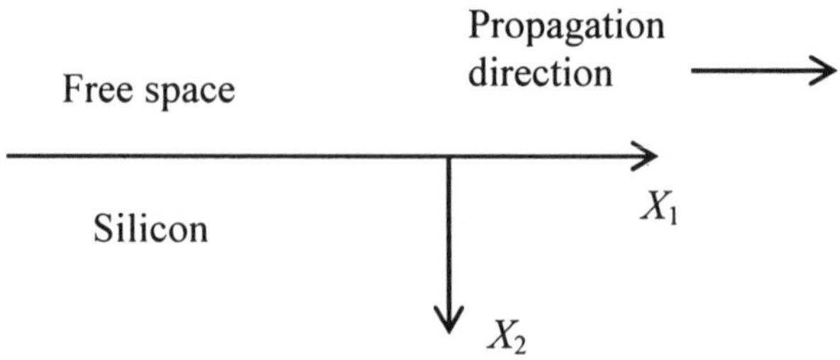

Fig. 9.5.1. An elastic half-space and coordinate system for surface waves.

The body is free from any deformations and fields. Suppose that the governing equations and boundary conditions allow the propagation of a small-amplitude Rayleigh surface wave with frequency ω, wave number ξ in the X_1 direction, phase speed $V_R = \omega/\xi$, and displacement u_α. Since the region is unbounded, we have an eigenvalue problem with a continuous spectrum. Given a wave number ξ, there always exists a Rayleigh wave with a frequency ω such that $\omega/\xi = V_R$ which depends on material properties only. In particular, silicon is of cubic symmetry (m3m) and allows plane-strain motions with $u_3 = 0$ and $\partial_3 = 0$. Rayleigh wave solutions in a half-space of m3m crystals with a traction-free boundary surface is given in [105]:

$$\begin{aligned} u_1 &= \exp(-2\pi\beta X_2/\lambda_R) \\ &\quad \cos\left(2\pi g\frac{X_2}{\lambda_R}+\alpha\right)\exp i\left[\omega\left(t-\frac{X_1}{V_R}\right)-\alpha\right], \\ u_2 &= ir\exp(-2\pi\beta X_2/\lambda_R) \\ &\quad \cos\left(2\pi g\frac{X_2}{\lambda_R}-\alpha\right)\exp i\left[\omega\left(t-\frac{X_1}{V_R}\right)-\alpha\right], \end{aligned} \tag{9.5.1}$$

where β, λ_R, g, α, V_R and r depend on material parameters. For silicon

$$\begin{aligned} &V_R = 4917\,m/s, \quad g = 0.4808, \\ &\beta = 0.4556, \quad r = 1.226, \quad \alpha = 58^\circ 1'. \end{aligned} \tag{9.5.2}$$

For germanium

$$V_R = 2924\,m/s, \quad g = 0.5122,$$
$$\beta = 0.4378, \quad r = 1.1938, \quad \alpha = 55^\circ 53'. \tag{9.5.3}$$

In real devices, instead of a half-space, a plate is used whose thickness is much larger than the surface wave length. The waves propagate along one surface of the plate, decay fast enough from that surface, and do not feel the other surface of the plate.

When an initial displacement field **w** with a uniform initial stress T_{KL}^0 and initial strain S_{KL}^0 is applied, the frequency of the wave of interest is perturbed a little and can be denoted by $\omega + \Delta\omega$. The change of wave frequency due to the initial fields is represented by the perturbation integral in Eq. (1.3.19). Equivalently, in terms of the wave speed,

$$\Delta V = \frac{1}{2V_R \xi^2} \frac{\int_V \hat{c}_{L\gamma M\alpha} u_{\gamma,L} u_{\alpha,M} dV}{\int_V \rho_0 u_\alpha u_\alpha dV}. \tag{9.5.4}$$

For initial deformations we consider pressure induced extension (compression) in the plate sealed in the shallow shell in Fig. 9.3.1. When the structure in the figure is subject to a surrounding pressure p, the following compressional force develops in the crystal plate:

$$Q = \frac{pa^2}{d}. \tag{9.5.5}$$

The strain fields in the plate are

$$S_1^0 = w_{1,1} = -\frac{Q}{2h\bar{c}_{11}},$$
$$S_2^0 = w_{2,2} = -\frac{c_{21}}{c_{22}} w_{1,1} = \frac{c_{21}}{c_{22}} \frac{Q}{2h\bar{c}_{11}}, \tag{9.5.6}$$

where

$$\bar{c}_{11} = c_{11} - c_{12}^2 / c_{22}. \tag{9.5.7}$$

All other displacement gradients vanish.

Pressure-induced change of wave speed is shown in Fig. 9.5.2. The relations are linear for small pressure. The sensitivity of Si is slightly larger than that of Ge.

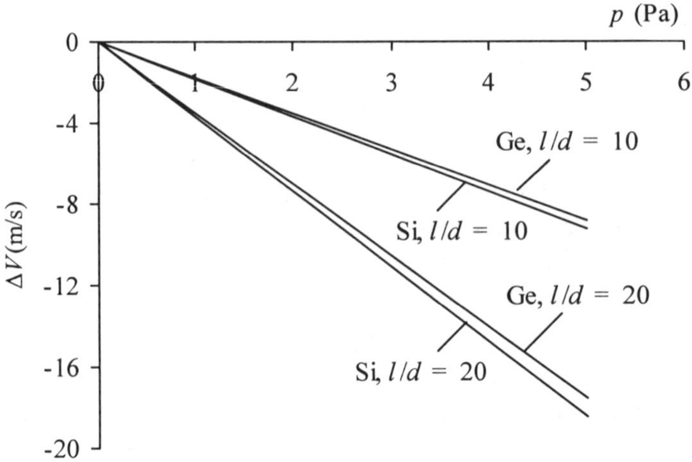

Fig. 9.5.2. Change of wave speed due to extension.

9.6. Surface Wave Pressure Sensors Based on Flexure

In this section we analyze surfaces waves propagating in a plate in flexure due to pressure [106]. The surface waves are the same as those in the previous section. For the biasing fields due to pressure induced flexure, we consider three cases of pure bending, bending under a uniform load, and bending under a concentrate load as shown in Fig. 9.6.1(a), (b) and (c), respectively.

9.6.1. Biasing deformations

9.6.1.1. Pure bending

In the case of pure bending, the middle part of the plate, i.e., the sensing portion, is subjected to a constant bending moment $M = -pb^2/2$. For plane-strain motions with $w_3 = 0$ and $\partial_3 = 0$, the deflection $w_2(X_1)$ is given by

$$-\frac{2}{3}h^3\gamma_{11}w_{2,11} = M(X_1) = -\frac{1}{2}pb^2, \tag{9.6.1}$$

and this leads to

$$\frac{2}{3}h^3\gamma_{11}w_2 = \frac{pb^2}{4}(X_1^2 - a^2), \tag{9.6.2}$$

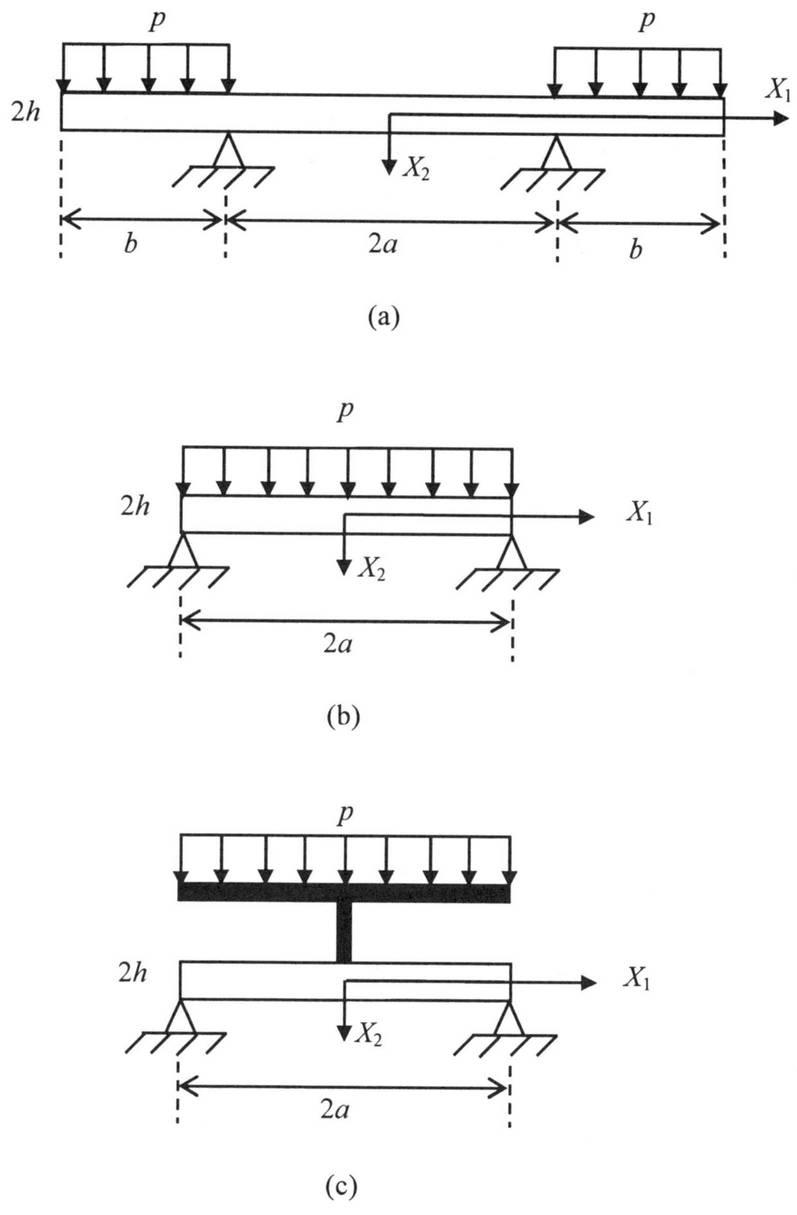

Fig. 9.6.1. A plate in different bending modes: (a) pure bending; (b) bending under a uniform load; (c) bending under a concentrated load.

where

$$\gamma_{11} = c_{11} - c_{12}^2 / c_{22}. \tag{9.6.3}$$

The displacement gradients needed in the perturbation integral can be found as follows. The axial displacement w_1 is

$$w_1(X_1, X_2) = -w_{2,1}(X_1)X_2, \tag{9.6.4}$$

from which $w_{1,1}$ and $w_{1,2}$ can be calculated. $w_{2,1}$ can be obtained from differentiating w_2, or from the fact that in the classical theory of flexure the shear strain

$$2S_{12}^0 = w_{1,2} + w_{2,1} = 0. \tag{9.6.5}$$

$w_{2,2}$ is obtained from the stress relaxation condition

$$T_{22}^0 = c_{21}S_{11}^0 + c_{22}S_{22}^0 + c_{23}S_{23}^0 = 0. \tag{9.6.6}$$

The results are

$$\begin{aligned} &w_{1,1} = -\frac{3pb^2}{4\gamma_{11}h^3}X_2, \quad w_{1,2} = -\frac{3pb^2}{4\gamma_{11}h^3}X_1, \quad w_{1,3} = 0, \\ &w_{2,1} = \frac{3pb^2}{4\gamma_{11}h^3}X_1, \quad w_{2,2} = \frac{c_{21}}{c_{11}}\frac{3pb^2}{4\gamma_{11}h^3}X_2, \quad w_{2,3} = 0, \\ &w_{3,1} = 0, \quad w_{3,2} = 0, \quad w_{3,3} = 0. \end{aligned} \tag{9.6.7}$$

The normal stress T_{11}^0 in pure bending is calculated from constitutive relations. To be fair for comparison with other cases, we will use $b = a$ in our calculations so that the pressure loaded areas in all of the three cases shown in Fig. 9.6.1 are the same.

9.6.1.2. Bending under a uniform load

For Case (b) of bending under a uniform load, the bending moment is given below:

$$-\frac{2}{3}h^3\gamma_{11}w_{2,11} = M(X_1) = \frac{p}{2}(a^2 - X_1^2). \tag{9.6.8}$$

Then

$$\frac{2}{3}h^3\gamma_{11}w_{2,1} = -\frac{p}{2}\left(a^2X_1 - \frac{1}{3}X_1^3\right), \tag{9.6.9}$$

from which $w_{1,1}$, $w_{1,2}$, $w_{2,1}$ and $w_{2,2}$ can be calculated as:

$$w_{1,1} = \frac{3p(a^2 - X_1^2)}{4\gamma_{11}h^3}X_2, \quad w_{1,2} = \frac{3p(a^2 X_1 - X_1^3/3)}{4\gamma_{11}h^3}, \quad w_{1,3} = 0,$$
$$w_{2,1} = -\frac{3p(a^2 X_1 - X_1^3/3)}{4\gamma_{11}h^3}, \quad w_{2,2} = -\frac{c_{21}}{c_{11}}\frac{3p(a^2 - X_1^2)}{4\gamma_{11}h^3}X_2, \quad (9.6.10)$$
$$w_{2,3} = 0, \quad w_{3,1} = 0, \quad w_{3,2} = 0, \quad w_{3,3} = 0.$$

In addition to the normal stress T_{11}^0 due to bending, in this case of non-pure bending there also exists a shear stress T_{12}^0. However, within the classical Kirchhoff flexural theory the shear strain S_{12}^0 vanishes. This imposes a constraint on the biasing deformation. As a consequence the corresponding shear stress T_{12}^0 cannot be obtained from the constitutive relations. Instead, we calculate the shear resultant force Q over a cross-section of the plate by the shear force-bending moment relation as follows:

$$Q(X_1) = M_{,1} = -pX_1 \,. \quad (9.6.11)$$

This shear force is due to a parabolic distribution of T_{12}^0 over a cross section. Therefore the shear stress is given by

$$T_{12}^0 = \frac{3Q}{4h^3}(h^2 - X_2^2) = -\frac{3pX_1}{4h^3}(h^2 - X_2^2)\,. \quad (9.6.12)$$

9.6.1.3. Bending under a concentrated load

For Case (c), bending under a concentrated load, we have

$$-\frac{2}{3}h^3\gamma_{11}w_{2,11} = M(X_1) = pa(a - X_1)\,. \quad (9.6.13)$$

The displacement gradients are

$$w_{1,1} = \frac{3pa(a - X_1)}{2\gamma_{11}h^3}X_2, \quad w_{1,2} = \frac{3pa(aX_1 - X_1^2/2)}{2\gamma_{11}h^3}, \quad w_{1,3} = 0,$$
$$w_{2,1} = -\frac{3pa(aX_1 - X_1^2/2)}{2\gamma_{11}h^3}, \quad w_{2,2} = -\frac{c_{21}}{c_{11}}\frac{3pa(a - X_1)}{2\gamma_{11}h^3}X_2, \quad (9.6.14)$$
$$w_{2,3} = 0, \quad w_{3,1} = 0, \quad w_{3,2} = 0, \quad w_{3,3} = 0.$$

In this case $Q = -pa$, and the shear stress is given, therefore, by

$$T_{12}^0 = \frac{3Q}{4h^3}(h^2 - X_2^2) = -\frac{3pa}{4h^3}(h^2 - X_2^2). \tag{9.6.15}$$

9.6.2. Frequency shifts

We are interested in comparing the sensitivities of the pressure sensors based on SAW of frequency f = 1 GHz, and the corresponding wave length is λ = 4.92×10^{-3} mm for Si and 2.92×10^{-3} mm for Ge. For geometric parameters we choose a = 20h, b = 20h, and 2h = 20λ. The plate thickness 2h is chosen to be sufficiently large as compared to the wave length so that the plate is effectively a half-space for the SAW under consideration.

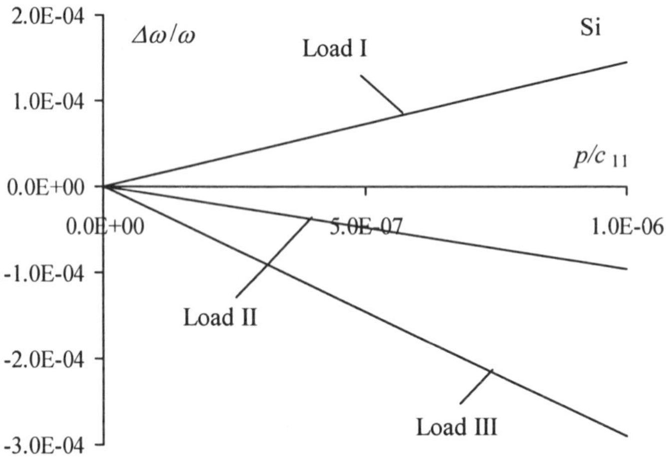

Fig. 9.6.2. Frequency shifts for different bending modes (silicon). Load I: pure bending. Load II: uniform load. Load III: concentrated load.

Pressure induced frequency shifts in silicon and germanium SAWs are plotted in Figs. 9.6.2 and 9.6.3, respectively. The frequency shifts are positive for the pure bending mode and are negative for the other two bending modes. This is due to that the upper surface of the plate, along which the acoustic waves propagate, is in extension for the pure bending mode, while it is in compression for the other two bending modes. These results show that, for both of these materials, the pressure sensitivity is the highest for bending under a concentrate load, the second for pure bending and the lowest for bending under a uniformly distributed load. The pressure p applied over the length span of $2a$ is set to be the same for the comparison.

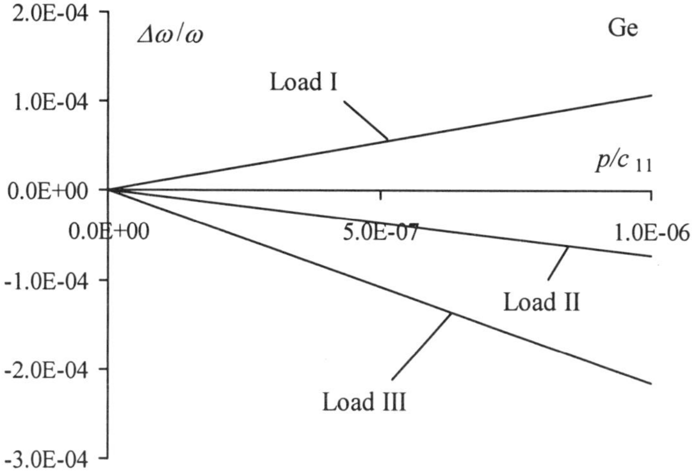

Fig. 9.6.3. Frequency shifts for different bending modes (germanium). Load I: pure bending. Load II: uniform load. Load III: concentrated load.

Table 9.6.1. Frequency shifts for different bending modes. Maximum $T_{11}^0 = 30$ MPa. Load I: pure bending. Load II: uniform load. Load III: concentrated load.

Bending modes	Pressure p (N/m)	Frequency shifts Si	Ge
Load I	1E5	8.77E - 05	8.41E - 05
Load II	1E5	- 5.84E - 05	- 5.61E - 05
Load III	5E4	- 8.74E - 05	- 8.37E - 05

Another way of comparing the sensitivity, which is a fairer comparison when the strength of the structures is taken into consideration, is to compare frequency shifts when the maximum tensile stresses in the three structures are the same while allowing the pressures on the structures to be different. The results for this comparison are shown in Table 9.6.1. In this comparison, for both materials, pure bending and bending under a concentrated load have about the same sensitivity which is considerably higher than that of bending under a uniform load.

Chapter 10

Temperature Sensors

Temperature sensors can be designed based on frequency shifts due to a temperature change. To model temperature sensors, in this chapter we first generalize the equations of electroelasticity in the first chapter to include thermal effects. Then equations for small fields superposed on a thermal bias are presented, followed by analyses of temperature sensors.

10.1. Thermoelectroelasticity

Thermal and viscous effects are treated together in this section [1,107,108]. The conservation of mass and the linear and angular momentum equations remain the same as in the first chapter. The energy equation including thermal effects and the second law of thermodynamics take the following form:

$$\frac{D}{Dt}\int_{v}\rho\left(\frac{1}{2}\mathbf{v}\cdot\mathbf{v}+e\right)dv$$
$$=\int_{v}[(\rho\mathbf{f}+\mathbf{F}^{E})\cdot\mathbf{v}+w^{E}+\rho\gamma]dv+\int_{s}(\mathbf{t}\cdot\mathbf{v}-\mathbf{n}\cdot\mathbf{q})ds, \qquad (10.1.1)$$
$$\frac{D}{Dt}\int_{v}\rho\eta dv\geq\int_{v}\frac{\rho\gamma}{\theta}dv-\int_{s}\frac{\mathbf{q}\cdot\mathbf{n}}{\theta}ds,$$

where $\mathbf{q}$ is the heat flux vector, η is the entropy per unit mass, γ is the body heat source per unit mass, and θ is the absolute temperature. The above integral balance laws can be localized to yield

$$\rho\dot{e}=\tau_{ij}v_{j,i}+\rho\gamma-q_{i,i}+\rho E_{i}\dot{\pi}_{i},$$
$$\rho\dot{\eta}\geq\frac{\rho\gamma}{\theta}-\left(\frac{q_{i}}{\theta}\right)_{,i}, \qquad (10.1.2)$$

where $\pi_i = P_i/\rho$. Eliminating γ in Eq. (10.1.2), we obtain the Clausius-Duhem inequality as

$$\rho(\theta\dot{\eta}-\dot{e})+\tau_{ij}v_{j,i}+\rho E_i\dot{\pi}_i-\frac{q_i\theta_{,i}}{\theta}\geq 0\,. \tag{10.1.3}$$

A free energy ψ can be introduced through the following Legendre transform:

$$\psi=e-\theta\eta-E_i\pi_i\,. \tag{10.1.4}$$

Then the energy equation in Eq. (10.1.2)$_1$ and the C-D inequality in Eq. (10.1.3) become

$$\rho(\dot{\psi}+\eta\dot{\theta}+\dot{\eta}\theta)=\tau_{ij}v_{j,i}-P_i\dot{E}_i+\rho\gamma-q_{i,i}\,, \tag{10.1.5}$$

and

$$-\rho(\dot{\psi}+\eta\dot{\theta})+\tau_{ij}v_{j,i}-P_i\dot{E}_i-\frac{q_i\theta_{,i}}{\theta}\geq 0\,. \tag{10.1.6}$$

Introducing the material heat flux and temperature gradient by

$$Q_K=JX_{K,k}q_k\,,\quad \Theta_K=\theta_{,K}=\theta_{,k}y_{k,K}\,, \tag{10.1.7}$$

we write the energy equation and the C-D inequality as

$$\begin{aligned}&\rho_0(\dot{\psi}+\eta\dot{\theta}+\dot{\eta}\theta)=T^S_{KL}E\dot{S}_{KL}-\mathcal{P}_K\dot{\mathcal{E}}_K+\rho_0\gamma-Q_{K,K}\,,\\&-\rho_0(\dot{\psi}+\eta\dot{\theta})+T^S_{KL}\dot{S}_{KL}-\mathcal{P}_K\dot{\mathcal{E}}_K-\frac{Q_K\Theta_K}{\theta}\geq 0.\end{aligned} \tag{10.1.8}$$

For constitutive relations we start with the following:

$$\begin{aligned}\psi&=\psi(S_{KL},\mathcal{E}_K,\theta,\Theta_K),\\T^S_{KL}&=T^S_{KL}(S_{KL},\mathcal{E}_K,\theta,\Theta_K,\dot{S}_{KL},\dot{\mathcal{E}}_K),\\\mathcal{P}_K&=\mathcal{P}_K(S_{KL},\mathcal{E}_K,\theta,\Theta_K,\dot{S}_{KL},\dot{\mathcal{E}}_K),\\Q_K&=Q_K(S_{KL},\mathcal{E}_K,\theta,\Theta_K,\dot{S}_{KL},\dot{\mathcal{E}}_K).\end{aligned} \tag{10.1.9}$$

Substitution of Eq. (10.1.9) into the C-D inequality in Eq. (10.1.8)$_2$ yields

$$\begin{aligned}&-\rho_0\frac{\partial\psi}{\partial\Theta_K}\dot{\Theta}_K-\rho_0\left(\eta+\frac{\partial\psi}{\partial\theta}\right)\dot{\theta}\\&+\left(T^S_{KL}-\rho_0\frac{\partial\psi}{\partial S_{KL}}\right)S_{KL}-\left(\mathcal{P}_K+\rho_0\frac{\partial\psi}{\partial\mathcal{E}_K}\right)\dot{\mathcal{E}}_K-\frac{1}{\theta}Q_K\Theta_K\geq 0.\end{aligned} \tag{10.1.10}$$

Since Eq. (10.1.10) is linear in $\dot{\Theta}_K$ and $\dot{\theta}$, for the inequality to hold ψ cannot depend on Θ_K, and η is related to ψ by

$$\eta = -\frac{\partial \psi}{\partial \theta}. \tag{10.1.11}$$

We break T_{KL}^S and $\mathcal{P}_K$ into reversible and dissipative parts as follows:

$$\begin{aligned}
&T_{KL}^S = T_{KL}^R + T_{KL}^D, \quad \mathcal{P}_K = \mathcal{P}_K^R + \mathcal{P}_K^D, \\
&T_{KL}^R = \rho_0 \frac{\partial \psi}{\partial S_{KL}}, \quad \mathcal{P}_K^R = -\rho_0 \frac{\partial \psi}{\partial \mathcal{E}_K}, \\
&T_{KL}^D = T_{KL}^D(S_{KL}, \mathcal{E}_K, \theta, \Theta_K, \dot{S}_{KL}, \dot{\mathcal{E}}_K), \\
&\mathcal{P}_K^D = \mathcal{P}_K^D(S_{KL}, \mathcal{E}_K, \theta, \Theta_K, \dot{S}_{KL}, \dot{\mathcal{E}}_K).
\end{aligned} \tag{10.1.12}$$

Then what is left for the C-D inequality in Eq. (10.1.10) is

$$T_{KL}^D \dot{S}_{KL} - \mathcal{P}_K^D \dot{\mathcal{E}}_K - \frac{1}{\theta} Q_K \Theta_K \ge 0. \tag{10.1.13}$$

From Eqs. (10.1.8)$_1$ and (10.1.11) we obtain the heat equation

$$\rho_0 \theta \dot{\eta} = T_{KL}^D \dot{S}_{KL} - \mathcal{P}_K^D \dot{\mathcal{E}}_K + \rho_0 \gamma - Q_{K,K}. \tag{10.1.14}$$

In summary, the nonlinear equations for thermoviscoelectroelasticity are

$$\begin{aligned}
&\rho_0 = \rho J, \\
&K_{Lk,L} + \rho_0 f_k = \rho_0 \ddot{y}_k, \\
&\mathcal{D}_{K,K} = \rho_E, \\
&\rho_0 \theta \dot{\eta} = T_{KL}^D \dot{S}_{KL} - \mathcal{P}_K^D \dot{\mathcal{E}}_K + \rho_0 \gamma - Q_{K,K},
\end{aligned} \tag{10.1.15}$$

with constitutive relations

$$\begin{aligned}
&\psi = \psi(S_{KL}, \mathcal{E}_K, \theta), \qquad \eta = -\frac{\partial \psi}{\partial \theta}, \\
&T_{KL}^S = T_{KL}^R + T_{KL}^D, \quad \mathcal{P}_K = \mathcal{P}_K^R + \mathcal{P}_K^D, \\
&T_{KL}^R = \rho_0 \frac{\partial \psi}{\partial S_{KL}}, \quad \mathcal{P}_K^R = -\rho_0 \frac{\partial \psi}{\partial \mathcal{E}_K}, \\
&T_{KL}^D = T_{KL}^D(S_{KL}, \mathcal{E}_K, \theta, \Theta_K, \dot{S}_{KL}, \dot{\mathcal{E}}_K), \\
&\mathcal{P}_K^D = \mathcal{P}_K^D(S_{KL}, \mathcal{E}_K, \theta, \Theta_K, \dot{S}_{KL}, \dot{\mathcal{E}}_K), \\
&Q_K = Q_K(S_{KL}, \mathcal{E}_K, \theta, \Theta_K, \dot{S}_{KL}, \dot{\mathcal{E}}_K),
\end{aligned} \tag{10.1.16}$$

which are restricted by

$$T^D_{KL}\dot{S}_{KL} - \mathcal{P}^D_K\dot{\mathcal{E}}_K - \frac{1}{\theta}\mathcal{Q}_K\Theta_K \geq 0\,. \qquad (10.1.17)$$

The equation for the conservation of mass in Eq. $(10.1.15)_1$ can be used to determine ρ separately from the other equations in (10.1.15). Equations $(10.1.15)_{2,3,4}$ can be written as five equations for $y_i(\mathbf{X},t)$, $\phi(\mathbf{X},t)$ and $\theta(\mathbf{X},t)$. On the boundary surface S, the thermal boundary conditions may be either prescribed temperature or heat flux

$$N_L\mathcal{Q}_L = \overline{\mathcal{Q}}\,. \qquad (10.1.18)$$

10.2. Linear Theory

For small deformations and weak electric fields

$$\begin{aligned} &D_{i,i} = \rho_e,\\ &T_{ji,j} + \rho_0 f_i = \rho_0\ddot{u}_i,\\ &\rho_0\theta\dot{\eta} = T^D_{ij}\dot{S}_{ij} - P^D_i\dot{E}_i - q_{i,i}. \end{aligned} \qquad (10.2.1)$$

The reversible part of the constitutive equations for small deformations and weak electric fields are determined by $\psi = \psi(S_{ij}, E_i, \theta)$ and

$$\eta = -\frac{\partial\psi}{\partial\theta}, \quad T^R_{ij} = \rho_0\frac{\partial\psi}{\partial S_{ij}}, \quad P^R_i = -\rho_0\frac{\partial\psi}{\partial E_i}. \qquad (10.2.2)$$

In order to linearize the constitutive relations for small temperature variations we expand ψ into a power series about $\theta = T_0$, $S_{ij} = 0$, and $E_i = 0$, where T_0 is a reference temperature. Denoting $T = \theta - T_0$, assuming $|T/T_0|<<1$, and keeping quadratic terms only, we can write

$$\begin{aligned} \rho_0\psi = &\frac{1}{2}c_{ijkl}S_{ij}S_{kl} - e_{kij}E_kS_{ij} - \frac{1}{2}\chi_{ij}E_iE_j\\ &-\frac{1}{2}\frac{\alpha}{T_0}T^2 - \beta_{kl}S_{kl}T - \lambda_kE_kT, \end{aligned} \qquad (10.2.3)$$

where β_{ij} are the thermoelastic constants, λ_k are the pyroelectric constants, and α is related to the specific heat. Equations (10.2.2) and (10.2.3) yield

$$\begin{aligned} T_{ij}^R &= c_{ijkl} S_{kl} - e_{kij} E_k - \beta_{ij} T, \\ D_i^R &= \varepsilon_0 E_i + P_i^R = e_{ijk} S_{jk} + \varepsilon_{ij} E_j + \lambda_k T, \\ \rho_0 \eta &= \frac{\alpha}{T_0} T + \beta_{kl} S_{kl} + \lambda_k E_k, \end{aligned} \tag{10.2.4}$$

which are the equations for linear thermopiezoelectricity. For the dissipative part of the constitutive relations we choose the following linear relations:

$$\begin{aligned} T_{ij}^D &= \mu_{ijkl} \dot{S}_{kl} - \alpha_{kij} \dot{E}_k, \\ D_i^D &= P_i^D = \beta_{ijk} \dot{S}_{jk} + \zeta_{ij} \dot{E}_j, \\ q_k &= -\kappa_{kl} \theta_{,l}. \end{aligned} \tag{10.2.5}$$

In the following we will assume $\beta_{ijk} = \alpha_{ijk}$. Equations (10.2.5) are restricted by

$$T_{ij}^D \dot{S}_{ij} - P_i^D \dot{E}_i - \frac{q_i \theta_{,i}}{\theta} \geq 0. \tag{10.2.6}$$

A dissipation function can be introduced as follows:

$$\chi(\dot{S}_{kl}, \dot{E}_j) = \frac{1}{2} \mu_{ijkl} \dot{S}_{kl} \dot{S}_{ij} - \alpha_{kij} \dot{E}_k \dot{S}_{ij} - \frac{1}{2} \zeta_{ij} \dot{E}_i \dot{E}_j \,. \tag{10.2.7}$$

Then Eqs. $(10.2.1)_3$, $(10.2.5)_{1,2}$ and (10.2.6) can be written as

$$\begin{aligned} &\rho_0 T_0 \dot{\eta} = 2\chi(\dot{S}_{ij}, \dot{E}_i) - q_{i,i}, \\ &T_{ij}^D = \frac{\partial \chi}{\partial \dot{S}_{ij}}, \quad P_i^D = -\frac{\partial \chi}{\partial \dot{E}_i}, \\ &2\chi(\dot{S}_{ij}, \dot{E}_i) + \frac{\kappa_{ij} \theta_{,i} \theta_{,j}}{\theta} \geq 0. \end{aligned} \tag{10.2.8}$$

Equation $(10.2.8)_4$ implies that

$$\chi(\dot{S}_{kl}, \dot{E}_j) \geq 0, \quad \kappa_{ij} \theta_{,i} \theta_{,j} \geq 0, \tag{10.2.9}$$

which further implies that μ_{ijkl}, $-\zeta_{ij}$, and κ_{ij} are positive definite. The formal similarity between Eq. (10.2.7) and the first three terms on the right-hand side of Eq. (10.2.3) suggests that the structures of μ_{ijkl}, ζ_{ij}, and α_{ijk} are the same as those of c_{ijkl}, χ_{ij}, and e_{ijk}, which are known for various crystal classes.

When the thermoelastic and pyroelectric effects are small, they can be neglected. Then the above equations for the linear theory reduce to two one-way coupled systems of equations. One represents the problem of viscopiezoelectricity with the following constitutive relations:

$$\begin{aligned} T_{ij}^R &= c_{ijkl}S_{kl} - e_{kij}E_k, \quad D_i^R = e_{ijk}S_{jk} + \varepsilon_{ij}E_j, \\ T_{ij}^D &= \mu_{ijkl}\dot{S}_{kl} - \alpha_{kij}\dot{E}_k, \quad D_i^D = \alpha_{ijk}\dot{S}_{jk} + \zeta_{ij}\dot{E}_j. \end{aligned} \tag{10.2.10}$$

The other governs the temperature field

$$\begin{aligned} \rho_0 T_0 \dot{\eta} &= T_{ij}^D \dot{S}_{ij} - D_i^D \dot{E}_i - q_{i,i}, \\ \rho_0 \eta &= \frac{\alpha}{T_0} T. \end{aligned} \tag{10.2.11}$$

Equations (10.2.10) can be substituted into Eqs. $(10.2.1)_{1,2}$ for four equations for y_i and ϕ. Once the mechanical and electric fields are found, they can be substituted into Eq. (10.2.11) to solve for the temperature field T.

Under harmonic excitation with an $\exp(i\omega t)$ factor, the linear constitutive relations in Eq. (10.2.10) can be written as

$$\begin{aligned} T_{ij} &= c_{ijkl}S_{kl} + \mu_{ijkl}\dot{S}_{kl} - e_{kij}E_k - \alpha_{kij}\dot{E}_k \\ &= (c_{ijkl} + i\omega\mu_{ijkl})S_{kl} - (e_{kij} + i\omega\alpha_{kij})E_k, \\ D_i &= e_{ijk}S_{jk} + \alpha_{ijk}\dot{S}_{jk} + \varepsilon_{ij}E_j + \zeta_{ij}\dot{E}_j \\ &= (e_{ijk} + i\omega\alpha_{ijk})S_{jk} + (\varepsilon_{ij} + i\omega\zeta_{ij})E_j, \end{aligned} \tag{10.2.12}$$

Formally, the material constants become complex and frequency-dependent.

10.3. Small Fields Superposed on a Thermal Bias

Consider the following three configurations of a thermoelastic body as shown in Fig. 10.3.1 [109].

(i) The reference configuration: At this state the body is undeformed with a reference temperature θ^R. A generic point is denoted by $\mathbf{X}$ with rectangular coordinates X_K. The mass density is denoted by ρ_0.

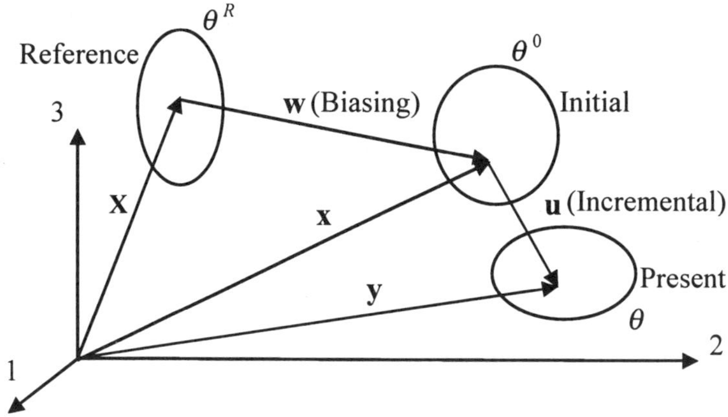

Fig. 10.3.1. Reference, initial and present configurations of a thermoelastic body.

(ii) The initial configuration: In this state the body is deformed finitely and statically, with an initial temperature of θ^0. The position of the material point associated with **X** is given by $x_\alpha = x_\alpha(\mathbf{X})$. The initial displacement is given by $\mathbf{w} = \mathbf{x} - \mathbf{X}$. The initial deformations satisfy the following equations of static nonlinear thermoelasticity:

$$
\begin{aligned}
&K^0_{LK,L} + \rho_0 f^0_k = 0, \\
&0 = -Q^0_{K,K} + \rho_0 \gamma^0.
\end{aligned}
\tag{10.3.1}
$$

The above equations are adjoined by constitutive relations defined by the specification of the free energy ψ and the heat flux vector Q_K

$$
\psi = \psi(S_{KL}, \theta), \quad Q_K = Q_K(S_{KL}, \theta, \Theta_K), \tag{10.3.2}
$$

where Θ_K is the material temperature gradient. For the initial state,

$$
\begin{aligned}
&K^0_{Lk} = x_{k,K}\rho_0 \left.\frac{\partial \psi}{\partial S_{KL}}\right|_{S^0_{KL},\theta^0}, \quad Q^0_K = Q_K(S^0_{KL}, \theta^0, \Theta^0_K), \\
&S^0_{KL} = (x_{\alpha,K}x_{\alpha,L} - \delta_{KL})/2, \quad \Theta^0_K = \theta^0_{,K}.
\end{aligned}
\tag{10.3.3}
$$

(iii) The present configuration: Time-dependent, small deformations are applied to the initial configuration of the deformed body. The final position of the material point associated with **X** is given by $y_i = y_i(\mathbf{X}, t)$.

The small, incremental displacement vector is denoted by **u**. The equations for the incremental fields are

$$\begin{aligned} & K^1_{K\alpha,K} + \rho_0 f^1_\alpha = \rho_0 \ddot{u}_\alpha, \\ & \rho_0(\theta^0 \dot{\eta}^1 + \theta^1 \dot{\eta}^0) = -Q^1_{K,K} + \rho_0 \gamma^1, \end{aligned} \tag{10.3.4}$$

where the effective linear constitutive relations for the small incremental stress tensor, entropy, and heat flux are

$$\begin{aligned} & K^1_{K\alpha} = G_{K\alpha L\gamma} u_{\gamma,L} - \rho_0 \Lambda_{K\alpha} \theta^1, \\ & \eta^1 = \Lambda_{L\gamma} u_{\gamma,L} + \alpha \theta^1, \\ & Q^1_K = A_{KM\alpha} u_{\alpha,M} + C_K \theta^1 + F_{KL} \theta^1_{,L}. \end{aligned} \tag{10.3.5}$$

In Eq. (10.3.5), the effective material constants are defined by

$$\begin{aligned} G_{K\alpha L\gamma} &= x_{\alpha,M} \rho_0 \left. \frac{\partial^2 \psi}{\partial S_{KM} \partial S_{LN}} \right|_{S^0_{KL}, \theta^0} x_{\gamma,N} \\ &\quad + \rho_0 \left. \frac{\partial \psi}{\partial S_{KL}} \right|_{S^0_{KL}, \theta^0} \delta_{\alpha\gamma} = G_{L\gamma K\alpha}, \\ \Lambda_{L\alpha} &= -\left. \frac{\partial^2 \psi}{\partial S_{KL} \partial \theta} \right|_{S^0_{KL}, \theta^0} x_{\alpha,K}, \quad \alpha = -\left. \frac{\partial^2 \psi}{\partial \theta^2} \right|_{S^0_{KL}, \theta^0}, \\ A_{KM\alpha} &= \left. \frac{\partial Q_K}{\partial S_{LM}} \right|_{S^0_{KL}, \theta^0, \Theta^0_K} x_{\alpha,L}, \quad C_K = \left. \frac{\partial Q_K}{\partial \theta} \right|_{S^0_{KL}, \theta^0, \Theta^0_K}, \\ F_{KL} &= \left. \frac{\partial Q_K}{\partial \Theta_L} \right|_{S^0_{KL}, \theta^0, \Theta^0_K}. \end{aligned} \tag{10.3.6}$$

High frequency incremental motions can be assumed to be isentropic for which $\eta^1 = 0$. Then, from Eq. $(10.3.5)_2$, the incremental temperature field θ^1 can be obtained and substituted into Eq. $(10.3.5)_1$, resulting in

$$K^1_{K\alpha} = \overline{G}_{K\alpha L\gamma} u_{\gamma,L}, \tag{10.3.7}$$

where the isentropic effective constants are

$$\overline{G}_{K\alpha L\gamma} = G_{K\alpha L\gamma} + \rho_0 \frac{\Lambda_{K\alpha} \Lambda_{L\gamma}}{\alpha} = \overline{G}_{L\gamma K\alpha}. \tag{10.3.8}$$

In many applications the biasing deformations are small. In this case, only their linear effects on the incremental fields need to be considered. Then the following free energy density is sufficient:

$$\rho_0\psi(S_{KL},\theta) = -a(\theta) - f_{AB}(\theta)S_{AB} + \frac{1}{2}c_{ABCD}(\theta)S_{AB}S_{CD} + \frac{1}{6}c_{ABCDEF}(\theta)S_{AB}S_{CD}S_{EF}, \tag{10.3.9}$$

where a and f_{AB} are related to the specific heat and the thermoelastic constants. Then

$$\begin{aligned} G_{K\alpha L\gamma}(\theta^0) &= c_{K\alpha L\gamma}(\theta^0) + \hat{c}_{K\alpha L\gamma}(\theta^0), \\ \hat{c}_{K\alpha L\gamma}(\theta^0) &= c_{K\alpha LN}(\theta^0)w_{\gamma,N} + c_{KNL\gamma}(\theta^0)w_{\alpha,N} \\ &\quad + c_{K\alpha L\gamma AB}(\theta^0)S^0_{AB} + T^0_{KL}\delta_{\alpha\gamma}, \\ T^0_{KL} &= c_{KLMN}(\theta^0)S^0_{MN} - f_{KL}(\theta^0), \quad S^0_{AB} = (w_{A,B} + w_{B,A})/2. \end{aligned} \tag{10.3.10}$$

For sensor applications, the linear effects of the initial temperature variation θ^0-θ^R are of main interest. Then, Eq. (10.3.10) can be approximated to the first order for all of the mechanical and thermal biasing fields by

$$\begin{aligned} G_{K\alpha L\gamma}(\theta^0) &\cong c_{K\alpha L\gamma}(\theta^R) + \left.\frac{\partial c_{K\alpha L\gamma}}{\partial\theta}\right|_{\theta^R}(\theta^0 - \theta^R) \\ &\quad + c_{K\alpha LN}(\theta^R)w_{\gamma,N} + c_{KNL\gamma}(\theta^R)w_{\alpha,N} + c_{K\alpha L\gamma AB}(\theta^R)S^0_{AB} \\ &\quad + [c_{KLMN}(\theta^R)S^0_{MN} - \lambda_{KL}(\theta^R)(\theta^0 - \theta^R)]\delta_{\alpha\gamma}, \end{aligned} \tag{10.3.11}$$

where

$$\lambda_{KL}(\theta) = \frac{df_{KL}}{d\theta} \tag{10.3.12}$$

are the thermoelastic constants.

10.4. Thickness-Shear Modes of a Free Plate

Consider a rotated Y-cut quartz plate whose reference state is shown in Fig. 10.4.1(a). The biasing deformation in Fig. 10.4.1(b) is a free thermal expansion due to a temperature change from θ^R to θ^0. The incremental motion is the thickness-shear vibration shown in Fig. 10.4.1(c).

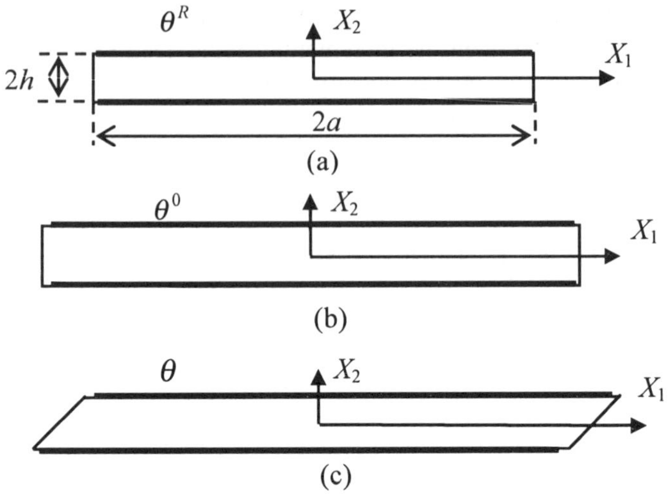

Fig. 10.4.1. (a) Reference, (b) initial, and (c) present configurations of a quartz plate.

Consider motions independent of X_3 with $\partial/\partial X_3 = 0$, $w_3 = 0$, and $u_3 = 0$. The biasing fields are assumed to be small and are governed by the linear theory of thermoelasticity. The relevant constitutive relations are

$$\begin{aligned} T_1^0 &= c_{11}S_1^0 + c_{12}S_2^0 - \lambda_1\Delta\theta = 0, \\ T_2^0 &= c_{21}S_1^0 + c_{22}S_2^0 - \lambda_1\Delta\theta = 0, \end{aligned} \tag{10.4.1}$$

where T_1^0 and T_2^0 are the initial extensional stresses in the X_1 and X_2 directions, and S_1^0 and S_2^0 are the corresponding initial strains. Shear deformation is not considered. In Eq. (10.4.1) we have denoted $\Delta\theta = \theta^0 - \theta^R$. Equation (10.4.1) can be solved for the strains

$$\begin{aligned} w_{1,1} &= S_1^0 = \alpha_1\Delta\theta, \\ w_{2,2} &= S_2^0 = -\frac{c_{21}}{c_{22}}S_1^0 + \frac{\lambda_1}{c_{22}}\Delta\theta, \end{aligned} \tag{10.4.2}$$

where

$$\begin{aligned} &\alpha_1 = \bar{\lambda}_1 / \bar{c}_{11}, \\ &\bar{c}_{11} = c_{11} - c_{12}^2 / c_{22}, \quad \bar{\lambda}_1 = \lambda_1(1 - c_{12}/c_{22}). \end{aligned} \tag{10.4.3}$$

The biasing deformations are uniform and the plate is long and thin. For the incremental motion we neglect edge effects and consider motions independent of X_1. For thickness-shear vibration in the X_1 direction

$$u_1 = u_1(X_2,t), \quad u_2 = u_3 = 0 . \tag{10.4.4}$$

The isentropic modification of the effective elastic constants is neglected. The relevant stress component for the incremental fields is found to be

$$\begin{aligned} K_{21}^1 &= \{c_{66}(\theta^0) + [2c_{66}(\theta^R) + c_{661}(\theta^R)]S_1^0 + c_{662}(\theta^R)S_2^0\}u_{1,2} \\ &= c_{66}(\theta^R)(1+2\overline{\alpha}\Delta\theta)u_{1,2}, \end{aligned} \tag{10.4.5}$$

where

$$\begin{aligned} 2\overline{\alpha} &= \frac{1}{c_{66}(\theta^R)}\left.\frac{dc_{66}}{d\theta}\right|_{\theta^R} + 2 + \frac{c_{661}(\theta^R)}{c_{66}(\theta^R)}\alpha_1(\theta^R) \\ &+ \frac{c_{662}(\theta^R)}{c_{66}(\theta^R)}\left[-\frac{c_{21}(\theta^R)}{c_{22}(\theta^R)}\alpha_1(\theta^R) + \frac{\lambda_1(\theta^R)}{c_{22}(\theta^R)}\right]. \end{aligned} \tag{10.4.6}$$

The equation of motion takes the following form:

$$c_{66}(\theta^R)(1+2\overline{\alpha}\Delta\theta)u_{1,22} = \rho_0\ddot{u}_1 . \tag{10.4.7}$$

For odd thickness-shear modes,

$$u_1(X_2,t) = u(t)\sin\frac{n\pi}{2b}X_2, \quad n = 1,3,5\ldots . \tag{10.4.8}$$

Equation (10.4.7) becomes the following ordinary differential equation for $u(t)$:

$$\ddot{u} + \omega_0^2(1+2\overline{\alpha}\Delta\theta)u = 0 , \tag{10.4.9}$$

where

$$\omega_0^2 = \frac{n^2\pi^2 c_{66}(\theta^R)}{4b^2\rho_0} . \tag{10.4.10}$$

Equation (10.4.9) implies the following frequency shift:

$$\frac{\omega - \omega_0}{\omega_0} = \overline{\alpha}\Delta\theta . \tag{10.4.11}$$

Equation (10.4.11) is linear in $\Delta\theta$, which is desirable for temperature sensing and is valid for small temperature changes only.

10.5. Thickness-Shear Modes of a Constrained Plate

Consider a rotated Y-cut quartz plate connected to two elastic layers by rigid end walls as shown in Fig. 10.5.1 [110].

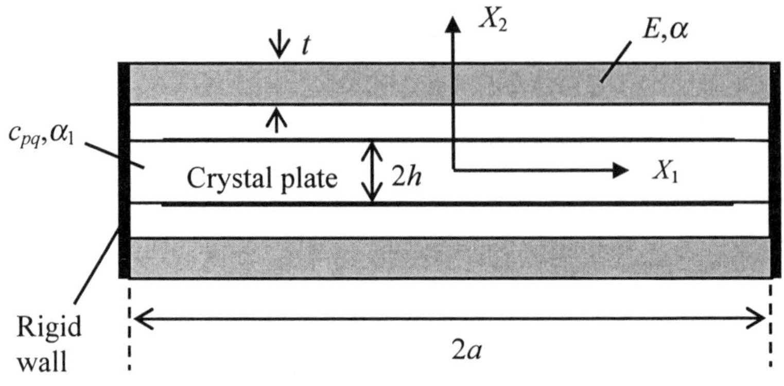

Fig. 10.5.1. A constrained crystal plate.

10.5.1. Analysis

The structure is long in the X_3 direction. The figure shows a cross-section. The elastic layers are isotropic with Young's modulus E and thermal expansion coefficient α. The thermal expansion coefficients of the crystal plate and the elastic layers are different. Under a temperature change they expand differently if allowed to do so freely. Due to the rigid end connections, they have to be of the same length for all temperatures. Thus, under a temperature change, strains and stresses are developed in both the crystal plate and the elastic layers. The operating mode of the crystal plate is the thickness-shear mode, with only one displacement component in the X_1 direction. The reference configuration of the crystal plate is shown in Fig. 10.5.2(a). The biasing deformation is due to a constrained thermal expansion which is shown in Fig. 10.5.2(b). The incremental motion is the thickness-shear vibration shown in Fig. 10.5.2(c).

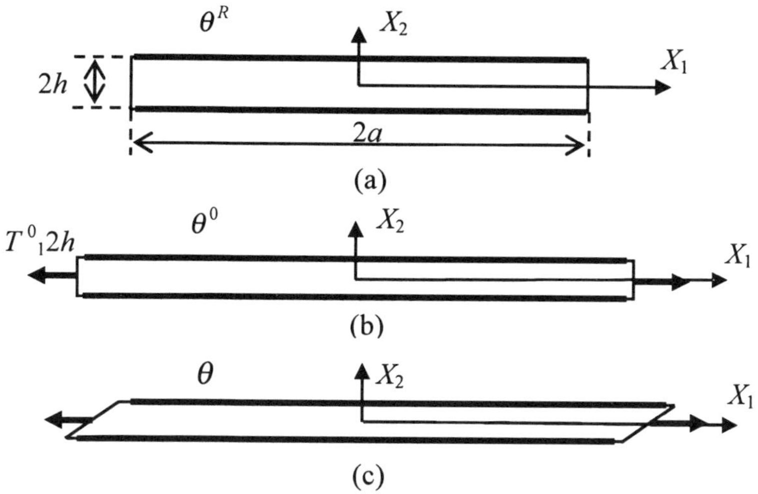

Fig. 10.5.2. (a) Reference, (b) initial , and (c) present configurations of the quartz plate.

Consider motions independent of X_3 with $\partial/\partial X_3 = 0$, $w_3 = 0$, and $u_3 = 0$. The relevant constitutive relations are

$$\begin{aligned} T_1^0 &= c_{11}S_1^0 + c_{12}S_2^0 - \lambda_1\Delta\theta, \\ T_2^0 &= c_{21}S_1^0 + c_{22}S_2^0 - \lambda_1\Delta\theta, \end{aligned} \tag{10.5.1}$$

where T_1^0 and T_2^0 are the initial extensional stresses in the X_1 and X_2 directions, and S_1^0 and S_2^0 are the corresponding initial strains. In Eq. (10.5.1), $\Delta\theta = \theta^0 - \theta^R$. The plate is assumed to be thin with $T_2^0 = 0$. We can then solve for S_2^0 from Eq. $(10.5.1)_2$ and substitute the resulting expression into Eq. $(10.5.1)_1$ to obtain the following one-dimensional constitutive relation:

$$\begin{aligned} S_1^0 &= \frac{1}{\bar{c}_{11}}T_1^0 + \alpha_1\Delta\theta, \\ \bar{c}_{11} &= c_{11} - c_{12}^2/c_{22}, \\ \bar{\lambda}_1 &= \lambda_1(1 - c_{12}/c_{22}), \quad \alpha_1 = \bar{\lambda}_1/\bar{c}_{11}. \end{aligned} \tag{10.5.2}$$

The relevant constitutive relation for the elastic layers is

$$S_1^0 = \frac{1}{E} T_1^0 + \alpha \Delta\theta . \tag{10.5.3}$$

The solution to the constrained thermal expansion problem of the structure in Fig. 10.5.1 is

$$\begin{aligned} T_1^0 &= \frac{Et\bar{c}_{11}}{Et + \bar{c}_{11}h}(\alpha - \alpha_1)\Delta\theta, \\ w_{1,1} &= S_1^0 = \frac{\alpha Et + \alpha_1 \bar{c}_{11} h}{Et + \bar{c}_{11} h}\Delta\theta, \\ w_{2,2} &= S_2^0 = -\frac{c_{21}}{c_{22}} S_1^0 + \frac{\lambda_1}{c_{22}}\Delta\theta. \end{aligned} \tag{10.5.4}$$

The relevant stress component is found to be

$$\begin{aligned} K_{21}^1 &= \{c_{66}(\theta^0) + [2c_{66}(\theta^R) + c_{661}(\theta^R)]S_1^0 + c_{662}(\theta^R)S_2^0\}u_{1,2} \\ &= c_{66}(\theta^R)(1 + 2\bar{\alpha}\Delta\theta)u_{1,2}, \end{aligned} \tag{10.5.5}$$

where

$$\begin{aligned} 2\bar{\alpha} &= \frac{1}{c_{66}(\theta^R)} \left.\frac{dc_{66}}{d\theta}\right|_{\theta^R} \\ &+ \frac{2c_{66}(\theta^R) + c_{661}(\theta^R)}{c_{66}(\theta^R)} \frac{\alpha(\theta^R)E(\theta^R)t + \alpha_1(\theta^R)\bar{c}_{11}(\theta^R)h}{E(\theta^R)t + \bar{c}_{11}(\theta^R)h} \\ &+ \frac{c_{662}(\theta^R)}{c_{66}(\theta^R)}\left[-\frac{c_{21}(\theta^R)}{c_{22}(\theta^R)} \frac{\alpha(\theta^R)E(\theta^R)t + \alpha_1(\theta^R)\bar{c}_{11}(\theta^R)h}{E(\theta^R)t + \bar{c}_{11}(\theta^R)h} + \frac{\lambda_1(\theta^R)}{c_{22}(\theta^R)}\right]. \end{aligned} \tag{10.5.6}$$

By setting $E = 0$ Eq. (10.5.6) reduces to Eq. (10.4.6) for an unconstrained plate. The frequency shift is given by

$$\frac{\omega - \omega_0}{\omega_0} = \bar{\alpha}\Delta\theta , \tag{10.5.7}$$

where

$$\omega_0^2 = \frac{n^2\pi^2 c_{66}(\theta^R)}{4b^2\rho_0} . \tag{10.5.8}$$

10.5.2. An example

For a numerical example, consider an AT-cut quartz plate. The temperature derivative of the relevant material constant c_{66} at 25°C is

$$\frac{1}{c_{66}}\frac{dc_{66}}{d\theta} = 126.7\times10^{-6}\,/^{\circ}\,\mathrm{C}\,. \tag{10.5.9}$$

The thermal expansion coefficient of quartz is

$$\alpha_1 = 13.71\times10^{-6}\,/^{\circ}\,\mathrm{C}\,. \tag{10.5.10}$$

For the plate thickness we choose h = 0.9696 mm. For thickness-shear vibration without biasing deformations, the fundamental thickness-shear frequency is $\omega_0/2\pi = 10^6$ Hz. We want to study effects of the constraints on the thermally induced frequency shift.

Figure 10.5.3 shows the frequency shift versus the temperature change for different values of α/α_1, the ratio between the thermal expansion coefficients of the crystal plate and the elastic layers. The slopes of the straight lines in the figure are related to the temperature sensitivity of the structure. The figure shows that the sensitivity is not sensitive to α/α_1.

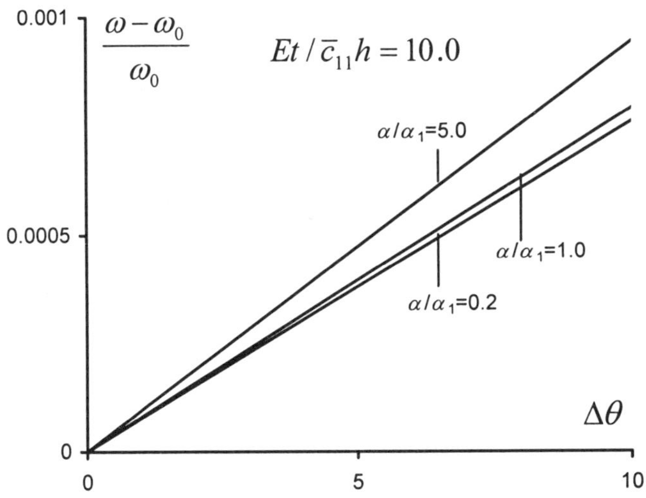

Fig. 10.5.3. Frequency shift versus temperature change for different α/α_1.

Figure 10.5.4 shows the frequency shift versus the temperature change for different values of the ratio between the extensional stiffness of the crystal plate and the elastic layers, $Et/\bar{c}_{11}h$. Again, the frequency shift is insensitive to the stiffness ratio.

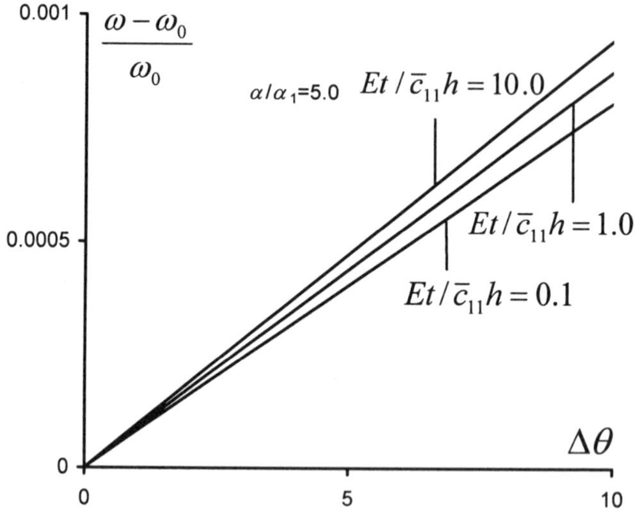

Fig. 10.5.4. Frequency shift versus temperature change for different $Et/\bar{c}_{11}h$.

A close examination of the numerical procedure shows that the first term on the right-hand side of Eq. (10.5.6), i.e. the dependence of the elastic constant c_{66} on temperature, is dominant. This term does not vary with the constraint. Therefore, the sensitivity of this particular structure does not vary much when a plate is in constrained or free thermal expansion.

Chapter 11
Piezoelectric Generators

Beginning from this chapter we study power handling devices. With electromechanical coupling, piezoelectric materials can be used to make generators for converting mechanical energy into electrical energy. Recently, due to the rapid development of wireless electronic devices, operating these devices without a wired power source has become an issue. One approach is to harvest power from the operating environment. Piezoelectric materials are natural candidates for scavenging ambient energy for powering small electronic devices of a very lower power requirement. Such a piezoelectric device is also called a piezoelectric power harvester.

11.1. Thickness-Stretch of a Ceramic Plate

In this section we present a simple analytical solution for a piezoelectric generator operating with thickness-stretch modes of a plate from the three-dimensional equations of linear piezoelectricity [111]. Consider an unbounded ceramic plate with thickness poling as shown in Fig. 11.1.1.

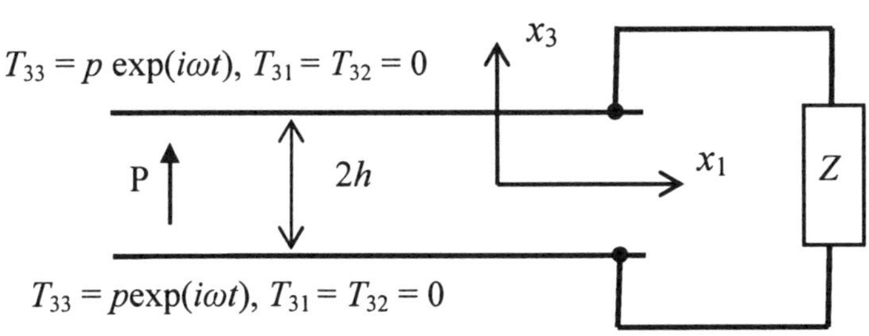

Fig. 11.1.1. A ceramic plate in thickness-stretch vibration.

11.1.1. Analysis

The major surfaces of the plate are electroded and the electrodes are connected by a circuit whose impedance is denoted by Z when the motion is time-harmonic. When the plate is driven by a time-harmonic surface normal stress, the mechanical boundary conditions are

$$T_{3j} = \delta_{3j} p \exp(i\omega t), \quad x_3 = \pm h . \tag{11.1.1}$$

On each of the two electrodes, the electric potential is spatially constant, varying with time. We denote the potential difference between the two electrodes by a voltage V

$$\phi(x_3 = h) - \phi(x_3 = -h) = V . \tag{11.1.2}$$

The free charge per unit area on the electrode at $x_3 = h$ and the current per unit area that flows out of the electrode at $x_3 = h$ are

$$Q = -D_3, \quad I = -\dot{Q} . \tag{11.1.3}$$

For time-harmonic motions, we use the following complex notation:

$$\{V, Q, I\} = \mathrm{Re}\{(\overline{V}, \overline{Q}, \overline{I})\exp(i\omega t)\} . \tag{11.1.4}$$

Then

$$\overline{I} = -i\omega\overline{Q} . \tag{11.1.5}$$

We also have the following relation for the output circuit:

$$\overline{I} = \overline{V} / Z . \tag{11.1.6}$$

Consider a possible complex solution in the following form:

$$\begin{aligned} u_3 &= u_3(x_3)\exp(i\omega t), \quad u_1 = u_2 = 0, \\ \phi &= \phi(x_3)\exp(i\omega t). \end{aligned} \tag{11.1.7}$$

The non-trivial components of strain and electric field are

$$S_{33} = u_{3,3}, \quad E_3 = -\phi_3 . \tag{11.1.8}$$

In Eq. (11.1.8) and from hereon, we have dropped the time-harmonic factor for simplicity. The non-trivial stress and electric displacement components are

$$\begin{aligned} T_{11} &= T_{22} = c_{13}u_{3,3} + e_{31}\phi_{,3} \\ T_{33} &= c_{33}u_{3,3} + e_{33}\phi_{,3}, \\ D_3 &= e_{33}u_{3,3} - \varepsilon_{33}\phi_{,3}. \end{aligned} \tag{11.1.9}$$

The equations to be satisfied are

$$\begin{aligned} c_{33}u_{3,33} + e_{33}\phi_{,33} &= -\rho\omega^2 u_3, \\ e_{33}u_{3,33} - \varepsilon_{33}\phi_{,33} &= 0. \end{aligned} \tag{11.1.10}$$

Equation $(11.1.10)_2$ can be integrated to yield

$$\phi = \frac{e_{33}}{\varepsilon_{33}} u_3 + B_1 x_3 + B_2, \tag{11.1.11}$$

where B_1 and B_2 are integration constants, and B_2 is immaterial. Substituting Eq. (11.1.11) into the expressions for T_{33}, D_3, and Eq. $(11.1.10)_1$ leads to

$$T_{33} = \bar{c}_{33}u_{3,3} + e_{33}B_1, \quad D_3 = -\varepsilon_{33}B_1, \tag{11.1.12}$$

$$\bar{c}_{33}u_{3,33} = -\rho\,\omega^2 u_3, \tag{11.1.13}$$

where

$$\bar{c}_{33} = c_{33}(1 + k_{33}^2), \quad k_{33}^2 = \frac{e_{33}^2}{\varepsilon_{33}c_{33}}. \tag{11.1.14}$$

The general solution to Eq. (11.1.13) and the corresponding expression for the electric potential are

$$\begin{aligned} u_3 &= A_1 \sin \xi x_3 + A_2 \cos \xi x_3, \\ \phi &= \frac{e_{33}}{\varepsilon_{33}}(A_1 \sin \xi x_3 + A_2 \cos \xi x_3) + B_1 x_3 + B_2, \end{aligned} \tag{11.1.15}$$

where A_1 and A_2 are integration constants, and

$$\xi^2 = \frac{\rho}{\bar{c}_{33}}\omega^2. \tag{11.1.16}$$

The expression for the relevant stress is then

$$T_{33} = \bar{c}_{33}(A_1\xi \cos \xi x_3 - A_2\xi \sin \xi x_3) + e_{33}B_1. \tag{11.1.17}$$

The boundary conditions require that

$$\begin{aligned} \bar{c}_{33}A_1\xi \cos \xi h - \bar{c}_{33}A_2\xi \sin \xi h + e_{33}B_1 &= p, \\ \bar{c}_{33}A_1\xi \cos \xi h + \bar{c}_{33}A_2\xi \sin \xi h + e_{33}B_1 &= p. \end{aligned} \tag{11.1.18}$$

Adding and subtracting the two equations in Eq. (11.1.18), respectively, yield the following:

$$\begin{aligned} &\bar{c}_{33} A_1 \xi \cos \xi h + e_{33} B_1 = p, \\ &\bar{c}_{33} A_2 \xi \sin \xi h = 0, \end{aligned} \tag{11.1.19}$$

which implies that $A_2 = 0$. From Eqs. (11.1.3), (11.1.5) and $(11.1.12)_2$,

$$\bar{Q} = \varepsilon_{33} B_1, \quad \bar{I} = -i\omega \varepsilon_{33} B_1. \tag{11.1.20}$$

Then the circuit condition in Eq. (11.1.6) can be written as

$$-i\omega \varepsilon_{33} B_1 = \bar{V} / Z. \tag{11.1.21}$$

The voltage across the electrodes takes the following form:

$$2\left(\frac{e_{33}}{\varepsilon_{33}} A_1 \sin \xi h + B_1 h\right) = \bar{V}. \tag{11.1.22}$$

Equations $(11.1.19)_1$, (11.1.21) and (11.1.22) are three equations for A_1, B_1 and $\bar{V}$. Solving these equations, we obtain

$$\begin{aligned} A_1 &= \frac{p}{\bar{c}_{33}} h \frac{1}{\xi h \cot \xi h - \dfrac{\bar{k}_{33}^2}{1 + Z / Z_0}} \frac{1}{\sin \xi h}, \\ B_1 &= -\frac{e_{33}}{\varepsilon_{33}} \frac{p}{\bar{c}_{33}} \frac{1}{(1 + Z / Z_0) \xi h \cot \xi h - \bar{k}_{33}^2}, \\ \bar{V} &= i\omega e_{33} \frac{p}{\bar{c}_{33}} \frac{Z}{(1 + Z / Z_0) \xi h \cot \xi h - \bar{k}_{33}^2}, \end{aligned} \tag{11.1.23}$$

where

$$\bar{k}_{33}^2 = \frac{e_{33}^2}{\varepsilon_{33} \bar{c}_{33}} = \frac{k_{33}^2}{1 + k_{33}^2}, \quad Z_0 = \frac{1}{i\omega C_0}, \quad C_0 = \frac{\varepsilon_{33}}{2h}. \tag{11.1.24}$$

In the above, C_0 is the static capacitance per unit area of the plate. From Eq. (11.1.23) the resonant frequencies are determined by:

$$\xi h \cot \xi h - \frac{\bar{k}_{33}^2}{1 + Z / Z_0} = 0. \tag{11.1.25}$$

From Eqs. (11.1.20) and (11.1.23) we have

$$\bar{I} = i\omega e_{33} \frac{p}{\bar{c}_{33}} \frac{1}{(1+Z/Z_0)\xi h \cot \xi h - \bar{k}_{33}^2}. \tag{11.1.26}$$

With the complex notation and the circuit condition, the average output electrical power per unit plate area over a period is given by

$$\begin{aligned} P_2 &= \frac{1}{4}(\bar{I}\bar{V}^* + \bar{I}^*\bar{V}) = \frac{1}{4}\bar{I}\bar{I}^*(Z+Z^*) = \frac{1}{2}|\bar{I}|^2 \operatorname{Re}\{Z\} \\ &= \frac{1}{2}p^2\omega^2 \frac{\varepsilon_{33}|\bar{k}_{33}^2|}{|\bar{c}_{33}|} \operatorname{Re}\{Z\} \left| \frac{1}{(1+Z/Z_0)\xi h \cot \xi h - \bar{k}_{33}^2} \right|^2, \end{aligned} \tag{11.1.27}$$

where an asterisk represents complex conjugate. Equation (11.1.27) shows that the output power P_2 depends on the input stress amplitude p quadratically, as expected. The output power also explicitly depends on ω^2, ε_{33} and $|\bar{k}_{33}^2|$ linearly, and is inversely proportional to $\bar{c}_{33}$. Therefore, a large dielectric constant ε_{33}, a large electromechanical coupling factor $\bar{k}_{33}$, and a small stiffness $\bar{c}_{33}$ are helpful for raising the output power. ω and $\bar{k}_{33}^2$, and hence ρ, ε_{33}, e_{33} and $\bar{c}_{33}$ are also present implicitly in the last factor in Eq. (11.1.27). When the driving frequency is near a resonant frequency, the denominator of this factor is very small, leading to a large output power. We further see from Eq. (11.1.27) that the output power P_2 depends on $\operatorname{Re}\{Z\}$ linearly for a small load impedance Z, and it diminishes as Z becomes large. The output power P_2 depends only on the real part of the load impedance (resistance), because its imaginary part represents a capacitance or inductance which takes away energy during half of a period and gives the energy back during the other half with no net energy consumption when averaged in a period.

To calculate the mechanical input power we need the velocities at the plate surfaces

$$\begin{aligned} v_3(x_3 = \pm h) &= \pm i\omega A_1 \sin \xi h \\ &= \pm i\omega \frac{p}{\bar{c}_{33}} h \frac{1}{\xi h \cot \xi h - \dfrac{\bar{k}_{33}^2}{1+Z/Z_0}}. \end{aligned} \tag{11.1.28}$$

Therefore the input mechanical power is

$$P_1 = 2\frac{1}{4}(pv_3^* + p^* v_3) = \frac{1}{2}p(v_3^* + v_3) = p\,\mathrm{Re}\{v_3\}$$

$$= \omega p^2 h \frac{\mathrm{Im}\left\{\bar{c}_{33}^*\left(\frac{\bar{k}_{33}^2}{1+Z/Z_0} - \xi h \cot \xi h\right)^*\right\}}{|\bar{c}_{33}|^2 \left|\frac{\bar{k}_{33}^2}{1+Z/Z_0} - \xi h \cot \xi h\right|^2}. \qquad (11.1.29)$$

The efficiency of the harvester, which measures its performance in converting the input mechanical power into the output electric power, is defined as

$$\eta = \frac{P_2}{P_1} = \frac{1}{2h}\frac{\omega \varepsilon_{33} |\bar{k}_{33}^2| \,\mathrm{Re}\{Z\}}{\frac{|(1+Z/Z_0)|^2}{|\bar{c}_{33}|}\mathrm{Im}\left\{\bar{c}_{33}^*\left(\frac{\bar{k}_{33}^2}{1+Z/Z_0} - \xi h \cot \xi h\right)^*\right\}}. \qquad (11.1.30)$$

The efficiency depends on the load impedance Z linearly for a small load. It diminishes with a large Z. It also becomes large at the resonant frequencies. However, the frequency dependence in Eq. (11.1.30) is different from that in Eq. (11.1.27) because Z is in general a function of the frequency. For example, if Z has the same frequency dependence as Z_0 in Eq. (11.1.24), then the low frequency limit of Eq. (11.1.27) is zero but that of Eq. (11.1.30) is a finite value.

Another quantity of practical interest is the power density defined as the output power per unit volume. In our case, it is given by

$$p_2 = \frac{P_2}{2h}. \qquad (11.1.31)$$

11.1.2. An example

For a numerical example, consider PZT-5H. The mechanical damping of the material is introduced through replacing c_{33} by $c_{33}(1+iQ^{-1})$, where $Q = 1\times 10^2$. The device thickness h is the only parameter that we have chosen, rather arbitrarily, as h = 1 cm. Figure 11.1.2 shows the output

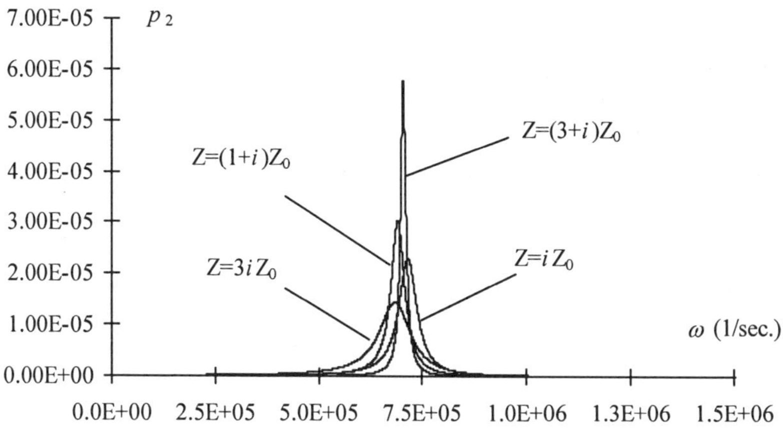

Fig. 11.1.2. Power density as a function of the driving frequency ω.

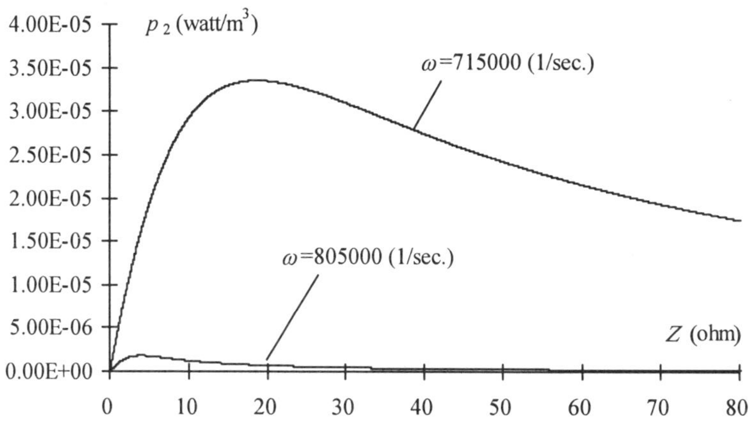

Fig. 11.1.3. Power density as a function of the load Z.

power density versus the driving frequency for different load impedance. The system has infinitely many resonant frequencies. Only the behavior near the first resonant frequency of 715 000 1/sec. is shown. Near resonance the output is maximal.

Output power density versus the load is shown in Fig. 11.1.3 for two different driving frequencies, 715 000 1/sec. and 805 000 1/sec. The figure shows that if the driving frequency is not close to a resonant frequency the output power is much less.

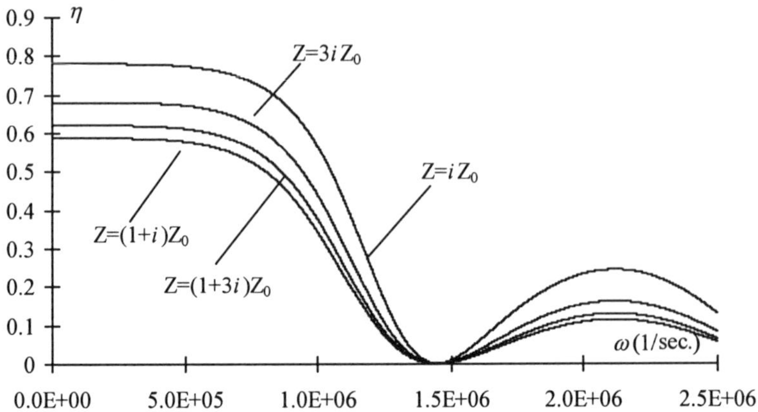

Fig. 11.1.4. Efficiency as a function of the driving frequency ω.

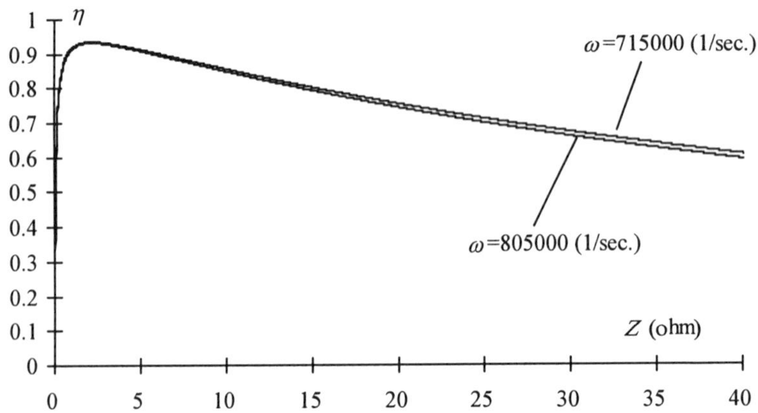

Fig. 11.1.5. Efficiency as a function of the load Z.

Efficiency versus the driving frequency is shown in Fig. 11.1.4 for different load impedance. The dependence of the efficiency on ω in Fig. 11.1.4 is more complicated than that of the output power in Fig. 11.1.2. In plotting Figs. 11.1.2 and 11.1.4, Z is chosen to be proportional to Z_0, or a function inversely proportional to ω. This has an effect on the behavior of the curves, especially for a small ω.

The efficiency as a function of Z is plotted in Fig. 11.1.5. In Figs. 11.1.4 and 11.1.5 the efficiency is higher than the electromechanical coupling factor of the material. This causes no contradiction because the coupling factor is defined for static processes and does not directly apply to our dynamic problem.

11.2. A Circular Shell

The thickness-stretch generator in the previous section is probably the simplest example of a piezoelectric generator. Its theoretical value is that it allows a simple analytical solution showing the basic behaviors of the device and the effects of various parameters. For real applications, other structural shapes and operating modes need to be studied. In this section we consider a circular cylindrical ceramic shell generator as shown in Fig. 11.2.1 [112].

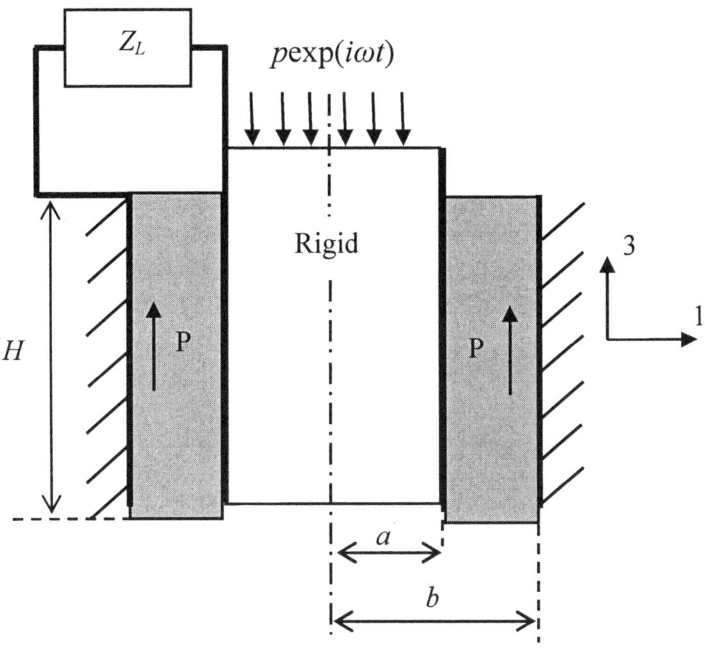

Fig. 11.2.1. A piezoelectric ceramic shell for power harvesting.

The ceramic is poled in the axial or 3 direction. The shell is electroded at its inner and outer surfaces, with electrodes shown by the thick lines in the figure. The electrodes are connected to a circuit whose impedance is Z_L when driven harmonically. The outer surface of the shell is rigidly attached to a fixed wall. The inner surface is driven harmonically by a pressure p at a given frequency ω. Since the ceramic layer is piezoelectric, across the thickness of the shell there exists a harmonic output voltage and a current due to the mechanically driven vibration of the shell.

Assume that $H >> b\text{-}a$ so that edge effects can be neglected. Consider axi-symmetric motions independent of z. In polar coordinates, let (r,θ,z) correspond to (1,2,3), we have

$$\begin{aligned} &u_z = u_z(r,t), \quad u_r = u_\theta = 0, \\ &\phi = \phi(r,t). \end{aligned} \tag{11.2.1}$$

The strain and electric field components are

$$2S_{rz} = u_{z,r}, \quad E_r = -\phi_{,r}\,. \tag{11.2.2}$$

The stress and electric displacement components are

$$T_{rz} = c_{44}u_{z,r} + e_{15}\phi_{,r}, \quad D_r = e_{15}u_{z,r} - \varepsilon_{11}\phi_{,r}\,. \tag{11.2.3}$$

The equation of motion and the charge equation of electrostatics take the following form:

$$\begin{aligned} &T_{rz,r} + \frac{1}{r}T_{rz} = \frac{1}{r}(rT_{rz})_{,r} = \rho\ddot{u}_z, \\ &D_{r,r} + \frac{1}{r}D_r = \frac{1}{r}(rD_r)_{,r} = 0. \end{aligned} \tag{11.2.4}$$

Equation $(11.2.4)_2$ can be integrated to yield

$$D_r = e_{15}u_{z,r} - \varepsilon_{11}\phi_{,r} = \frac{c_1(t)}{r}, \tag{11.2.5}$$

where c_1 is an integration constant. Substitution of Eq. (11.2.5) into Eqs. $(11.2.3)_1$ and $(11.2.4)_1$ gives

$$\begin{aligned} &T_{rz} = \bar{c}_{44}u_{z,r} - \frac{e_{15}}{\varepsilon_{11}}\frac{c_1}{r}, \\ &\frac{1}{r}\left(r\bar{c}_{44}u_{z,r} - \frac{e_{15}}{\varepsilon_{11}}c_1\right)_{,r} = \rho\ddot{u}_z, \end{aligned} \tag{11.2.6}$$

where

$$\bar{c}_{44} = c_{44}(1 + k_{15}^2), \quad k_{15}^2 = \frac{e_{15}^2}{\varepsilon_{11} c_{44}}. \tag{11.2.7}$$

For time-harmonic motions let

$$\{u_z, \phi, c_1, c_2\} = \mathrm{Re}\{\{U, \Phi, C_1, C_2\}\exp(i\omega t)\}. \tag{11.2.8}$$

Then Eq. (11.2.6)$_2$ becomes

$$U_{,rr} + \frac{1}{r}U_{,r} + \frac{\rho\omega^2}{\bar{c}_{44}}U = 0. \tag{11.2.9}$$

The general solution to Eq. (11.2.9) is

$$U = C_3 J_0(\xi r) + C_4 Y_0(\xi r), \tag{11.2.10}$$

where J_0 and Y_0 are the first and second kind Bessel functions of order zero, and

$$\xi^2 = \frac{\rho\omega^2}{\bar{c}_{44}}. \tag{11.2.11}$$

Then,

$$\begin{aligned} \Phi &= \frac{e_{15}}{\varepsilon_{11}}[C_3 J_0(\xi r) + C_4 Y_0(\xi r)] - \frac{C_1}{\varepsilon_{11}}\ln r - \frac{C_2}{\varepsilon_{11}}, \\ T_{rz} &= \bar{c}_{44}[C_3 \xi J_0'(\xi r) + C_4 \xi Y_0'(\xi r)] - \frac{e_{15}}{\varepsilon_{11}}\frac{C_1}{r}, \\ D_r &= \frac{C_1}{r}, \end{aligned} \tag{11.2.12}$$

which will be needed for prescribing boundary conditions. In Eq. (11.2.12), a prime represents differentiation with respect to the whole argument of the Bessel functions, i.e., ξr. For boundary conditions we also need the charge or current on the electrodes. For the outer electrode at $r = b$, we have the charge as the following integral over the electrode and the current as the time derivative of the charge:

$$\begin{aligned} Q_e &= \int -D_r dA = -c_1 2\pi H, \\ I &= -\dot{Q}_e. \end{aligned} \tag{11.2.13}$$

In the complex notation, the current flowing out of this electrode is given by

$$\bar{I} = i\omega C_1 2\pi H. \tag{11.2.14}$$

We denote the voltage across the shell by V and its complex amplitude by $\bar{V}$. The electromechanical boundary conditions and the equation of the output circuit are

$$
\begin{aligned}
&T_{rz}(a) = \bar{c}_{44}[C_3\xi J_0'(\xi a) + C_4\xi Y_0'(\xi a)] - \frac{e_{15}}{\varepsilon_{11}}\frac{C_1}{a} = \frac{p\pi a^2}{2\pi a H} = \frac{pa}{2H},\\
&U(b) = C_3 J_0(\xi b) + C_4 Y_0(\xi b) = 0,\\
&\Phi(b) - \Phi(a) = \frac{e_{15}}{\varepsilon_{11}}[C_3 J_0(\xi b) + C_4 Y_0(\xi b)\\
&\qquad - C_3 J_0(\xi a) - C_4 Y_0(\xi a)] - \frac{C_1}{\varepsilon_{11}}\ln\frac{b}{a} = \bar{V},\\
&\bar{I} = i\omega C_1 2\pi H = \frac{\bar{V}}{Z_L},
\end{aligned}
\tag{11.2.15}
$$

which are four equations for C_1, C_3, C_4 and $\bar{V}$, with p as the driving force. The solution to Eq. (11.2.15) can be found as

$$
\begin{aligned}
&C_1 = \frac{pa^2\varepsilon_{11}e_{15}[J_0(\xi b)Y_0(\xi a) - J_0(\xi a)Y_0(\xi b)]}{\Delta},\\
&C_3 = \frac{pa^2\varepsilon_{11}Y_0(\xi b)\ln(b/a)(1 + Z_L/Z_0)}{\Delta},\\
&C_4 = -\frac{pa^2\varepsilon_{11}J_0(\xi b)\ln(b/a)(1 + Z_L/Z_0)}{\Delta},\\
&\bar{V} = \frac{pa^2 e_{15}[J_0(\xi b)Y_0(\xi a) - J_0(\xi a)Y_0(\xi b)]\ln(b/a)Z_L/Z_0}{\Delta},
\end{aligned}
\tag{11.2.16}
$$

where

$$
\begin{aligned}
&\Delta = 2H\{e_{15}^2[J_0(\xi a)Y_0(\xi b) - J_0(\xi b)Y_0(\xi a)]\\
&\qquad + a\bar{c}_{44}\varepsilon_{11}\xi\ln(b/a)\\
&\qquad [Y_0(\xi b)J_0'(\xi a) - J_0(\xi b)Y_0'(\xi a)(\xi a)](1 + Z_L/Z_0)\},\\
&Z_0 = \frac{1}{i\omega C_0},\quad C_0 = \frac{\varepsilon_{11}2\pi H}{\ln(b/a)}.
\end{aligned}
\tag{11.2.17}
$$

$\Delta = 0$ yields an equation that determines the resonance frequencies. The output electrical power is

$$P_2 = \frac{1}{4}(\bar{I}\,\bar{V}^* + \bar{I}^*\bar{V}), \tag{11.2.18}$$

where an asterisk represents complex conjugate. The total shear force acting on the inner surface is:

$$F = p\pi a^2 \exp(i\omega t). \tag{11.2.19}$$

The velocity of the inner surface is

$$\begin{aligned} \dot{u}_3 &= i\omega U(a)\exp(i\omega t) \\ &= i\omega[C_3 J_0(\xi a) + C_4 Y_0(\xi a)]\exp(i\omega t). \end{aligned} \tag{11.2.20}$$

Therefore, the input mechanical power is

$$P_1 = -\frac{1}{4}p\pi a^2\{[i\omega U(a)]^* + i\omega U(a)\}. \tag{11.2.21}$$

The efficiency of the device is given by

$$\eta = \frac{P_2}{P_1}. \tag{11.2.22}$$

The power density is

$$p_2 = \frac{P_2}{\pi b^2 H}. \tag{11.2.23}$$

The basic behavior of this shell generator is similar to what is shown in Figs. 11.1.2 through 11.1.5.

11.3. A Beam Bimorph

The generators analyzed in the previous two sections operate with thickness-stretch and thickness-shear modes, respectively. These are the so-called high frequency modes whose frequencies depend on the plate or shell thickness. The frequencies of energy sources may vary. Therefore low frequency generators are also needed. Flexure of a beam is a common low frequency mode. In this section we analyze a beam bimorph generator as shown in Fig. 11.3.1 [113]. It consists of a pair of identical piezoelectric layers poled along the thickness direction and are separated by a metallic layer in the middle. It operates with the lowest flexural mode. The end mass m_0 is for adjusting the vibration characteristics of the structure through design for optimal performance.

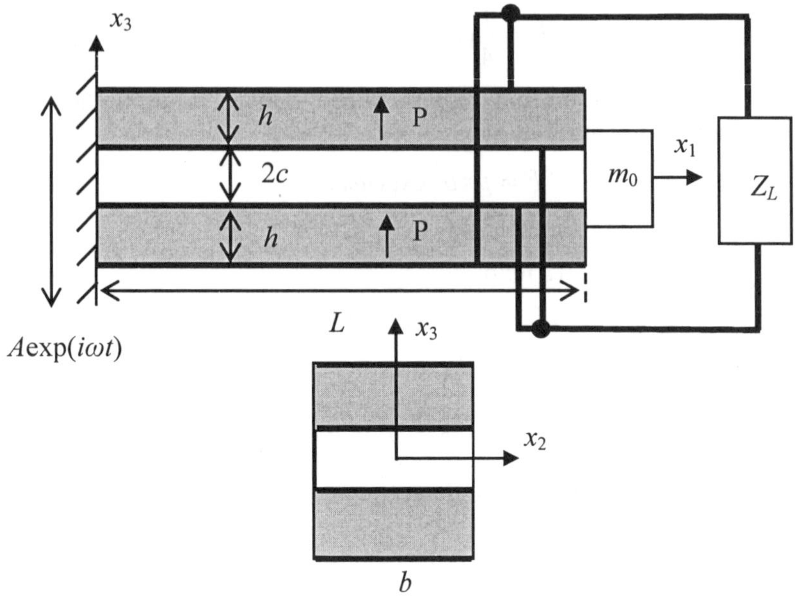

Fig. 11.3.1. A piezoelectric bimorph with an elastic layer and an end mass.

We assume $L >> b >> h$. The left end of the bimorph is cantilevered into a wall that vibrates along the vertical direction harmonically with a known amplitude A at a given frequency ω. This gives the input mechanical power. The two electrodes placed upon the upper and lower surfaces of the bimorph and the middle metallic layer are connected to the load circuit whose impedance is denoted by Z_L when driven harmonically.

11.3.1. Analysis

Consider the flexural motion of the bimorph in the x_3 direction. The structure is a laminated beam. Since it is essentially symmetric about the middle surface, flexure can be produced with little coupling to extension. To derive equations for classical flexure, we use a global approximation of the fields along the plate thickness. The flexural displacement and the axial strain are given by

$$u_3 = u_3^{(0)},$$
$$S_1 = x_3 S_1^{(1)} = -x_3 u_{3,11}^{(0)} = -x_3 u_{3,11}. \tag{11.3.1}$$

The electric field in the ceramic layers corresponding to the electrode configuration shown in the figure has the following components:

$$E_1 = 0, \quad E_2 = 0, \quad E_3 = E_3^{(0)} = -\frac{V}{h}, \tag{11.3.2}$$

where V denotes the voltage across each of the two piezoelectric layers. The stress components are one-dimensional:

$$T_1 = T_1(x_1, t),$$
$$T_2 = T_3 = T_4 = T_5 = T_6 = 0. \tag{11.3.3}$$

Then the relevant constitutive relations for the ceramic layers can be written as

$$S_1 = s_{11} T_1 + d_{31} E_3,$$
$$D_3 = d_{31} T_1 + \varepsilon_{33} E_3. \tag{11.3.4}$$

From Eq. (11.3.4) we solve for the axial stress T_1 and the transverse electric displacement D_3 and obtain

$$T_1 = s_{11}^{-1}(-x_3 u_{3,11}) - s_{11}^{-1} d_{31} E_3,$$
$$D_3 = s_{11}^{-1} d_{31}(-x_3 u_{3,11}) + \bar{\varepsilon}_{33} E_3, \tag{11.3.5}$$

where

$$\bar{\varepsilon}_{33} = \varepsilon_{33}(1 - k_{31}^2), \quad k_{31}^2 = d_{31}^2 / (\varepsilon_{33} s_{11}). \tag{11.3.6}$$

The metallic layer in the middle is assumed to be elastic and its constitutive relation is:

$$T_1 = E S_1 = E(-x_3 u_{3,11}), \tag{11.3.7}$$

where E is the Young's modulus. The bending moment M is defined by the following integral over a cross section of the beam, which can be integrated explicitly with the expressions for T_1 in Eqs. $(11.3.5)_1$ and (11.3.7):

$$M = \int x_3 T_1 dx_2 dx_3 = -D u_{3,11} + s_{11}^{-1} d_{31} \frac{V}{h} 2G, \tag{11.3.8}$$

where the bending stiffness D is defined as

$$D = \left\{ \frac{2}{3} E c^3 + \frac{2}{3} s_{11}^{-1} [(c+h)^3 - c^3] \right\} b,$$
$$G = \left(c + \frac{h}{2} \right) hb, \tag{11.3.9}$$

and G is the first moment of the cross-sectional area of one of the ceramic layers about the x_2-axis. The shear force Q in the beam is given by the shear force-bending moment relation

$$Q = \int T_{13} dx_2 dx_3 = M_{,1} = -D u_{3,111}. \tag{11.3.10}$$

The flexural motion is governed by

$$M_{,11} = m \ddot{u}_3, \tag{11.3.11}$$

where m is the mass per unit length of the beam

$$m = \rho 2cb + 2\rho' hb. \tag{11.3.12}$$

In the above, we have denoted by ρ and ρ', respectively, the mass densities of the metallic and piezoelectric layers. The electric charge on the top electrode at $x_3 = c+h$ is given by

$$Q_e = -b \int_0^L D_3(x_3 = c+h) dx_1$$
$$= b \left\{ s_{11}^{-1} d_{31} (c+h) [u_{3,1}(L,t) - u_{3,1}(0,t)] + \bar{\varepsilon}_{33} \frac{V}{h} L \right\}, \tag{11.3.13}$$

where Eq. $(11.3.5)_2$ has been used. The current flowing out of this electrode is

$$I = -\dot{Q}_e. \tag{11.3.14}$$

When the motion is time-harmonic, under the complex notation, with $I = \mathrm{Re}\{\bar{I} \exp(i\omega t)\}$ and $V = \mathrm{Re}\{\bar{V} \exp(i\omega t)\}$, the output voltage and current are related by

$$2\bar{I} = \bar{V} / Z. \tag{11.3.15}$$

Substitution of Eq. (11.3.8) into Eq. (11.3.11) yields

$$-D u_{3,1111} = m \ddot{u}_3. \tag{11.3.16}$$

At the left end of the beam, the boundary conditions are

$$u_3(0,t) = A \exp(i\omega t),$$
$$u_{3,1}(0,t) = 0. \tag{11.3.17}$$

At the right end, the bending moment vanishes and the shear force balances the inertial force of the attached mass, i.e.,

$$\begin{aligned} &M(L,t)=0, \\ &-Q(L,t)=m_0\ddot{u}_3(L,t). \end{aligned} \tag{11.3.18}$$

Let

$$u_3=\mathrm{Re}\{U\exp(i\omega t)\}. \tag{11.3.19}$$

Then Eqs. (11.3.16) through (11.3.18) become

$$\begin{aligned} &-DU_{,1111}=-\omega^2 mU, \quad 0<x_1<L, \\ &U(0)=A, \\ &U_{,1}(0)=0, \\ &-DU_{,11}(L)+s_{11}^{-1}d_{31}\frac{\overline{V}}{h}2G=0, \\ &DU_{,111}(L)=-\omega^2 m_0 U(L). \end{aligned} \tag{11.3.20}$$

The general solution to Eq. $(11.3.20)_1$ can be written as

$$U=B_1\sin\alpha x_1+B_2\cos\alpha x_1+B_3\sinh\alpha x_1+B_4\cosh\alpha x_1, \tag{11.3.21}$$

where B_1 through B_4 are undetermined constants, and

$$\alpha=\left(\frac{m}{D}\omega^2\right)^{1/4}. \tag{11.3.22}$$

Substitution of Eq. (11.3.21) into Eqs. (11.3.13) and (11.3.14) gives the following expression for the complex current:

$$\begin{aligned} \overline{I}=-i\omega b\Big\{&\frac{d_{31}}{s_{11}}(c+h)[B_1\alpha\cos\alpha L-B_2\alpha\sin\alpha L \\ &+B_3\alpha\cosh\alpha L+B_4\alpha\sinh\alpha L]+\overline{\varepsilon}_{33}\frac{\overline{V}}{h}L\Big\}. \end{aligned} \tag{11.3.23}$$

We then substitute Eqs. (11.3.21) and (11.3.23) into the four boundary conditions in Eqs. $(11.3.20)_{2\text{-}5}$ and the circuit condition in Eq. (11.3.15). This yields the following five equations for B_1 through B_4 and $\overline{V}$:

$$\begin{aligned}
&B_2 + B_4 = A,\\
&B_1\alpha + B_3\alpha = 0,\\
&-D(-B_1\alpha^2 \sin\alpha L - B_2\alpha^2 \cos\alpha L\\
&\quad + B_3\alpha^2 \sinh\alpha L + B_4\alpha^2 \cosh\alpha L) + s_{11}^{-1} d_{31}\frac{\bar{V}}{h}2G = 0,\\
&D(-B_1\alpha^3 \cos\alpha L + B_2\alpha^3 \sin\alpha L + B_3\alpha^3 \cosh\alpha L + B_4\alpha^3 \sinh\alpha L)\\
&\quad = -\omega^2 m_0(B_1 \sin\alpha L + B_2 \cos\alpha L + B_3 \sinh\alpha L + B_4 \cosh\alpha L),\\
&-2i\omega b\left\{\frac{d_{31}}{s_{11}}(c+h)[B_1\alpha\cos\alpha L - B_2\alpha\sin\alpha L\right.\\
&\quad \left. + B_3\alpha\cosh\alpha L + B_4\alpha\sinh\alpha L] + \bar{\varepsilon}_{33}\frac{\bar{V}}{h}L\right\} = \frac{\bar{V}}{Z_L},
\end{aligned} \tag{11.3.24}$$

where the wall amplitude A is the driving term. The solution to Eq. (11.3.24) is found to be

$$\begin{aligned}
B_1 = \frac{A}{\Delta}\Bigg[&-2\alpha\left(\frac{2}{Z_0}+\frac{1}{Z_L}\right)\cos\alpha L\cdot\cosh\alpha L\cdot\omega^2 m_0 D\\
&-\alpha^4\left(\frac{2}{Z_0}+\frac{1}{Z_L}\right)(\cos\alpha L\cdot\sinh\alpha L+\sin\alpha L\cdot\cosh\alpha L)D^2\\
&-4i(\cos\alpha L\cdot\sinh\alpha L+\sin\alpha L\cdot\cosh\alpha L)(c+h)d_{31}^2 s_{11}^{-2}h^{-1}\omega^3 m_0 bG\\
&-8i\alpha^3\sin\alpha L\cdot\sinh\alpha L\cdot d_{31}^2 s_{11}^{-2}h^{-1}\omega bDG(c+h)\Bigg],
\end{aligned} \tag{11.3.25}$$

$$\begin{aligned}
B_2 = \frac{A}{\Delta}\Bigg[&2\alpha\left(\frac{2}{Z_0}+\frac{1}{Z_L}\right)\sin\alpha L\cdot\cosh\alpha L\cdot\omega^2 m_0 D\\
&+\alpha^4\left(\frac{2}{Z_0}+\frac{1}{Z_L}\right)(\sin\alpha L\cdot\sinh\alpha L-\cos\alpha L\cdot\cosh\alpha L-1)D^2\\
&-4i(-1-\sin\alpha L\cdot\sinh\alpha L+\cos\alpha L\cdot\cosh\alpha L)\\
&(c+h)d_{31}^2 s_{11}^{-2}h^{-1}\omega^3 m_0 bG\\
&-8i\alpha^3\cos\alpha L\cdot\sinh\alpha L\cdot d_{31}^2 s_{11}^{-2}h^{-1}\omega bDG(c+h)\Bigg],
\end{aligned} \tag{11.3.26}$$

$$B_3 = \frac{A}{\Delta}\left[2\alpha\left(\frac{2}{Z_0}+\frac{1}{Z_L}\right)\cos\alpha L\cdot\cosh\alpha L\cdot\omega^2 m_0 D\right.$$
$$+\alpha^4\left(\frac{2}{Z_0}+\frac{1}{Z_L}\right)(\cos\alpha L\cdot\sinh\alpha L+\sin\alpha L\cdot\cosh\alpha L)D^2$$
$$+4i(\cos\alpha L\cdot\sinh\alpha L+\sin\alpha L\cdot\cosh\alpha L)(c+h)d_{31}^2 s_{11}^{-2}h^{-1}\omega^3 m_0 bG$$
$$\left.+8i\alpha^3\sin\alpha L\cdot\sinh\alpha L\cdot d_{31}^2 s_{11}^{-2}h^{-1}\omega bDG(c+h)\right], \tag{11.3.27}$$

$$B_4 = \frac{A}{\Delta}\left[-2\alpha\left(\frac{2}{Z_0}+\frac{1}{Z_L}\right)\cos\alpha L\cdot\sinh\alpha L\cdot\omega^2 m_0 D\right.$$
$$-\alpha^4\left(\frac{2i\omega b\bar{\varepsilon}_{33}L}{h}+\frac{1}{Z_L}\right)(\sin\alpha L\cdot\sinh\alpha L+\cos\alpha L\cdot\cosh\alpha L+1)D^2$$
$$+4i(1-\sin\alpha L\cdot\sinh\alpha L-\cos\alpha L\cdot\cosh\alpha L)(c+h)d_{31}^2 s_{11}^{-2}h^{-1}\omega^3 m_0 bG$$
$$\left.-8i\alpha^3\sin\alpha L\cdot\cosh\alpha L\cdot d_{31}^2 s_{11}^{-2}h^{-1}\omega bDG(c+h)\right], \tag{11.3.28}$$

$$\bar{V} = \frac{A}{\Delta}4i\alpha^2 d_{31}s_{11}^{-1}\omega bD(c+h)\{-\cos\alpha L\cdot\omega^2 m_0$$
$$+\cos^2\alpha L(\omega^2 m_0\cosh\alpha L+\alpha^3 D\sinh\alpha L) \tag{11.3.29}$$
$$+\sin\alpha L[-\alpha^3 D+\sin\alpha L(\omega^2 m_0\cosh\alpha L+\alpha^3 D\sinh\alpha L)]\},$$

where

$$\Delta = -2\alpha\left(\frac{2}{Z_0}+\frac{1}{Z_L}\right)(\cos\alpha L\cdot\sinh\alpha L-\sin\alpha L\cdot\cosh\alpha L)\omega^2 m_0 D$$
$$-2\alpha^4 D^2\left(\frac{2}{Z_0}+\frac{1}{Z_L}\right)(1+\cos\alpha L\cdot\cosh\alpha L)$$
$$+8i(1-\cos\alpha L\cdot\cosh\alpha L)(c+h)d_{31}^2 s_{11}^{-2}h^{-1}\omega^3 m_0 bG \tag{11.3.30}$$
$$-8i(\cos\alpha L\cdot\sinh\alpha L+\sin\alpha L\cdot\cosh\alpha L)(c+h)\alpha^3 d_{31}^2 s_{11}^{-2}h^{-1}\omega bDG,$$

$$Z_0 = \frac{1}{i\omega C_0}, \quad C_0 = \frac{\bar{\varepsilon}_{33}bL}{h}.$$

$\Delta = 0$ yields an equation that determines the resonant frequencies of the structure. With the complex notation, the output electrical power is given by

$$P_2 = \frac{1}{4}[(2\bar{I})\bar{V}^* + (2\bar{I})^*\bar{V}]. \tag{11.3.31}$$

To calculate the mechanical input power, we need the shear force in the beam at the left end. From Eqs. (11.3.21) and (11.3.10) we have

$$\begin{aligned} Q(x_1) = -D\alpha^3(&-B_1\cos\alpha x_1 + B_2\sin\alpha x_1 \\ &+ B_3\cosh\alpha x_1 + B_4\sinh\alpha x_1). \end{aligned} \tag{11.3.32}$$

The velocity of the left end of the beam is given by

$$\dot{u}_3 = i\omega A\exp(i\omega t). \tag{11.3.33}$$

This leads to the following expression for the input mechanical power:

$$P_1 = -\frac{1}{4}[Q(0)\cdot(i\omega A)^* + Q^*(0)(i\omega A)]. \tag{11.3.34}$$

The efficiency is

$$\eta = \frac{P_2}{P_1}. \tag{11.3.35}$$

The power density for the bimorph is given by

$$p_2 = \frac{P_2}{b(2c+2h)L}. \tag{11.3.36}$$

11.3.2. An example

For numerical results we use PZT-5H. To include damping, the real elastic constant s_{11} is replaced by $s_{11}(1 - iQ^{-1})$. We fix Q to be 10^2. $L = 25$ mm, $b = 8$ mm, $h = 0.31$ mm, and the acceleration amplitude $\omega^2 A = 1.0$ m/s^2 are fixed in all calculations. The elastic layer is taken to be aluminum alloy with the Young's modulus $E = 70$ GPa and mass density $\rho = 2700$ kg/m^3. The basic behavior of the bimorph generator is similar to that in Figs. 11.1.2 through 11.1.5. Figures 11.3.2 and 11.3.3 show the effects of some design parameters of this specific generator.

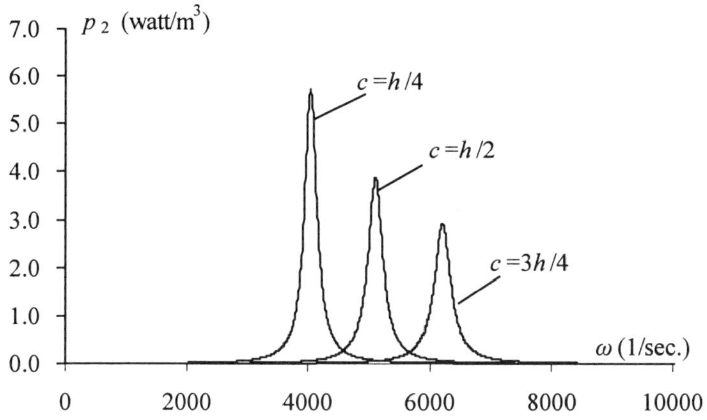

Fig. 11.3.2. Power density versus frequency for different thickness ratios. $m_0 = 0$, $Z_L = iZ_0$

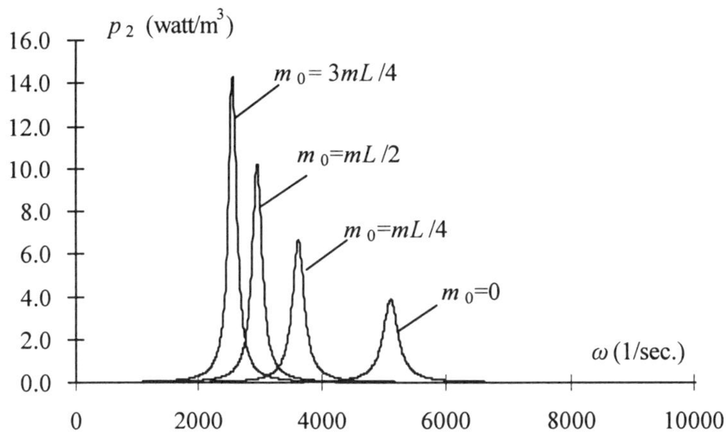

Fig. 11.3.3. Power density versus frequency for different end mass. $c = h/2$, $Z_L = iZ_0$.

11.4. A Spiral Bimorph

For energy sources at low frequencies, to further lower the resonant frequencies of a generator without increasing its dimension, a spiral bimorph (see Fig. 11.4.1) can be considered [114].

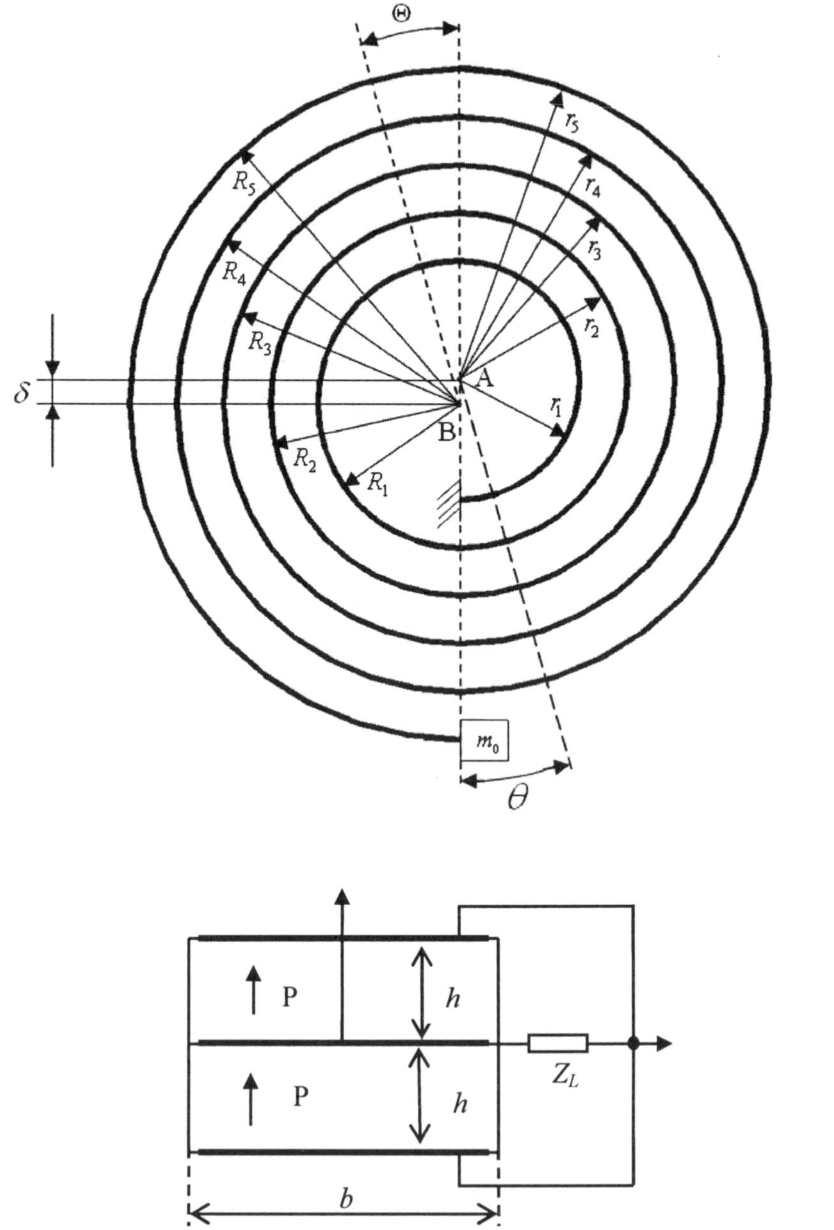

Fig. 11.4.1. A spiral-shaped piezoelectric bimorph and its cross section.

The spiral bimorph is essentially a long bimorph beam. It can be modeled by the one-dimensional equations of a piezoelectric ring. For a ring, the low frequency modes of extension and flexure are naturally coupled. Numerical results [114] show that the mode with the highest output may not be the lowest mode (see Fig. 11.4.2). The optimal design of this generator is relatively more complicated.

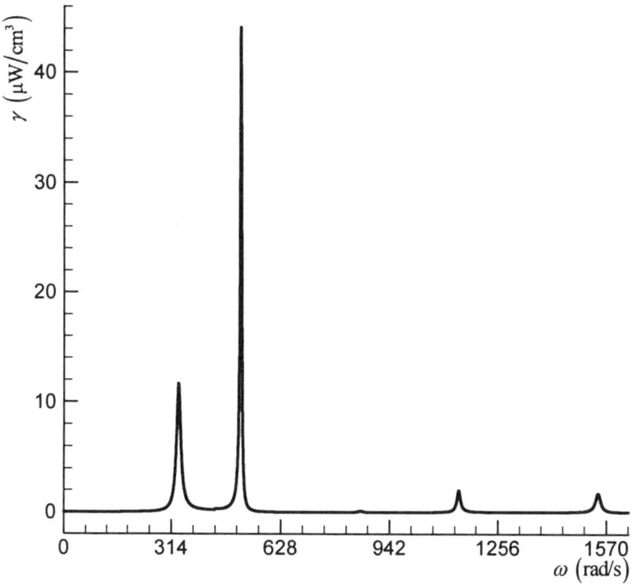

Fig. 11.4.2. Power density as a function of the driving frequency ω for $m_0 = 0$.

11.5. Nonlinear Behavior near Resonance

Piezoelectric generators are resonant devices. A small driving force may produce a strong resonance with relatively large mechanical deformation which brings in nonlinearity. In this section we examine nonlinear effects in a generator operating with thickness-shear modes of a plate under surface traction (see Fig. 11.5.1) [115].

Limited by the availability of nonlinear material constants, we consider quartz for which the relevant third- and fourth-order elastic constants are known. Quartz is a material with weak piezoelectric coupling and is not a good candidate for power harvesting, but the qualitative nonlinear behavior predicted is expected to be shared by power harvesters made from other materials for which the nonlinear material constants still need to be measured.

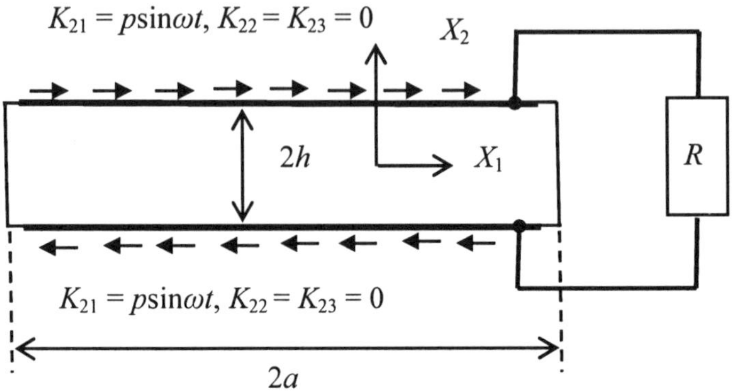

Fig. 11.5.1. A thickness-shear piezoelectric generator.

11.5.1. Analysis

For thickness-shear vibration of a plate of rotated Y-cut quartz, the relevant stress component and electric displacement component are [9]

$$\begin{aligned} K_{21} &= c_{66}u_{1,2} + e_{26}\phi_{,2} + \beta(u_{1,2})^2 + \gamma(u_{1,2})^3, \\ \mathcal{D}_2 &= e_{26}u_{1,2} - \varepsilon_{22}\phi_{,2}, \end{aligned} \tag{11.5.1}$$

where

$$\begin{aligned} \beta &= 0, \\ \gamma &= \frac{1}{2}c_{22} + c_{266} + \frac{1}{6}c_{6666}. \end{aligned} \tag{11.5.2}$$

In Eq. (11.5.1), we have neglected electrical nonlinearity which is caused by electromechanical coupling and is small. The plate is assumed to be thin with $a >> h$. Therefore, we neglect edge effects and consider pure thickness-shear modes $u_1(X_2,t)$ and $\phi(X_2,t)$ governed by

$$\begin{aligned} K_{21,2} &= c_{66}u_{1,22} + e_{26}\phi_{,22} + 3\gamma(u_{1,2})^2 u_{1,22} = \rho_0\ddot{u}_1, \\ \mathcal{D}_{2,2} &= e_{26}u_{1,22} - \varepsilon_{22}\phi_{,22} = 0. \end{aligned} \tag{11.5.3}$$

The free charge on the electrode at $X_2 = h$ and the current that flows out of the electrode are given by

$$Q = \int_{-a}^{a} -\mathcal{D}_2 dX_1, \quad i = -\dot{Q}. \tag{11.5.4}$$

Equation $(11.5.3)_2$ can be integrated to give

$$\phi = \frac{e_{26}}{\varepsilon_{22}} u_1 + C_1(t) X_2 + C_2(t), \tag{11.5.5}$$

where C_1 and C_2 are integration constants. Without loss of generality we take $C_2 = 0$. With Eq. (11.5.5), we can write Eq. (11.5.1) as

$$\begin{aligned} K_{21} &= \bar{c}_{66} u_{1,2} + \gamma (u_{1,2})^3 + e_{26} C_1, \\ \mathcal{D}_2 &= -\varepsilon_{22} C_1, \end{aligned} \tag{11.5.6}$$

and Eq. $(11.5.3)_1$ becomes

$$K_{21,2} = \bar{c}_{66} u_{1,22} + 3\gamma (u_{1,2})^2 u_{1,22} = \rho_0 \ddot{u}_1, \tag{11.5.7}$$

where

$$\bar{c}_{66} = c_{66}(1 + k_{26}^2), \quad k_{26}^2 = \frac{e_{26}^2}{\varepsilon_{22} c_{66}}. \tag{11.5.8}$$

The device operates with the fundamental thickness-shear mode. Near resonance, we use a one-mode Galerkin approximation with the fundamental thickness-shear mode from the linear theory as the base function, i.e.,

$$u_1(X_2, t) \cong u(t) \sin \frac{\pi}{2h} X_2. \tag{11.5.9}$$

Substituting (11.5.9) into the variational formulation in Eq. (1.1.20):

$$\int_{-h}^{h} (K_{21,2} - \rho_0 \ddot{u}_1) \delta u_1 dX_2 = 0, \tag{11.5.10}$$

making use of the traction boundary conditions, we obtain the following equation for u:

$$\ddot{u} + \omega_0^2 u + e C_1(t) + \mu u^3 = F \sin \omega t, \tag{11.5.11}$$

where

$$\omega_0^2 = \frac{\pi^2}{4\rho_0 h^2} \bar{c}_{66}, \quad e = \frac{2e_{26}}{\rho_0 h}, \quad \mu = \frac{3\pi^4}{64 \rho_0 h^4} \gamma, \quad F = \frac{2p}{\rho_0 h}. \tag{11.5.12}$$

Let

$$\begin{aligned} u &= U_1 \sin \omega t + U_2 \cos \omega t, \\ C_1 &= A_1 \sin \omega t + A_2 \cos \omega t, \end{aligned} \tag{11.5.13}$$

where U_1, U_2, A_1 and A_2 are undetermined constants. Substituting Eq. (11.5.13) into Eq. (11.5.9) and then into Eq. (11.5.5), we have

$$\begin{aligned} \phi = &\left(\frac{e_{26}}{\varepsilon_{22}} U_1 \sin \frac{\pi}{2h} X_2 + A_1 X_2 \right) \sin \omega t \\ &+ \left(\frac{e_{26}}{\varepsilon_{22}} U_2 \sin \frac{\pi}{2h} X_2 + A_2 X_2 \right) \cos \omega t. \end{aligned} \tag{11.5.14}$$

The voltage across the plate thickness is given by

$$\begin{aligned} V &= \phi(h) - \phi(-h) \\ &= 2\left[\frac{e_{26}}{\varepsilon_{22}} U_1 + A_1 h \right] \sin \omega t + 2\left[\frac{e_{26}}{\varepsilon_{22}} U_2 + A_2 h \right] \cos \omega t. \end{aligned} \tag{11.5.15}$$

From Eqs. (11.5.4) and $(11.5.6)_2$ we have

$$i = -2a\varepsilon_{22}\omega(A_1 \cos \omega t - A_2 \sin \omega t) = I \sin(\omega t + \phi), \tag{11.5.16}$$

where

$$I = 2a\varepsilon_{22}\omega\sqrt{A_1^2 + A_2^2}, \quad \tan\phi = -A_1 / A_2. \tag{11.5.17}$$

Substituting Eq. (11.5.13) into Eq. (11.5.11), collecting coefficients of the resulting sine and cosine terms separately, neglecting higher harmonics of $2\omega t$ and $3\omega t$ which is a common approximation, we obtain the following two equations:

$$\begin{aligned} &\frac{3}{4}\mu U_1^3 + (\omega_0^2 - \omega^2)U_1 + \frac{3}{4}\mu U_1 U_2^2 + eA_1 - F = 0, \\ &\frac{3}{4}\mu U_2^3 + (\omega_0^2 - \omega^2)U_2 + \frac{3}{4}\mu U_2 U_1^2 + eA_2 = 0. \end{aligned} \tag{11.5.18}$$

For the electric output circuit we consider the simple case of a load resistor so that $i = V/R$, leading to the following two equations:

$$\begin{aligned} &\frac{e_{26}}{\varepsilon_{22}R} U_1 + \frac{h}{R} A_1 - a\varepsilon_{22}\omega A_2 = 0, \\ &\frac{e_{26}}{\varepsilon_{22}R} U_2 + \frac{h}{R} A_2 + a\varepsilon_{22}\omega A_1 = 0. \end{aligned} \tag{11.5.19}$$

Equations (11.5.18) and (11.5.19) are four equations for U_1, U_2, A_1 and A_2. We rewrite Eq. (11.5.19) by solving it for A_1 and A_2

$$A_1 = -\frac{e_{26}hU_1 + a\varepsilon_{22}e_{26}R\omega U_2}{\varepsilon_{22}(h^2 + a^2\varepsilon_{22}^2R^2\omega^2)},$$
$$A_2 = \frac{-e_{26}hU_2 + a\varepsilon_{22}e_{26}R\omega U_1}{\varepsilon_{22}(h^2 + a^2\varepsilon_{22}^2R^2\omega^2)}. \tag{11.5.20}$$

Equations (11.5.18) and (11.5.20) are solved numerically. The output power P is given by:

$$P = \frac{\omega}{2\pi}\int_t^{t+2\pi/\omega} iVdt = \frac{1}{2}RI^2. \tag{11.5.21}$$

11.5.2. Numerical results

The fourth-order elastic constants c_{6666} is needed for numerical calculation. The only piezoelectric crystal for which c_{6666} is known seems to be an AT-cut quartz plate. For geometric parameters we consider $h = 1$ mm and $a = 10$ mm. Behaviors of the output current versus the driving

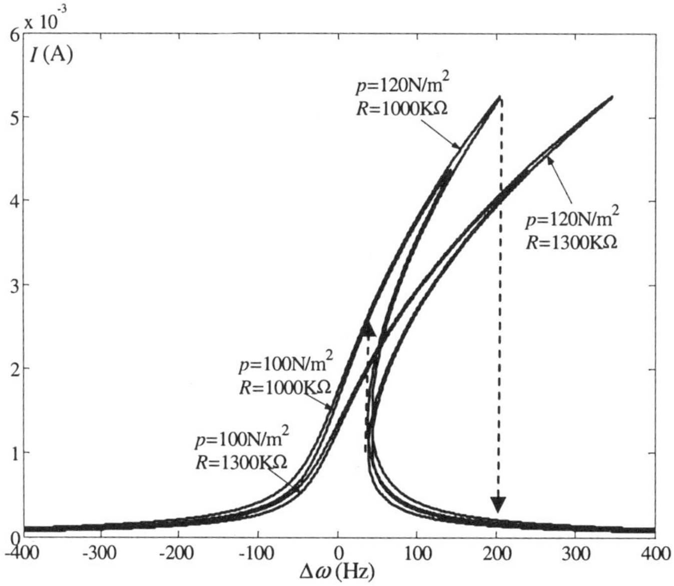

Fig. 11.5.2. Amplitude of the output current near resonance.

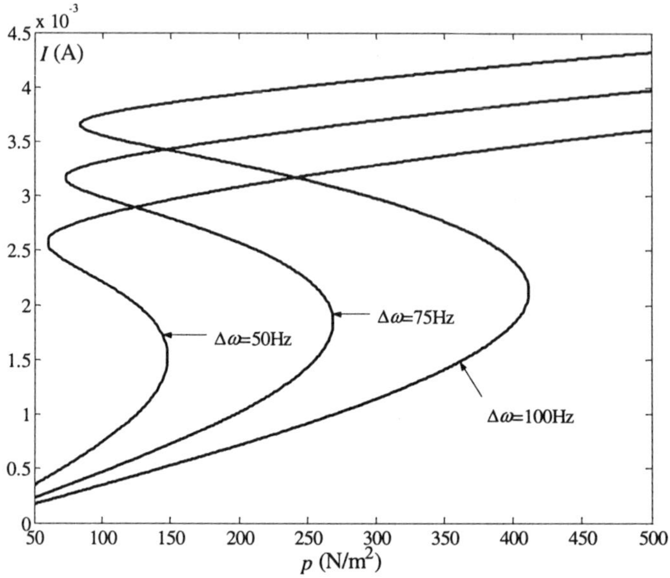

Fig. 11.5.3. Amplitude of the output current versus the amplitude of the driving force.

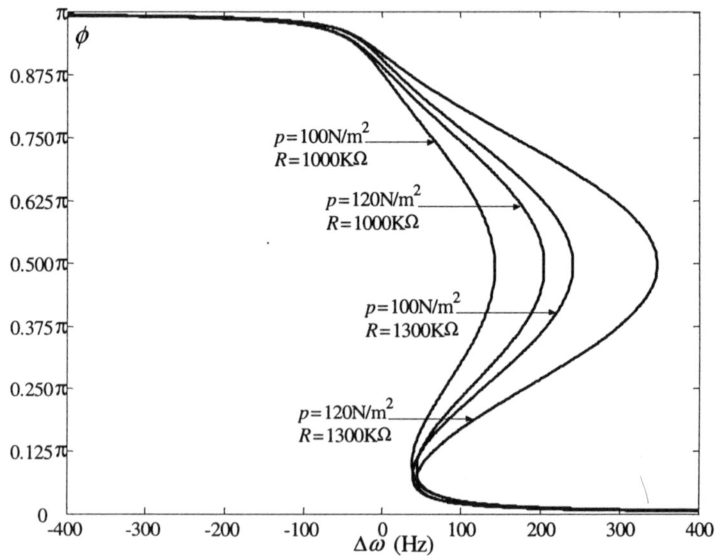

Fig. 11.5.4. Phase angle of the output current near resonance.

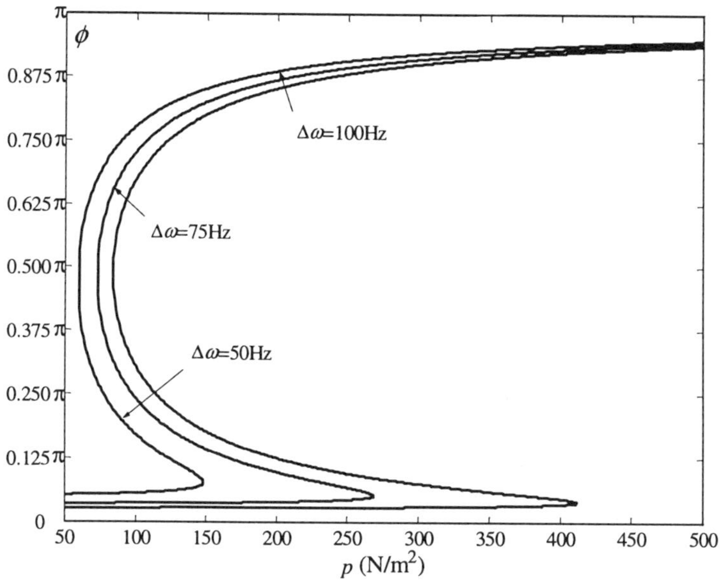

Fig. 11.5.5. Phase angle of the output current versus the amplitude of the driving force.

frequency and the applied force are shown in Figs. 11.5.2 and 11.5.3, respectively. Near nonlinear resonance the current becomes multi-valued. A jump phenomenon is usually associated with this type of behavior, as indicated by the dotted lines in Fig. 11.5.2. From Eq. (11.5.21), the output power is quadratic in the current amplitude and should also exhibit multi-valuedness and jumps.

The phase angle of the output current versus the driving frequency and the force is shown in Figs. 11.5.4 and 11.5.5, respectively, which also show multi-valuedness and jumps as expected.

Chapter 12
Piezoelectric Transformers

Piezoelectric transformers for raising or lowering electric voltage are widely used in various electronic equipment. In a piezoelectric transformer energy converts twice: from electrical to mechanical and then back to electrical. The basic behavior of a piezoelectric transformer is governed by the linear theory of piezoelectricity.

12.1. A Thickness-Stretch Mode Plate Transformer

As a simple example, consider a transformer consisting of two piezoelectric ceramic plates poled in the thickness direction. It has two traction-free surfaces and an interface, which are all electroded (see Fig. 12.1.1) [116].

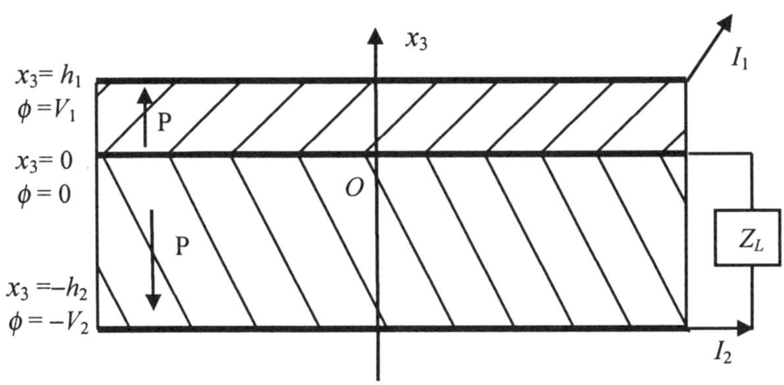

Fig. 12.1.1. A thickness-stretch mode transformer.

12.1.1. Governing equations

The driving portion $0 \le x_3 \le h_1$ is poled in the positive x_3 direction and the receiving portion $-h_2 \le x_3 \le 0$ is poled in the negative x_3 direction.

Under a time-harmonic driving voltage $V_1(t)$ with a proper frequency, the transformer can be driven into thickness-stretch vibration and produce an output voltage $V_2(t)$. The output electrodes are connected by a circuit whose impedance per unit area is denoted by Z_L. Since the transformer is inhomogeneous with ceramics poled in different directions, we need to model each layer separately with interface continuity conditions. The plates are thin and edge effects are neglected.

For the driving portion the governing equations for thickness-stretch vibrations are

$$\begin{aligned} &c_{33}u_{3,33} + e_{33}\phi_{,33} = \rho\ddot{u}_3, \\ &e_{33}u_{3,33} - \varepsilon_{33}\phi_{,33} = 0, \\ &E_3 = -\phi_{,3}, \\ &T_{33} = c_{33}u_{3,3} - e_{33}E_3, \\ &D_3 = e_{33}u_{3,3} + \varepsilon_{33}E_3. \end{aligned} \tag{12.1.1}$$

The boundary conditions at the top surface are

$$\begin{aligned} &T_{3j} = 0, \quad x_3 = h_1, \\ &\phi = V_1(t), \quad x_3 = h_1. \end{aligned} \tag{12.1.2}$$

The current density per unit area flowing out of the driving electrode at $x_3 = h_1$ is given by

$$I_1 = -\dot{Q}_1, \quad Q_1 = -D_3. \tag{12.1.3}$$

For the receiving portion the piezoelectric constants change their signs due to reversed polarization, and the governing equations are

$$\begin{aligned} &c_{33}u_{3,33} - e_{33}\phi_{,33} = \rho\ddot{u}_3, \\ &-e_{33}u_{3,33} - \varepsilon_{33}\phi_{,33} = 0, \\ &E_3 = -\phi_{,3}, \\ &T_{33} = c_{33}u_{3,3} + e_{33}E_3, \\ &D_3 = -e_{33}u_{3,3} + \varepsilon_{33}E_3. \end{aligned} \tag{12.1.4}$$

The mechanical boundary condition at the bottom surface is

$$T_{3j} = 0, \quad x_3 = -h_2. \tag{12.1.5}$$

Equation (12.1.4) implies that

$$D_3(x_3, t) = f(t), \tag{12.1.6}$$

where $f(t)$ is an integration constant which may still be a function of time. The current density flowing out of the receiving electrode at $x_3 = -h_2$ is

$$I_2 = -\dot{Q}_2, \quad Q_2 = D_3 = f(t). \tag{12.1.7}$$

At the interface $x_3 = 0$, the following continuity and boundary conditions need to be prescribed:

$$\begin{aligned} &u_3(0^+,t) = u_3(0^-,t), \quad \phi(0^+,t) = \phi(0^-,t) = 0, \\ &T_{33}(0^+,t) = [c_{33}u_{3,3} + e_{33}\phi_{,3}]\big|_{0^+} \\ &\quad = [c_{33}u_{3,3} - e_{33}\phi_{,3}]\big|_{0^-} = T_{33}(0^-,t). \end{aligned} \tag{12.1.8}$$

12.1.2. Analytical solution

Let

$$\begin{aligned} &[u_3, \phi, V_1(t), V_2(t), I_1(t), I_2(t), f(t)] \\ &\quad = \mathrm{Re}\{[U_3(x_3), \Phi(x_3), \overline{V}_1, \overline{V}_2, \overline{I}_1, \overline{I}_2, F]\exp(i\omega t)\}. \end{aligned} \tag{12.1.9}$$

Then Eq. (12.1.8) becomes

$$\begin{aligned} &U_3(0^+) = U_3(0^-), \quad \Phi(0^+) = \Phi(0^-) = 0, \\ &[c_{33}U_{3,3} + e_{33}\Phi_{,3}]\big|_{0^+} = [c_{33}U_{3,3} - e_{33}\Phi_{,3}]\big|_{0^-}. \end{aligned} \tag{12.1.10}$$

The equations of motion and the electric potential can be written as

$$\begin{aligned} &\bar{c}_{33}U_{3,33} + \rho\omega^2 U_3 = 0, \qquad -h_2 \le x_3 \le h_1, \\ &\Phi(x_3) = \begin{cases} \dfrac{x_3}{h_1}\overline{V}_1 + e_{33}\varepsilon_{33}^{-1}\{[U_3(x_3) - U_3(0^+)] \\ \quad -\dfrac{x_3}{h_1}[U_3(h_1) - U_3(0^+)]\}, \quad 0^+ \le x_3 \le h_1, \\ -e_{33}\varepsilon_{33}^{-1}[U_3(x_3) - U_3(0^-)] + Fx_3, \quad -h_2 \le x_3 \le 0^-, \end{cases} \end{aligned} \tag{12.1.11}$$

where

$$\bar{c}_{33} = c_{33} + e_{33}^2 / \varepsilon_{33}. \tag{12.1.12}$$

The circuit condition takes the following form:

$$\overline{I}_2 = \overline{V}_2 / Z_L. \tag{12.1.13}$$

The solution for the transforming ratio, the output current and the input admittance can be written as

$$\frac{\bar{V}_2}{\bar{V}_1}=\frac{\Gamma_1 Z_L}{Z_L+Z_2},\quad \bar{I}_2=\frac{\Gamma_1}{Z_L+Z_2}\bar{V}_1,\quad -\frac{\bar{I}_1}{\bar{V}_1}=\frac{1}{Z_1}-\frac{\Gamma_1\Gamma_2}{Z_L+Z_2},\tag{12.1.14}$$

where

$$\begin{aligned}\Gamma_1 &= \frac{e_{33}\varepsilon_{33}^{-1}}{\Delta}[\gamma_{31}\sin\eta h_2+\gamma_{21}(1-\cos\eta h_2)],\\ Z_2 &= \frac{e_{33}\varepsilon_{33}^{-1}}{i\omega\Delta}[\gamma_{32}\sin\eta h_2+\gamma_{22}(1-\cos\eta h_2)]-\frac{1}{i\omega}h_2,\end{aligned}\tag{12.1.15}$$

$$\begin{aligned}1/Z_1 &= -\frac{i\omega}{h_1\Delta}[-\varepsilon_{33}\Delta+e_{33}\gamma_{11}\sin\eta h_1+e_{33}\gamma_{21}(\cos\eta h_1-1)],\\ \Gamma_2 &= -\frac{e_{33}}{h_1\Delta}[\gamma_{12}\sin\eta h_1+\gamma_{22}(\cos\eta h_1-1)],\end{aligned}\tag{12.1.16}$$

$$\begin{aligned}&\gamma_{11}=\alpha_{12}\mu_1-h_1^{-1}e_{33}\alpha_{22},\quad \gamma_{12}=\alpha_{12}\mu_2,\\ &\gamma_{21}=h_1^{-1}e_{33}\alpha_{21}-\alpha_{11}\mu_1,\quad \gamma_{22}=-\alpha_{11}\mu_2,\\ &\gamma_{31}=\gamma_{11}\tau_1+\gamma_{21}\tau_2+\Delta\tau_3,\quad \gamma_{32}=\gamma_{12}\tau_1+\gamma_{22}\tau_2+\Delta\tau_3 h_1,\\ &\Delta=\alpha_{11}\alpha_{22}-\alpha_{12}\alpha_{21},\end{aligned}\tag{12.1.17}$$

$$\begin{aligned}\alpha_{11} &= \bar{c}_{33}\eta\cos\eta h_1-h_1^{-1}e_{33}^2\varepsilon_{33}^{-1}\sin\eta h_1,\\ \alpha_{12} &= -[\bar{c}_{33}\eta\sin\eta h_1+h_1^{-1}e_{33}^2\varepsilon_{33}^{-1}(\cos\eta h_1-1)],\\ \alpha_{21} &= \bar{c}_{33}\eta\tau_1\cos\eta h_2,\\ \alpha_{22} &= \bar{c}_{33}\eta(\tau_2\cos\eta h_2+\sin\eta h_2),\\ \mu_1 &= \bar{c}_{33}\eta\tau_3\cos\eta h_2,\quad \mu_2=\bar{c}_{33}\eta\tau_3 h_1\cos\eta h_2-e_{33}.\end{aligned}\tag{12.1.18}$$

Equation (12.1.14) is formally valid for a transformer in general. Equations (12.1.15) and (12.1.16) are specific expressions depending on the specific transformer under consideration.

The dependence of the transforming ratio and the input admittance on the load Z_L is of interest in transformer design, which is explicitly shown in Eq. (12.1.14). For small loads the transforming ratio is linear in Z_L and the input admittance approaches a constant. For large Z_L the transforming ratio approaches Γ_1 (saturation) and the input admittance approaches $1/Z_1$. The input and output powers are given by

$$P_1 = \frac{1}{4}(\bar{I}_1\bar{V}_1^* + \bar{I}_1^*\bar{V}_1), \quad P_2 = \frac{1}{4}(\bar{I}_2\bar{V}_2^* + \bar{I}_2^*\bar{V}_2). \tag{12.1.19}$$

The efficiency of the transformer is

$$\nu = \frac{P_2}{P_1}$$

$$= \frac{Z_1 Z_1^* \Gamma_1 \Gamma_1^* (Z_L + Z_L^*)}{Z_1 Z_1^* [\Gamma_1^* \Gamma_2^* (Z_L + Z_2) + \Gamma_1 \Gamma_2 (Z_L^* + Z_2^*)] - (Z_L + Z_2)(Z_L^* + Z_2^*)(Z_1 + Z_1^*)}. \tag{12.1.20}$$

Equation (12.1.20) shows that for small loads ν depends on Z_L linearly, and for large loads ν decreases to zero. When the load is a pure resistor, Z_L is real. In this case Eq. (12.1.20) can be written in a simpler form

$$\nu = \frac{\xi_1 Z_L}{1 + \xi_2 Z_L + \xi_3 Z_L^2}, \tag{12.1.21}$$

where ξ_n, $n = 1, 2, 3$, are real functions of Γ_1, Γ_2, Z_1 and Z_2. Equation (12.1.21) implies that the efficiency as a function of a resistor load has a maximum. The above discussions are based on the general relations in (12.1.14). They are valid for piezoelectric transformers in general.

12.1.3. Numerical results

Figure 12.1.2 shows the transforming ratio as a function of the driving frequency. The transforming ratio assumes its maxima at resonant frequencies as expected. This shows that the transformer is a resonant device operating at a particular frequency.

Figure 12.1.3 shows the transforming ratio as a function of the load Z_L. As the load increases from zero, the transforming ratio increases from zero essentially linearly. For large loads, the transforming ratio is essentially a constant, exhibiting a saturation behavior. Physically, for very large values of the load, the output electrodes are essentially open. The output voltage is saturated and the output current essentially vanishes.

Admittance versus the driving frequency and the load are plotted in Figs. 12.1.4 and 12.1.5, respectively.

Efficiency versus the driving frequency and the load are shown in Figs. 12.1.6 and 12.1.7, respectively.

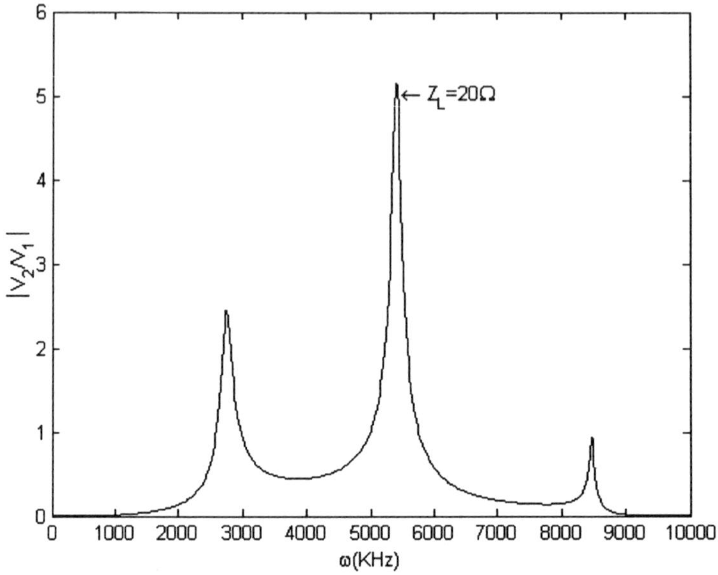

Fig. 12.1.2. Transforming ratio versus the driving frequency.

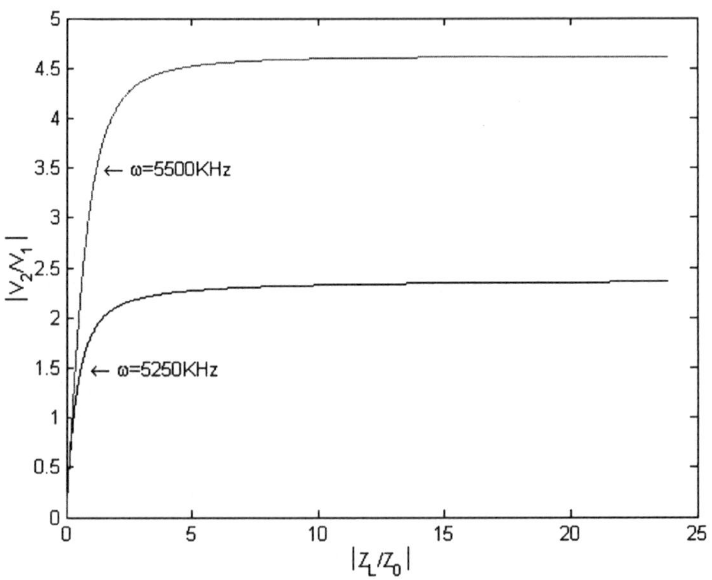

Fig. 12.1.3. Transforming ration versus the load.

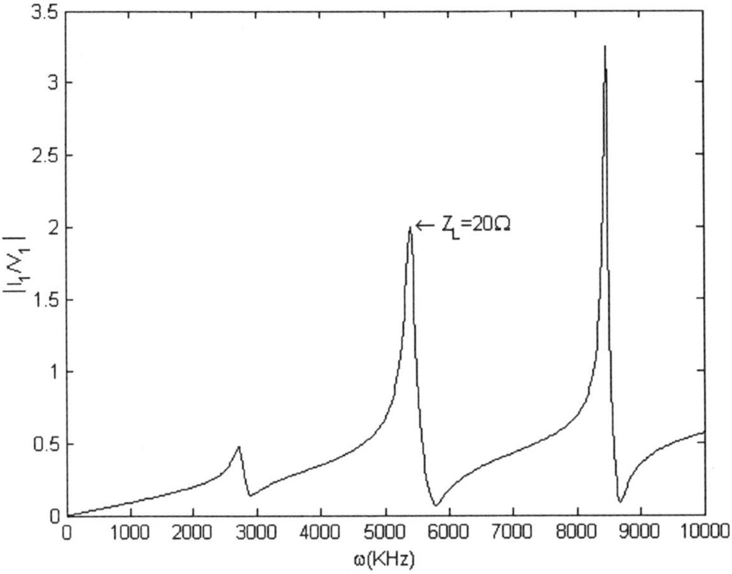

Fig. 12.1.4. Admittance versus the driving frequency.

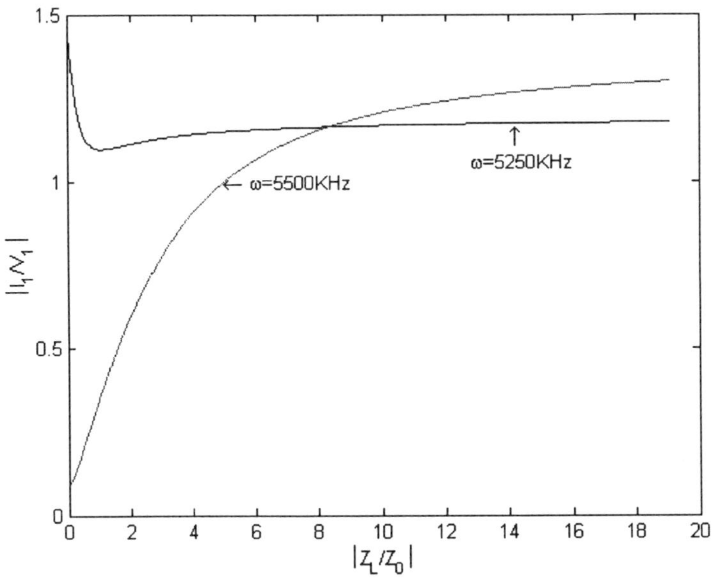

Fig. 12.1.5. Admittance versus the load.

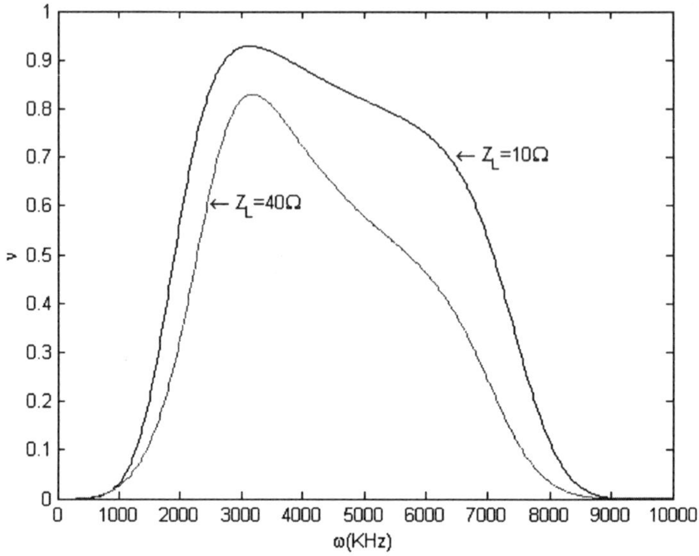

Fig. 12.1.6. Efficiency versus the driving frequency.

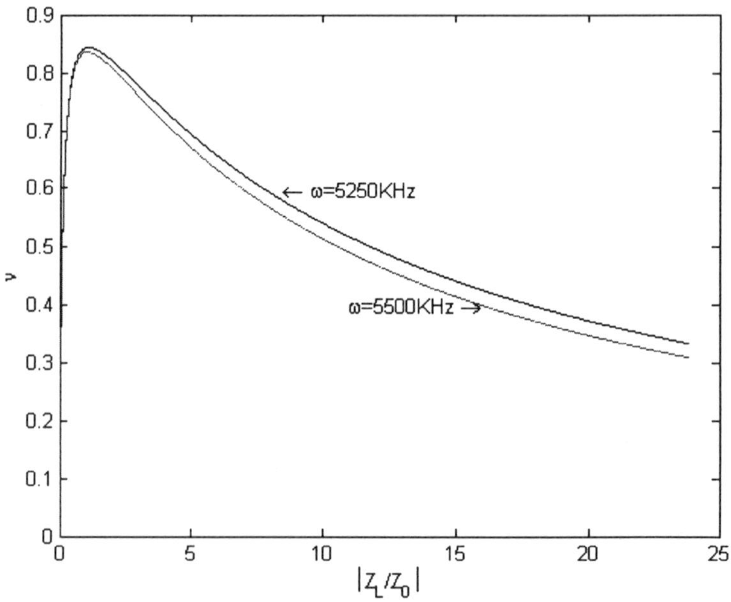

Fig. 12.1.7. Efficiency versus the load.

12.2. Rosen Transformer

Piezoelectric transformers are classified as low-voltage and high-voltage transformers. The thickness-stretch transformer analyzed in the previous section is a low-voltage one. Driving and receiving are both across the plate thickness, the smallest dimension. In this section, we analyze Rosen transformers operating with extensional modes of rods (see Fig. 12.2.1) [117]. A Rosen transformer is driven across its thickness and the output is taken from its length, the largest dimension. Therefore, a high voltage can be generated and it is a so-called high-voltage transformer. The transformer has a total length $a+b$, width w and thickness h.

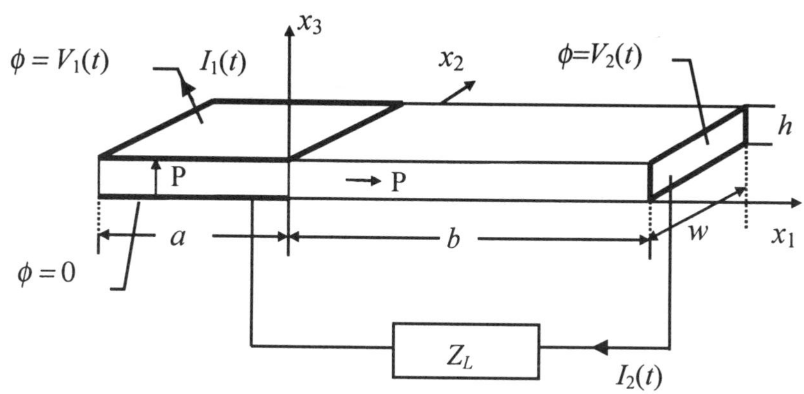

Fig. 12.2.1. A Rosen ceramic transformer.

The transformer has a slender shape with a, $b >> w >> h$. The driving portion $-a < x_1 < 0$ is poled in the x_3 direction and is electroded at $x_3 = 0$ and $x_3 = h$, with electrodes in the areas bounded by the thick lines shown in the figure. In the receiving portion $0 < x_1 < b$, the ceramic is poled in the x_1 direction with one output electrode at the end $x_1 = b$. The other output electrode is shared with the driving portion (where $\phi = 0$).

12.2.1. One-dimensional model

Since the rod is slender and we are considering extensional motions only, we make the usual assumption of a unidirectional state of stress throughout the entire transformer:

$$T_2 = T_3 = T_4 = T_5 = T_6 = 0. \tag{12.2.1}$$

For the driving portion $-a < x_1 < 0$, based on the electrode configuration, we approximately have

$$\phi = \frac{x_3}{h} V_1, \tag{12.2.2}$$

which implies that

$$E_1 = E_2 = 0, \quad E_3 = -\frac{1}{h} V_1. \tag{12.2.3}$$

The relevant equation of motion and constitutive relations are

$$\begin{aligned} &T_{1,1} = \rho \ddot{u}_1, \\ &S_1 = s_{11} T_1 + d_{31} E_3, \quad D_3 = d_{31} T_1 + \varepsilon_{33} E_3. \end{aligned} \tag{12.2.4}$$

From Eqs. (12.2.1) to (12.2.4) we obtain the equation for u_1 and the expressions for T_1 and D_3 as

$$\begin{aligned} &\frac{1}{s_{11}} u_{1,11} = \rho \ddot{u}_1, \\ &T_1 = \frac{1}{s_{11}}\left(u_{1,1} + \frac{d_{31}}{h} V_1\right), \quad D_3 = \frac{d_{31}}{s_{11}} u_{1,1} - \bar{\varepsilon}_{33} \frac{V_1}{h}, \end{aligned} \tag{12.2.5}$$

where

$$\bar{\varepsilon}_{33} = \varepsilon_{33}(1 - k_{31}^2), \quad k_{31}^2 = \frac{d_{31}^2}{\varepsilon_{33} s_{11}}. \tag{12.2.6}$$

At the left end, we have the following boundary condition:

$$T_1 = \frac{1}{s_{11}}\left(u_{1,1} + \frac{d_{31}}{h} V_1\right) = 0, \qquad x_1 = -a. \tag{12.2.7}$$

The current flowing out of the driving electrode at $x_3 = h$ is given by

$$I_1 = -\dot{Q}_1, \quad Q_1 = -w \int_{-a}^{0} D_3 dx_1, \tag{12.2.8}$$

where Q_1 is the charge on the driving electrode at $x_3 = h$.

For the receiving portion $0 < x_1 < b$, the stress assumption in Eq. (12.2.1) is still valid. For the electric field, since the portion is not electroded on its lateral surfaces, we approximately have

$$D_2 = D_3 = 0, \tag{12.2.9}$$

which implies, from the constitutive equations, that

$$E_2 = E_3 = 0. \tag{12.2.10}$$

Hence, in the receiving portion, the dominating electric field component is

$$E_1 = -\phi_{,1}, \quad \phi = \phi(x_1, t). \tag{12.2.11}$$

The relevant equation of motion, the charge equation, and the relevant constitutive relations are

$$\begin{aligned} &T_{1,1} = \rho \ddot{u}_1, \quad D_{1,1} = 0, \\ &S_1 = s_{33} T_1 + d_{33} E_1, \quad D_1 = d_{33} T_1 + \varepsilon_{33} E_1. \end{aligned} \tag{12.2.12}$$

Then the equation for u_1 and the expressions for T_1, D_1 and ϕ are

$$\begin{aligned} &\frac{1}{\bar{s}_{33}} u_{1,11} = \rho \ddot{u}_1, \quad T_1 = \frac{1}{\bar{s}_{33}}(u_{1,1} - k_{33}^2 c_1), \\ &D_1 = \frac{d_{33}}{s_{33}} c_1, \quad \phi = \frac{1}{\bar{d}_{33}}(c_1 x_1 - u_1) + c_2, \end{aligned} \tag{12.2.13}$$

where

$$\bar{s}_{33} = s_{33}(1 - k_{33}^2), \qquad \bar{d}_{33} = d_{33}\left(1 - \frac{1}{k_{33}^2}\right), \quad k_{33}^2 = \frac{d_{33}^2}{s_{33}\varepsilon_{33}}, \tag{12.2.14}$$

and $c_1(t)$ and $c_2(t)$ are two integration constants which may still be functions of time. The following boundary conditions need to be satisfied at the right end:

$$T_1 = \frac{1}{\bar{s}_{33}}(u_{1,1} - k_{33}^2 c_1) = 0, \quad \phi = V_2(t), \quad x_1 = b. \tag{12.2.15}$$

Physically, c_1 is related to the electric charge Q_2 and, hence, the current I_2 on the receiving electrode at $x_1 = b$

$$D_1 = \frac{d_{33}}{s_{33}} c_1 = -\frac{Q_2}{wh}, \quad I_2 = -\dot{Q}_2. \tag{12.2.16}$$

At the junction of the two portions $x_1 = 0$, the following continuity and boundary conditions need to be prescribed:

$$\begin{aligned} &u_1(0^-) = u_1(0^+), \\ &T_1(0^-) = \frac{1}{s_{11}}(u_{1,1} + \frac{d_{31}}{h} V_1)|_{0^-} = \frac{1}{\bar{s}_{33}}(u_{1,1} - k_{33}^2 c_1)|_{0^+} = T_1(0^+), \end{aligned} \tag{12.2.17}$$

$$\phi(0^+) = \frac{1}{2}V_1 .$$

We need to solve the two second-order Eqs. (12.2.5)$_1$ and (12.2.13)$_1$ for u_1 in two regions. Let

$$\begin{aligned}&\{u_1,\phi,V_1,V_2 I_1,I_2,c_1,c_2\}\\&=\mathrm{Re}\{\{U,\Phi,\overline{V}_1,\overline{V}_2\overline{I}_1,\overline{I}_{2,}C_1,C_2\}\exp(i\omega t)\}.\end{aligned} \tag{12.2.18}$$

For the output electrodes, we have the following circuit condition:

$$\overline{I}_2 = \overline{V}_2 / Z_L . \tag{12.2.19}$$

When $Z_L = 0$ or ∞, we have shorted or open output electrodes.

12.2.2. Free vibration analysis

The basic mechanism of a high voltage transformer can be shown by its vibration modes from a free vibration analysis. For free vibrations we set $V_1 = 0$ and $c_1 = 0$. Physically this means that the driving electrodes are shorted and the receiving electrodes are open. Mathematically the equations and boundary conditions all become homogeneous. From Eqs. (12.2.5)$_1$, (12.2.13)$_1$, (12.2.7), (12.2.15) and (12.2.17), the equations and boundary conditions we need to solve reduce to

$$\begin{aligned}&-\frac{1}{s_{11}}U_{,11} = \rho\omega^2 U , \quad -a < x_1 < 0 ,\\&-\frac{1}{\overline{s}_{33}}U_{,11} = \rho\omega^2 U , \quad 0 < x_1 < b ,\\&\frac{1}{s_{11}}U_{,1}(-a) = 0 , \quad \frac{1}{\overline{s}_{33}}U_{,1}(b) = 0 ,\\&U(0^-) = U(0^+) , \quad \frac{1}{s_{11}}U_{,1}(0^-) = \frac{1}{\overline{s}_{33}}U_{,1}(0^+) ,\end{aligned} \tag{12.2.20}$$

where ω is an unknown resonant frequency at which nontrivial solutions of U to the above equations exist. Mathematically, Eq. (12.2.20) is an eigenvalue problem. The solution with $\omega = 0$ is a rigid body displacement with U being a constant, which is not of interest. The solution to Eq. (12.2.20) is

$$\omega = \omega_{(n)} , \qquad n = 1,2,3,\ldots , \tag{12.2.21}$$

$$U^{(n)} = \begin{cases} \left(\dfrac{s_{11}}{\bar{s}_{33}}\right)^{\frac{1}{2}} \sin k_{(n)} x_1 + \dfrac{1}{\tan \bar{k}_{(n)} b} \cos k_{(n)} x_1, & -a < x_1 < 0, \\ \sin \bar{k}_{(n)} x_1 + \dfrac{1}{\tan \bar{k}_{(n)} b} \cos \bar{k}_{(n)} x_1 \quad , & 0 < x_1 < b, \end{cases}$$

(12.2.22)

where

$$k_{(n)}^2 = \rho s_{11} \omega_{(n)}^2, \quad \bar{k}_{(n)}^2 = \rho \bar{s}_{33} \omega_{(n)}^2, \tag{12.2.23}$$

and $\omega_{(n)}$ is the n-th root of the following frequency equation:

$$\frac{\tan ka}{\tan \bar{k} b} = -\left(\frac{s_{11}}{\bar{s}_{33}}\right)^{\frac{1}{2}}. \tag{12.2.24}$$

The width w and thickness h do not appear in the frequency equation. This is as expected from a one-dimensional model. Once U is known, Φ in the receiving portion can be found from Eq. $(12.2.13)_4$. We have

$$\Phi^{(n)} = \begin{cases} 0, \quad -a < x_1 < 0, \\ -\dfrac{1}{\bar{d}_{33}} [\sin \bar{k}_{(n)} x_1 + \dfrac{1}{\tan \bar{k}_{(n)} b} (\cos \bar{k}_{(n)} x_1 - 1)], \quad 0 < x_1 < b. \end{cases}$$

(12.2.25)

12.2.3. Forced vibration analysis

For forced vibration driven by $V_1(t) = \bar{V}_1 \exp(i\omega t)$, from Eqs. $(12.2.5)_1$, $(12.2.13)_1$, (12.2.7), (12.2.15) and (12.2.17), we need to solve the following problem for U:

$$-\frac{1}{s_{11}} U_{,11} = \rho \omega^2 U \ , \quad -a < x_1 < 0 \, ,$$

$$-\frac{1}{\bar{s}_{33}} U_{,11} = \rho \omega^2 U \, , \quad 0 < x_1 < b \, ,$$

$$\frac{1}{s_{11}} \left(U_{,1} + \frac{d_{31}}{h} \bar{V}_1 \right) = 0, \qquad x_1 = -a \, ,$$

$$\frac{1}{\bar{s}_{33}}(U_{,1} - k_{33}^2 C_1) = 0, \quad x_1 = b,$$

$$U(0^-) = U(0^+), \quad \frac{1}{s_{11}}\left[U_{,1}(0^-) + \frac{d_{31}}{h}\bar{V}_1\right] = \frac{1}{\bar{s}_{33}}[U_{,1}(0^+) - k_{33}^2 C_1]. \tag{12.2.26}$$

The solution to Eq. (12.2.26) can be written as

$$U = \begin{cases} \left(\alpha_{11}\dfrac{d_{31}\bar{V}_1}{h} + \beta_{11}k_{33}^2 C_1\right)\sin kx_1 \\ \quad + \left(\alpha_{12}\dfrac{d_{31}\bar{V}_1}{h} + \beta_{12}k_{33}^2 C_1\right)\cos kx_1, \quad -a < x_1 < 0, \\ \left(\alpha_{22}\dfrac{d_{31}\bar{V}_1}{h} + \beta_{22}k_{33}^2 C_1\right)\sin \bar{k}x_1 \\ \quad + \left(\alpha_{12}\dfrac{d_{31}\bar{V}_1}{h} + \beta_{12}k_{33}^2 C_1\right)\cos \bar{k}x_1, \quad 0 < x_1 < b, \end{cases} \tag{12.2.27}$$

where

$$\begin{aligned} & k = \omega\sqrt{\rho s_{11}}, \quad \bar{k} = \omega\sqrt{\rho \bar{s}_{33}}, \\ & \alpha_{11} = -\frac{1}{\Delta\cos ka}\bar{s}_{33}\cos\bar{k}b\sin ka(\cos ka - 1) - \frac{1}{k\cos ka}, \\ & \beta_{11} = \frac{1}{\Delta}s_{11}\sin ka(1-\cos\bar{k}b), \quad \alpha_{12} = \frac{1}{\Delta}\bar{s}_{33}\cos\bar{k}b(\cos ka - 1), \\ & \beta_{12} = -\frac{1}{\Delta}s_{11}\cos ka(1-\cos\bar{k}b), \alpha_{22} = \frac{1}{\Delta}\bar{s}_{33}\sin\bar{k}b(\cos ka - 1), \\ & \beta_{22} = -\frac{1}{\Delta\cos\bar{k}b}s_{11}\sin\bar{k}b\cos ka(1-\cos\bar{k}b) + \frac{1}{\bar{k}\cos\bar{k}b}, \\ & \Delta = s_{11}\bar{k}\sin\bar{k}b\cos ka + \bar{s}_{33}k\sin ka\cos\bar{k}b. \end{aligned} \tag{12.2.28}$$

$\Delta = 0$ implies the frequency equation in Eq. (12.2.24).

With Eqs. (12.2.27), from $(12.2.13)_4$ and $(12.2.17)_3$, we obtain the voltage distribution in the receiving portion as

$$\Phi = \frac{1}{2}\bar{V}_1 + \frac{1}{\bar{d}_{33}}\left[C_1 x_1 - \left(\alpha_{22}\frac{d_{31}\bar{V}_1}{h} + \beta_{22}k_{33}^2 C_1\right)\sin\bar{k}x_1 \right.$$
$$\left. + \left(\alpha_{12}\frac{d_{31}\bar{V}_1}{h} + \beta_{12}k_{33}^2 C_1\right)(1-\cos\bar{k}x_1)\right], \qquad 0 < x_1 < b. \tag{12.2.29}$$

Although $\bar{V}_1$ is considered given, Eqs. (12.2.27) and (12.2.29) still have an unknown constant C_1. From Eqs. (12.2.29) and $(12.2.15)_2$ the output voltage is

$$\bar{V}_2 = \Phi(b) = \Gamma_1\bar{V}_1 - Z_2\bar{I}_2, \tag{12.2.30}$$

where

$$\Gamma_1 = \frac{1}{2} + \frac{d_{31}}{\bar{d}_{33}}\frac{1}{h}[\alpha_{12}(1-\cos\bar{k}b) - \alpha_{22}\sin\bar{k}b],$$
$$Z_2 = \frac{1}{1-k_{33}^2}\frac{1}{b}[b - \beta_{22}k_{33}^2\sin\bar{k}b + \beta_{12}k_{33}^2(1-\cos\bar{k}b)]\frac{b}{i\omega\varepsilon_{33}wh}, \tag{12.2.31}$$

and Eq. (12.2.16) has been used to replace C_1 by $\bar{I}_2$. From Eqs. (12.2.8), $(12.2.5)_3$, and (12.2.27), the driving current is

$$\bar{I}_1 = -\bar{V}_1/Z_1 + \Gamma_2\bar{I}_2, \tag{12.2.32}$$

where

$$1/Z_1 = \frac{1}{a}\left[a - \frac{k_{31}^2}{1-k_{31}^2}(\alpha_{12} + \alpha_{11}\sin ka - \alpha_{12}\cos ka)\right]\frac{i\omega\bar{\varepsilon}_{33}wa}{h},$$
$$\Gamma_2 = k_{33}^2\frac{s_{33}d_{31}}{s_{11}d_{33}}\frac{1}{h}(\beta_{12} + \beta_{11}\sin ka - \beta_{12}\cos ka). \tag{12.2.33}$$

Solving Eqs. (12.2.30), (12.2.32) and the circuit condition (12.2.19), we obtain the transforming ratio, the output current, and the input admittance as

$$\frac{\bar{V}_2}{\bar{V}_1} = \frac{\Gamma_1 Z_L}{Z_L + Z_2}, \quad \bar{I}_2 = \frac{\Gamma_1}{Z_L + Z_2}\bar{V}_1,$$
$$-\frac{\bar{I}_1}{\bar{V}_1} = \frac{1}{Z_1} - \frac{\Gamma_1\Gamma_2}{Z_L + Z_2}. \tag{12.2.34}$$

The input and output powers are given by

$$P_1 = \frac{1}{4}(\bar{I}_1\bar{V}_1^* + \bar{I}_1^*\bar{V}_1), \quad P_2 = \frac{1}{4}(\bar{I}_2\bar{V}_2^* + \bar{I}_2^*\bar{V}_2). \tag{12.2.35}$$

Then the efficiency of the transformer can be written as

$$\eta = \frac{P_2}{P_1}. \tag{12.2.36}$$

When the load is a pure resistor

$$\eta = \frac{\lambda Z_L}{1 + \mu Z_L + \nu Z_L^2}, \tag{12.2.37}$$

where λ, μ, and ν are real functions of Γ_1, Γ_1, Z_1 and Z_2.

12.2.4. Numerical results

As an example, consider a transformer made from polarized ceramics PZT-5H. For forced vibration analysis, damping is introduced by replacing s_{11} and s_{33} by $s_{11}(1-iQ^{-1})$ and $s_{33}(1-iQ^{-1})$. For free vibration $Q = 0$. For forced vibration $Q = 1000$. Consider a transformer with $a = b = 22$ mm, $w = 10$ mm, and $h = 2$ mm.

12.2.4.1. Free vibration solution

With the above data, the first root of Eq. (12.2.24), i.e., the resonant frequency of the operating mode, is found to be

$$f = \omega_{(1)} / 2\pi = 36.35 \text{ kHz}. \tag{12.2.38}$$

The mode shape of $U^{(1)}$ is shown in Fig. 12.2.2, and is normalized by its maximum. The location of the nodal point is not simply in the middle of the transformer, although $a = b$ in this example. This is because the material is not uniform. The location of the nodal point depends on the compliances s_{11} and $\bar{s}_{33}$. In this example we have $\bar{s}_{33} < s_{11}$. The material of the left half of is more compliant in the x_1 direction than the material of the right half. Therefore the left half is stretched more. Hence, the nodal point (which is also the center of mass of the deformed shape, which cannot move according to Newton's law) appears in the left half. In Fig. 12.2.2, $U^{(1)}$ is continuous at $x_1 = 0$ but $U_{,1}^{(1)}$ is not, as dictated by Eq. (12.2.20)$_{5,6}$. Accurate prediction of vibration modes and their nodal points is particularly important for Rosen transformers. Theoretically, Rosen transformers can be mounted at their nodal points. Then mounting will not affect the vibration and the performance of these transformers.

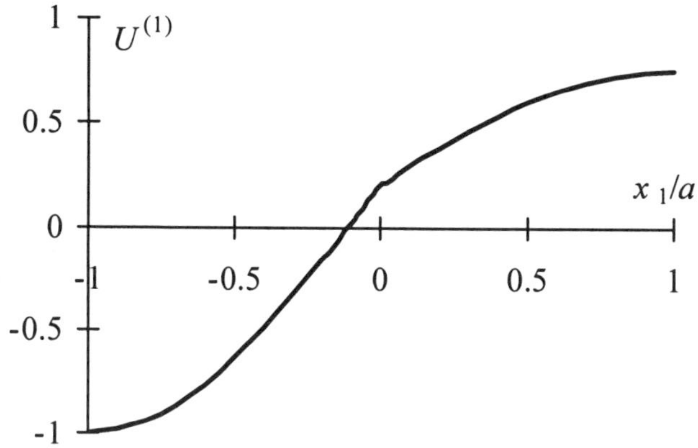

Fig. 12.2.2. Normalized mechanical displacement $U^{(1)}$ of the first extensional mode.

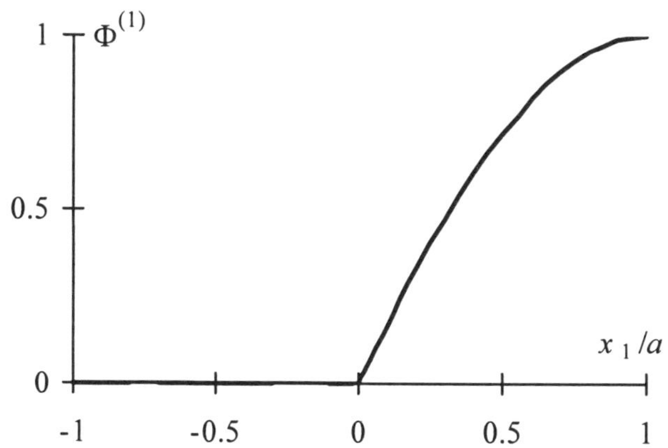

Fig. 12.2.3. Normalized electric potential $\Phi^{(1)}$ of the first extensional mode.

Normalized $\Phi^{(1)}$ as a function of x_1 is shown in Fig. 12.2.3. $\Phi^{(1)}$ rises in the receiving portion $[0,b]$. This is why a non-uniform ceramic rod can work as a transformer. The rate of change of $\Phi^{(1)}$ is large near $x_1 = 0$ where the extensional stress is large. The rate of change is small near $x_1 = b$, which is a free end with vanishing extensional stress.

12.2.4.2. Forced vibration solution

The basic behavior of the force vibration solution of the Rosen transformer is similar to what is shown in Figs. 12.1.2 through 12.1.7. The difference between a low-voltage transformer and a high-voltage one can be seen from Fig. 12.2.4 for the transforming ratio versus the aspect ratio b/h. The transforming ratio increases essentially linearly with b/h. Therefore, a long and thin Rosen transformer can produce a high voltage.

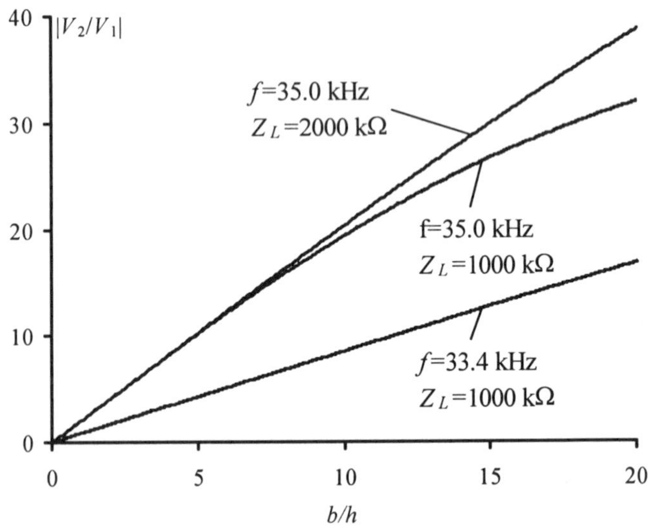

Fig. 12.2.4. Transforming ratio versus b/h.

12.3. A Thickness-Shear Mode Transformer — Free Vibration

In this section we analyze another high-voltage transformer (see Fig. 12.3.1) [118].

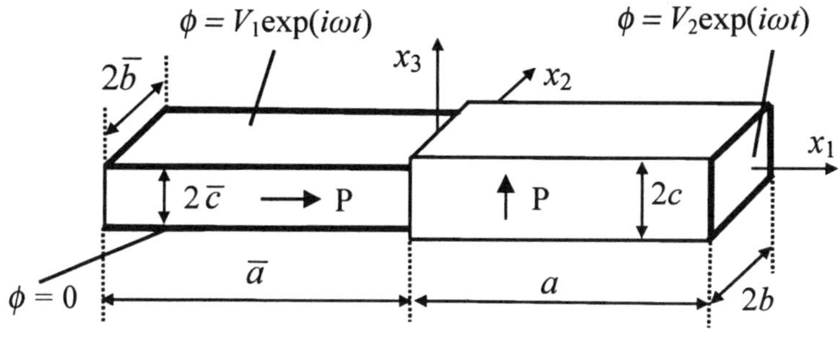

Fig. 12.3.1. A thickness-shear mode transformer.

12.3.1. Governing equations

It is assumed that $a,\bar{a} >> b,\bar{b} >> c,\bar{c}$. The driving portion $-\bar{a} < x_1 < 0$ is electroded at $x_3 = \pm\bar{c}$, with electrodes in the areas bounded by the thick lines. In the receiving portion $0 < x_1 < a$, the beam is electroded at the end $x_1 = a$. The driving portion and the receiving portion may have slightly different thickness $2\bar{c}$ and $2c$, and width $2\bar{b}$ and $2b$. V_1 is the input voltage and V_2 is the output voltage. The transformer is assumed to be made of an arbitrary piezoelectric material. When it is made of polarized ceramics, the polarizations in the driving and receiving portions are as shown in the figure.

12.3.1.1. Driving portion

For the driving portion $-\bar{a} < x_1 < 0$, the mechanical displacement and the electric potential are approximated by

$$
\begin{aligned}
&u_1 = x_3 u_1^{(0,1)}(x_1,t), \quad u_2 = u_3 = 0, \\
&\phi = \phi^{(0,0)}(x_1,t) + x_3\phi^{(0,1)}(x_1,t) = \frac{x_3+\bar{c}}{2\bar{c}} V_1 \exp(i\omega t), \qquad (12.3.1) \\
&\phi^{(0,0)} = \frac{1}{2} V_1 \exp(i\omega t), \qquad \phi^{(0,1)} = \frac{1}{2\bar{c}} V_1 \exp(i\omega t).
\end{aligned}
$$

The one-dimensional equation of motion is [118,20]

$$\frac{1}{\bar{s}_{11}^*}u_{1,11}^{(0,1)} - \frac{3\bar{\kappa}^2}{\bar{c}^2\bar{s}_{55}}u_1^{(0,1)} = \bar{\rho}\ddot{u}_1^{(0,1)}, \quad -\bar{a} < x_1 < 0, \tag{12.3.2}$$

where we have denoted all the geometric and material parameters of the driving portion by an over bar. The boundary condition of vanishing shear force is

$$T_{13}^{(0,0)} = \frac{4\bar{b}\bar{c}}{\bar{s}_{55}}(\bar{\kappa}^2 u_1^{(0,1)} + \bar{\kappa}\bar{d}_{35}\phi^{(0,1)}) = 0, \quad x_1 = -\bar{a}. \tag{12.3.3}$$

12.3.1.2. Receiving portion

For the receiving portion $0 < x_1 < a$, $\phi^{(0,0)}$ is the major part of the electric potential. The equation of motion is

$$\frac{1}{s_{11}^*}u_{1,11}^{(0,1)} - \frac{3\kappa^2}{c^2 s_{55}}u_1^{(0,1)} - \frac{3\kappa d_{15}}{c^2 s_{55}}\phi_{,1}^{(0,0)} = \rho\ddot{u}_1^{(0,1)}, \quad 0 < x_1 < a. \tag{12.3.4}$$

The one-dimensional electrostatic equation is

$$\frac{d_{15}\kappa}{s_{55}}u_{1,1}^{(0,1)} - \tilde{\varepsilon}_{11}\phi_{,11}^{(0,0)} = 0, \quad 0 < x_1 < a, \tag{12.3.5}$$

where the term $D^{(0,0)}$ for the electric charge on the lateral surfaces at $x_2 = \pm b$ and $x_3 = \pm c$ has been dropped for unelectroded surfaces. From Eq. (12.3.5) we obtain

$$\phi_{,1}^{(0,0)} = \frac{d_{15}\kappa}{\tilde{\varepsilon}_{11}s_{55}}u_1^{(0,1)} + C_1, \tag{12.3.6}$$

where C_1 is an integration constant. Physically C_1 is related to the charge and hence the current on the electrode at $x_1 = a$. Substituting Eq. (12.3.6) into Eq. (12.3.4), we obtain

$$\frac{1}{s_{11}^*}u_{1,11}^{(0,1)} - \frac{3\kappa^2}{c^2 s_{55}^*}u_1^{(0,1)} - \frac{3\kappa d_{15}}{c^2 s_{55}}C_1 = \rho\ddot{u}_1^{(0,1)}, \qquad 0 < x_1 < a,$$

$$\frac{1}{s_{55}^*} = \frac{1}{s_{55}}\left(1 + \frac{d_{15}^2}{\tilde{\varepsilon}_{11}s_{55}}\right). \tag{12.3.7}$$

For boundary conditions we need

$$\begin{aligned} T_{13}^{(0,0)} &= \frac{4bc}{s_{55}}(\kappa^2 u_1^{(0,1)} + \kappa d_{15}\phi_{,1}^{(0,0)}) \\ &= \frac{4bc}{\overset{*}{s}_{55}}\left(\kappa^2 u_1^{(0,1)} + \frac{\overset{*}{s}_{55}\kappa d_{15}}{s_{55}}C_1\right), \\ D_1^{(0,0)} &= 4bc\left(\frac{d_{15}\kappa}{s_{55}}u_1^{(0,1)} - \tilde{\varepsilon}_{11}\phi_{,1}^{(0,0)}\right) = -\tilde{\varepsilon}_{11}4bcC_1. \end{aligned} \tag{12.3.8}$$

Then the boundary conditions take the following form:

$$\begin{aligned} &T_{13}^{(0,0)} = \frac{4bc}{\overset{*}{s}_{55}}\left(\kappa^2 u_1^{(0,1)} + \frac{\overset{*}{s}_{55}\kappa d_{15}}{s_{55}}C_1\right) = 0, \quad x_1 = a, \\ &\phi^{(0,0)} = V_2\exp(i\omega t), \qquad x_1 = a, \\ &\phi^{(0,0)}(0^+) = \frac{1}{2}V_1\exp(i\omega t). \end{aligned} \tag{12.3.9}$$

12.3.1.3. Continuity conditions

For continuity conditions we impose

$$u_1^{(0,1)}(0^-) = u_1^{(0,1)}(0^+), \quad u_{1,1}^{(0,1)}(0^-) = u_{1,1}^{(0,1)}(0^+), \tag{12.3.10}$$

in which the second equation is an approximation.

12.3.2. Free vibration analysis

Consider free vibrations for which the driving voltage V_1 or $\phi^{(0,0)}$ and $\phi^{(0,1)}$ vanish in the driving portion, which means that the driving electrodes are shorted. The receiving electrodes are open with no current or charge on the electrodes ($D_1^{(0,0)} = 0$). Hence, C_1 is zero. This is the simplest circuit condition for the receiving electrodes. Then all the equations and boundary conditions become homogeneous. We need to solve an eigenvalue problem for ω^2:

$$\begin{aligned}
&-u_{1,11}^{(0,1)}+\frac{3\bar{\kappa}^2\bar{s}_{11}^*}{\bar{c}^2\bar{s}_{55}}u_1^{(0,1)}=\bar{\rho}\bar{s}_{11}^*\omega^2u_1^{(0,1)}, \quad -\bar{a}<x_1<0,\\
&-u_{1,11}^{(0,1)}+\frac{3\kappa^2 s_{11}^*}{c^2 s_{55}^*}u_1^{(0,1)}=\rho s_{11}^*\omega^2u_1^{(0,1)}, \quad 0<x_1<a,\\
&u_1^{(0,1)}(-\bar{a})=0, \quad u_1^{(0,1)}(a)=0,\\
&u_1^{(0,1)}(0^-)=u_1^{(0,1)}(0^+), \quad u_{1,1}^{(0,1)}(0^-)=u_{1,1}^{(0,1)}(0^+).
\end{aligned} \tag{12.3.11}$$

We try the following solution which already satisfies the two boundary conditions in Eq. (12.3.11)$_{3,4}$ at the two ends of the transformer:

$$u_1^{(0,1)}=\begin{cases} A\sin\bar{k}(x_1+\bar{a}), & -\bar{a}<x_1<0,\\ B\sin k(a-x_1), & 0<x_1<a,\end{cases} \tag{12.3.12}$$

where A , B , $\bar{k}$ and k are undetermined constants. Substitution of Eq. (12.3.12) into Eq. (12.3.11)$_{1,2}$ yields

$$\bar{k}^2=\bar{s}_{11}^*\left(\bar{\rho}\omega^2-\frac{3\bar{\kappa}^2}{\bar{c}^2\bar{s}_{55}}\right), \qquad k^2=s_{11}^*\left(\rho\omega^2-\frac{3\kappa^2}{c^2 s_{55}^*}\right), \tag{12.3.13}$$

which shows that the solution of $u_1^{(0,1)}$ may have exponential or sinusoidal behaviors depending on the signs of $\bar{k}^2$ and k^2. This is related to the energy trapping phenomenon of thickness-shear modes. For the operating mode of a transformer, sinusoidal behavior in both portions is desired. Hence, we consider the case when both $\bar{k}^2$ and k^2 are positive:

$$\bar{\rho}\omega^2-\frac{3\bar{\kappa}^2}{\bar{c}^2\bar{s}_{55}}>0, \quad \rho\omega^2-\frac{3\kappa^2}{c^2 s_{55}^*}>0. \tag{12.3.14}$$

Substituting Eq. (12.3.12) into the continuity conditions in Eq. (12.3.11)$_{5,6}$, we have

$$\begin{aligned}
&A\sin(\bar{k}\bar{a})=B\sin(ka),\\
&\bar{k}A\cos(\bar{k}\bar{a})=-Bk\cos(ka),
\end{aligned} \tag{12.3.15}$$

which, for non-trivial solutions of A and B, yields the frequency equation

$$\frac{\tan(\bar{k}\bar{a})}{\tan(ka)}=-\frac{\bar{k}}{k}. \tag{12.3.16}$$

The corresponding mode shape function is

$$u_1^{(0,1)} = \begin{cases} \dfrac{\sin \bar{k}(x_1 + \bar{a})}{\sin(\bar{k}\bar{a})}, & -\bar{a} < x_1 < 0, \\ \dfrac{\sin k(a - x_1)}{\sin(ka)}, & 0 < x_1 < a. \end{cases} \tag{12.3.17}$$

Then the electric potential can be found as

$$\phi^{(0,0)} = \begin{cases} 0, & -\bar{a} < x_1 < 0, \\ \cos k(a - x_1) - \cos(ka), & 0 < x_1 < a. \end{cases} \tag{12.3.18}$$

For a transformer it is also desired that the wave number k of the operating mode in the receiving portion satisfies

$$k^2 = s_{11}^* \left(\rho\omega^2 - \frac{3\kappa^2}{c^2 s_{55}^*} \right) \le \left(\frac{\pi}{a} \right)^2, \tag{12.3.19}$$

so that the receiving portion is not longer than one-half of the wave length in the x_1 direction of the thickness-shear mode. Then the shear deformation does not change its sign and the voltage generated by the shear deformation accumulates spatially without cancellation. Solutions satisfying this condition will be obtained for a ceramic transformer below.

12.3.3. Ceramic transformers

Consider a ceramic transformer of constant width $b = \bar{b}$. The driving portion is polarized in the x_1 direction and the receiving portion is polarized in the x_3 direction. For ceramics, usually

$$|d_{15}| > |d_{31}|, \quad |d_{15}| > |d_{33}|. \tag{12.3.20}$$

For example,

	d_{15}	d_{31}	d_{33}
PZT − 2	440	− 60	152
PZT − 5H	741	− 274	593
PZT − 7A	362	− 60	150

(12.3.21)

This suggests that the thickness-shear transformer discussed here which operates with d_{15} may be more effective in transforming than Rosen extensional transformers which use d_{31} and d_{33}. For a ceramic transformer, the equations derived above take specific forms. We have

$$\begin{aligned} \bar{\omega}_\infty^2 &= \frac{\pi^2}{4\bar{c}^2 \rho s_{44}}, \quad -\bar{a} < x_1 < 0, \\ \omega_\infty^2 &= \frac{\pi^2}{4c^2 \rho s_{44}(1-k_{15}^2)}, \quad 0 < x_1 < a, \end{aligned} \tag{12.3.22}$$

because for free vibrations the electrodes in the driving portion are shorted and the electrodes for the receiving portion are open. In addition,

$$\begin{aligned} \frac{1}{\bar{s}_{11}^*} &= \frac{1}{s_{33}} + \frac{3\bar{\kappa}^4}{\rho\bar{\omega}_\infty^2 \bar{c}^2 s_{44}^2} = \frac{1}{s_{33}} + \frac{\pi^2}{12}\frac{1}{s_{44}}, \\ \frac{1}{s_{11}^*} &= \frac{1}{s_{11}} + \frac{3\kappa^4}{\rho\omega_\infty^2 c^2 s_{44}^2} = \frac{1}{s_{11}} + \frac{\pi^2}{12}\frac{1-k_{15}^2}{s_{44}}, \end{aligned} \tag{12.3.23}$$

$$\begin{aligned} \tilde{\varepsilon}_{11} &= \varepsilon_{11}^T - \frac{d_{15}d_{15}}{s_{44}} = \varepsilon_{11}(1-k_{15}^2), \\ \frac{1}{s_{55}^*} &= \frac{1}{s_{44}}\left(1 + \frac{d_{15}^2}{\tilde{\varepsilon}_{11}s_{44}}\right) = \frac{1}{s_{44}}\frac{1}{1-k_{15}^2}, \end{aligned} \tag{12.3.24}$$

$$\begin{aligned} \bar{k}^2 &= \bar{s}_{11}^*\left(\rho\omega^2 - \frac{3\bar{\kappa}^2}{\bar{c}^2 s_{44}}\right) = \bar{s}_{11}^*\rho(\omega^2 - \bar{\omega}_\infty^2), \\ k^2 &= s_{11}^*\left(\rho\omega^2 - \frac{3\kappa^2}{c^2 s_{55}^*}\right) = s_{11}^*\rho(\omega^2 - \omega_\infty^2). \end{aligned} \tag{12.3.25}$$

12.3.3.1. A transformer of uniform thickness

First consider the case of a transformer with constant thickness $\bar{c} = c$ and equal length of the driving and receiving portions ($\bar{a} = a$). We have the inequality

$$\bar{\omega}_\infty^2 = \frac{\pi^2}{4c^2 \rho s_{44}} < \omega_\infty^2 = \frac{\pi^2}{4c^2 \rho s_{44}(1-k_{15}^2)}. \tag{12.3.26}$$

Before we start solving the frequency equation, we note that the transformer is assumed to be long and, for voltage accumulation, the entire receiving portion of the transformer should be vibrating in phase. Therefore, the transformer should be vibrating at a frequency very close to and slightly higher than the infinite beam thickness-shear frequency of the receiving portion, with a very small wave number k. Hence, we should approximately have

$$\omega^2 \approx \omega_\infty^2 = \frac{\pi^2}{4c^2 \rho s_{44}(1-k_{15}^2)}. \tag{12.3.27}$$

Then, at this frequency, for the driving portion, a finite wave number can be approximately determined by

$$\bar{k}^2 = \bar{s}_{11}^* \rho(\omega^2 - \bar{\omega}_\infty^2) \approx \bar{s}_{11}^* \rho(\omega_\infty^2 - \bar{\omega}_\infty^2). \tag{12.3.28}$$

For PZT-5H, by trial and error, it can be quickly found that, when a = 20 cm and c = 1 cm, the first root of the frequency equation is 59.66 kHz. It is indeed very close to and slightly higher than the infinite beam frequency ω_∞ of the receiving portion, which is found to be 59.31 kHz. We then have $\bar{k}a = 5.4\pi$ and $ka = 0.86\pi$. The corresponding shear and potential distributions are shown in Fig. 12.3.2. The electric potential rises in the receiving portion and a voltage is generated between $x_1 = 0$ and $x_1 = a$. Hence, this mode can be used as an operating mode of the transformer. While in the receiving portion the shear deformation does not change sign along the beam and voltage is accumulated, in the driving portion, the shear changes sign a few times. This is fine because this mode can be excited by a few pairs of electrodes in the driving portion with alternating signs of driving voltages among pairs of driving electrodes. Transformers with a short driving portion can also be designed, with fewer pairs of driving electrodes.

12.3.3.2 A transformer of non-uniform thickness ($\bar{c} < c$)

The reason that the mode in Fig. 12.3.2 needs several pairs of driving electrodes is the change of sign of the shear deformation in the driving portion. This is a result of the fact that the infinite beam frequency of the driving portion is lower than that of the receiving portion. The infinite beam frequencies of the driving and receiving portions can be made close by adjusting $\bar{c}$ and c. This can lead to a reduction of the wave number in the driving portion and, hence, fewer pairs of driving electrodes. To lower the infinite beam shear frequency of the receiving

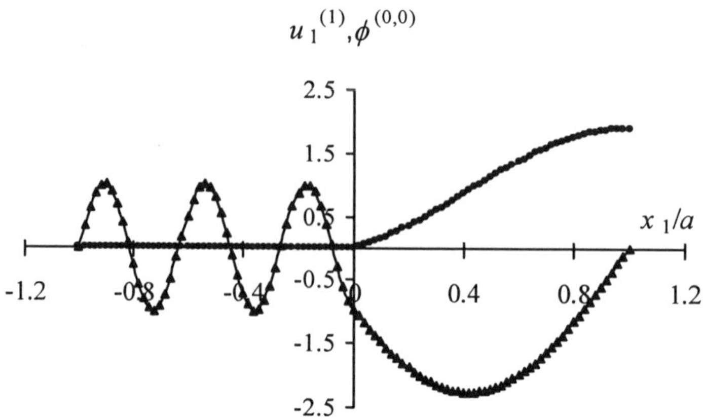

Fig. 12.3.2. Shear and potential distributions of the operating mode ($\bar{c} = c$). $u_1^{(1)}$ is marked by triangles, and $\phi^{(0,0)}$ by circles.

portion, we increase the beam thickness c of the receiving portion such that $\bar{c} < c$. Consider the case of equal length of the driving and receiving portions ($\bar{a} = a$). It can be verified that the following is a limit solution to the frequency equation:

$$\begin{aligned} \bar{k}a &= \frac{\pi^+}{2}, \quad \tan(\bar{k}a) = -\infty, \\ ka &= \frac{\pi^-}{2}, \quad \tan(ka) = \infty. \end{aligned} \tag{12.3.29}$$

Then

$$\begin{aligned} \bar{k}^2 a^2 &= a^2 \bar{s}_{11}^* \rho(\omega^2 - \bar{\omega}_\infty^2) = \pi^2 / 4, \\ k^2 a^2 &= a^2 s_{11}^* \rho(\omega^2 - \omega_\infty^2) = \pi^2 / 4, \end{aligned} \tag{12.3.30}$$

which leads to the frequency

$$\omega^2 = \bar{\omega}_\infty^2 + \frac{\pi^2}{4a^2 \bar{s}_{11}^* \rho} = \omega_\infty^2 + \frac{\pi^2}{4a^2 s_{11}^* \rho}. \tag{12.3.31}$$

From Eqs. (12.3.31) and (12.3.22) we have the following condition:

$$\frac{1}{\bar{c}^2 s_{44}} + \frac{1}{a^2 \bar{s}_{11}^*} = \frac{1}{c^2 s_{44}(1 - k_{15}^2)} + \frac{1}{a^2 s_{11}^*}, \tag{12.3.32}$$

that can be satisfied by adjusting the geometrical parameters $\bar{c}$ and c. The mode shapes of $u_1^{(1)}$ and $\phi^{(0,0)}$ are shown in Fig. 12.3.3. The mode can be driven by one pair of electrodes.

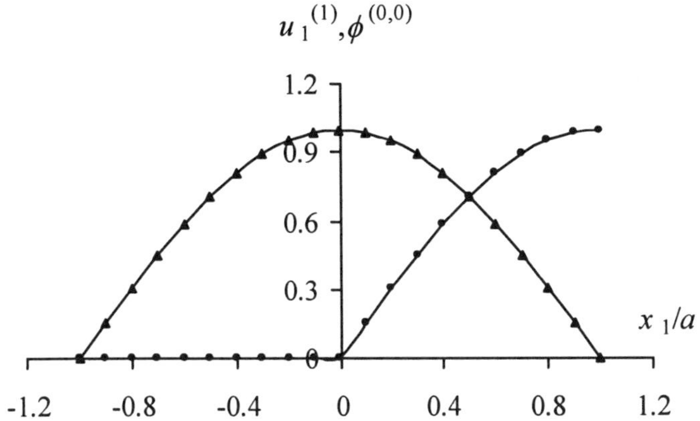

Fig. 12.3.3. Shear and potential distributions of the operating mode ($\bar{c} < c$). $u_1^{(1)}$ is marked by triangles, and $\phi^{(0,0)}$ by circles.

Similarly, it can be verified that the following is also a limit solution to the frequency equation:

$$\bar{k}a = \pi^+, \quad \tan(\bar{k}a) = 0^+,$$
$$ka = \pi^-, \quad \tan(ka) = 0^-, \tag{12.3.33}$$

$$\bar{k}^2a^2 = a^2\bar{s}_{11}^*\rho(\omega^2 - \bar{\omega}_\infty^2) = \pi^2,$$
$$k^2a^2 = a^2 s_{11}^*\rho(\omega^2 - \omega_\infty^2) = \pi^2, \tag{12.3.34}$$

$$\omega^2 = \bar{\omega}_\infty^2 + \frac{\pi^2}{a^2\bar{s}_{11}^*\rho} = \omega_\infty^2 + \frac{\pi^2}{a^2 s_{11}^*\rho}, \tag{12.3.35}$$

$$\frac{1}{4\bar{c}^2 s_{44}} + \frac{1}{a^2\bar{s}_{11}^*} = \frac{1}{4c^2 s_{44}(1-k_{15}^2)} + \frac{1}{a^2 s_{11}^*}. \tag{12.3.36}$$

The mode shapes for the shear deformation and electric potential in this case is shown in Fig. 12.3.4, which can also be driven by one pair of electrodes.

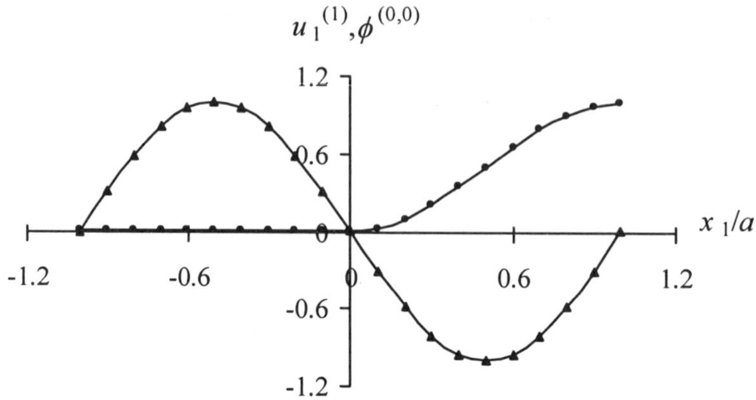

Fig. 12.3.4. Shear and potential distributions of the operating mode ($\bar{c} < c$). $u_1^{(1)}$ is marked by triangles, and $\phi^{(0,0)}$ by circles.

12.4 A Thickness-Shear Mode Transformer — Forced Vibration

For a complete understanding of a piezoelectric transformer, a forced vibration analysis is necessary. The thickness-shear transformer analyzed in the previous section is a thin rod. Its operating modes are slowly varying in the driving and receiving portions when the transformer is long. Slowly varying modes can be approximately described by zero-dimensional equations of piezoelectric parallelepipeds [20]. In this section we perform a forced vibration analysis of the thickness-shear transformer in the previous using zero-dimensional equations [119].

12.4.1 Governing equations

The driving portion is in $-2\bar{l} < X_1 < 0$. The receiving portion is $0 < X_1 < 2l$. V_1 is the input voltage and V_2 is the output voltage. For each portion of the transformer there exists a local Cartesian coordinate system x_i with its origin at the center of the portion and directions along the global system X_i. The thickness-shear motion we are considering can be approximately represented by

$$u_1 \cong x_3 u_1^{(0,0,1)}, \quad u_2 \cong 0, \quad u_3 \cong 0\,. \tag{12.4.1}$$

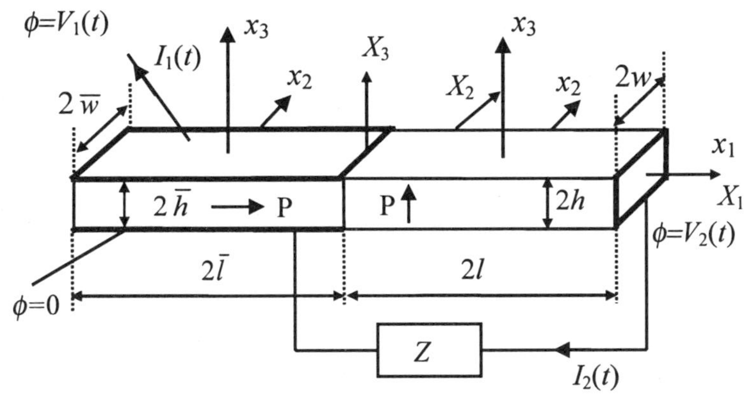

Fig. 12.4.1. A thickness-shear piezoelectric transformer.

In the driving portion the electric potential is the known driving potential with

$$\phi \cong \phi^{(0,0,0)} + x_3\phi^{(0,0,1)}, \quad \phi^{(0,0,0)} = V_1/2, \quad \phi^{(0,0,1)} = V_1/(2\bar{h}). \qquad (12.4.2)$$

The zero-dimensional equation of motion is [20]

$$-T_{31}^{(0,0,0)} + F_1^{(0,0,1)} = \rho\frac{8\bar{l}\bar{w}\bar{h}^3}{3}\ddot{u}_1^{(0,0,1)}, \qquad (12.4.3)$$

where

$$F_1^{(0,0,1)} = \int_{-\bar{w}}^{\bar{w}}\int_{-\bar{h}}^{\bar{h}} T_{11}(X_1 = 0^-)X_3 dX_2 dX_3. \qquad (12.4.4)$$

Relevant zero-dimensional constitutive relations are

$$\begin{aligned} T_{31}^{(0,0,0)} &= 8\bar{l}\bar{w}\bar{h}(c_{44}\kappa^2 S_5^{(0,0,0)} - e_{15}\kappa E_3^{(0,0,0)}), \\ D_3^{(0,0,0)} &= 8\bar{l}\bar{w}\bar{h}(e_{15}\kappa S_5^{(0,0,0)} + \varepsilon_{11}E_3^{(0,0,0)}). \end{aligned} \qquad (12.4.5)$$

In Eq. (12.4.5) we have introduced a thickness-shear correction factor κ. For a ceramic parallelepiped, we use $\kappa^2 = \pi^2/12$. For the charge and current on the electrode at $X_3 = \bar{h}$, we have

$$Q_1 = -D_3^{(0,0,0)}/(2\bar{h}), \quad I_1 = -\dot{Q}_1. \qquad (12.4.6)$$

In the receiving portion, the electric potential can be written as

$$\phi \cong \phi^{(0,0,0)} + x_1\phi^{(1,0,0)}, \quad \phi^{(0,0,0)} = V_2/2, \quad \phi^{(1,0,0)} = V_2/(2l), \qquad (12.4.7)$$

where V_2 is unknown. The equation of motion is

$$-T_{31}^{(0,0,0)} + F_1^{(0,0,1)} = \rho\frac{8lwh^3}{3}\ddot{u}_1^{(0,0,1)}, \tag{12.4.8}$$

where

$$F_1^{(0,0,1)} = \int_{-\bar{w}}^{\bar{w}}\int_{-\bar{h}}^{\bar{h}} -T_{11}(X_1 = 0^+)X_3 dX_2 dX_3. \tag{12.4.9}$$

Relevant constitutive relations are

$$\begin{aligned} T_{31}^{(0,0,0)} &= 8lwh(c_{44}\kappa^2 S_5^{(0,0,0)} - e_{15}\kappa E_1^{(0,0,0)}), \\ D_1^{(0,0,0)} &= 8lwh(e_{15}\kappa S_5^{(0,0,0)} + \varepsilon_{11}E_1^{(0,0,0)}). \end{aligned} \tag{12.4.10}$$

The electric charge on the electrode at $X_1 = 2l$ and the electric current flows out of the electrode are given by

$$Q_2 = -D_1^{(0,0,0)}/(2l), \quad I_2 = -\dot{Q}_2. \tag{12.4.11}$$

In deriving the above we have assumed that there is no body force and made use of the traction free boundary conditions at $X_1 = -2\bar{l}$ and $2l$, $X_2 = \pm w$, and $X_3 = \pm h$.

Substituting Eq. (12.4.10)$_1$ into Eq. (12.4.8), and Eq. (12.4.5)$_1$ into Eq. (12.4.3), adding the resulting equations and making use of the continuity conditions

$$\begin{aligned} &F_1^{(0,0,1)}(X_1 = 0^-) = -F_1^{(0,0,1)}(X_1 = 0^+), \\ &u_1(X_1 = 0^-) = u_1(X_1 = 0^+), \end{aligned} \tag{12.4.12}$$

we obtain

$$\begin{aligned} &-8lwh(c_{44}\kappa^2 S_5^{(0,0,0)} - e_{15}\kappa E_1^{(0,0,0)}) \\ &-8\bar{l}\bar{w}\bar{h}(c_{44}\kappa^2 S_5^{(0,0,0)} - e_{15}\kappa E_3^{(0,0,0)}) \\ &= \frac{8}{3}\rho(lwh^3 + \bar{l}\bar{w}\bar{h}^3)\ddot{u}_1^{(0,0,1)}. \end{aligned} \tag{12.4.13}$$

Under the known time-harmonic driving voltage $V_1 = \bar{V}_1\exp(i\omega t)$, for time-harmonic solutions we write

$$\begin{aligned} &u_1^{(0,0,1)} = \bar{u}\exp(i\omega t), \quad V_2 = \bar{V}_2\exp(i\omega t), \\ &I_1 = \bar{I}_1\exp(i\omega t), \quad I_2 = \bar{I}_2\exp(i\omega t). \end{aligned} \tag{12.4.14}$$

The receiving electrodes are connected by an output circuit which, when the motion is time-harmonic, has an impedance Z_L. We have the following circuit condition:

$$\bar{I}_2 = \bar{V}_2 / Z_L . \tag{12.4.15}$$

With successive substitutions, we obtain the following two equations for $\bar{u}$ and $\bar{V}_2$, driven by $\bar{V}_1$:

$$\begin{aligned} &-8lwh\left(c_{44}\kappa^2\bar{u} + e_{15}\kappa\frac{\bar{V}_2}{2l}\right) - 8\bar{l}\bar{w}\bar{h}\left(c_{44}\kappa^2\bar{u} + e_{15}\kappa\frac{\bar{V}_1}{2\bar{h}}\right) \\ &= -\frac{8}{3}\rho(lwh^3 + \bar{l}\bar{w}\bar{h}^3)\omega^2\bar{u}, \\ &4whi\omega\left(e_{15}\kappa\bar{u} - \varepsilon_{11}\frac{\bar{V}_2}{2l}\right) = \frac{\bar{V}_2}{Z_L}. \end{aligned} \tag{12.4.16}$$

Once $\bar{u}$ and $\bar{V}_2$ are obtained, the currents are given by

$$\begin{aligned} \bar{I}_2 &= 4whi\omega\left(e_{15}\kappa\bar{u} - \varepsilon_{11}\frac{\bar{V}_2}{2l}\right), \\ \bar{I}_1 &= 4\bar{l}\bar{w}i\omega\left(e_{15}\kappa\bar{u} - \varepsilon_{11}\frac{\bar{V}_1}{2\bar{h}}\right). \end{aligned} \tag{12.4.17}$$

12.4.2 Forced vibration analysis

We consider the case of $l = \bar{l}$ and $w = \bar{w}$. Solving Eq. (12.4.16) for $\bar{u}$ and $\bar{V}_2$, we obtain the transforming ratio and normalized input and output currents as

$$\begin{aligned} \frac{\bar{V}_2}{\bar{V}_1} &= \frac{k_{15}^2}{(1+h/\bar{h})(1+Z_2/Z_L)(\omega^2/\omega_0^2-1) - k_{15}^2 h/\bar{h}}\frac{l}{\bar{h}}, \\ \frac{\bar{I}_2}{(\bar{V}_1/Z_2)} &= \frac{k_{15}^2}{(1+h/\bar{h})(1+Z_L/Z_2)(\omega^2/\omega_0^2-1) - k_{15}^2 hZ_L/(\bar{h}Z_2)}\frac{l}{\bar{h}}, \\ -\frac{\bar{I}_1}{(\bar{V}_1/Z_1)} &= 1 - \frac{k_{15}^2(1+Z_2/Z_L)}{(1+h/\bar{h})(1+Z_2/Z_L)(\omega^2/\omega_0^2-1) - k_{15}^2 h/\bar{h}}, \end{aligned} \tag{12.4.18}$$

where

$$k_{15}^2 = \frac{e_{15}^2}{\varepsilon_{11} c_{44}}, \quad \omega_0^2 = \frac{3\kappa^2 c_{44}}{\rho(h^2 - h\bar{h} + \bar{h}^2)},$$
$$Z_1 = \frac{1}{i\omega C_1}, \quad C_1 = \frac{\varepsilon_{11} 4\bar{l}\bar{w}}{2\bar{h}}, \quad Z_2 = \frac{1}{i\omega C_2}, \quad C_2 = \frac{\varepsilon_{11} 4wh}{2l}. \tag{12.4.19}$$

In Eq. (12.4.19), ω_0 is the thickness-shear resonant frequency for shorted receiving electrodes ($Z_L = 0$) as predicted by the zero-dimensional theory, C_1 and C_2 are the static capacitance of the driving and receiving portions, and Z_1 and Z_2 are the impedance of the two portions.

As a numerical example, we consider PZT-5H. Material damping is included by allowing c_{pq} to assume complex values. c_{44} is replaced by $c_{44}(1+iQ^{-1})$, where c_{44} and Q are real and the value of Q is fixed to be 10^2 in the calculation.

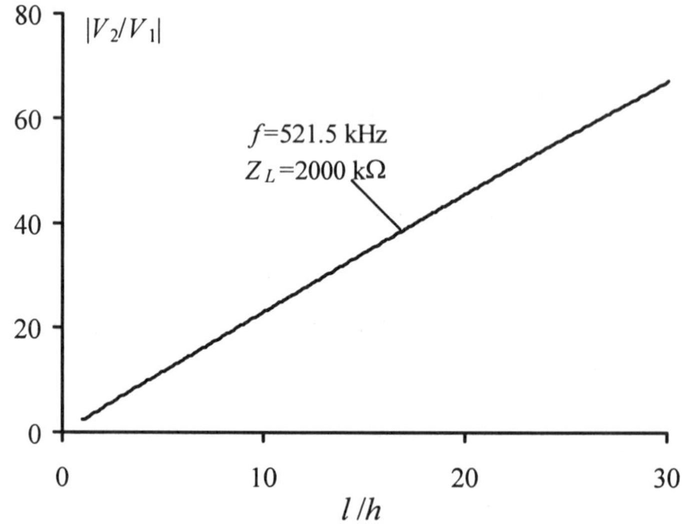

Fig. 12.4.2. Transforming ratio versus aspect ratio ($\bar{h} = h$).

The basic behavior of the transformer is similar to what is shown in Figs. 12.1.2 to 12.1.7. To show that the present transformer is a high-voltage transformer, $|V_2/V_1|$ versus the aspect ratio $l/\bar{h}$ (the length of the receiving portion over the thickness of the driving portion) is plotted in Fig. 12.4.2. An essentially linearly increasing behavior is observed. The figure exhibits the voltage raising capability of the transformer. For large aspect ratios or long and thin transformers, high-voltage output can be achieved. The zero-dimensional equations are particularly suitable for long and thin transformers with almost uniform fields. The dependence of the transforming ration on $l/\bar{h}$ can also be seen from the factor of $l/\bar{h}$ in Eq. $(12.4.18)_1$.

Chapter 13

Power Transmission through an Elastic Wall

In certain applications there are electronic devices sealed in an armor for operations in hazardous environments where physical access to the devices is prohibited. There is a need for periodically recharging the batteries inside the armor that power the devices. Piezoelectric transducers may be used to generate acoustic waves propagating through the armor for transmitting a small amount of power to the devices inside the armor. In this section we examine the performance of such a power transmission technique.

13.1. Formulation of the Problem

Consider the structure illustrated in Fig. 13.1.1, in which a metal plate representing the armor is sandwiched by two piezoelectric layers. These piezoelectric layers model two piezoelectric transducers, one for generating acoustic waves driven by a prescribed electric voltage source and the other for converting the acoustic energy into electric energy to power a load circuit, characterized by its impedance Z_L as indicated in the figure.

Consider thickness-stretch vibrations. For motions of the ceramic layers we have

$$\begin{aligned} &c_{33}u_{3,33} + e_{33}\phi_{,33} = \rho\ddot{u}_3, \\ &e_{33}u_{3,33} - \varepsilon_{33}\phi_{,33} = 0, \\ &T_{33} = c_{33}u_{3,3} - e_{33}E_3, \\ &D_3 = e_{33}u_{3,3} + \varepsilon_{33}E_3, \\ &E_3 = -\phi_{,3}. \end{aligned} \tag{13.1.1}$$

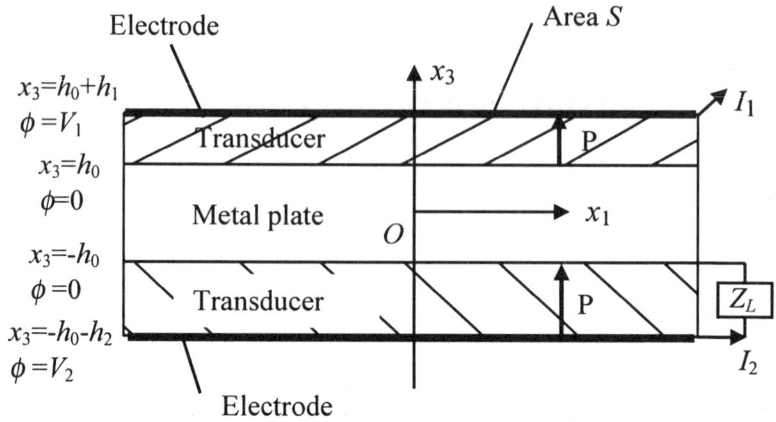

Fig. 13.1.1. An elastic plate sandwiched by two piezoelectric transducers.

The driving transducer is electroded on its outer surface at $x_3 = h_0+h_1$. The electrode is subjected to a time-harmonic driving voltage $V_1(t)$, and the mechanical boundary condition is traction-free. These require that

$$\begin{aligned} T_{3j} &= 0, \quad x_3 = h_0 + h_1, \\ \phi &= V_1(t), \quad x_3 = h_0 + h_1. \end{aligned} \tag{13.1.2}$$

The current density flowing out of the driving electrode at $x_3 = h_0 + h_1$ is given by

$$I_1 = -\dot{Q}_1, \quad Q_1 = -D_3(h_0 + h_1, t), \tag{13.1.3}$$

where Q_1 is the charge per unit area on the electrode.

The receiving transducer is electroded on its outer surface at $x_3 = -(h_0+h_2)$. This electrode has an output voltage $V_2(t)$, and is traction-free. Thus, we have the following boundary conditions:

$$\begin{aligned} T_{3j} &= 0, \quad x_3 = -(h_0 + h_2), \\ \phi &= V_2(t), \quad x_3 = -(h_0 + h_2). \end{aligned} \tag{13.1.4}$$

The electric displacement D_3 in the receiving transducer is spatially constant. We denote it by

$$D_3(x_3, t) = f(t), \tag{13.1.5}$$

where $f(t)$ is an undetermined function of time. The current density on the receiving electrode at $x_3 = -(h_0 + h_2)$ is given by

$$I_2 = -\dot{Q}_2, \quad Q_2 = D_3(-h_0 - h_2, t) = f(t), \tag{13.1.6}$$

where Q_2 is the charge per unit area on the electrode at $x_3 = -(h_0 + h_2)$.

The middle layer occupying $-h_0 \le x_3 \le h_0$ is metallic and conducting and, hence, the electric potential ϕ is spatially constant. We let the middle layer be grounded and, thus, the electric potential ϕ vanishes identically within the middle layer. Therefore, we have the following governing equations for this elastic layer in thickness-stretch vibrations:

$$\begin{aligned} &c_{33}^0 u_{3,33} = \rho_0 \ddot{u}_3, \\ &T_{33} = c_{33}^0 u_{3,3}, \end{aligned} \tag{13.1.7}$$

where c_{33}^0 is the elastic constant of the elastic layer and ρ_0 is its mass density.

On the two interfaces, we require that the displacement u_3, the electric potential ϕ and the mechanical tractions T_{3i} be continuous, i.e.,

$$\begin{aligned} &u_3(h_0^+, t) = u_3(h_0^-, t), \quad u_3(-h_0^+, t) = u_3(-h_0^-, t), \\ &\phi(h_0^+, t) = \phi(-h_0^-, t) = 0, \\ &T_{33}(h_0^+, t) = [c_{33}u_{3,3} + e_{33}\phi_{,3}]\big|_{h_0^+} = [c_{33}^0 u_{3,3}]\big|_{h_0^-} = T_{33}(h_0^-, t), \\ &T_{33}(-h_0^-, t) = [c_{33}u_{3,3} + e_{33}\phi_{,3}]\big|_{-h_0^-} = [c_{33}^0 u_{3,3}]\big|_{-h_0^+} = T_{33}(-h_0^+, t). \end{aligned} \tag{13.1.8}$$

Under a time-harmonic driving voltage, we denote

$$\begin{aligned} &[u_3, \phi, V_1(t), V_2(t), I_1(t), I_2(t), f(t)] \\ &\quad = \mathrm{Re}\{[U_3(x_3), \Phi(x_3), \overline{V}_1, \overline{V}_2, \overline{I}_1, \overline{I}_2, F]\exp(i\omega t)\}. \end{aligned} \tag{13.1.9}$$

Then the continuity conditions in Eq. (13.1.8) takes the following form:

$$\begin{aligned} &U_3(h_0^+) = U_3(h_0^-), \quad U_3(-h_0^+) = U_3(-h_0^-), \\ &\Phi(h_0^+) = \Phi(-h_0^-) = 0, \\ &[c_{33}U_{3,3} + e_{33}\Phi_{,3}]\big|_{h_0^+} = [c_{33}^0 U_{3,3}]\big|_{h_0^-}, \\ &[c_{33}^0 U_{3,3}]\big|_{-h_0^+} = [c_{33}U_{3,3} + e_{33}\Phi_{,3}]\big|_{-h_0^-}. \end{aligned} \tag{13.1.10}$$

Next we integrate Eq. (13.1.1)$_2$ with respect to x_3 twice over $h_0 \le x_3 \le h_0+h_1$ and $-(h_0+h_2) \le x_3 \le -h_0$, respectively, to obtain expressions of the electric potential Φ in terms of U_3, within the driving and receiving transducers, respectively. These expressions are then substituted into Eq. (13.1.1)$_1$ to eliminate the electric potential Φ and to obtain the differential equations for the thickness displacement U_3. We collect, in the following, the equation of motion for the metal plate and the transducers as well as the expressions of the electric potential:

$$
\begin{aligned}
&c_{33}^0 U_{3,33} + \rho_0 \omega^2 U_3 = 0, \quad -h_0 \le x_3 \le h_0, \\
&\bar{c}_{33} U_{3,33} + \rho \omega^2 U_3 = 0, \quad -h_0 - h_2 \le x_3 \le -h_0, \\
&\bar{c}_{33} U_{3,33} + \rho \omega^2 U_3 = 0, \quad h_0 \le x_3 \le h_0 + h_1, \\
&\Phi = \frac{x_3 - h_0}{h_1} \bar{V}_1 + \frac{e_{33}}{\varepsilon_{33}} \{[U_3(x_3) - U_3(h_0^+)] \\
&\quad - \frac{x_3 - h_0}{h_1} [U_3(h_0 + h_1) - U_3(h_0^+)]\}, \quad h_0 \le x_3 \le h_0 + h_1, \\
&\Phi = 0, \quad -h_0 \le x_3 \le h_0, \\
&\Phi = e_{33} \varepsilon_{33}^{-1} [U_3(x_3) - U_3(-h_0^-)] \\
&\quad + \varepsilon_{33}^{-1} (x_3 + h_0) F, \quad -h_0 - h_2 \le x_3 \le -h_0,
\end{aligned}
\tag{13.1.11}
$$

where $\bar{c}_{33} = c_{33} + e_{33}^2 / \varepsilon_{33}$. We now introduce the output current $J_2 = \bar{I}_2 S$, where S denotes the electroded surface area of the receiving transducer. The receiving transducer is connected to a load circuit. Under a time-harmonic excitation, the effect of this load circuit can be characterized by its impedance

$$Z_L = \bar{V}_2 / J_2 . \tag{13.1.12}$$

13.2. Theoretical Analysis

The solution to the boundary value problem in the previous section is found to be

$$
\begin{aligned}
&\bar{V}_2 = \Phi(-h_2 - h_0) = \Gamma_1 \bar{V}_1 - Z_2 \bar{I}_2, \\
&\bar{I}_1 = -\bar{V}_1 / Z_1 + \Gamma_2 \bar{I}_2,
\end{aligned}
\tag{13.2.1}
$$

where

$$\begin{aligned}\Gamma_1 &= -e_{33}\varepsilon_{33}^{-1}\{\chi_{31}[\sin\eta(h_2+h_0)-\sin\eta h_0]\\ &\quad+\chi_{41}[\cos\eta h_0-\cos\eta(h_2+h_0)]\},\\ Z_2 &= \frac{h_2}{i\omega\varepsilon_{33}S}\{1+e_{33}h_2^{-1}\chi_{32}[\sin\eta(h_2+h_0)-\sin\eta h_0]\\ &\quad+e_{33}h_2^{-1}\chi_{42}[\cos\eta h_0-\cos\eta(h_2+h_0)]\},\end{aligned}\tag{13.2.2}$$

$$\begin{aligned}1/Z_1 &= \frac{i\omega\varepsilon_{33}S}{h_1}\{1-e_{33}\varepsilon_{33}^{-1}\chi_{11}[\sin\eta(h_1+h_0)-\sin\eta h_0]\\ &\quad-e_{33}\varepsilon_{33}^{-1}\chi_{21}[\cos\eta(h_1+h_0)-\cos\eta h_0]\},\\ \Gamma_2 &= \frac{e_{33}}{h_1}\{\chi_{12}[\sin\eta(h_1+h_0)-\sin\eta h_0]\\ &\quad+\chi_{22}[\cos\eta(h_1+h_0)-\cos\eta h_0]\},\end{aligned}\tag{13.2.3}$$

$$\begin{aligned}\chi_{11} &= \alpha_{12}\Delta^{-1}h_1^{-1}e_{33}\\ &\quad+\Delta^{-1}\Delta_2^{-1}\vartheta_1[\mu_{22}(\gamma_{11}\alpha_{22}-\gamma_{21}\alpha_{12})-\mu_{21}(\gamma_{12}\alpha_{22}-\gamma_{22}\alpha_{12})],\\ \chi_{12} &= \Delta^{-1}\Delta_2^{-1}\vartheta_2[\mu_{11}(\gamma_{12}\alpha_{22}-\gamma_{22}\alpha_{12})-\mu_{12}(\gamma_{11}\alpha_{22}-\gamma_{21}\alpha_{12})],\\ \chi_{21} &= -\alpha_{11}\Delta^{-1}h_1^{-1}e_{33}\\ &\quad+\Delta^{-1}\Delta_2^{-1}\vartheta_1[\mu_{22}(\gamma_{21}\alpha_{11}-\gamma_{11}\alpha_{21})-\mu_{21}(\gamma_{22}\alpha_{11}-\gamma_{12}\alpha_{21})],\\ \chi_{22} &= \Delta^{-1}\Delta_2^{-1}\vartheta_2[\mu_{11}(\gamma_{22}\alpha_{11}-\gamma_{12}\alpha_{21})-\mu_{12}(\gamma_{21}\alpha_{11}-\gamma_{11}\alpha_{21})],\end{aligned}\tag{13.2.4}$$

$$\begin{aligned}\chi_{31} &= \Delta_1^{-1}\Delta_2^{-1}\vartheta_1[\mu_{22}(\gamma'_{11}\alpha'_{22}-\gamma'_{21}\alpha'_{12})-\mu_{21}(\gamma'_{12}\alpha'_{22}-\gamma'_{22}\alpha'_{12})],\\ \chi_{32} &= \alpha'_{12}\Delta_1^{-1}\varepsilon_{33}^{-1}e_{33}\\ &\quad+\Delta_1^{-1}\Delta_2^{-1}\vartheta_2[\mu_{11}(\gamma'_{12}\alpha'_{22}-\gamma'_{22}\alpha'_{12})-\mu_{12}(\gamma'_{11}\alpha'_{22}-\gamma'_{21}\alpha'_{12})],\\ \chi_{41} &= \Delta_1^{-1}\Delta_2^{-1}\vartheta_1[\mu_{22}(\gamma'_{21}\alpha'_{11}-\gamma'_{11}\alpha'_{21})-\mu_{21}(\gamma'_{22}\alpha'_{11}-\gamma'_{12}\alpha'_{21})],\\ \chi_{42} &= -\alpha'_{11}\Delta_1^{-1}\varepsilon_{33}^{-1}e_{33}\\ &\quad+\Delta_1^{-1}\Delta_2^{-1}\vartheta_2[\mu_{11}(\gamma'_{22}\alpha'_{11}-\gamma'_{12}\alpha'_{21})-\mu_{12}(\gamma'_{21}\alpha'_{11}-\gamma'_{11}\alpha'_{21})],\end{aligned}\tag{13.2.5}$$

$$\begin{aligned}
&\alpha_{11} = \sin\eta h_0, \\
&\alpha_{21} = \bar{c}_{33}\eta\cos\eta h_0 - h_1^{-1}e_{33}^2\varepsilon_{33}^{-1}[\sin\eta(h_0+h_1) - \sin\eta h_0], \\
&\alpha_{12} = \cos\eta h_0, \\
&\alpha_{22} = -\bar{c}_{33}\eta\sin\eta h_0 - h_1^{-1}e_{33}^2\varepsilon_{33}^{-1}[\cos\eta(h_0+h_1) - \cos\eta h_0], \\
&\alpha'_{11} = \sin\eta h_0, \quad \alpha'_{12} = -\cos\eta h_0, \\
&\alpha'_{21} = \bar{c}_{33}\eta\cos\eta h_0, \quad \alpha'_{22} = \bar{c}_{33}\eta\sin\eta h_0,
\end{aligned} \tag{13.2.6}$$

$$\begin{aligned}
&\gamma_{11} = \sin\eta' h_0, \quad \gamma_{12} = \cos\eta' h_0, \\
&\gamma_{21} = c_{33}^0\eta'\cos\eta' h_0, \quad \gamma_{22} = -c_{33}^0\eta'\sin\eta' h_0, \\
&\gamma'_{11} = \sin\eta' h_0, \quad \gamma'_{12} = -\cos\eta' h_0, \\
&\gamma'_{21} = c_{33}^0\eta'\cos\eta' h_0, \quad \gamma'_{22} = c_{33}^0\eta'\sin\eta' h_0,
\end{aligned} \tag{13.2.7}$$

$$\begin{aligned}
&\eta = (\rho\omega^2/\bar{c}_{33})^{\frac{1}{2}}, \; \eta' = (\rho_0\omega^2/c_{33}^0)^{\frac{1}{2}}, \\
&\Delta = \alpha_{11}\alpha_{22} - \alpha_{12}\alpha_{21}, \\
&\Delta_1 = \alpha'_{11}\alpha'_{22} - \alpha'_{12}\alpha'_{21}, \quad \Delta_2 = \mu_{11}\mu_{22} - \mu_{12}\mu_{21},
\end{aligned} \tag{13.2.8}$$

$$\begin{aligned}
&\mu_{11} = \Delta^{-1}[\tau_{11}(\gamma_{11}\alpha_{22} - \gamma_{21}\alpha_{12}) + \tau_{12}(-\gamma_{11}\alpha_{21} + \gamma_{21}\alpha_{11})], \\
&\mu_{12} = \Delta^{-1}[\tau_{11}(\gamma_{12}\alpha_{22} - \gamma_{22}\alpha_{12}) + \tau_{12}(-\gamma_{12}\alpha_{21} + \gamma_{22}\alpha_{11})], \\
&\mu_{21} = \Delta_1^{-1}[\tau_{21}(\gamma'_{11}\alpha'_{22} - \gamma'_{21}\alpha'_{12}) + \tau_{22}(-\gamma'_{11}\alpha'_{21} + \gamma'_{21}\alpha'_{11})], \\
&\mu_{22} = \Delta_1^{-1}[\tau_{21}(\gamma'_{12}\alpha'_{22} - \gamma'_{22}\alpha'_{12}) + \tau_{22}(-\gamma'_{12}\alpha'_{21} + \gamma'_{22}\alpha'_{11})], \\
&\vartheta_1 = -\Delta^{-1}h_1^{-1}e_{33}(\Delta + \tau_{11}\alpha_{12} - \tau_{12}\alpha_{11}), \\
&\vartheta_2 = -\Delta_1^{-1}\varepsilon_{33}^{-1}e_{33}(\Delta_1 + \tau_{21}\alpha'_{12} - \tau_{22}\alpha'_{11}),
\end{aligned} \tag{13.2.9}$$

$$\begin{aligned}
&\tau_{11} = \bar{c}_{33}\eta\cos\eta(h_1+h_0) \\
&\quad - h_1^{-1}e_{33}^2\varepsilon_{33}^{-1}[\sin\eta(h_1+h_0) - \sin\eta h_0], \\
&\tau_{12} = -\bar{c}_{33}\eta\sin\eta(h_1+h_0) \\
&\quad - h_1^{-1}e_{33}^2\varepsilon_{33}^{-1}[\cos\eta(h_1+h_0) - \cos\eta h_0], \\
&\tau_{21} = \bar{c}_{33}\eta\cos\eta(h_2+h_0), \; \tau_{22} = \bar{c}_{33}\eta\sin\eta(h_2+h_0).
\end{aligned} \tag{13.2.10}$$

In Eq. (13.2.1), the coefficients Γ_1 and Γ_2 are dimensionless numbers, while the coefficients Z_1 and Z_2 have the dimension of impedance. Physically, Z_1 may be considered as the input impedance of the system $(-\overline{V}_1/\overline{I}_1)$ when the output circuit is open (I_2 = 0). Since the driving voltage $\overline{V}_1$ is prescribed for the system, solving Eq. (13.2.1) and the load circuit equation, we obtain the normalized output voltage, the output current, and the input admittance as follows:

$$\frac{\overline{V}_2}{\overline{V}_1}=\frac{\Gamma_1 Z_L}{Z_L+Z_2},\quad \overline{I}_2=\frac{\Gamma_1}{Z_L+Z_2}\overline{V}_1,\quad -\frac{\overline{I}_1}{\overline{V}_1}=\frac{1}{Z_1}-\frac{\Gamma_1\Gamma_2}{Z_L+Z_2}. \tag{13.2.11}$$

The input and output powers are given by

$$P_1=\frac{1}{4}(\overline{I}_1\overline{V}_1^*+\overline{I}_1^*\overline{V}_1),\quad P_2=\frac{1}{4}(\overline{I}_2\overline{V}_2^*+\overline{I}_2^*\overline{V}_2). \tag{13.2.12}$$

Then the efficiency of the power transmission is

$$\nu=P_2/P_1$$
$$=\frac{Z_1Z_1^*\Gamma_1\Gamma_1^*(Z_L+Z_L^*)}{Z_1Z_1^*[\Gamma_1^*\Gamma_2^*(Z_L+Z_2)+\Gamma_1\Gamma_2(Z_L^*+Z_2^*)]-(Z_L+Z_2)(Z_L^*+Z_2^*)(Z_1+Z_1^*)}. \tag{13.2.13}$$

In the case that the load is a pure resistor, the load impedance Z_L is real and the system efficiency ν has the following simple expression:

$$\nu=\frac{\xi_1 Z_L}{1+\xi_2 Z_L+\xi_3 Z_L^2}, \tag{13.2.14}$$

where ξ_n, n = 1, 2, 3, are real and are dependent on the coefficients Γ_1, Γ_2, Z_1 and Z_2.

13.3. Numerical Results

For a numerical example, considering PZT-5H for the piezoelectric transducers. For the elastic plate, consider steel with ρ_0 = 7850 kg/m^3 and c_{33}^0 = 2.69×10^{11} N/m^2. c_{33} and c_{33}^0 are replaced by $c_{33}(1+iQ^{-1})$ and $c_{33}^0(1+iQ^{-1})$, where c_{33}, c_{33}^0 and Q are real numbers. Q = 10^2. h_0 = 3 mm, h_1 = 1 mm, h_2 = 2 mm, and S = 0.01 m^2. The numerical results are normalized by

$$Z_0=\frac{h_2}{i\omega\varepsilon_{33}S}. \tag{13.3.1}$$

Figure 13.3.1 shows the normalized output voltage $|\overline{V}_2/\overline{V}_1|$ versus the driving frequency ω. The output voltage peaks sharply at the thickness-stretch resonant frequencies. This indicates that the system may be effective near its resonant frequencies. The highest peak in this illustrative example occurs at the fourth resonance. The peak value increases monotonically with the first several modes and then decreases. In the lowest mode, the output transducer is not severely stressed because of the traction-free boundary condition at the outer surface. Correspondingly, the electric field within the output transducer generated through piezoelectric coupling is not particularly strong. This leads to a moderate output voltage. As the mode number increases, the stress and the corresponding electric field in the output transducer become increasingly stronger, leading to an increasing output voltage. This trend, however, reverses when the mode becomes so high that there are nodal points present within the output transducer, across which the stress reverses its sign and so does the electric field. This causes voltage cancellation across the nodal points. Therefore, the voltage within the output transducer no longer increases monotonically. This leads to a reduced output voltage.

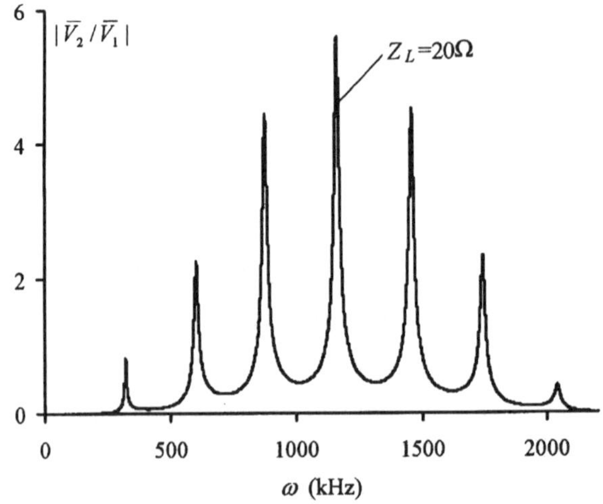

Fig. 13.3.1. Output voltage versus the driving frequency.

The dependence of the output voltage $|\overline{V}_2/\overline{V}_1|$ on the normalized load $|Z_L/Z_0|$ is shown in Fig. 13.3.2, in which the two curves correspond to two driving frequencies near the second resonant frequency. As indicated by Eq. (13.2.11)$_1$, these curves are essentially linear for small loads and the output voltages approach different (saturation) constants for large loads.

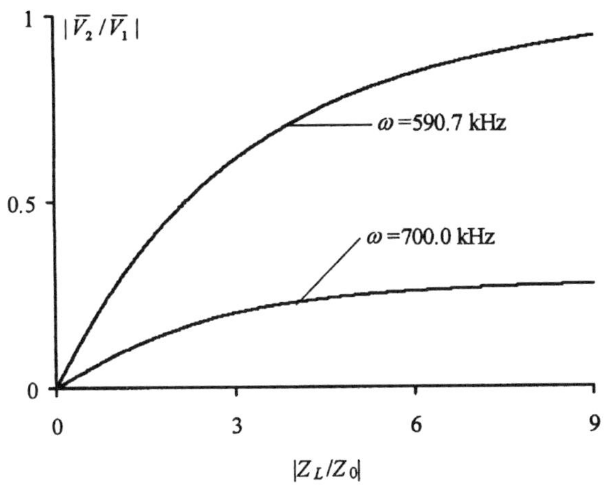

Fig. 13.3.2. Output voltage versus the load impedance.

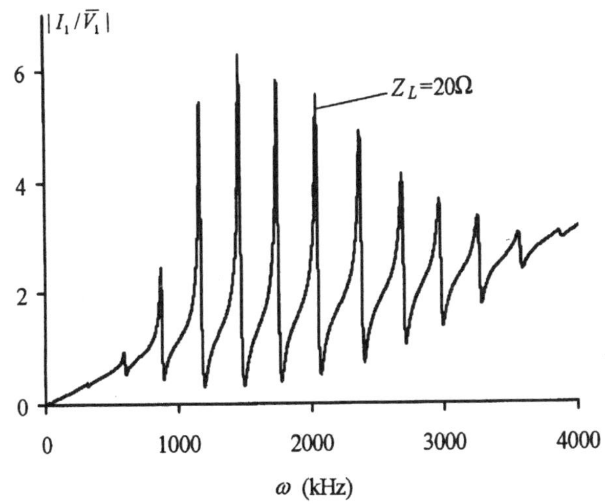

Fig. 13.3.3. Input admittance versus the driving frequency.

Figure 13.3.3 shows the input admittance $|I_1/\overline{V}_1|$ versus the driving frequency ω. The input admittance peaks at the resonant frequencies.

The input admittance versus the normalized impedance of the load circuit for two driving frequencies near the second resonant frequency is shown in Fig. 13.3.4.

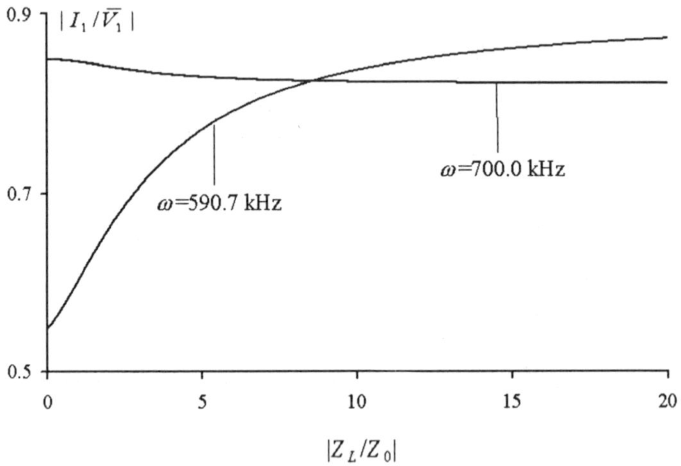

Fig. 13.3.4. Input admittance versus the load impedance.

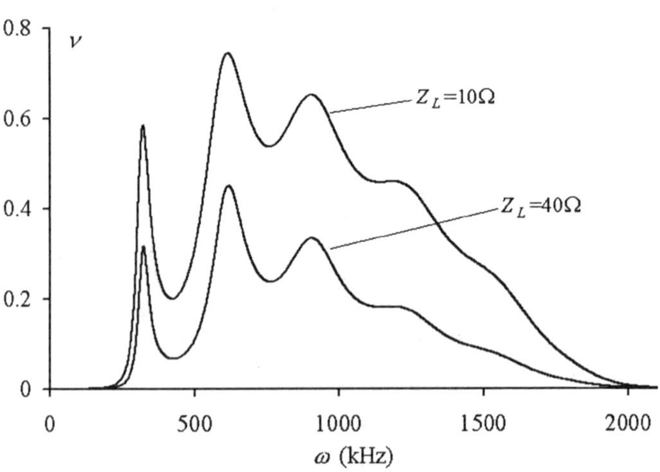

Fig. 13.3.5. Efficiency versus the driving frequency.

Figure 13.3.5 shows the efficiency ν versus the driving frequency ω. The frequency dependence of the efficiency is relatively complicated.

Figure 13.3.6 shows the efficiency ν versus the normalized impedance $|Z_L/Z_0|$. As seen from Eq. (13.2.14), ν increases with the load impedance for small loads. For large loads it approaches zero.

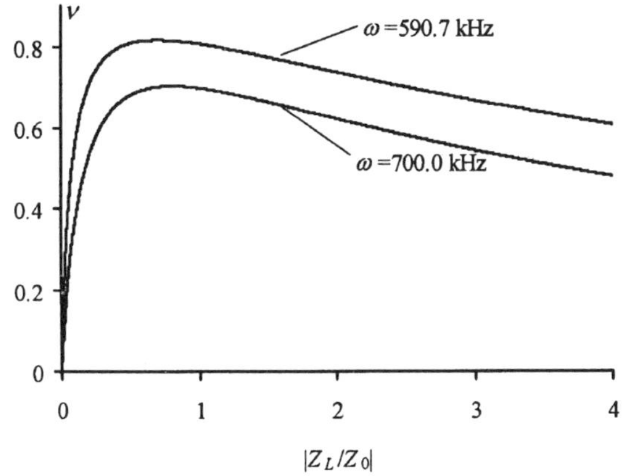

Fig. 13.3.6. Efficiency versus the load impedance.

Chapter 14
Acoustic Wave Amplifiers

Piezoelectric materials are either dielectrics or semiconductors. Mechanical fields and mobile charges in piezoelectric semiconductors can interact, and this is called the acoustoelectric effect. An acoustic wave traveling in a piezoelectric semiconductor can be amplified by the application of a DC electric field. The acoustoelectric effect and the acoustoelectric amplification of acoustic waves have led to piezoelectric semiconductor devices. The basic behavior of piezoelectric semiconductors can be described by a simple extension of the theory of piezoelectricity [121]. Structural theories for plates and shells were derived in [122–124]. More references can be found in a review article [125].

14.1. Equations for Piezoelectric Semiconductors

Consider a homogeneous, one-carrier piezoelectric semiconductor under a uniform DC electric field $\overline{E}_j$. The steady state current is $\overline{J}_i = q\overline{n}\mu_{ij}\overline{E}_j$, where q is the carrier charge which may be the electronic charge or its opposite, $\overline{n}$ is the steady state carrier density which produces electrical neutrality, and μ_{ij} is the carrier mobility. When an acoustic wave $\mathbf{u}(\mathbf{x},t)$ propagates through the material, perturbations of the electric field, the carrier density and the current are denoted by E_j, n and J_i. The linear theory for small signals consists of the equations of motion, Gauss's law, and conservation of charge [121]

$$\begin{aligned} T_{ji,j} &= \rho\ddot{u}_i, \\ D_{i,i} &= qn, \\ q\dot{n} + J_{i,i} &= 0. \end{aligned} \tag{14.1.1}$$

The above equations are accompanied by the following linearized constitutive relations:

$$\begin{aligned} T_{ij} &= c_{ijkl}S_{kl} - e_{kij}E_k, \\ D_i &= e_{ijk}S_{jk} + \varepsilon_{ij}E_j, \\ J_i &= q\bar{n}\mu_{ij}E_j + qn\mu_{ij}\bar{E}_j - qd_{ij}n_{,j}, \end{aligned} \tag{14.1.2}$$

where d_{ij} are the carrier diffusion constants. Equations (14.1.1) can be written as five equations for **u**, ϕ and n

$$\begin{aligned} &c_{ijkl}u_{k,lj} + e_{kij}\phi_{,kj} + f_i = \rho\ddot{u}_i, \\ &e_{ikl}u_{k,li} - \varepsilon_{ij}\phi_{,ij} = qn, \\ &\dot{n} - \bar{n}\mu_{ij}\phi_{,ij} + \mu_{ij}\bar{E}_j n_{,i} - d_{ij}n_{,ij} = 0. \end{aligned} \tag{14.1.3}$$

On the boundary of a finite body with a unit outward normal n_i, the mechanical displacement u_i or the traction vector $T_{ij}n_i$, the electric potential ϕ or the normal component of the electric displacement vector D_in_i, and the carrier density n or the normal current J_in_i may be prescribed.

The acoustoelectric effect and amplification of acoustic waves can also be achieved through composite structures of piezoelectric dielectrics and nonpiezoelectric semiconductors. In these composites the acoustoelectric effect is due to the combination of the piezoelectric effect and semiconduction in each component phase.

14.2. Equations for a Thin Film

For later use we derive two-dimensional equations for the extensional motion of a thin piezoelectric semiconductor film. Consider such a film of thickness $2h$ as shown in Fig. 14.2.1, along with the coordinate system.

The film is assumed to be very thin in the sense that its thickness is much smaller than the wavelength of the waves we are interested in. For thin films the following stress components can be approximately taken to vanish:

$$T_{2j} = 0, \quad j = 1,2,3. \tag{14.2.1}$$

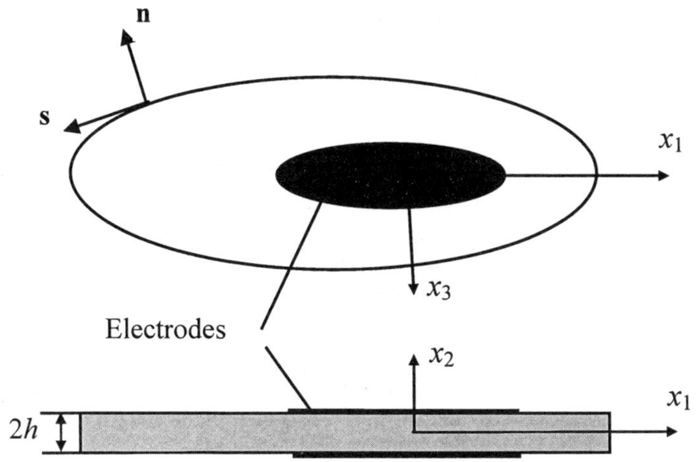

Fig. 14.2.1. Plan view and cross section of a thin film of a piezoelectric semiconductor.

According to the compact matrix notation, with the range of p, q as 1, 2, … and 6, Eq. (14.2.1) can be written as

$$T_q = 0, \quad q = 2,4,6\,. \tag{14.2.2}$$

For convenience we introduce a convention that subscripts u, v, w take the values 2, 4, 6 while subscripts r, s, t take the remaining values 1, 3, 5. Then Eq. $(14.1.2)_{1,2}$ can be written as

$$\begin{aligned} T_r &= c_{rs}S_s + c_{ru}S_u - e_{kr}E_k, \\ T_v &= c_{vs}S_s + c_{vw}S_w - e_{kv}E_k = 0, \\ D_i &= e_{is}S_s + e_{iu}S_u + \varepsilon_{ij}E_j, \end{aligned} \tag{14.2.3}$$

where Eq. (14.2.2) has been used. From Eq. $(14.2.3)_2$ we have

$$S_u = -c_{uv}^{-1}c_{vs}S_s + c_{uv}^{-1}e_{kv}E_k\,. \tag{14.2.4}$$

Substitution of Eq. (14.2.4) into Eq. $(14.2.3)_{1,3}$ gives the constitutive relations for the film

$$\begin{aligned} T_r &= c_{rs}^p S_s - e_{kr}^p E_k, \\ D_i &= e_{is}^p S_s + \varepsilon_{ij}^p E_j, \end{aligned} \tag{14.2.5}$$

where the film material constants are

$$c_{rs}^{p} = c_{rs} - c_{rv}c_{vw}^{-1}c_{ws}, \quad e_{ks}^{p} = e_{ks} - e_{kw}c_{wv}^{-1}c_{vs},$$
$$\varepsilon_{kj}^{p} = \varepsilon_{kj} + e_{kv}c_{vw}^{-1}e_{jw}. \tag{14.2.6}$$

We now introduce another convention that subscripts a, b, c and d assume 1 and 3 but not 2. Then Eq. (14.2.5) can be written as

$$T_{ab} = c_{abcd}^{p}S_{cd} - e_{kab}^{p}E_{k},$$
$$D_{i} = e_{iab}^{b}S_{ab} + \varepsilon_{ij}^{p}E_{j}. \tag{14.2.7}$$

Integrating the equations in $(14.1.1)_1$ for i = 1, 3 and $(14.1.1)_{2,3}$ with respect to x_2 through the film thickness, we obtain the following two-dimensional equations of motion, Gauss's law and conservation of charge:

$$T_{ab,a} + \frac{1}{2h}[T_{2b}(x_2 = h) - T_{2b}(x_2 = -h)] = \rho\ddot{u}_b,$$
$$D_{a,a} + \frac{1}{2h}[D_2(x_2 = h) - D_2(x_2 = -h)] = qn, \tag{14.2.8}$$
$$q\dot{n} + J_{a,a} + \frac{1}{2h}[J_2(x_2 = h) - J_2(x_2 = -h)] = 0,$$

where u_a, T_{ab}, D_a, J_a and n are averages of the corresponding three-dimensional quantities along the film thickness.

14.3. Surface Waves

Consider the propagation of anti-plane surface waves in a piezoelectric dielectric half-space carrying a thin, non-piezoelectric semiconductor film of silicon (see Fig. 14.3.1) [126].

14.3.1. Analytical solution

For the ceramic half-space, anti-plane motions are governed by [27]

$$\bar{c}_{44}\nabla^2 u_3 = \rho\ddot{u}_3,$$
$$\nabla^2\psi = 0, \tag{14.3.1}$$
$$\psi = \phi - \frac{e_{15}}{\varepsilon_{11}}u_3,$$

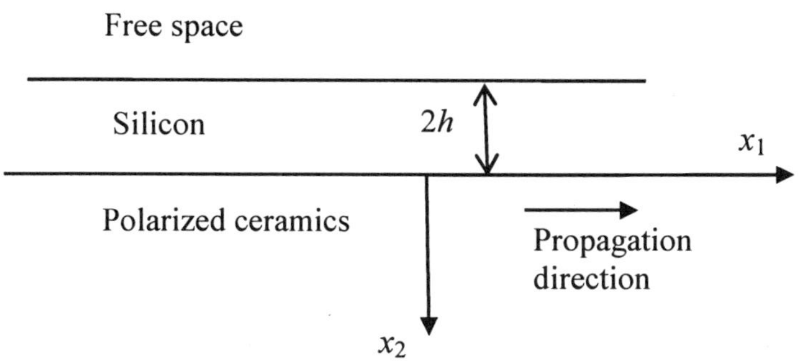

Fig. 14.3.1. A ceramic half-space with a semiconductor film (silicon).

and

$$\begin{aligned} T_{23} &= \bar{c}_{44}u_{3,2} + e_{15}\psi_{,2}, \\ T_{31} &= \bar{c}_{44}u_{3,1} + e_{15}\psi_{,1}, \\ D_1 &= -\varepsilon_{11}\psi_{,1}, \\ D_2 &= -\varepsilon_{11}\psi_{,2}, \end{aligned} \tag{14.3.2}$$

where

$$\bar{c}_{44} = c_{44} + \frac{e_{15}^2}{\varepsilon_{11}} = c_{44}(1 + k_{15}^2), \quad k_{15}^2 = \frac{e_{15}^2}{\varepsilon_{11}c_{44}}. \tag{14.3.3}$$

For a surface wave solution we must have

$$u_3, \phi \to 0, \quad x_2 \to +\infty . \tag{14.3.4}$$

Consider the possibility of solutions in the following form:

$$\begin{aligned} u_3 &= A\exp(-\xi_2 x_2)\exp[i(\xi_1 x_1 - \omega t)], \\ \psi &= B\exp(-\xi_1 x)\exp[i(\xi_1 x_1 - \omega t)], \end{aligned} \tag{14.3.5}$$

where A and B are undetermined constants, and ξ_2 should be positive for decaying behavior away from the surface. Equation $(14.3.5)_2$ already satisfies Eqs. $(14.3.1)_2$. For Eq. $(14.3.5)_1$ to satisfy Eq. $(14.3.1)_1$ we must have

$$\bar{c}_{44}(\xi_1^2 - \xi_2^2) = \rho\omega^2 , \tag{14.3.6}$$

which leads to the following expression for ξ_2:

$$\xi_2^2 = \xi_1^2 - \frac{\rho\omega^2}{\bar{c}_{44}} = \xi_1^2\left(1 - \frac{v^2}{v_T^2}\right) > 0\,, \tag{14.3.7}$$

where

$$v^2 = \frac{\omega^2}{\xi_1^2}, \quad v_T^2 = \frac{\bar{c}_{44}}{\rho}. \tag{14.3.8}$$

The following are needed for prescribing boundary and continuity conditions:

$$\begin{aligned} \phi &= \left[B\exp(-\xi_1 x_2) + \frac{e_{15}}{\varepsilon_{11}} A\exp(-\xi_2 x_2)\right]\exp[i(\xi_1 x_1 - \omega t)], \\ T_{23} &= -[A\bar{c}_{44}\xi_2\exp(-\xi_2 x_2) \\ &\quad + e_{15}B\xi_1\exp(-\xi_1 x_2)]\exp[i(\xi_1 x_1 - \omega t)], \\ D_2 &= \varepsilon_{11}B\xi_1\exp(-\xi_1 x_2)\exp[i(\xi_1 x_1 - \omega t)]. \end{aligned} \tag{14.3.9}$$

Electric fields can also exist in the free space of $x_2 < 0$, which is governed by

$$\begin{aligned} &\nabla^2\phi = 0, \quad x_2 < 0, \\ &\phi \to 0, \quad x_2 \to -\infty. \end{aligned} \tag{14.3.10}$$

A surface wave solution to Eq. (14.3.10) is

$$\phi = C\exp(\xi_1 x_2)\exp[i(\xi_1 x_1 - \omega t)]\,, \tag{14.3.11}$$

where C is an undetermined constant. From Eq. (14.3.11), in the free space,

$$D_2 = -\varepsilon_0\xi_1 C\exp(\xi_1 x_2)\exp[i(\xi_1 x_1 - \omega t)]\,. \tag{14.3.12}$$

The semiconductor film is one-dimensional with $n = n(x_1, t)$. Consider the case when the DC biasing electric field is in the x_1 direction. Let

$$\begin{aligned} &u_3 = A\exp[i(\xi_1 x_1 - \omega t)], \quad \phi = C\exp[i(\xi_1 x_1 - \omega t)], \\ &n = N\exp[i(\xi_1 x_1 - \omega t)], \end{aligned} \tag{14.3.13}$$

where N is an undetermined constant. Equation (14.3.13) already satisfies the continuity of displacement between the film and the ceramic half-space, and the continuity of electric potential between the film and the free space. We use a prime to indicate the elastic and dielectric constants as well as the mass density of the film. Silicon is a cubic crystal with m3m symmetry and does not have piezoelectric coupling. The elastic and dielectric constants are given by

$$\begin{pmatrix} c'_{11} & c'_{12} & c'_{12} & 0 & 0 & 0 \\ c'_{12} & c'_{11} & c_{12} & 0 & 0 & 0 \\ c'_{12} & c'_{12} & c'_{11} & 0 & 0 & 0 \\ 0 & 0 & 0 & c'_{44} & 0 & 0 \\ 0 & 0 & 0 & 0 & c'_{44} & 0 \\ 0 & 0 & 0 & 0 & 0 & c'_{44} \end{pmatrix}, \begin{pmatrix} \varepsilon'_{11} & 0 & 0 \\ 0 & \varepsilon'_{11} & 0 \\ 0 & 0 & \varepsilon'_{11} \end{pmatrix}. \tag{14.3.14}$$

From Eq. (14.2.7) and $(14.1.2)_3$ we obtain:

$$\begin{aligned} T_{13} &= c^p_{55} S_{13} = c^p_{44} u_{3,1} = c^p_{44} i\xi_1 A \exp[i(\xi_1 x_1 - \omega t)], \\ D_1 &= \varepsilon^p_{11} E_1 = -\varepsilon^p_{11} \phi_{,1} = -\varepsilon^p_{11} i\xi_1 C \exp[i(\xi_1 x_1 - \omega t)], \\ J_1 &= -q\bar{n}\mu_{11}\phi_{,1} + qn\mu_{11}\bar{E}_1 - qd_{11}n_{,1} \\ &= (-q\bar{n}\mu_{11} i\xi_1 C + qN\mu_{11}\bar{E}_1 - qd_{11} i\xi_1 N)\exp[i(\xi_1 x_1 - \omega t)]. \end{aligned} \tag{14.3.15}$$

Substitution of Eqs. (14.3.9), (14.3.11), (14.3.12), (14.3.13) and (14.3.15) into the continuity condition of the electric potential between the ceramic half-space and the film, Eq. $(14.2.8)_1$ for $b = 3$, and Eq. $(14.2.8)_{2,3}$ yields

$$\begin{aligned} & B + \frac{e_{15}}{\varepsilon_{11}} A = C, \\ & -c^p_{44}\xi_1^2 A - \frac{1}{2h}(A\bar{c}_{44}\xi_2 + e_{15}B\xi_1) = -\rho'\omega^2 A, \\ & \varepsilon^p_{11}\xi_1^2 C + \frac{1}{2h}(\varepsilon_{11}B\xi_1 + \varepsilon_0\xi_1 C) = qN, \\ & -qi\omega N + i\xi_1(-q\bar{n}\mu_{11} i\xi_1 C + qN\mu_{11}\bar{E}_1 - qd_{11} i\xi_1 N) = 0, \end{aligned} \tag{14.3.16}$$

which is a system of linear, homogeneous equations for A, B, C and N. The equations of the thin film appear as interface continuity conditions between the ceramic half-space and the free space. For non-trivial solutions the determinant of the coefficient matrix has to vanish:

$$\begin{vmatrix} \dfrac{e_{15}}{\varepsilon_{11}} & 1 & -1 & 0 \\ \rho'\omega^2 - c_{44}^p\xi_1^2 - \dfrac{\bar{c}_{44}\xi_2}{2h} & -\dfrac{e_{15}\xi_1}{2h} & 0 & 0 \\ 0 & \dfrac{\varepsilon_{11}\xi_1}{2h} & \dfrac{\varepsilon_0\xi_1}{2h} + \varepsilon_{11}^p\xi_1^2 & -q \\ 0 & 0 & q\bar{n}\mu_{11}\xi_1^2 & \begin{matrix} -qi\omega + i\xi_1 q\mu_{11}\bar{E}_1 \\ + qd_{11}\xi_1^2 \end{matrix} \end{vmatrix} = 0, \tag{14.3.17}$$

which determines the dispersion relation, a relation between ω and ξ_1, of the surface wave. In terms of the surface wave speed $v = \omega / \xi_1$, Eq. (14.3.17) can be written in the following form:

$$\left(\frac{v^2}{v_T'^2} - 1\right)\frac{c_{44}^p}{\bar{c}_{44}}2h\xi_1 - \sqrt{1 - \frac{v^2}{v_T^2}} + \bar{k}_{15}^2 = \frac{\bar{k}_{15}^2}{1 + \dfrac{\varepsilon_0}{\varepsilon_{11}} + \dfrac{\varepsilon_{11}^p}{\varepsilon_{11}}\xi_1 2h + \dfrac{q\bar{n}\mu_{11}2h}{\varepsilon_{11}[d_{11}\xi_1 + i(\mu_{11}\bar{E}_1 - v)]}}, \tag{14.3.18}$$

where

$$v_T'^2 = \frac{c_{44}^p}{\rho'}, \quad \bar{k}_{15}^2 = \frac{e_{15}^2}{\varepsilon_{11}\bar{c}_{44}}. \tag{14.3.19}$$

14.3.2. Discussion

When h = 0, i.e. the semiconductor film does not exist, Eq. (14.3.18) reduces to

$$v^2 = v_T^2\left[1 - \frac{\bar{k}_{15}^4}{(1 + \varepsilon_{11}/\varepsilon_0)^2}\right] = v_{B-G}^2, \tag{14.3.20}$$

which is the speed of the Bleustein-Gulyaev surface wave.

When $\bar{k}_{15}^2 = 0$, i.e. the half-space is non-piezoelectric, electromechanical coupling disappears and the wave is purely elastic. In this case Eq. (14.3.18) reduces to

$$\left(\frac{v^2}{v_T'^2}-1\right)\frac{c_{44}^p}{\bar{c}_{44}}2h\xi_1-\sqrt{1-\frac{v^2}{v_T^2}}=0\,, \tag{14.3.21}$$

which is the equation that determines the speed of the Love wave (an anti-plane surface wave in an elastic half-space carrying an elastic layer) in the limit when the film is very thin compared to the wavelength ($\xi_1 h \ll 1$). Love waves are known to exist when the elastic stiffness of the layer is smaller that that of the half-space.

The denominator of the right hand side of Eq. (14.3.18) indicates that a complex wave speed may be expected and the imaginary part of the complex wave speed may change its sign (transition from a damped wave to a growing wave) when $\mu_{11}\bar{E}_1\xi_1-\omega$ changes sign or

$$v=\frac{\omega}{\xi_1}=\mu_{11}\bar{E}_1\,, \tag{14.3.22}$$

i.e. the acoustic wave speed is equal to the carrier drift speed under the biasing electric field.

When the semiconduction is small, Eq. (14.3.18) can be solved by an iteration or perturbation procedure. As the lowest (zero) order of approximation, we neglect the small semiconduction and denote the zero-order solution by $v_{(0)}$. Then, from Eq. (14.3.18),

$$\begin{aligned}&\left(\frac{v_{(0)}^2}{v_T'^2}-1\right)\frac{c_{44}^p}{\bar{c}_{44}}2h\xi_1-\sqrt{1-\frac{v_{(0)}^2}{v_T^2}}+\bar{k}_{15}^2\\&=\frac{\bar{k}_{15}^2}{1+\dfrac{\varepsilon_0}{\varepsilon_{11}}+\dfrac{\varepsilon_{11}^p}{\varepsilon_{11}}\xi_1\,2h}\,,\end{aligned} \tag{14.3.23}$$

which is dispersive. For the next order, we substitute $v_{(0)}$ into the right-hand side of Eq. (14.3.18) and obtain the following equation for $v_{(1)}$

$$\begin{aligned}&\left(\frac{v_{(1)}^2}{v_T'^2}-1\right)\frac{c_{44}^p}{\bar{c}_{44}}2h\xi_1-\sqrt{1-\frac{v_{(1)}^2}{v_T^2}}+\bar{k}_{15}^2\\&=\frac{\bar{k}_{15}^2}{1+\dfrac{\varepsilon_0}{\varepsilon_{11}}+\dfrac{\varepsilon_{11}^p}{\varepsilon_{11}}\xi_1\,2h+\dfrac{q\bar{n}\mu_{11}2h}{\varepsilon_{11}[d_{11}\xi_1+i(\mu_{11}\bar{E}_1-v_{(0)})]}}\,,\end{aligned} \tag{14.3.24}$$

which suggests a wave that is both dispersive and dissipative.

14.3.3. Numerical results

For numerical results consider PZT-5H. Since $c'_{44} > c_{44}$, the counterpart of the elastic Love wave does not exist, but a modified Bleustein-Gulyaev wave is expected. We plot the real parts of $v_{(0)}$ and $v_{(1)}$ versus ξ_1 in Fig. 14.3.2. The dimensionless wave number X and the dimensionless wave speed Y of different orders are defined by

$$X = \xi_1 \Big/ \frac{\pi}{2h},$$
$$Y_{(0)} = v_{(0)} / v_{B-G}, \quad Y_{(1)} = \mathrm{Re}\{v_{(1)}\} / v_{B-G}. \tag{14.3.25}$$

γ is a dimensionless number given by

$$\gamma = \mu_{11} \overline{E}_1 / v_{B-G}, \tag{14.3.26}$$

which may be considered as a normalized electric field. It represents the ratio of the carrier drift velocity and the speed of the Bleustein-Gulyave

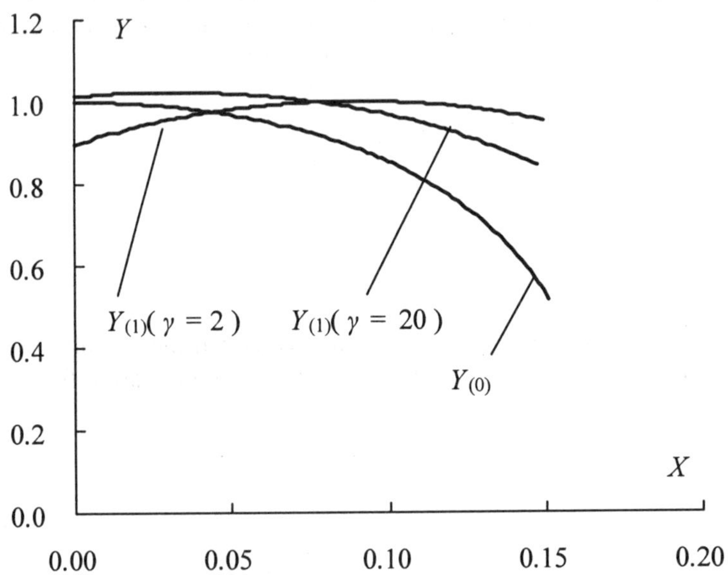

Fig. 14.3.2. Dispersion relations.

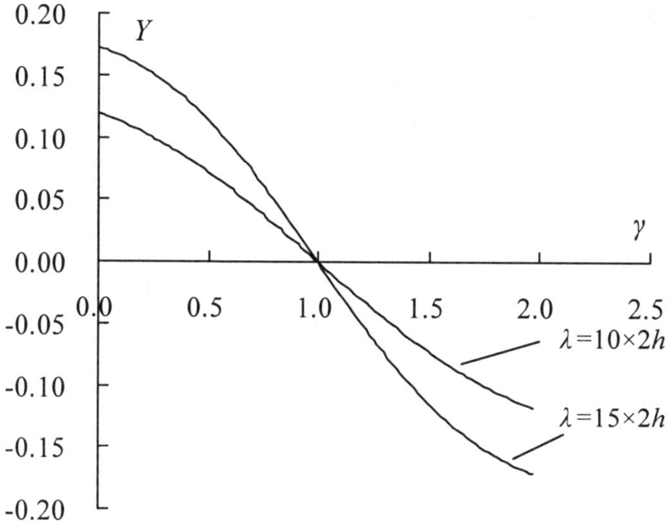

Fig. 14.3.3. Dissipation as a function of the DC bias.

wave. Because of the use of thin film equations for the semiconductor film, the solution is valid only when the wavelength is much larger than the film thickness ($X << 1$). The figure shows that semiconduction causes additional dispersion. This conduction induced dispersion varies according to the DC biasing electric field.

Figure 14.3.3 shows the imaginary part of $v_{(1)}$ versus γ. The dimensionless number describing the decaying behavior of the waves is defined by

$$Y = \mathrm{Im}\{v_{(1)}\}/v_{B-G}\ . \tag{14.3.27}$$

When the DC bias is large enough (approximately $\gamma > 1$) the decay constant becomes negative, indicating wave amplification. The transition from damped waves to growing waves occurs when Eq. (14.3.22) is true for $v_{(0)}$.

14.4. Interface Waves

Next consider anti-plane waves propagating between two piezoelectric half-spaces of polarized ceramics with a semiconductor film of silicon between them (see Fig. 14.4.1) [127].

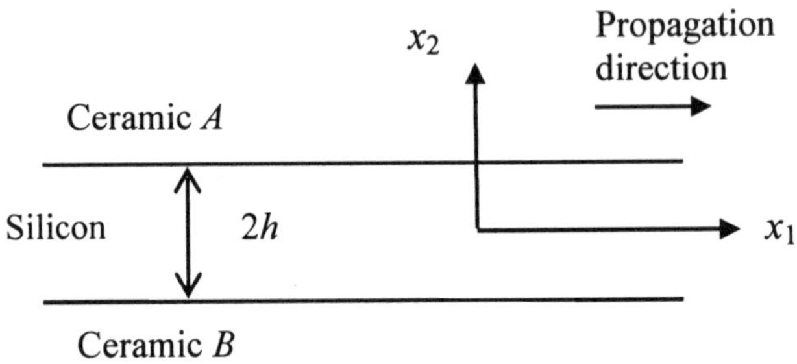

Fig. 14.4.1. Two ceramic half-spaces with a semiconductor film.

14.4.1. Analytical solution

For the upper half-space, from the relevant equations in the previous section, the solution can be written as

$$\begin{aligned} u_A &= U\exp(-\eta_A x_2)\exp i(\xi x_1 - \omega t), \\ \psi_A &= \Psi_A \exp(-\xi x_2)\exp i(\xi x_1 - \omega t), \end{aligned} \tag{14.4.1}$$

where U and Ψ_A are undetermined constants,

$$\eta_A^2 = \xi^2 - \frac{\rho_A \omega^2}{\bar{c}_A} = \xi^2 \left(1 - \frac{v^2}{v_A^2}\right) > 0, \tag{14.4.2}$$

and

$$v = \frac{\omega}{\xi}, \quad v_A^2 = \frac{\bar{c}_A}{\rho_A}. \tag{14.4.3}$$

For continuity conditions, we need ϕ_A, T_{23} and D_2 in ceramic A. We denote T_{23} and D_2 by T_A and D_A:

$$\begin{aligned}
\phi_A &= \psi_A + \frac{e_A}{\varepsilon_A} u_A \\
&= \left[\Psi_A \exp(-\xi x_2) + \frac{e_A}{\varepsilon_A} U \exp(-\eta_A x_2) \right] \exp i(\xi x_1 - \omega t), \\
T_A &= \bar{c}_A u_{A,2} + e_A \psi_{A,2} \\
&= -[\bar{c}_A \eta_A U \exp(-\eta_A x_2) + e_A \xi \Psi_A \exp(-\xi x_2)] \exp i(\xi x_1 - \omega t), \\
D_A &= -\varepsilon_A \psi_{A,2} \\
&= \varepsilon_A \xi \Psi_A \exp(-\xi x_2) \exp i(\xi x_1 - \omega t).
\end{aligned} \tag{14.4.4}$$

Similarly, for the lower half-space, the solutions can be written as

$$\begin{aligned}
u_B &= U \exp(\eta_B x_2) \exp i(\xi x_1 - \omega t), \\
\psi_B &= \Psi_B \exp(\xi x_2) \exp i(\xi x_1 - \omega t),
\end{aligned} \tag{14.4.5}$$

where

$$\eta_B^2 = \xi^2 - \frac{\rho_B \omega^2}{\bar{c}_B} = \xi^2 \left(1 - \frac{v^2}{v_B^2} \right) > 0 , \tag{14.4.6}$$

$$v_B^2 = \frac{\bar{c}_B}{\rho_B} . \tag{14.4.7}$$

The continuity of u between the two half-spaces is already satisfied. For the other continuity conditions, we need ϕ_B, T_{23} and D_2 in ceramic B. We denote T_{23} and D_2 by T_B and D_B:

$$\begin{aligned}
\phi_B &= \psi_B + \frac{e_B}{\varepsilon_B} u_B \\
&= \left[\Psi_B \exp(\xi x_2) + \frac{e_B}{\varepsilon_B} U \exp(\eta_B x_2) \right] \exp i(\xi x_1 - \omega t), \\
T_B &= \bar{c}_B u_{B,2} + e_B \psi_{B,2} \\
&= [\bar{c}_B \eta_B U \exp(\eta_B x_2) + e_B \xi \Psi_B \exp(\xi x_2)] \exp i(\xi x_1 - \omega t), \\
D_B &= -\varepsilon_B \psi_{B,2} \\
&= -\varepsilon_B \xi \Psi_B \exp(\xi x_2) \exp i(\xi x_1 - \omega t).
\end{aligned} \tag{14.4.8}$$

The fields in the semiconductor film can be treated as functions of x_1 and time only. Denote the fields in the film by

$$\begin{aligned} u &= U \exp i(\xi x_1 - \omega t), \quad \phi = \Phi \exp i(\xi x_1 - \omega t), \\ n &= N \exp i(\xi x_1 - \omega t), \end{aligned} \tag{14.4.9}$$

where Φ and N are undetermined constants. Equation (14.4.9) already satisfies the continuity of mechanical displacement between the film and the ceramic half-spaces. We use a prime to indicate the elastic and dielectric constants as well as the mass density of the film (see Eq. (14.3.14)). When the DC biasing electric field is in the x_1 direction

$$\begin{aligned} T_{13} &= \bar{c}'_{55} S_{13} = \bar{c}'_{44} u_{3,1} = \bar{c}'_{44} i\xi U \exp i(\xi x_1 - \omega t), \\ D_1 &= \bar{\varepsilon}'_{11} E_1 = -\bar{\varepsilon}'_{11} \phi_{,1} = -\bar{\varepsilon}'_{11} i\xi \Phi \exp i(\xi x_1 - \omega t), \\ J_1 &= -q\bar{n}\mu_{11}\phi_{,1} + qn\mu_{11}\bar{E}_1 - qd_{11}n_{,1} \\ &= (-q\bar{n}\mu_{11} i\xi\Phi + qN\mu_{11}\bar{E}_1 - qd_{11} i\xi N) \exp i(\xi x_1 - \omega t). \end{aligned} \tag{14.4.10}$$

From the continuity of the electric potential and the equations of the semiconductor film, we have

$$\begin{aligned} &\Psi_A + \frac{e_A}{\varepsilon_A} U = \Phi, \\ &\Psi_B + \frac{e_B}{\varepsilon_B} U = \Phi, \\ &-\bar{c}'_{44}\xi^2 U + \frac{1}{2h}[-(\bar{c}_A\eta_A U + e_A\xi\Psi_A) - (\bar{c}_B\eta_B U + e_B\xi\Psi_B)] \\ &\quad = -\rho'\omega^2 U, \\ &\bar{\varepsilon}'_{11}\xi^2\Phi + \frac{1}{2h}(\varepsilon_A\xi\Psi_A + \varepsilon_B\xi\Psi_B) = qN, \\ &-qi\omega N + i\xi\,(-q\bar{n}\mu_{11} i\xi\,\Phi + qN\mu_{11}\bar{E}_1 - qd_{11} i\xi N) = 0. \end{aligned} \tag{14.4.11}$$

Equation (14.4.11) is a system of linear homogeneous equations for U, Ψ_A, Ψ_B, Φ and N. For non-trivial solutions the determinant of the coefficient matrix has to vanish, which gives the following frequency equation that determines the dispersion relations of the waves:

$$\rho' \frac{\omega^2}{\xi^2} \xi 2h - \bar{c}'_{44} \xi 2h - \bar{c}_A \frac{\eta_A}{\xi} - \bar{c}_B \frac{\eta_B}{\xi} + \frac{e_A^2}{\varepsilon_A} + \frac{e_B^2}{\varepsilon_B}$$
$$= \frac{(e_A + e_B)^2}{\bar{\varepsilon}'_{11} \xi 2h + \varepsilon_A + \varepsilon_B + \dfrac{q\bar{n}\mu_{11} 2h}{d_{11}\xi + i\left(\bar{E}_1 \mu_{11} - \dfrac{\omega}{\xi}\right)}}. \tag{14.4.12}$$

Or, in terms of the wave speed v, Eq. (14.4.12) can be written as

$$\bar{c}'_{44}\left(\frac{v^2}{v_T'^2} - 1\right)\xi 2h - \bar{c}_A \sqrt{1 - \frac{v^2}{v_A^2}} - \bar{c}_B \sqrt{1 - \frac{v^2}{v_B^2}} + \frac{e_A^2}{\varepsilon_A} + \frac{e_B^2}{\varepsilon_B}$$
$$= \frac{(e_A + e_B)^2}{\bar{\varepsilon}'_{11} \xi 2h + \varepsilon_A + \varepsilon_B + \dfrac{q\bar{n}\mu_{11} 2h}{d_{11}\xi + i(\bar{E}_1 \mu_{11} - v)}}, \tag{14.4.13}$$

where

$$v_T'^2 = \frac{c'_{44}}{\rho'}. \tag{14.4.14}$$

14.4.2. Discussion

We make the following observations from Eq. (14.4.13):

(i) As a special case, when $h = 0$, i.e. the semiconductor film does not exist, Eq. (14.4.14) reduces to the equation that determines the speed of interface waves between two ceramic half-spaces

$$\bar{c}_A \sqrt{1 - \frac{v^2}{v_A^2}} + \bar{c}_B \sqrt{1 - \frac{v^2}{v_B^2}} = \frac{e_A^2}{\varepsilon_A} + \frac{e_B^2}{\varepsilon_B} - \frac{(e_A + e_B)^2}{\varepsilon_A + \varepsilon_B}; \tag{14.4.15}$$

(ii) Different from the interface waves determined by Eq. (14.4.15) which are not dispersive, waves determined by Eq. (14.4.13) are dispersive due to the presence of the film which introduces a length parameter h into the problem;

(iii) If the two half-spaces are of the same ceramics with opposite poling directions, we have

$$\bar{c}_B = \bar{c}_A = \bar{c}, \quad e_B = -e_A = -e, \quad \varepsilon_B = \varepsilon_A = \varepsilon. \tag{14.4.16}$$

Then Eq. (14.4.13) simplifies to

$$\frac{\bar{c}'_{44}}{\bar{c}}\left(\frac{v^2}{v_T'^2}-1\right)\xi h-\sqrt{1-\frac{v^2}{v_T^2}}+k^2=0\,, \qquad (14.4.17)$$

where

$$v_T^2=\frac{\bar{c}}{\rho},\quad k^2=\frac{e^2}{\varepsilon\bar{c}}\,. \qquad (14.4.18)$$

Equation (14.4.17) represents a dispersive but non-dissipative wave. In this case the electric fields produced by the two ceramic half-spaces cancel with each other in the semiconductor film. Hence, there is no conduction and dissipation in the film.

(iv) If the two half-spaces are of the same ceramics with the same poling direction, Eq. (14.4.13) reduces to

$$\frac{\bar{c}'_{44}}{\bar{c}}\left(\frac{v^2}{v_T'^2}-1\right)\xi h-\sqrt{1-\frac{v^2}{v_T^2}}+k^2 = \frac{k^2}{\dfrac{\bar{\varepsilon}'_{11}}{\varepsilon}\xi h+1+\dfrac{q\bar{n}\mu_{11}h}{\varepsilon[d_{11}\xi+i(\bar{E}_1\mu_{11}-v)]}}\,. \qquad (14.4.19)$$

The denominator of the right-hand side of Eq. (14.4.19) indicates that a complex wave speed may be expected and the imaginary part of the complex wave speed may change its sign when $\mu_{11}\bar{E}_1-v$ changes its sign or

$$v=\frac{\omega}{\xi}=\mu_{11}\bar{E}_1\,, \qquad (14.4.20)$$

i.e., the acoustic wave speed is equal to the carrier drift speed.

14.5. Waves in a Plate

In this section we study anti-plane waves in a piezoelectric plate of polarized ceramics carrying two identical semiconductor layers of silicon on each of its major surfaces (see Fig. 14.5.1) [128].

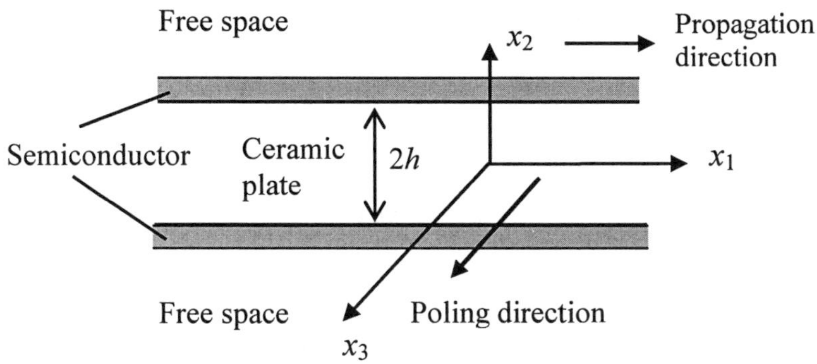

Fig. 14.5.1. A ceramic plate with semiconductor films.

Waves in the structure can be classified as symmetric and anti-symmetric.

14.5.1. Symmetric waves

First consider symmetric waves. From Eqs. (14.3.1) and (14.3.2), fields in the plate can be written as

$$\begin{aligned} u_3 &= A\cos\xi_2 x_2 \exp i(\xi x_1 - \omega t), \\ \psi &= B\cosh\xi_1 x_2 \exp i(\xi x_1 - \omega t), \end{aligned} \tag{14.5.1}$$

$$\bar{c}_{44}(\xi_1^2 + \xi_2^2) = \rho\omega^2, \tag{14.5.2}$$

$$\xi_2^2 = \frac{\rho\omega^2}{\bar{c}_{44}} - \xi_1^2 = \xi_1^2\left(\frac{v^2}{v_T^2} - 1\right), \tag{14.5.3}$$

$$v^2 = \frac{\omega^2}{\xi_1^2}, \quad v_T^2 = \frac{\bar{c}_{44}}{\rho}, \tag{14.5.4}$$

$$\begin{aligned} \phi &= \left(\frac{e_{15}}{\varepsilon_{11}} A\cos\xi_2 x_2 + B\cosh\xi_1 x_2\right)\exp i(\xi x_1 - \omega t), \\ T_{23} &= (-A\bar{c}_{44}\xi_2 \sin\xi_2 x_2 + e_{15}B\xi_1 \sinh\xi_1 x_2)\exp i(\xi x_1 - \omega t), \\ D_2 &= -\varepsilon_{11}B\xi_1 \sinh\xi_1 x_2 \exp i(\xi x_1 - \omega t). \end{aligned} \tag{14.5.5}$$

The electric field in the free space of $x_2 > h$ is governed by

$$\nabla^2\phi = 0, \quad x_2 > h, \qquad \phi \to 0, \quad x_2 \to +\infty. \tag{14.5.6}$$

The relevant solution is

$$\phi = C\exp\xi_1(h - x_2)\exp i(\xi x_1 - \omega t), \qquad D_2 = -\varepsilon_0\phi_{,2} = \varepsilon_0\xi_1 C\exp\xi_1(h - x_2)\exp i(\xi x_1 - \omega t), \tag{14.5.7}$$

where C is an undetermined constant.

Fields in the semiconductor film at $x_2 = h$ depends approximately on x_1 and time only. Denote the fields in the film by

$$u_3 = A\cos\xi_2 h\exp i(\xi x_1 - \omega t), \quad \phi = C\exp i(\xi x_1 - \omega t), \qquad n = N\exp i(\xi x_1 - \omega t), \tag{14.5.8}$$

where N is an undetermined constant. At $x_2 = h$, Eq. (14.5.8) already satisfies the continuity of mechanical displacement between the film and the ceramic plate, and the continuity of electric potential between the film and the free space. We use a prime to indicate the elastic and dielectric constants as well as the mass density of the film (see Eq. (14.3.14)). When the DC biasing electric field is in the x_1 direction,

$$u_3 = A\exp(-\xi_2 x_2)\exp i(\xi_1 x_1 - \omega t), \qquad \psi = B\exp(-\xi_1 x_2)\exp i(\xi_1 x_1 - \omega t, \tag{14.5.9}$$

The continuity condition of the electric potential between the ceramic plate and the equations of the film yield

$$\begin{aligned}
&\frac{e_{15}}{\varepsilon_{11}} A\cos\xi_2 h + B\cosh\xi_1 h = C, \\
&-\bar{c}'_{44}\xi_1^2 A\cos\xi_2 h - \frac{1}{2b}(-A\bar{c}_{44}\xi_2\sin\xi_2 h + e_{15}B\xi_1\sinh\xi_1 h) \\
&\quad = -\rho'\omega^2 A\cos\xi_2 h, \\
&\bar{\varepsilon}'_{11}\xi_1^2 C + \frac{1}{2b}(\varepsilon_0\xi_1 C + \varepsilon_{11}B\xi_1\sinh\xi_1 h) = qN, \\
&-qi\omega N + i\xi_1(-q\bar{n}\mu_{11}i\xi_1 C + qN\mu_{11}\bar{E}_1 - qd_{11}i\xi_1 N) = 0,
\end{aligned} \tag{14.5.10}$$

which is a system of linear homogeneous equations for A, B, C and N. For nontrivial solutions the determinant of the coefficient matrix has to vanish:

$$\begin{vmatrix} \dfrac{e_{15}}{\varepsilon_{11}}\cos\xi_2 h & \cosh\xi_1 h & -1 & 0 \\ \begin{array}{l}(\rho'\omega^2-\bar{c}'_{44}\xi_1^2)\cos\xi_2 h \\ +\dfrac{\bar{c}_{44}\xi_2}{2b}\sin\xi_2 h\end{array} & -\dfrac{e_{15}\xi_1}{2b}\sinh\xi_1 h & 0 & 0 \\ 0 & \dfrac{\varepsilon_{11}\xi_1}{2b}\sinh\xi_1 h & \bar{\varepsilon}'_{11}\xi_1^2+\dfrac{\varepsilon_0\xi_1}{2b} & -q \\ 0 & 0 & q\bar{n}\mu_{11}\xi_1^2 & \begin{array}{c}-qi\omega+i\xi_1 q\mu_{11}\bar{E}_1 \\ +qd_{11}\xi_1^2\end{array} \end{vmatrix}$$

$$=0, \tag{14.5.11}$$

which determines the dispersion relation. Equation (14.5.11) can be further written as

$$\bar{k}_{15}^2\tanh\xi_1 h+\frac{\xi_2}{\xi_1}\tan\xi_2 h+\left(\frac{\omega^2}{\xi_1^2}\frac{\rho'}{\bar{c}'_{44}}-1\right)\frac{\bar{c}'_{44}}{\bar{c}_{44}}\xi_1 2b$$
$$=\frac{\bar{k}_{15}^2\tanh^2\xi_1 h}{\tanh\xi_1 h+\dfrac{\varepsilon_0}{\varepsilon_{11}}+\dfrac{\bar{\varepsilon}'_{11}}{\varepsilon_{11}}\xi_1 2b-\dfrac{2bq\bar{n}\mu_{11}}{\varepsilon_{11}\left[i\left(\dfrac{\omega}{\xi_1}-\mu_{11}\bar{E}_1\right)-d_{11}\xi_1\right]}}, \tag{14.5.12}$$

where $\bar{k}_{15}^2=e^2/(\varepsilon_{11}\bar{c}_{44})$. The symmetric fields for $x_2 < 0$ and the continuity conditions at $x_2 = -h$ yield the same dispersion relation. For symmetric fields, all fields in the lower film are the same as those in the upper film.

14.5.2. Anti-symmetric waves

For anti-symmetric waves, consider the following fields in the plate

$$\begin{aligned} u_3 &= A\sin\xi_2 x_2 \exp i(\xi x_1 - \omega t), \\ \psi &= B\sinh\xi_1 x_2 \exp i(\xi x_1 - \omega t), \end{aligned} \qquad (14.5.13)$$

$$\begin{aligned} \phi &= \left(\frac{e_{15}}{\varepsilon_{11}} A\sin\xi_2 x_2 + B\sinh\xi_1 x_2\right)\exp i(\xi x_1 - \omega t), \\ T_{23} &= (A\bar{c}_{44}\xi_2\cos\xi_2 x_2 + e_{15}B\xi_1\cosh\xi_1 x_2)\exp i(\xi x_1 - \omega t), \\ D_2 &= -\varepsilon_{11}B\xi_1\cosh\xi_1 x_2 \exp i(\xi x_1 - \omega t). \end{aligned} \qquad (14.5.14)$$

The fields in the free space of for $x_2 > 0$ are still given by Eq. (14.5.7). For the fields in the film at $x_2 = h$, Eq. $(14.5.8)_1$ should be replaced by

$$u_3 = A\sin\xi_2 h \exp i(\xi x_1 - \omega t)\,, \qquad (14.5.15)$$

and Eq. $(14.5.9)_1$ should be replaced by

$$T_{13} = \bar{c}'_{44} i\xi_1 A\sin\xi_2 h \exp i(\xi x_1 - \omega t)\,. \qquad (14.5.16)$$

The continuity condition of the electric potential between the ceramic plate and the film and the equations of the film then imply that

$$\begin{aligned} &\frac{e_{15}}{\varepsilon_{11}} A\sin\xi_2 h + B\sinh\xi_1 h = C, \\ &-\bar{c}'_{44}\xi_1^2 A\sin\xi_2 h - \frac{1}{2b}(A\bar{c}_{44}\xi_2\cos\xi_2 h + e_{15}B\xi_1\cosh\xi_1 h) \\ &\quad = -\rho'\omega^2 A\sin\xi_2 h, \\ &\bar{\varepsilon}'_{11}\xi_1^2 C + \frac{1}{2b}(\varepsilon_0\xi_1 C + \varepsilon_{11}B\xi_1\cosh\xi_1 h) = qN, \\ &-qi\omega N + i\xi_1(-q\bar{n}\mu_{11}i\xi_1 C + qN\mu_{11}\bar{E}_1 - qd_{11}i\xi_1 N) = 0. \end{aligned} \qquad (14.5.17)$$

For non-trivial solutions,

$$
\begin{vmatrix}
\frac{e_{15}}{\varepsilon_{11}}\sin\xi_2 h & \sinh\xi_1 h & -1 & 0 \\
\begin{matrix}(\rho'\omega^2-\bar{c}'_{44}\xi_1^2)\sin\xi_2 h \\ -\frac{\bar{c}_{44}\xi_2}{2b}\cos\xi_2 h\end{matrix} & -\frac{e_{15}\xi_1}{2b}\cosh\xi_1 h & 0 & 0 \\
0 & \frac{\varepsilon_{11}\xi_1}{2b}\cosh\xi_1 h & \bar{\varepsilon}'_{11}\xi_1^2+\frac{\varepsilon_0\xi_1}{2b} & -q \\
0 & 0 & q\bar{n}\mu_{11}\xi_1^2 & \begin{matrix}-qi\omega+i\xi_1 q\mu_{11}\bar{E}_1 \\ +qd_{11}\xi_1^2\end{matrix}
\end{vmatrix}
$$

$$
=0. \tag{14.5.18}
$$

Equation (14.5.8) can be further written as

$$
\bar{k}_{15}^2\tan\xi_2 h-\frac{\xi_2}{\xi_1}\tanh\xi_1 h+\left(\frac{\omega^2}{\xi_1^2}\frac{\rho'}{\bar{c}'_{44}}-1\right)\frac{\bar{c}'_{44}}{\bar{c}_{44}}\xi_1 2b\tan\xi_2 h\tanh\xi_1 h
$$

$$
=\frac{\bar{k}_{15}^2\tan\xi_2 h}{1+\left(\frac{\varepsilon_0}{\varepsilon_{11}}+\frac{\bar{\varepsilon}'_{11}}{\varepsilon_{11}}\xi_1 2b\right)\tanh\xi_1 h-\dfrac{2bq\bar{n}\mu_{11}\tanh\xi_1 h}{\varepsilon_{11}\left[i\left(\frac{\omega}{\xi_1}-\mu_{11}\bar{E}_1\right)-d_{11}\xi_1\right]}}. \tag{14.5.19}
$$

The anti-symmetric fields for $x_2 < 0$ and the continuity conditions at $x_2 = -h$ yield the same dispersion relation. For anti-symmetric fields the carrier density perturbation n of the lower film has an opposite sign to that of the upper film, but q, $\bar{n}$ and $\bar{E}_1$ are the same as those of the upper film.

14.5.3. Discussion

Equations (14.5.12) and (14.5.19) show clearly that the waves are dispersive. When $b = 0$, i.e., the semiconductor films do not exist, Eqs. (14.5.12) and (14.5.19) reduce to

$$\bar{k}_{15}^2 \tanh \xi_1 h + \frac{\xi_2}{\xi_1} \tan \xi_2 h = \frac{\bar{k}_{15}^2 \tanh^2 \xi_1 h}{\tanh \xi_1 h + \dfrac{\varepsilon_0}{\varepsilon_{11}}} , \tag{14.5.20}$$

$$\bar{k}_{15}^2 \tan \xi_2 h - \frac{\xi_2}{\xi_1} \tanh \xi_1 h = \frac{\bar{k}_{15}^2 \tan \xi_2 h}{1 + \dfrac{\varepsilon_0}{\varepsilon_{11}} \tanh \xi_1 h} , \tag{14.5.21}$$

or

$$\xi_2 h (1 + \frac{\varepsilon_{11}}{\varepsilon_0} \tanh \xi_1 h) \tan \xi_2 h = -\bar{k}_{15}^2 \xi_1 h \tanh \xi_1 h , \tag{14.5.22}$$

$$\bar{k}_{15}^2 \xi_1 h \tan \xi_2 h = \xi_2 h \left(\frac{\varepsilon_{11}}{\varepsilon_0} + \tanh \xi_1 h \right) , \tag{14.5.23}$$

which are the frequency equations for thickness-twist waves in an unelectroded ceramic plate given by Eq. (3.1.23).

14.6. Gap Waves

Structures of acoustoelectric devices often have air gap(s). Since electric fields can exist in free space, components of a device on both sides of an air gap are electrically coupled, although they are not mechanically so. This offers some options in design. In this section we study the propagation of anti-plane waves in a piezoelectric dielectric half-space with a thin film of a non-piezoelectric semiconductor and a thin air gap (see Fig. 14.6.1) [129].

14.6.1. Analytical solution

From the previous sections of this chapter, fields in the ceramic half-space are

$$\begin{aligned} u_3 &= A \exp(-\xi_2 x_2) \exp i(\xi_1 x_1 - \omega t), \\ \psi &= B \exp(-\xi_1 x_2) \exp i(\xi_1 x_1 - \omega t, \end{aligned} \tag{14.6.1}$$

$$\bar{c}_{44} (\xi_1^2 - \xi_2^2) = \rho \omega^2 , \tag{14.6.2}$$

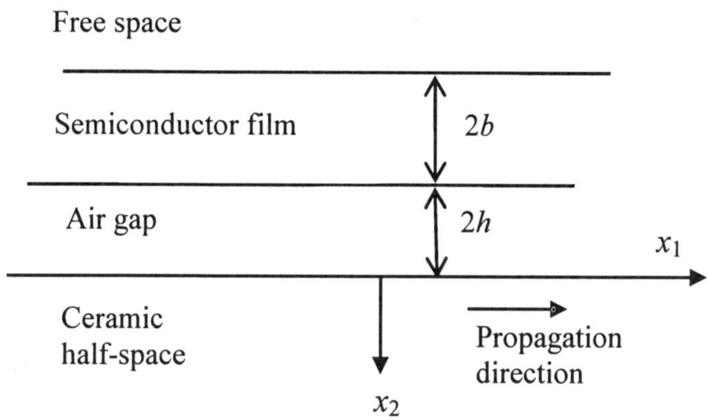

Fig. 14.6.1. A ceramic half-space with a semiconductor film and an air gap.

$$\xi_2^2 = \xi_1^2 - \frac{\rho\omega^2}{\bar{c}_{44}} = \xi_1^2\left(1 - \frac{v^2}{v_T^2}\right) > 0\,, \tag{14.6.3}$$

$$v^2 = \frac{\omega^2}{\xi_1^2}, \quad v_T^2 = \frac{\bar{c}_{44}}{\rho}, \tag{14.6.4}$$

$$\begin{aligned} \phi &= \left[B\exp(-\xi_1 x_2) + \frac{e_{15}}{\varepsilon_{11}}A\exp(-\xi_2 x_2)\right]\exp i(\xi_1 x_1 - \omega t), \\ T_{23} &= -[A\bar{c}_{44}\xi_2\exp(-\xi_2 x_2) + e_{15}B\xi_1\exp(-\xi_1 x_2)]\exp i(\xi_1 x_1 - \omega t), \\ D_2 &= \varepsilon_{11}B\xi_1\exp(-\xi_1 x_2)\exp i(\xi_1 x_1 - \omega t). \end{aligned} \tag{14.6.5}$$

Fields in the free space $x_2 < -2h - 2b$ can be represented by

$$\phi = C\exp(\xi_1 x_2)\exp i(\xi_1 x_1 - \omega t)\,, \tag{14.6.6}$$

$$D_2 = -\varepsilon_0\xi_1 C\exp(\xi_1 x_2)\exp i(\xi_1 x_1 - \omega t)\,. \tag{14.6.7}$$

When the DC biasing electric field is in the x_1 direction, the fields in the semiconductor film are

$$\begin{aligned} &u_3 = D\exp i(\xi_1 x_1 - \omega t), \quad \phi = C\exp i(\xi_1 x_1 - \omega t), \\ &n = N\exp i(\xi_1 x_1 - \omega t). \end{aligned} \tag{14.6.8}$$

Equation (14.6.8) already satisfies the continuity of electric potential between the film and the free space. From Eq. (14.6.8),

$$\begin{aligned} T_{13} &= \bar{c}'_{55} S_{13} = \bar{c}'_{44} u_{3,1} = \bar{c}'_{44} i\xi_1 D \exp i(\xi_1 x_1 - \omega t), \\ D_1 &= \bar{\varepsilon}'_{11} E_1 = -\bar{\varepsilon}'_{11} \phi_{,1} = -\bar{\varepsilon}'_{11} i\xi_1 C \exp i(\xi_1 x_1 - \omega t), \\ J_1 &= -q\bar{n}\mu_{11}\phi_{,1} + qn\mu_{11}\bar{E}_1 - qd_{11} n_{,1} \\ &= (-q\bar{n}\mu_{11} i\xi_1 C + qN\mu_{11}\bar{E}_1 - qd_{11} i\xi_1 N)\exp i(\xi_1 x_1 - \omega t). \end{aligned} \tag{14.6.9}$$

Fields in the air gap can be treated as functions of x_1 and t only. Let

$$\phi = C \exp i(\xi_1 x_1 - \omega t). \tag{14.6.10}$$

Equation (14.6.10) already satisfies the continuity of electric potential between the air gap and the semiconductor film. In the air gap,

$$D_1 = \varepsilon_0 E_1 = -\varepsilon_0 \phi_{,1} = -\varepsilon_0 i\xi_1 C \exp i(\xi_1 x_1 - \omega t). \tag{14.6.11}$$

Substitution of the above fields into the continuity condition of the electric potential between the ceramic half-space and the air gap, the traction-free boundary condition of the half-space surface, and the equations of the film yields

$$\begin{aligned} & B + \frac{e_{15}}{\varepsilon_{11}} A = C, \\ & A\bar{c}_{44}\xi_2 + e_{15} B\xi_1 = 0, \\ & -\bar{c}'_{44}\xi_1^2 D = -\rho'\omega^2 D, \\ & \bar{\varepsilon}'_{11}\xi_1^2 C + \frac{1}{2b}[D_2(\text{film} - \text{gap interface}) + \varepsilon_0 \xi_1 C] = qN, \\ & -qi\omega N + i\xi_1(-q\bar{n}\mu_{11} i\xi_1 C + qN\mu_{11}\bar{E}_1 - qd_{11} i\xi_1 N) = 0, \\ & \varepsilon_0 \xi_1^2 C + \frac{1}{2h}[\varepsilon_{11} B\xi_1 - D_2(\text{film} - \text{gap interface})] = 0, \end{aligned} \tag{14.6.12}$$

which is a system of linear, homogeneous equations for A, B, C, N, D_2(film-gap interface), and D. Since the film and the air gap are very thin, the continuity conditions are applied at $x_2 = 0$. D appears in Eq. $(14.6.12)_3$ only, which represents a pure elastic wave in the film without mechanical coupling to the substrate. This mode is not our focus. In the following analysis we study Eq. $(14.6.12)_{1,2,4\text{-}6}$. Or we only study waves with $D = 0$. This means that the film only interacts with the structure electrically, but does not move. This is possible when the film is semiconducting but is not piezoelectric. For nontrivial solutions the determinant of the coefficient matrix of Eq. $(14.6.12)_{1,2,4\text{-}6}$ for A, B, C, N and D_2(film-gap interface) has to vanish, which yields

$$\begin{vmatrix} \dfrac{e_{15}}{\varepsilon_{11}} & 1 & -1 & 0 & 0 \\ \bar{c}_{44}\xi_2 & e_{15}\xi_1 & 0 & 0 & 0 \\ 0 & 0 & \bar{\varepsilon}'_{11}\xi_1^2 + \dfrac{\varepsilon_0\xi_1}{2b} & -q & \dfrac{1}{2t} \\ 0 & 0 & q\bar{n}\mu_{11}\xi_1^2 & -qi\omega + i\xi_1 q\mu_{11}\bar{E}_1 + qd_{11}\xi_1^2 & 0 \\ 0 & \dfrac{\varepsilon_{11}\xi_1}{2h} & \varepsilon_0\xi_1^2 & 0 & \dfrac{-1}{2h} \end{vmatrix} = 0. \tag{14.6.13}$$

Equation (14.6.13) determines the dispersion relation. It can also be written as

$$\begin{aligned} &\left(1+\xi_1\,2h+\frac{\bar{\varepsilon}'_{11}}{\varepsilon_0}\xi_1\,2b\right)\bar{k}_{15}^2-\frac{\xi_2}{\xi_1}\left(1+\frac{\bar{\varepsilon}'_{11}}{\varepsilon_0}\xi_1 2b+\frac{\varepsilon_{11}}{\varepsilon_0}+\xi_1 2h\right) \\ &\quad =\frac{q\bar{n}\mu_{11}\xi_1\,2b\left(\dfrac{\xi_2}{\xi_1}-\bar{k}_{15}^2\right)}{\varepsilon_0[d_{11}\xi_1^2+i(\bar{E}_1\mu_{11}\xi_1-\omega)]}, \end{aligned} \tag{14.6.14}$$

or, in terms of the wave speed defined by $v=\omega/\xi_1$, Eq. (14.6.14) can be written in the following form:

$$\begin{aligned} &\left(1+\xi_1\,2h+\frac{\bar{\varepsilon}'_{11}}{\varepsilon_0}\xi_1\,2b\right)\bar{k}_{15}^2-\sqrt{1-\frac{v^2}{v_T^2}}\left(1+\frac{\bar{\varepsilon}'_{11}}{\varepsilon_0}\xi_1 2b+\frac{\varepsilon_{11}}{\varepsilon_0}+\xi_1 2h\right) \\ &\quad =\frac{q\bar{n}\mu_{11}2b\left(\sqrt{1-\dfrac{v^2}{v_T^2}}-\bar{k}_{15}^2\right)}{\varepsilon_0[d_{11}\xi_1+i(\bar{E}_1\mu_{11}-v)]}. \end{aligned} \tag{14.6.15}$$

14.6.2. Discussion

Note that setting $h = 0$ in Eq. (14.6.15) to remove the air gap does not yield the dispersion relation in the third section of this chapter because the semiconductor film in this section does not have mechanical interaction with the half-space and in the third section of this chapter the film is mechanically attached to the half-space with an ideal bonding.

When $b = h = 0$, i.e. the semiconductor film and the air gap do not exist, Eq. (14.6.15) reduces to

$$v^2 = v_T^2\left[1 - \frac{\bar{k}_{15}^4}{(1+\varepsilon_{11}/\varepsilon_0)^2}\right] = v_{B-G}^2 \,. \tag{14.6.16}$$

When the semiconduction is small, Eq. (14.6.15) can be solved by an iteration procedure. As the lowest (zero) order of approximation, we neglect the small semiconduction and denote the zero-order solution by $v_{(0)}$. Then, from Eq. (14.6.15),

$$\frac{v_{(0)}^2}{v_T^2} = 1 - \bar{k}_{15}^4\left(\frac{1+\xi_1 2h + \dfrac{\bar{\varepsilon}_{11}'}{\varepsilon_0}\xi_1 2b}{1+\dfrac{\bar{\varepsilon}_{11}'}{\varepsilon_0}\xi_1 2b + \dfrac{\varepsilon_{11}}{\varepsilon_0} + \xi_1 2h}\right)^2, \tag{14.6.17}$$

which determines the speed of piezoelectric gap waves.

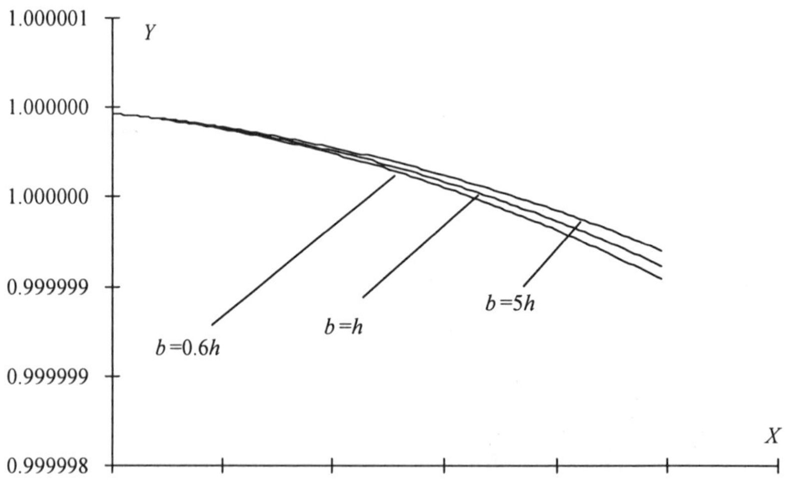

Fig. 14.6.2. Dispersion relations of gap waves for different values of film-thickness/air gap ratio.

We examine Eq. (14.6.17) numerically. For the half-space we consider PZT-5H. Figure 14.6.2 shows the wave speed for different values of b/h. The dimensionless wave number X and the dimensionless wave speed Y are defined by

$$X = \xi_1 \Big/ \frac{\pi}{2h}, \quad Y = v_{(0)} \big/ v_{B-G} \,. \tag{14.6.18}$$

The figure shows that the waves are dispersive. This is due to the presence of the film and the gap, introducing two geometric parameters b and h. The dispersion is less for long waves with a small X. The dispersion also depends on the ratio of b/h, but is not sensitive to it. For relatively thick films with $b >> h$, the dispersion is less.

References

[1] H. F. Tiersten, On the nonlinear equations of thermoelectroelasticity, *Int. J. Eng. Sci.*, **9**, 587–604, 1971.

[2] H. F. Tiersten, Electroelastic interactions and the piezoelectric equations, *J. Acoust. Soc. Am.*, **70**, 1567–1576, 1981.

[3] A. H. Meitzler, H. F. Tiersten, A. W. Warner, D. Berlincourt, G. A. Couqin and F. S. Welsh, III, *IEEE Standard on Piezoelectricity*, IEEE, New York, 1988.

[4] H. F. Tiersten, *Linear Piezoelectric Plate Vibrations, Plenum*, New York, 1969.

[5] J. C. Baumhauer and H. F. Tiersten, Nonlinear electroelastic equations for small fields superposed on a bias, *J. Acoust. Soc. Am.*, **54**, 1017–1034, 1973.

[6] H. F. Tiersten, On the accurate description of piezoelectric resonators subject to biasing deformations, *Int. J. Eng. Sci.*, **33**, 2239–2259, 1995.

[7] H. F. Tiersten, Perturbation theory for linear electroelastic equations for small fields superposed on a bias, *J. Acoust. Soc. Am.*, **64**, 832–837, 1978.

[8] H. F. Tiersten, Nonlinear electroelastic equations cubic in the small field variables, *J. Acoust. Soc. Am.*, **57**, 660–666, 1975.

[9] H. F. Tiersten, Analysis of intermodulation in thickness-shear and trapped energy resonators, *J. Acoust. Soc. Am.*, **57**, 667–681, 1975.

[10] J. S. Yang, *An Introduction to the Theory of Piezoelectricity*, Springer, New York, 2005.

[11] J. S. Yang and S. H. Guo, Effects of nonlinear elastic constants on electromechanical coupling factors, *IEEE Trans. on Ultrasonics, Ferroelectrics, and Frequency Control*, **52**, 2303–2305, 2005.

[12] Y. T. Hu, J. S. Yang, S. N. Jiang and Q. Jiang, Effects of biasing fields on electromechanical coupling factors, *World Journal of Engineering*, **2**, 63–69, 2005.

[13] H. F. Tiersten, Thickness vibrations of piezoelectric plates, *J. Acoust. Soc. Am.*, **35**, 53–58, 1963.

[14] A. Ballato and T. J. Lukaszek, Mass-loading of thickness-excited crystal resonators having arbitrary piezo-coupling, *IEEE Trans. on Sonics and Ultrasonics*, **21**, 269–274, 1974.

[15] J. A. Kosinski, Thickness vibration of flat piezoelectric plates with massy electrodes of unequal thickness, *Proc. IEEE Ultrasonics Symp.*, 70–73, 2003.

[16] J. S. Yang, Y. T. Hu, Y. Zeng and H. Fan, Thickness-shear vibration of rotated Y-cut quartz plates with imperfectly bounded surface mass layers, *IEEE Trans. on Ultrasonics, Ferroelectrics, and Frequency Control*, **53**, 241–245, 2006.

[17] J. Wang and L. J. Shen, Exact thickness-shear resonance frequency of electroded piezoelectric crystal plates, *J. Zhejiang Univ. SCI*, **6A**, 980–985, 2005.

[18] J. S. Yang, H. G. Zhou and W. P. Zhang, Thickness-shear vibration of rotated Y-cut quartz plates with relatively thick electrodes of unequal thickness, *IEEE Trans. on Ultrasonics, Ferroelectrics, and Frequency Control*, **52**, 918–922, 2005.

[19] J. S. Yang, X. M. Yang, J. A. Turner, J. A. Kosinski and R. A. Pastore, Jr., Two-dimensional equations for electroelastic plates with relatively large shear deformations, *IEEE Trans. on Ultrasonics, Ferroelectrics, and Frequency Control*, **50**, 765772, 2003.

[20] J. S. Yang, *Mechanics of Piezoelectric Structures*, World Scientific, Singapore, 2006.

[21] P. C. Y. Lee and M. S. H. Tang, Thickness vibrations of doubly rotated crystal plates under initial deformations, *IEEE Trans. on Ultrasonics, Ferroelectrics, and Frequency Control*, **34**, 659–666, 1987.

[22] P. C. Y. Lee and Y. K. Yong, Temperature derivatives of elastic stiffness derived from the frequency-temperature behavior of quartz plates, *J. Appl. Phys.*, **56**, 1514–1521, 1984.

[23] H. Ekstein, High frequency vibrations of thin crystal plates, *Phys. Rev.*, **68**, 11–23, 1945.

[24] E. G. Newman and R. D. Mindlin, Vibrations of a monoclinic crystal plate, *J. Acoust. Soc. Am.*, **29**, 1206–1218, 1957.

[25] R. K. Kaul and R. D. Mindlin, Vibrations of an infinite, monoclinic crystal plate at high frequencies and long wavelengths, *J. Acoust. Soc. Am.*, **34**, 1895–1901, 1962.

[26] R. D. Mindlin, Thickness-twist vibrations of an infinite, monoclinic, crystal plate, *Int. J. Solids Structures*, **1**, 141–145, 1965.

[27] J. L. Bleustein, Some simple modes of wave propagation in an infinite piezoelectric plate, *J. Acoust. Soc. Am.*, **45**, 614–620, 1969.

[28] H. F. Tiersten, Wave propagation in an infinite piezoelectric plate, *J. Acoust. Soc. Am.*, **35**, 234–239, 1963.

[29] S. Syngellakis and P. C. Y. Lee, Piezoelectric wave dispersion curves for infinite anisotropic plates, *J. Appl. Phys.*, **73**, 7152–7161, 1993.

[30] J. Wang, J. S. Yang and J. Y. Li, Energy trapping of thickness-shear vibration modes of elastic plates with functionally graded materials,

IEEE Trans. on Ultrasonics, Ferroelectrics, and Frequency Control, submitted.

[31] R. D. Mindlin, Bechmann's number for harmonic overtones of thickness/twist vibrations of rotated Y-cut quartz plates, *J. Acoust. Soc. Am.*, **41**, 969–973, 1966.

[32] R. D. Mindlin and M. Forray, Thickness-shear and flexural vibrations of contoured crystal plates, *J. Appl. Phys.*, **25**, 12–20, 1954.

[33] J. Bleustein, Thickness-twist and face-shear vibrations of a contoured crystal plate, *Int. J. of Solids Structures*, **2**, 351–360, 1966.

[34] G. T. Pearman, Thickness-twist vibrations in beveled AT-cut quartz plates, J. *Acoust. Soc. Am.*, **45**, 928–934, 1968.

[35] P. C. Y. Lee and S.-S. Chen, Vibrations of contoured and partially plated, contoured, rectangular, AT-cut quartz plates, *J. Acoust. Soc. Am.*, **46**, 1193–1202, 1969.

[36] P. C. Y. Lee and J. Wang, Thickness-shear and flexural vibrations of contoured crystal strip resonators, *Proc. IEEE Ultrasonics Symp.*, 559–564, 1993.

[37] P. C. Y. Lee and J. Wang, Piezoelectrically forced thickness-shear and flexural vibrations of contoured quartz resonators, *J. Appl. Phys.*, **79**, 3411–3422, 1996.

[38] J. Wang and P. C. Y. Lee, The effect of cubically varying contours on the thickness-shear and flexural vibrations of crystal plates, *Proc. IEEE Ultrasonics Symp.*, 977–980, 1996.

[39] J. Wang, Thickness-shear and flexural vibrations of linearly contoured crystal strips with multiprecision computation, *Computers and Structures*, **70**, 437–445, 1999.

[40] J. S. Yang, Shear horizontal vibrations of a piezoelectric/ferroelectric wedge, *Acta Mechanica*, **173**, 13–17, 2004.

[41] J. S. Yang and H. G. Zhou, Vibrations of an isotropic elastic wedge, *IEEE Trans. on Ultrasonics, Ferroelectrics, and Frequency Control*, **52**, 139–144, 2005.

[42] R. D. Mindlin and P. C. Y. Lee, Thickness-shear and flexural vibrations of partially plated, crystal plates, *Int. J. Solids Struct.*, **2**, 125–139, 1966.

[43] R. D. Mindlin, High frequency vibrations of plated, crystal plates, *Progress in Applied Mechanics*, Macmillan, New York, 73–84, 1963.

[44] J. S. Yang, H. Y. Fang, J. Wang and L. J. Shen, *Effects of Electrodes with Varying Thickness on Energy Trapping in Thickness-Shear Quartz Resonators*, to be submitted.

[45] J. S. Yang and J. A. Kosinski, Effects of piezoelectric coupling on energy trapping of thickness-shear modes, *IEEE Trans. on Ultrasonics, Ferroelectrics, and Frequency Control*, **51**, 1047–1049, 2004.

[46] J. L. Bleustein and H. F. Tiersten, Forced thickness-shear vibrations of discontinuously plated piezoelectric plates, *J. Acoust. Soc. Am.*, **43**, 1311–1318, 1968.

[47] R. D. Mindlin, Thickness-shear and flexural vibrations of crystal plates, *J. Appl. Phys.*, **22**, 316–323, 1951.

[48] R. D. Mindlin, Forced thickness-shear and flexural vibrations of piezoelectric crystal plates, *J. Appl. Phys.*, **23**, 83–88, 1952.

[49] R. D. Mindlin, *An Introduction to the Mathematical Theory of Vibrations of Plates*, the US Army Signal Corps Engineering Laboratories, Fort Monmouth, NJ, 1955.

[50] J. S. Yang, X. M. Yang, J. A. Turner, J. A. Kosinski and R. A. Pastore, Jr., Effects of middle surface curvature on vibrations of a thickness-shear mode crystal resonator, *Int. J. of Solids and Structures*, accepted.

[51] J. S. Yang, Frequency shifts in a piezoelectric body due to small amounts of additional mass on its surface, *IEEE Trans. on Ultrasonics, Ferroelectrics, and Frequency Control*, **51**, 1199–1202, 2004.

[52] J. S. Yang and S. H. Guo, Mass sensitivity of anti-plane (SH) modes in an isotropic elastic wedge, *IEEE Trans. on Ultrasonics, Ferroelectrics, and Frequency Control*, **52**, 2306–2307, 2005.

[53] J. S. Yang and S. H. Guo, A mass sensor based on torsional modes in a conical shell, *Applied Physics Letters*, **87**, 214108, 2005.

[54] K. Kraus, *Thin Elastic Shells*, John Wiley and Sons, New York, 1967.

[55] J. S. Yang and S. H. Guo, Frequency shifts in a piezoelectric body due to a surface mass layer with consideration of the layer stiffness, *IEEE Trans. on Ultrasonics, Ferroelectrics, and Frequency Control*, **52**, 1200–1203, 2005.

[56] J. S. Yang and S. H. Guo, Frequency shifts in a piezoelectric body due to a surface mass layer: variational analysis, unpublished.

[57] J. S. Yang and S. H. Guo, Vibrations of a crystal body with a shear-deformable surface mass layer, *Acta Mechanica*, accepted.

[58] J. S. Yang, An ill-posed problem in elasticity, *World Journal of Engineering*, **2**, 34–36, 2005.

[59] K. F. Graff, *Wave Motion in Elastic Solids*, Dover, New York, 1991.

[60] A. C. Eringen and E. S. Suhubi, *Elastodynamics*, Vol. II, Academic Press, New York, 1975.

[61] J. S. Yang and R. C. Batra, Thickness shear vibrations of a circular cylindrical piezoelectric shell, *J. Acoust. Soc. Am.*, **97**, 309–312, 1995.

[62] J. S. Yang, Frequency shifts in a crystal resonator due to submersion in a fluid, *IEEE Trans. on Ultrasonics, Ferroelectrics, and Frequency Control*, **53**, 662–664, 2006.

[63] R. L. Panton, *Incompressible Flow*, John Wiley & Sons, New York, 1984.

[64] J. S. Yang and S. H. Guo, *Torsional waves in a Circular Cylindrical Shell in Contact with a Viscous Fluid*, unpublished.

[65] B. Y. Lao, Gyroscopic effect in surface acoustic waves, *Proc. IEEE Ultrasonics Symp.*, 687–691, 1980.

[66] H. F. Tiersten, D. S. Stevens and P. K. Das, Acoustic surface wave accelerometer and rotation rate sensor, *Proc. IEEE Ultrasonics Symp.*, 692–695, 1980.

[67] H. F. Tiersten, D. S. Stevens and P. K. Das, Circulating flexural wave rotation rate sensor, *Proc. IEEE Ultrasonics Symp.*, 163–166, 1981.

[68] N. S. Clarke and J. S. Burdess, A rotation rate sensor based upon a Rayleigh resonator, *ASME Journal of Applied Mechanics*, **61**, 139–143, 1994.

[69] N. S. Clarke and J. S. Burdess, Rayleigh waves on a rotating surface, *ASME Journal of Applied Mechanics*, **61**, 724–726, 1994.

[70] W. D. Gates, Vibrating angular rate sensor may threaten the gyroscope, *Electronics*, **41**, 130–134, 1968.

[71] J. Soderkvist, *Micromachined gyroscopes, Sensors and Actuators A*, **43**, 65–71, 1994.

[72] J. S. Yang, A review of analyses related to vibrations of rotating piezoelectric bodies and gyroscopes, *IEEE Trans. on Ultrasonics, Ferroelectrics, and Frequency Control*, **52**, 698–706, 2005.

[73] J. S. Yang, H. Fang and Q. Jiang, Thickness vibrations of rotating piezoelectric plates, *J. Acoust. Soc. Am.*, **104**, 1427–1435, 1998.

[74] J. A. Kosinski, R. A. Pastore, Jr., H. Y. Fang and J. S. Yang, Thickness vibrations of a rotating AT-cut quartz plate, *Proc. IEEE Int. Ultrasonics Symp.*, 795–798, 2001.

[75] H. Y. Fang, J. S. Yang and Q. Jiang, Rotation sensitivity of waves propagating in a rotating piezoelectric plate, *Int. J. of Solids and Structures*, **39**, 5241–5251, 2002.

[76] H. Y. Fang, J. S. Yang and Q. Jiang, Rotation perturbed surface acoustic waves propagating in piezoelectric crystals, *Int. J. of Solids and Structures*, **37**, 4933–4947, 2000.

[77] H. Y. Fang, J. S. Yang and Q. Jiang, Surface waves propagating over a rotating piezoelectric half-space, *IEEE Trans. on Ultrasonics, Ferroelectrics, and Frequency Control*, **48**, 998–1004, 2001.

[78] J. S. Yang and H. Y. Fang, Analysis of a rotating elastic beam with piezoelectric films as an angular rate sensor, *IEEE Trans. on Ultrasonics, Ferroelectrics, and Frequency Control*, **49**, 798–804, 2002.

[79] J. S. Yang and H. Y. Fang, A new ceramic tube piezoelectric gyroscope, *Sensors and Actuators A*, **107**, 42–49, 2003.

[80] J. S. Yang, H. Y. Fang and Q. Jiang, Analysis of a ceramic bimorph piezoelectric gyroscope, *Int. J. of Applied Electromagnetics and Mechanics*, **10**, 459–473, 1999.

[81] J. S. Yang, H. Y. Fang and Q. Jiang, A vibrating piezoelectric ceramic shell as a rotation sensor, *Smart Materials and Structures*, **9**, 445–451, 2000.

[82] J. S. Yang, H. Y. Fang and Q. Jiang, One-dimensional equations for a piezoelectric ring and applications in a gyroscope, *IEEE Trans. on Ultrasonics, Ferroelectrics, and Frequency Control*, **48**, 1275–1282, 2001.

[83] J. S. Yang and H. Y. Fang, A piezoelectric gyroscope based on extensional vibrations of rods, *Int. J. of Applied Electromagnetics and Mechanics*, **17**, 289–300, 2003.

[84] J. S. Yang, Analysis of ceramic thickness shear piezoelectric gyroscopes, *J. Acoust. Soc. Am.*, **102**, 3542–3548, 1997.

[85] J. S. Yang, H. Y. Fang and Q. Jiang, Equations for a piezoelectric parallelepiped and applications in a gyroscope, *Int. J. of Applied Electromagnetics and Mechanics*, **10**, 337–350, 1999.

[86] P. C. Y. Lee and K.-M. Wu, The influence of support-configuration on the acceleration sensitivity of quartz resonator plates, *IEEE Trans. on Sonics and Ultrasonics*, **25**, 220–223, 1978.

[87] P. C. Y. Lee and K.-M. Wu, In-plane accelerations and forces on frequency changes in doubly rotated quartz plates, *J. Acoust. Soc. Am.*, **75**, 1105–1117, 1984.

[88] H. F. Tiersten and D. V. Shick, On the normal acceleration sensitivity of ST-cut quartz surface wave resonators supported along rectangular edges, *J. Appl. Phys.*, **64**, 4334–4341, 1988.

[89] D. V. Shick, Y. S. Zhou and H. F. Tiersten, An analysis of the in-plane acceleration sensitivity of quartz surface-wave resonators rigidly supported along the edges, *J. Appl. Phys.*, **65**, 35–40, 1989.

[90] R. L. Filler, The acceleration sensitivity of quartz crystal oscillators: a review, *IEEE Trans. on Ultrasonics, Ferroelectrics, and Frequency Control*, **35**, 297–305, 1988.

[91] J. S. Yang and Y. T. Hu, Mechanics of electroelastic bodies under biasing fields, *Applied Mechanics Reviews*, **57**, 173–189, 2004.

[92] J. A. Kosinski, R. A. Pastore, Jr. and J. S. Yang, Implications of in-plane stretch and thickness compression coupled to flexure on the lower bound of BAW acceleration sensitivity, *Proc. IEEE Frequency Control Symposium*, 345–347, 2000.

[93] J. S. Yang and S. H. Guo, An estimate on the second-order normal acceleration sensitivity of a quartz resonator, *IEEE Trans. on Ultrasonics, Ferroelectrics, and Frequency Control*, accepted.

[94] M. Nakazawa, T. Lukaszek and A. Ballato, Force- and acceleration-frequency effects in grooved and ring-supported resonators, *Proc. 35th Annual Symp. on Frequency Control*, 71–91, 1981.

[95] J. A. Kosinski, R. A. Pastore, Jr., J. S. Yang, X. M. Yang and J. A. Turner, Second-order frequency shifts in crystal resonators under relatively large biasing fields, *Proc. IEEE Int. Frequency Control Symp.*, 103–110, 2002.

[96] J. S. Yang, X. Zhang, J. A. Kosinski and R. A. Pastore, Jr., Thickness-shear vibrations of a quartz plate under time-dependent biasing deformations, *IEEE Trans. on Ultrasonics, Ferroelectrics, and Frequency Control,* **50**, 1114–1123, 2003.

[97] E. P. EerNisse, R. W. Ward and R. B. Wiggins, Survey of quartz bulk resonator sensor technologies, *IEEE Trans. on Ultrasonics, Ferroelectrics, and Frequency Control*, **35**, 323–330, 1988.

[98] E. P. EerNisse and R. B. Wiggins, Review of thickness-shear mode quartz resonator sensors for temperature and pressure, *IEEE Sensors Journal*, **1**, 79–87, 2001.

[99] E. P. EerNisse, Quartz resonators vs their environment: time base or sensor? *Jpn. J. Appl. Phys.*, **40**, Part 1, 3479–3483, 2001.

[100] Y. T. Hu, J. S. Yang, S. N. Jiang and Q. Jiang, Analysis of a thickness-shear crystal resonator in a circular cylindrical shell as a pressure sensor, *World Journal of Engineering*, **2**, 26–32, 1005.

[101] Y. T. Hu, Z. J. Cui, S. N. Jiang and J. S. Yang, Thickness-shear vibrations of a circular crystal plate in a cylindrical shell as a pressure sensor, *Applied Mathematics and Mechanics*, **27**, 749–755, 2006.

[102] J. S. Yang and X. Zhang, A high sensitivity pressure sensor, *Sensors and Actuators A*, **101**, 332–337, 2002.

[103] Y. T. Hu, J. S. Yang and Y. Zeng, A high-sensitivity, dual-plate, thickness-shear mode pressure sensor, *IEEE Trans. on Ultrasonics, Ferroelectrics, and Frequency Control*, accepted.

[104] Q. Jiang, X. M. Yang, H. G. Zhou and J. S. Yang, Analysis of surface acoustic wave pressure sensors, *Sensors and Actuators A*, **118**, 1–5, 2005.

[105] R. Stoneley, The propagation of surface elastic waves in a cubic crystal, *Proc. of the Royal Society of London, Series A*, **232**, 447–458, 1955.

[106] J. S. Yang, X. M. Yang and Q. Jiang, Bending mode effect on sensitivity of plate surface acoustic wave pressure sensors. *IEEE Trans. on Ultrasonics, Ferroelectrics, and Frequency Control*, **52**, 1748–1753, 2005.

[107] H. F. Tiersten, On the influence of material objectivity on electroelastic dissipation in polarized ferroelectric ceramics, *Mathematics and Mechanics of Solids*, **1**, 45–55 1996.

[108] J. S. Yang, Nonlinear equations of thermoviscoelectroelasticity, *Mathematics and Mechanics of Solids*, **3**, 113–124, 1998.

[109] J. S. Yang, Equations for small fields superposed on finite biasing fields in a thermoelectroelastic body, *IEEE Trans. on Ultrasonics, Ferroelectrics, and Frequency Control*, **50**, 187–192, 2003.

[110] J. S. Yang and X. Zhang, Vibrations of a crystal plate under a thermal bias, *J. Thermal Stresses*, **26**, 467–477, 2003.

[111] J. S. Yang, H. G. Zhou and Q. Jiang, Performance of a piezoelectric harvester in thickness-stretch mode of a plate, *IEEE Trans. on Ultrasonics, Ferroelectrics, and Frequency Control*, **52**, 1872–1876, 2005.

[112] S. N. Jiang, Q. Jiang, Y. T. Hu, X. F. Li, S. H. Guo and J. S. Yang, Analysis of a piezoelectric ceramic shell in thickness-shear vibration as a power harvester, *Int. J. of Applied Electromagnetics and Mechanics*, accepted.

[113] S. N. Jiang, X. F. Li, S. H. Guo, Y. T. Hu, J. S. Yang and Q. Jiang, Performance of a piezoelectric bimorph for scavenging vibration energy, *Smart Materials and Structures*, **14**, 769–774, 2005.

[114] Y. T. Hu, H. P. Hu and J. S. Yang, A low frequency piezoelectric power harvester using a spiral-shaped bimorph, *Sci. in China*, submitted.

[115] Y. T. Hu, H. Xu, J. S. Yang and Q. Jiang, Nonlinear behavior of a piezoelectric power harvester near resonance, *IEEE Trans. on Ultrasonics, Ferroelectrics, and Frequency Control*, **53**, 1387–1391, 2006.

[116] Y. T. Hu, C. Y. Chen, X. H. Yang, Q. G. Du and Z. J. Cui, Electric energy transmission between two piezoelectric transducers, *Acta Mechanica Solida Sinica*, **24**, 304–312, 2003.

[117] J. S. Yang and X. Zhang, Extensional vibration of a nonuniform piezoceramic rod and high voltage generation, *Int. J. of Applied Electromagnetics and Mechanics*, **16**, 29–42, 2002.

[118] J. S. Yang and W. Zhang, A thickness-shear high voltage piezoelectric transformer, *Int. J. of Applied Electromagnetics and Mechanics*, **10**, 105–121, 1999.

[119] J. S. Yang and X. Zhang, Analysis of a thickness-shear piezoelectric transformer, *Int. J. of Applied Electromagnetics and Mechanics*, **21**, 131–141, 2005.

[120] Y. T. Hu, X. Zhang, J. S. Yang and Q. Jiang, Transmitting electric energy through a metal wall by acoustic waves using piezoelectric transducers, *IEEE Trans. on Ultrasonics, Ferroelectrics, and Frequency Control*, **50**, 773–781, 2003.

[121] D. L. White, Amplification of ultrasonic waves in piezoelectric semiconductors, *J. Appl. Phys.*, **33**, 2547–2554, 1962.

[122] J. S. Yang and H. G. Zhou, Amplification of acoustic waves in piezoelectric semiconductor plates, *Int. J. of Solids Structures*, **42**, 3171–3183, 2005.

[123] J. S. Yang, X. M. Yang and J. A. Turner, Amplification of acoustic waves in laminated piezoelectric semiconductor plates, *Archive of Applied Mechanics*, **74**, 288–298, 2004.

[124] J. S. Yang, X. M. Yang and J. A. Turner, Amplification of acoustic waves in piezoelectric semiconductor shells, *J. of Intelligent Material Systems and Structures*, **16**, 613–621, 2005.

[125] F. S. Hickernell, The piezoelectric semiconductor and acoustoelectronic device development in the sixties, *IEEE Trans. on Ultrasonics, Ferroelectrics, and Frequency Control*, **52**, 737–745, 2005.

[126] J. S. Yang and H. G. Zhou, Acoustoelectric amplification of piezoelectric surface waves, *Acta Mechanica*, **172**, 113–122, 2004.

[127] J. S. Yang and H. G. Zhou, Interface waves between two piezoelectric half-spaces with a semiconductor film, *J. Zhejiang University-Science*, **6A**, 90–96, 2005.

[128] J. S. Yang and H. G. Zhou, Wave propagation in a piezoelectric ceramic plate sandwiched between two semiconductor layers, *Int. J. of Applied Electromagnetics and Mechanics*, **22**, 97–109, 2005.

[129] J. S. Yang and H. G. Zhou, Propagation and amplification of gap waves between a piezoelectric half-space and a semiconductor film, *Acta Mechanica*, **176**, 83–93, 2005.

[130] B. A. Auld, Acoustic *Fields and Waves in Solids*, *Vol. 1*, John Wiley and Sons, New York, 1973.

[131] H. Jaffe and D. A. Berlincourt, Piezoelectric transducer materials, *Proc. of IEEE*, **53**, 1372–1386, 1965.

[132] R. Bechmann, Elastic and piezoelectric constants of alpha-quartz, *Phys. Rev.*, **110**, 1060–1061, 1958.

[133] B. K. Sinha and H. F. Tiersten, First temperature derivatives of the fundamental elastic constants of quartz, *J. Appl. Phys.*, **50**, 2732–2739, 1979.

[134] R. N. Thurston, H. J. McSkimin and P. Andreatch, Jr., Third-order elastic constants of quartz, *J. Appl. Phys.*, **37**, 267–275, 1966.

[135] D. F. Nelson, *Electric, Optic and Acoustic Interactions in Crystals*, John Wiley and Sons, New York, 1979.

[136] J. J. Gagnepain and R. Besson, Nonlinear effects in piezoelectric quartz crystals, *Physical Acoustics*, *Vol. XI*, W. P. Mason and R. N. Thurston, ed., Academic Press, New York, 1975.

[137] B. P. Sorokin, P. P. Turchin, S. I. Burkov, D. A. Glushkov and K. S. Alexandrov, Influence of static electric field, mechanical pressure and temperature on the propagation of acoustic waves in $La_3Ga_5SiO_{14}$ piezoelectric single crystals, *Proc. IEEE Int. Frequency Control Symp.*, 161–169, 1996.

[138] A. W. Warner, M. Onoe and G. A. Couqin, Determination of elastic and piezoelectric constants for crystals in class (3m), *J. Acoust. Soc. Am.*, **42**, 1223–1231, 1967.

[139] Y. Cho and K. Yamanouchi, Nonlinear, elastic, piezoelectric, electrostrictive, and dielectric constants of lithium niobate, *J. Appl. Phys.*, **61**, 875–887, 1987.

Appendix 1
Notation

δ_{ij}, δ_{KL}	– Kronecker delta
δ_{iK}, δ_{Ki}	– Shifter
ε_{ijk}, ε_{IJK}	– Permutation tensor
X_K	– Reference position of a material point
y_i	– Present position of a material point
u_K	– Mechanical displacement vector
J	– Jacobian
C_{KL}	– Deformation tensor
S_{KL}	– Finite strain tensor
S_{kl}	– Linear strain tensor
v_i	– Velocity vector
d_{ij}	– Deformation rate tensor
ω_{ij}	– Spin tensor
D/Dt	– Material time derivative
ρ_0	– Reference mass density (scalar)
ρ	– Present mass density
ρ_e	– Free charge density per unit present volume (scalar)
ρ_E	– Free charge density per unit reference volume (scalar)
σ_e	– Surface free charge per unit present area (scalar)
σ_E	– Surface free charge per unit reference area (scalar)
Q_e	– Free charge (scalar)
Q	– Quality factor
I	– Current
V	– Voltage
Z	– Impedance
ε_0	– Permittivity of free space (scalar)
ϕ	– Electrostatic potential
E_i	– Electric field
P_i	– Electric polarization per unit present volume

π_i – Electric polarization per unit mass

D_i – Electric displacement vector

$\mathcal{E}_K$ – Reference electric field vector

$\mathcal{P}_K$ – Reference electric polarization vector

$\mathcal{D}_K$ – Reference electric displacement vector

f_j – Mechanical body force per unit mass

σ_{ij} – Cauchy stress tensor

σ_{ij}^E – Electrostatic stress tensor

$\sigma_{ij}^S, F_{Lj}, T_{KL}^S$ – Symmetric stress tensor in spatial, two-point, and material form

$\sigma_{ij}^M, M_{Lj}, T_{KL}^M$ – Symmetric Maxwell stress tensor in spatial, two-point, and material form

$\tau_{ij}, K_{Lj}, \hat{T}_{KL}$ – Total stress tensor in spatial, two-point, and material form

K_{LM} – $K_{Lj}\delta_{jM}$

T_{kl} – Linear stress tensor

T_k – Mechanical surface traction per unit reference area

t_k – Mechanical surface traction per unit present area

ψ – Free energy per unit mass

H – Electric enthalpy per unit volume

$*$ – Complex conjugate

i – Imaginary unit

Appendix 2
Electroelastic Material Constants

Material constants for a few common piezoelectrics are summarized below. Numerical results given in this book are calculated from these constants.

Permittivity of free space $\varepsilon_0 = 8.854 \times 10^{-12}\,\text{F/m}$.

Polarized ceramics

The material matrices for PZT-5H are [130]

$$\rho = 7500\ \text{kg/m}^3,$$

$$[c_{pq}] = \begin{pmatrix} 12.6 & 7.95 & 8.41 & 0 & 0 & 0 \\ 7.95 & 12.6 & 8.41 & 0 & 0 & 0 \\ 8.41 & 8.41 & 11.7 & 0 & 0 & 0 \\ 0 & 0 & 0 & 2.3 & 0 & 0 \\ 0 & 0 & 0 & 0 & 2.3 & 0 \\ 0 & 0 & 0 & 0 & 0 & 2.325 \end{pmatrix} \times 10^{10}\,\text{N/m}^2,$$

$$[e_{ip}] = \begin{pmatrix} 0 & 0 & 0 & 0 & 17 & 0 \\ 0 & 0 & 0 & 17 & 0 & 0 \\ -6.5 & -6.5 & 23.3 & 0 & 0 & 0 \end{pmatrix} \text{C/m}^2,$$

$$[\varepsilon_{ij}] = \begin{pmatrix} 1700\varepsilon_0 & 0 & 0 \\ 0 & 1700\varepsilon_0 & 0 \\ 0 & 0 & 1470\varepsilon_0 \end{pmatrix}$$

$$= \begin{pmatrix} 1.505 & 0 & 0 \\ 0 & 1.505 & 0 \\ 0 & 0 & 1.302 \end{pmatrix} \times 10^{-8}\,\text{C/(V - m)}.$$

For PZT-5H, an equivalent set of material constants are [130]

$$s_{11} = 16.5, \quad s_{33} = 20.7, \quad s_{44} = 43.5,$$
$$s_{12} = -4.78, \quad s_{13} = -8.45 \times 10^{-12}\,\mathrm{m^2/N},$$
$$d_{31} = -274, \quad d_{15} = 741, \quad d_{33} = 593 \times 10^{-12}\,\mathrm{C/N},$$
$$\varepsilon_{11} = 3130\varepsilon_0, \quad \varepsilon_{33} = 3400\varepsilon_0.$$

When poling is along other directions, the material matrices can be obtained by tensor transformations. For PZT-5H, when poling is along the x_1-axis, we have

$$[c_{pq}] = \begin{pmatrix} 11.7 & 8.41 & 8.41 & 0 & 0 & 0 \\ 8.41 & 12.6 & 7.95 & 0 & 0 & 0 \\ 8.41 & 7.95 & 12.6 & 0 & 0 & 0 \\ 0 & 0 & 0 & 2.325 & 0 & 0 \\ 0 & 0 & 0 & 0 & 2.3 & 0 \\ 0 & 0 & 0 & 0 & 0 & 2.3 \end{pmatrix} \times 10^{10}\,\mathrm{N/m^2},$$

$$[e_{ip}] = \begin{pmatrix} 23.3 & -6.5 & -6.5 & 0 & 0 & 0 \\ 0 & 0 & 0 & 0 & 0 & 17 \\ 0 & 0 & 0 & 0 & 17 & 0 \end{pmatrix} \mathrm{C/m^2},$$

$$[\varepsilon_{ij}] = \begin{pmatrix} 1.302 & 0 & 0 \\ 0 & 1.505 & 0 \\ 0 & 0 & 1.505 \end{pmatrix} \times 10^{-8}\,\mathrm{C/Vm}.$$

When poling is along the x_2-axis,

$$[c_{pq}] = \begin{pmatrix} 12.6 & 8.41 & 7.95 & 0 & 0 & 0 \\ 8.41 & 11.7 & 8.41 & 0 & 0 & 0 \\ 7.95 & 8.41 & 12.6 & 0 & 0 & 0 \\ 0 & 0 & 0 & 2.3 & 0 & 0 \\ 0 & 0 & 0 & 0 & 2.325 & 0 \\ 0 & 0 & 0 & 0 & 0 & 2.3 \end{pmatrix} \times 10^{10}\,\mathrm{N/m^2},$$

$$[e_{ip}] = \begin{pmatrix} 0 & 0 & 0 & 0 & 0 & 17 \\ -6.5 & 23.3 & -6.5 & 0 & 0 & 0 \\ 0 & 0 & 0 & 17 & 0 & 0 \end{pmatrix} \text{ C/m}^2,$$

$$[\varepsilon_{ij}] = \begin{pmatrix} 1.505 & 0 & 0 \\ 0 & 1.302 & 0 \\ 0 & 0 & 1.505 \end{pmatrix} \times 10^{-8} \text{C/Vm}.$$

For PZT-G1195

$$\rho = 7500\,\text{kg/m}^3, \quad c_{11}^E = c_{22}^E = 148, \quad c_{33}^E = 131, \quad c_{12}^E = 76.2,$$

$$c_{13}^E = c_{23}^E = 74.2, \quad c_{44}^E = c_{55}^E = 25.4, \quad c_{66}^E = 35.9\ \text{GPa},$$

$$e_{15} = 9.2, \quad e_{31} = -2.1, \quad e_{33} = 9.5\ \text{C/m}^2.$$

Material constants of a few other polarized ceramics are given in the following tables [131]:

Material	c_{11}	c_{12}	c_{13}	c_{33}	c_{44}	c_{66}
PZT-4	13.9	7.78	7.40	11.5	2.56	3.06
PZT-5A	12.1	7.59	7.54	11.1	2.11	2.26
PZT-6B	16.8	8.47	8.42	16.3	3.55	4.17
PZT-5H	12.6	7.91	8.39	11.7	2.30	2.35
PZT-7A	14.8	7.61	8.13	13.1	2.53	3.60
PZT-8	13.7	6.99	7.11	12.3	3.13	3.36
$BaTiO_3$	15.0	6.53	6.62	14.6	4.39	4.24
	$\times 10^{10}$ N/m^2					

Material	e_{31}	e_{33}	e_{15}	ε_{11}	ε_{33}
PZT-4	–5.2	15.1	12.7	0.646	0.562
PZT-5A	–5.4	15.8	12.3	0.811	0.735
PZT-6B	–0.9	7.1	4.6	0.360	0.342
PZT-5H	–6.5	23.3	17.0	1.505	1.302
PZT-7A	–2.1	9.5	9.2	0.407	0.208
PZT-8	–4.0	13.2	10.4	0.797	0.514
$BaTiO_3$	–4.3	17.5	11.4	0.987	1.116
	C/m^2			$\times 10^{-8}$ C/Vm	

Density	PZT-5H	PZT-5A	PZT-6B	PZT-4
kg/m^3	7500	7750	7550	7500

Density	PZT-7A	PZT-8	$BaTiO_3$
kg/m^3	7600	7600	5700

Quartz

When referred to the crystal axes, the second-order material constants for left-hand quartz have the following values [132]:

$$\rho = 2649 \text{ kg/m}^3,$$

$$[c_{pq}] = \begin{pmatrix} 86.74 & 6.99 & 11.91 & -17.91 & 0 & 0 \\ 6.99 & 86.74 & 11.91 & 17.91 & 0 & 0 \\ 11.91 & 11.91 & 107.2 & 0 & 0 & 0 \\ -17.91 & 17.91 & 0 & 57.94 & 0 & 0 \\ 0 & 0 & 0 & 0 & 57.94 & -17.91 \\ 0 & 0 & 0 & 0 & -17.91 & 39.88 \end{pmatrix}$$

$$\times 10^9 \text{ N/m}^2,$$

$$[e_{ip}] = \begin{pmatrix} 0.171 & -0.171 & 0 & -0.0406 & 0 & 0 \\ 0 & 0 & 0 & 0 & 0.0406 & -0.171 \\ 0 & 0 & 0 & 0 & 0 & 0 \end{pmatrix} \mathrm{C/m^2},$$

$$[\varepsilon_{ij}] = \begin{pmatrix} 39.21 & 0 & 0 \\ 0 & 39.21 & 0 \\ 0 & 0 & 41.03 \end{pmatrix} \times 10^{-12}\,\mathrm{C/Vm}.$$

Temperature derivatives of the elastic constants of quartz at 25 °C are [133]

pq	11	33	12	13
$(1/c_{pq})(dc_{pq}/dT)$ (10^{-6}/°C)	18.16	−66.60	−1222	−178.6

pq	44	66	14
$(1/c_{pq})(dc_{pq}/dT)$ (10^{-6}/°C)	−89.72	126.7	−49.21

For quartz there are 31 non-zero third-order elastic constants. 14 are given in the following table. These values, at 25 °C, and based on a least-squares fit, are all in 10^{11} N/m² [134]

Constant	Value	Standard error
c_{111}	−2.10	0.07
c_{112}	−3.45	0.06
c_{113}	+0.12	0.06
c_{114}	−1.63	0.05
c_{123}	−2.94	0.05
c_{124}	−0.15	0.04
c_{133}	−3.12	0.07

Constant	Value	Standard error
c_{134}	+0.02	0.04
c_{144}	−1.34	0.07
c_{155}	−2.00	0.08
c_{222}	−3.32	0.08
c_{333}	−8.15	0.18
c_{344}	−1.10	0.07
c_{444}	−2.76	0.17

In addition, there are 17 relations among the third-order elastic constants of quartz [135]

$$c_{122} = c_{111} + c_{112} - c_{222}, \quad c_{156} = \frac{1}{2}(c_{114} + 3c_{124}),$$

$$c_{166} = \frac{1}{4}(-2c_{111} - c_{112} + 3c_{222}),$$

$$c_{224} = -c_{114} - 2c_{124}, \quad c_{256} = \frac{1}{2}(c_{114} - c_{124}),$$

$$c_{266} = \frac{1}{4}(2c_{111} - c_{112} - c_{222}),$$

$$c_{366} = \frac{1}{2}(c_{113} - c_{123}), \quad c_{456} = \frac{1}{2}(-c_{144} + c_{155}),$$

$$c_{223} = c_{113}, \quad c_{233} = c_{133}, \quad c_{234} = -c_{134}, \quad c_{244} = c_{155}, \quad c_{255} = c_{144},$$

$$c_{355} = c_{344}, \quad c_{356} = c_{134}, \quad c_{455} = -c_{444}, \quad c_{466} = c_{124}.$$

For the fourth-order elastic constants there are 69 non-zero ones of which 23 are independent [136]

$$c_{1111}, \quad c_{3333}, \quad c_{4444}, \quad c_{6666}, \quad c_{1112}, \quad c_{1113}, \quad c_{1123}, \quad c_{2214}, \quad c_{3331},$$

$$c_{4456}, \quad c_{5524}, \quad c_{4443}, \quad c_{1133}, \quad c_{3344}, \quad c_{1456}, \quad c_{1155}, \quad c_{1134}, \quad c_{2356},$$

$$c_{4423}, \quad c_{4413}, \quad c_{3314}, \quad c_{6614}, \quad c_{6624}.$$

There are 46 relations [136]

$$c_{2222}=c_{1111},\quad c_{2266}=\frac{1}{6}(c_{1111}-c_{1112}),\quad c_{2223}=c_{1113},$$

$$c_{2221}=c_{1112},\quad c_{6612}=\frac{1}{6}(c_{1111}-4c_{6666}-c_{1112}),\quad c_{2213}=c_{1123},$$

$$c_{1166}=c_{2266},\quad c_{1122}=\frac{1}{3}(-c_{1111}+4c_{1112}+8c_{6666}),$$

$$c_{6613}=\frac{1}{4}(c_{1113}-c_{1123}),$$

$$c_{5555}=c_{4444},\quad c_{4455}=\frac{1}{3}c_{4444},\quad c_{6623}=c_{6613},$$

$$c_{1124}=-c_{2214}+c_{6614}+c_{6624},$$

$$c_{3312}=-c_{1133},\quad c_{1114}=3(-c_{2214}+2c_{6614}-2c_{6624}),\quad c_{2233}=c_{1133},$$

$$c_{2256}=\frac{1}{2}(-2c_{2214}+3c_{6614}-5c_{6624}),\quad c_{6633}=c_{1133},$$

$$c_{2224}=3(c_{2214}-3c_{6614}+c_{6624}),$$

$$c_{3355}=c_{3344},\quad c_{1156}=\frac{1}{2}(-2c_{2214}+7c_{6614}-c_{6624}),\quad c_{3332}=c_{3331},$$

$$c_{1256}=\frac{1}{2}(-2c_{2214}+3c_{6614}-c_{6624}),\quad c_{5534}=-c_{4443},$$

$$c_{6665}=\frac{3}{2}(c_{6614}-c_{6624}),$$

$$c_{4442}=-4c_{4456}-c_{5524},\quad c_{1234}=c_{1134}-2c_{2356},\quad c_{2255}=c_{4412},$$

$$c_{5514}=2c_{4456}+c_{5524},\quad c_{1356}=2c_{1134}-3c_{2356},\quad c_{5566}=c_{1456},$$

$$c_{5556}=3c_{4456},\quad c_{2234}=4c_{2356}-3c_{1134},\quad c_{3324}=-c_{3314},$$

$$c_{4441}=2c_{4456}-c_{5524},\quad c_{6634}=c_{1234},\quad c_{3356}=c_{3314},\quad c_{5512}=c_{4412},$$

$$c_{1144}=c_{4412},\quad c_{5523}=c_{4413},\quad c_{2456}=c_{1456},\quad c_{2244}=c_{1155},\quad c_{5513}=c_{4423},$$

$$c_{4466}=c_{1456},\quad c_{4412}=c_{1155}-4c_{1456},\quad c_{3456}=\frac{1}{2}(c_{4423}-c_{4413}).$$

The fourth-order elastic constants are usually unknown. Some scattered results are [136]

$$c_{1111} = 1.59 \times 10^{13}\,\mathrm{N/m^2} \pm 20\%,$$
$$c_{3333} = 1.84 \times 10^{13}\,\mathrm{N/m^2} \pm 20\%,$$

and [9]

$$c^{E}_{6666} = 77 \times 10^{11}\,\mathrm{N/m^2}\,.$$

AT-cut quartz is a special case of rotated Y-cut quartz ($\theta = 35.25°$) whose material constants are [4]

$$[c_{pq}] = \begin{pmatrix} 86.74 & -8.25 & 27.15 & -3.66 & 0 & 0 \\ -8.25 & 129.77 & -7.42 & 5.7 & 0 & 0 \\ 27.15 & -7.42 & 102.83 & 9.92 & 0 & 0 \\ -3.66 & 5.7 & 9.92 & 38.61 & 0 & 0 \\ 0 & 0 & 0 & 0 & 68.81 & 2.53 \\ 0 & 0 & 0 & 0 & 2.53 & 29.01 \end{pmatrix} \times 10^{9}\,\mathrm{N/m^2}\,,$$

$$[e_{ip}] = \begin{pmatrix} 0.171 & -0.152 & -0.0187 & 0.067 & 0 & 0 \\ 0 & 0 & 0 & 0 & 0.108 & -0.095 \\ 0 & 0 & 0 & 0 & -0.0761 & 0.067 \end{pmatrix} \mathrm{C/m^2}\,,$$

$$[\varepsilon_{ij}] = \begin{pmatrix} 39.21 & 0 & 0 \\ 0 & 39.82 & 0.86 \\ 0 & 0.86 & 40.42 \end{pmatrix} \times 10^{-12}\,\mathrm{C/Vm}\,.$$

Langasite

The second-order material constants of $La_3Ga_5SiO_{14}$ are [137]

$$\rho = 5743\ \mathrm{kg/m^3},$$

$$[c_{pq}] = \begin{pmatrix} 18.875 & 10.475 & 9.589 & -1.412 & 0 & 0 \\ 10.475 & 18.875 & 9.589 & 1.412 & 0 & 0 \\ 9.589 & 9.589 & 26.14 & 0 & 0 & 0 \\ -1.412 & 1.412 & 0 & 5.35 & 0 & 0 \\ 0 & 0 & 0 & 0 & 5.35 & -1.412 \\ 0 & 0 & 0 & 0 & -1.412 & 4.2 \end{pmatrix}$$

$$\times 10^{10}\,\mathrm{N/m^2},$$

$$[e_{ip}] = \begin{pmatrix} -0.44 & 0.44 & 0 & -0.08 & 0 & 0 \\ 0 & 0 & 0 & 0 & 0.08 & 0.44 \\ 0 & 0 & 0 & 0 & 0 & 0 \end{pmatrix} \text{C/m}^2,$$

$$[\varepsilon_{ij}] = \begin{pmatrix} 18.92\varepsilon_0 & 0 & 0 \\ 0 & 18.92\varepsilon_0 & 0 \\ 0 & 0 & 50.7\varepsilon_0 \end{pmatrix}$$

$$= \begin{pmatrix} 167.5 & 0 & 0 \\ 0 & 167.5 & 0 \\ 0 & 0 & 448.9 \end{pmatrix} \times 10^{-12} \text{C/Vm}.$$

The third-order material constants of $La_3Ga_5SiO_{14}$ at 20°C are given in [137]. The third-order elastic constants c_{pqr} (in 10^{10} N/m^2) are

c_{111}	–97.2	c_{134}	–4.1
c_{112}	0.7	c_{144}	–4.0
c_{113}	–11.6	c_{155}	–19.8
c_{114}	–2.2	c_{222}	–96.5
c_{123}	0.9	c_{333}	–183.4
c_{124}	–2.8	c_{344}	–38.9
c_{133}	–72.1	c_{444}	20.2

The third-order piezoelectric constants e_{ipq} (in C/m^2) are

e_{111}	9.3	e_{124}	–4.8
e_{113}	–3.5	e_{134}	6.9
e_{114}	1.0	e_{144}	–1.7
e_{122}	0.7	e_{315}	–4

The third-order electrostrictive constants H_{pq} (in 10^{-11}N/V^2) are

H_{11}	–26	H_{31}	–24
H_{12}	65	H_{33}	–40
H_{13}	20	H_{41}	–170
H_{14}	–43	H_{44}	–44

The third-order dielectric permeability ε_{111} (in 10^{-20} F/V) are

ε_{111}	–0.5

Lithium Niobate

The second-order material constants for lithium niobate are [138]

$$\rho = 4700 \text{ kg/m}^3,$$

$$[c_{pq}] = \begin{pmatrix} 2.03 & 0.53 & 0.75 & 0.09 & 0 & 0 \\ 0.53 & 2.03 & 0.75 & -0.09 & 0 & 0 \\ 0.75 & 0.75 & 2.45 & 0 & 0 & 0 \\ 0.09 & -0.09 & 0 & 0.60 & 0 & 0 \\ 0 & 0 & 0 & 0 & 0.60 & 0.09 \\ 0 & 0 & 0 & 0 & 0.09 & 0.75 \end{pmatrix} \times 10^{11}\text{N/m}^2,$$

$$[e_{ip}] = \begin{pmatrix} 0 & 0 & 0 & 0 & 3.70 & -2.50 \\ -2.50 & 2.50 & 0 & 3.70 & 0 & 0 \\ 0.20 & 0.20 & 1.30 & 0 & 0 & 0 \end{pmatrix} \text{C/m}^2,$$

$$[\varepsilon_{ij}] = \begin{pmatrix} 38.9 & 0 & 0 \\ 0 & 38.9 & 0 \\ 0 & 0 & 25.7 \end{pmatrix} \times 10^{-11}\text{C/Vm}.$$

The third-order material constants of lithium niobate are given in [139]. The third-order elastic constants c_{pqr} (in 10^{11} N/m^2) are

Constant	Value	Standard error
c_{111}	–21.2	4.0
c_{112}	–5.3	1.2
c_{113}	–5.7	1.5
c_{114}	2.0	0.8
c_{123}	–2.5	1.0
c_{124}	0.4	0.3
c_{133}	–7.8	1.9
c_{134}	1.5	0.3
c_{144}	–3.0	0.2
c_{155}	–6.7	0.3
c_{222}	–23.3	3.4
c_{333}	–29.6	7.2
c_{344}	–6.8	0.7
c_{444}	–0.3	0.4

The third-order piezoelectric constants e_{ipq} $(= -k_{1\,ipq})$ are

Constant	Value	Standard error
e_{115}	17.1	6.6
e_{116}	–4.7	6.4
e_{125}	19.9	2.1
e_{126}	–15.9	5.3
e_{135}	19.6	2.7
e_{136}	–0.9	2.7
e_{145}	20.3	5.7

Constant	Value	Standard error
e_{311}	14.7	6.0
e_{312}	13.0	11.4
e_{313}	-10.0	8.7
e_{314}	11.0	4.6
e_{333}	-17.3	5.9
e_{344}	-10.2	5.6
	C/m^2	

The third-order electrostirctive constants l_{pq} (compressed from b_{ijkl} + $\varepsilon_0\delta_{ij}\delta_{kl}$ - $\varepsilon_0\delta_{ik}\delta_{jl}$ - $\varepsilon_0\delta_{il}\delta_{kl}$) (in 10^{-9} F/m^2) are

Constant	Value	Standard error
l_{11}	1.11	0.39
l_{12}	2.19	0.56
l_{13}	2.32	0.67
l_{31}	0.19	0.61
l_{33}	–2.76	0.41
l_{14}	1.51	0.17
l_{41}	1.85	0.17
l_{44}	–1.83	0.11

The third-order dielectric constants ε_{ip} (in 10^{-19} F/V) are

Constant	Value	Standard error
ε_{31}	–2.81	0.06
ε_{22}	-2.40	0.09
ε_{33}	-2.91	0.06

Lithium Tantalate

The second-order material constants for lithium tantalate are [138]

$$\rho = 7450 \text{ kg/m}^3,$$

$$[c_{pq}] = \begin{pmatrix} 2.33 & 0.47 & 0.80 & -0.11 & 0 & 0 \\ 0.47 & 2.33 & 0.80 & 0.11 & 0 & 0 \\ 0.80 & 0.80 & 2.45 & 0 & 0 & 0 \\ -0.11 & -0.11 & 0 & 0.94 & 0 & 0 \\ 0 & 0 & 0 & 0 & 0.94 & -0.11 \\ 0 & 0 & 0 & 0 & -0.11 & 0.93 \end{pmatrix} \times 10^{11} \text{N/m}^2,$$

$$[e_{ip}] = \begin{pmatrix} 0 & 0 & 0 & 0 & 2.6 & -1.6 \\ -1.6 & 1.6 & 0 & 2.6 & 0 & 0 \\ 0 & 0 & 1.9 & 0 & 0 & 0 \end{pmatrix} \text{C/m}^2,$$

$$[\varepsilon_{ij}] = \begin{pmatrix} 36.3 & 0 & 0 \\ 0 & 36.3 & 0 \\ 0 & 0 & 38.2 \end{pmatrix} \times 10^{-11} \text{C/Vm}.$$

Cadmium Sulfide

$\rho = 4820 \text{ kg/m}^3$. The second-order material constants are [130]

$$\begin{aligned} &c_{11} = 9.07, \quad c_{33} = 9.38, \quad c_{44} = 1.504, \\ &c_{12} = 5.81, \quad c_{13} = 5.10 \times 10^{10} \text{N/m}^2, \\ &e_{15} = -0.21, \quad e_{31} = -0.24, \quad e_{33} = -0.44 \text{ C/m}^2, \\ &\varepsilon_{11} = 9.02\varepsilon_0, \quad \varepsilon_{33} = 9.53\varepsilon_0. \end{aligned}$$

Silicon

$\rho_0 = 2332 \text{ kg/m}^3$. The second-order material constants are [130]

$$c_{11} = 16.57, \quad c_{44} = 7.956, \quad c_{12} = 6.39 \times 10^{10} \text{N/m}^2.$$

For the third-order elastic constants there are 20 non-zero ones among which six are independent

$$c_{111} = -825, \quad c_{112} = -451, \quad c_{123} = -64,$$
$$c_{144} = 12, \quad c_{155} = -310, \quad c_{456} = -64 \text{ GPa}.$$

The other 14 are determined from the following relations

$$c_{113} = c_{112}, \quad c_{122} = c_{112}, \quad c_{133} = c_{112}, \quad c_{166} = c_{155},$$
$$c_{222} = c_{111}, \quad c_{223} = c_{112}, \quad c_{233} = c_{112}, \quad c_{244} = c_{155},$$
$$c_{255} = c_{144}, \quad c_{266} = c_{155}, \quad c_{333} = c_{111}, \quad c_{344} = c_{155},$$
$$c_{355} = c_{155}, \quad c_{366} = c_{144}.$$

Germanium

$\rho = 5332$ kg/m^3. The second-order material constants are [130]

$$c_{11} = 12.9, \quad c_{44} = 6.68, \quad c_{12} = 4.9 \times 10^{10}\,\text{N/m}^2.$$

The third-order material constants are

$$c_{111} = -710, \quad c_{112} = -389, \quad c_{123} = -18,$$
$$c_{144} = -23, \quad c_{155} = -292, \quad c_{456} = -53 \text{ GPa}.$$

Index

absolute temperature 371
abstract notation 128
acceleration sensitivity 301, 311
acoustoelectric
    amplification 463
    effect 463
adiabatic elimination 378
admittance 423, 431, 459
air gap 355, 484
aluminum 155
amplifier 463
amplitude modulation 340
angular rate sensor 209, 245
ANSYS 275
anti-plane 132, 182, 186, 200
apparent material constant 19, 24, 378
aspect ratio 360, 434, 448
BAW v
beam 245, 258, 264, 407
Bechmann's number 96, 100, 171
Bessel
    equation 193
    function 133, 161, 193, 197, 397
    function, modified 193, 197
bias 16, 36, 209, 376
    frequency perturbation 21
    small 20
    thermal 376
bi-convex 97
bi-mesa 94
bimorph
    beam 264, 399
    plate 355
    spiral 407
Bleustein-Gulyaev wave 78, 470
boundary
    integral equation 185
    partition 5, 9
    value problem 5, 9, 19
bulk wave v
cadmium sulfide 515
capacitance
    motional 48
    static 290, 299
centrifugal 210
ceramic 12, 78, 167, 186, 204, 215, 223, 232, 238, 282, 288, 291, 296, 387, 395, 399, 407, 417, 439, 503
charge 10
circuit equation 10, 249, 290, 388, 398, 402, 419, 454
Clausius-Duhem inequality 371
compatibility 343, 348
concave-convex 120, 330
concentrated mass 288, 400
constrained 382
continuity condition 270, 278, 427, 453
convex
    bi-convex 97
    plano-convex 119
Coriolis force 210
correction factor 67
cubic
    crystal 362
    theory 22
current 10
curvature 119, 329
cut
    AT-cut 110, 123, 333, 510
    rotated Y-cut 14, 85, 217
    Y-cut 40, 42, 43, 110, 123, 359
cutoff frequency 78, 106
cylinder 159, 160, 184, 196
damping coefficient 68
degenerate 287
diffusion constant 464
dispersion relation (curve) 77, 81, 87, 91, 94, 112, 117, 212, 217, 219, 472, 488

dissipation 473
    function 375
double resonance 245, 295
drift 471
effective material constant 19, 24, 378
eigenvalue problem 74, 128, 138, 158, 188, 320
electrode 9
    asymmetric 49, 52, 60
    imperfectly bonded 52
    inertia 49, 60
    mass ratio 132
    nonuniform 104
    open 71
    partial 103, 106, 171
    shear deformable 60
    shear stiffness 60
    shorted 72
electromechanical coupling 28, 31, 33, 35, 38, 41, 106
energy trapping 93, 98, 103, 106
enthalpy 7
entropy 371
extension 78, 124, 288, 303, 361, 425, 464
exterior problem 195, 198
face-shear 78, 114
FGM 98
film 464, 467, 474, 485
finite element 275
finite plate 112, 117
flexure 78, 110, 114, 247, 260, 265, 301, 364
fluid sensor 181
four-vector 128, 138
free energy 4, 7, 379
frequency
    amplitude dependence 67, 70
    equation 438
    modulation 339
    perturbation 21
    shift 22, 139, 142, 153, 191, 332, 368
    spectrum 114, 118, 125, 126
    split 225, 227
functionally graded 98
gap wave 484
generator 387
germanium 363, 369, 516
gold 65, 155
gyroscope 209, 245
gyroscopic mode
    primary 245
    secondary 245
half-space 165, 166, 201, 202, 236, 361, 364, 466
harvester (see generator)
heat equation 373
hoop stress 353
ill-posed 157, 181
impedance 249, 388, 456
inertia 127
initial 70, 209, also see bias
inner product 130, 322
interface 417, 452
    wave 474
interior problem 194, 197
inverted mesa 94
isentropic 378
jump phenomenon 70, 413
Kronecker delta 1
langasite 13, 43, 110, 123, 346, 351, 359, 360, 510
Legendre transform 372
lithium niobate 512
lithium tantalate 515
Love wave 471
mass ratio 51, 132, 135, 157
mass sensor 127
material constant
    apparent 19, 24, 378
    complex 376
    effective 19, 24, 378
    fourth-order 4, 34, 67, 409, 508
    frequency dependence 376
    second-order 4
    temperature derivative 379, 381, 384
    thermoelastic 374, 379
    third-order 4, 40, 67, 343, 409, 507, 511, 512, 516
    time derivative 3
Mathieu equation 336
matrix notation 11
mesa resonator 93
    bi-mesa 94

inverted mesa 94
plano-mesa 94
mobility 464
mode
high frequency 25
low frequency 25
monoclinic 14, 49, 66
motional capacitance 48
natural boundary condition 6, 10, 20
nodal point 433, 442
nonlinear
cubic 20, 22
resonance 409
theory 1
thickness-shear 31, 66
weak 22
one-dimensional 143, 245, 258, 264, 282, 407
open problem 159, 185
overhang plate 315
parallelepiped 296, 444
perturbation 21, 127, 135, 138, 187, 318, 471
integral 22, 191, 311
phase angle 414
piezoelectric stiffening 47, 49
plano-convex 97, 119
plano-mesa 94
plate 73, 159, 183, 191, 291, 296, 341, 347, 352, 379, 382, 479
Poisson's effect 39, 335
positive definite 7, 375
power harvester (see generator)
pressure sensor 341
primary mode 245
pure bending 363
pyroelectric 374
quality factor 68, 252
quartz 13, 42, 85, 110, 123, 175, 187, 217, 227, 346, 351, 355, 359, 360, 506
quasistatic 335
Rayleigh
quotient 140, 151
wave 78, 362
resonance 46, 390, 398, 406
resonator
contoured 97
ring 143, 282
rod 288, 425
Rosen transformer 425
saturation 254, 421
SAW v
second law of thermodynamics 371
second order
acceleration sensitivity 313
perturbation 318
solution 336
secondary mode 245
semiconductor 463
sensor
angular rate 209, 245
fluid 181
mass 127
pressure 341
temperature 371
SH modes (see anti-plane)
shear horizontal (see SH)
shear-lag model 52
shell
bending with shear 148
circular cylindrical 192, 274, 341, 347, 352
conical 134
shallow 119, 352
membrane 137
shifter 2
silicon 362, 368, 515
silver 65, 155
specific heat 374
spectrum 114, 118, 125, 126
sphere 185
spiral 407
statically
determinate 353
indeterminate 343, 348, 356
stiffening 47, 49
straight-crested wave 74, 229, 236
strain energy 343
stress relaxation 247, 383, 426
surface
clamped 30
free 29
wave 163, 199, 236, 361, 364, 466
temperature sensor 371
thermal bias 376

thermoelastic 374, 379
thermoelectroelasticity 371
thermopiezoelectricity 375
thickness
  contraction 303
  vibration 25, 221
thickness-shear 25, 78, 410, 434
  approximation 122
  axial 162, 199
  mode 47, 131, 159, 191, 196, 379
  nonlinear 31, 66
  slowly varying 73
  static 25, 47
  tangential 162, 197
  vibration 43, 91
  wave 87
thickness-stretch 78, 387, 451
thickness-twist 78, 85, 171, 175, 204
torsion 134, 185, 192
transformer 417
transverse isotropy 12
tube 258
two-dimensional 66, 110, 114, 119, 134, 192, 274, 291, 395, 464
variation 5, 10, 20, 89, 139, 144, 150
vibration
  forced 47, 68, 250, 261, 270, 279, 287, 290, 299, 429, 447
  free 45, 67, 108, 271, 286, 298, 428
  sensitivity 334
viscopiezoelectricity 376
viscosity 188
wave
  Bleustein-Gulyaev 78, 470
  bulk v
  dissipative 471
  exact 73
  gap 484
  interface 474
  long 87, 471
  longitudinal 212
  Love 471
  plane 210
  plate 73, 228, 479
  Rayleigh 78, 362
  straight-crested 74, 229, 236
  surface 163, 199, 236, 361, 364, 466
  transverse 212
wedge 97, 132, 184
Young's modulus 345, 350, 358
zero-dimensional 288, 296, 444